Wolfgang Runge

Technology Entrepreneurship

A Treatise on Entrepreneurs and Entrepreneurship for and in Technology Ventures

VOLUME 2

Technology Entrepreneurship

A Treatise on Entrepreneurs and Entrepreneurship
for and in Technology Ventures

VOLUME 2

by
Wolfgang Runge

Copyright Acknowledgements

Acknowledgements are due to the Zentrum für Europäische Wirtschaftsforschung GmbH, Mannheim (ZEW) for permissions to reproduce a large number of figures and tables from its publications.

Permission to use figures from the Global Entrepreneurship Monitor (GEM) 2008 Global Report, which appear here, has been granted by the copyright holders. The GEM is an international consortium and this report was produced from data collected in, and received from, 43 countries in 2008. Our thanks go to the authors, national teams, researchers, funding bodies and other contributors who have made this possible.

Impressum

Karlsruher Institut für Technologie (KIT)
KIT Scientific Publishing
Straße am Forum 2
D-76131 Karlsruhe

KIT Scientific Publishing is a registered trademark of Karlsruhe Institute of Technology. Reprint using the book cover is not allowed.

www.ksp.kit.edu

Print on Demand 2014

ISBN 978-3-7315-0107-7 (Set)
DOI: 10.5445/KSP/1000036457

ISBN 978-3-7315-0108-4 (Band 1)
DOI: 10.5445/KSP/1000036459

ISBN 978-3-7315-0109-1 (Band 2)
DOI: 10.5445/KSP/1000036460

CONTENTS

4.3 Approaches to New Technology Venture Growth

Venture growth is currently an issue of strategic management or entrepreneurship – and of particular interest to policy. For the field of strategy the emphasis was and is on existing firms and business growth, but this makes insights to be transferred to firm's foundation and early growth of new (technology) ventures problematic.

Understanding the origins of new venture growth and why and how startups grow and become successful or even very successful firms ("promising startups") is still one of the least understood aspects of entrepreneurship research.

> It is our conviction, in line with Bhidé [2000:11],
> "We cannot expect to derive a fool-proof formula or a 'complete' description for starting a profitable business."

Furthermore, entrepreneurs or the leadership team of young technology-based ventures are being presented with a much larger "menu" or array of choices for decision-making (opportunities, financing, etc.), at many points in time, than established firms.

From a choice-taking perspective the critical questions become the following: a) Are there identifiable patterns of growth, technology, and financial choices for a particular industry or even industry segment? b) What factors shape or determine which choices are taken? c) Are these patterns related to venture productivity and performance (Equation I.2)?

Furthermore, are the above questions to be differentiated for the various types of NTBF – RBSUs, academic NTBFs, EBSUs (ch. 1.2.6.1)? Do they differ with regard to initial (startup) financing, such as financing by bootstrapping, mixed private/loan financing, financing by private investors or venture capital investors? What is the role of the different types of markets for NTBFs (ch. 1.2.5.3; "economic markets" "policy-driven markets", etc. – Table I.15)?

Under which circumstances do we have a transition of a growing young firm from "strategy logics" to formal "strategy-making"? How is the evolvement of strategy bound to the change of entrepreneurial leadership to management, if at all?

> Following Drucker's third pitfall (4.2.3) one can say: *Growth creates problems*!

There is evidence that the young firms' growth dynamics, as a teleological process targeted at a given vision, mission and goal, may not be governed by a random process and this has stimulated scholars to investigate factors and put forward a large number of growth models.

Each of these theories focus on explaining different aspects. There is no commonly agreed framework or theory for growth of startups. The frameworks are not mutually compatible or consistent. They do not agree on which (endogenous and/or exogenous) variables are the essential drivers of startup growth. Bhidé [2000:210] suggested, for instance, that "the transition to a large enterprise requires greater heterogeneity of assets and functions and investment in administrative infrastructure."

Most of the literature on the theory of the firm assumes a firm to be already in operation, but "mainstream economics has little to tell us about how and why some firms survive and grow and others do not." [Bhidé 2000:242]

What is needed is a conceptually grounded *explanatory model* allowing making sense of the extremely broad spectrum of findings in the technology entrepreneurship area. Such an approach to tackle the developmental processes found among new technology firms as they generate and increase revenues and build their employees' base can only be fully illuminated by *micro-data from case histories*.

In this line it is important to find out how the overall GST framework for technology entrepreneurship as the conceptual basis of this book can make use of gross aspects of common venture growth models and add insights on the structural and processual levels – and raise further research questions.

Teleological, future-oriented GST models would imply that it is the purpose or final goal of the founder(s) and their views of "success" that guides the development process of a new firm. Hence, the developing entity *is purposeful and adaptive* to internal and external forces and events, and the process can be seen as a repetitive sequence of goal formulation, implementation, execution, evaluation and modification of intermediate goals or milestones or even distinct changes of goals (*purposeful enactment*).

For our context, the *explicit consideration of economic recession for firms' growth* will illustrate important broader conceptual implications for reasoning and exemplify the general paths followed to generate statements like hypotheses, proposals, propositions or explanations.

Figure I.2 and surrounding text has outlined that for statements with conditionals there is a need for care in making clear whether "reason why" or "reason for thinking that" relations illustrated in terms of antecedents and consequents are being stated. For instance, "prediction" in chemistry refers often to properties and reactivities of molecular classes and *exceptions* ("single events," "single data points") explained *a posteriori* by a particular condition.

In a related way, our growth models targeting expectations are often statements of that kind, for instance: the particular startup configuration let expect strong growth (> 25 percent) over the next 4-5 years unless an economic recession prevents that. Or, the startup can be expected to show good growth (around 10 percent) provided the oil price will not fall significantly below $80 per barrel.

Discussing NTBF growth we shall focus essentially on the time periods surrounding firm birth (Figure I.122) and its first eight to ten years of existence. The entrepreneurs' reasons and motivations to found a firm (ch. 2.1.2.5) and the ideas and opportunities they pursue (ch. 3) have been discussed in previous sub-chapters.

However, growth of a new firm is not only a matter of the drive for achieving the goal, but also of the pace of development (Figure I.122) and attitudes toward resources. "But founders, not outside investors, should determine the proper pace of growth for a company. And a founder who is about to lose his or her life savings is far more likely to drive a company towards profitability." "Landing equity money early on quickly leads to bad habits" and "bringing in outside money usually creates expectations of very rapid growth." In the end, creating a culture that emphasizes long-term profitability over rapid growth is critical for success." [Wadhwa 2009]

The author encountered the description of the above situation several times with founders as guest lecturers for his Technology Entrepreneurship curriculum (Box I.20). All the cited firms were LLCs and for the first ten years of existence under full control of the founder-owners who had invested much of their private money.

Box I.20: How founders of NTBF with preferences for financial sources determine the pace of their firms' developments.

As Wadhwa [2009] says "Hungry companies figure out ways to keep eating because they don't know whether there will ever be another meal."

We hear the same from WITec GmbH (B.2). WITec was funded with private money, which is not necessarily a bad thing, one of the founders (Klaus Weishaupt) said, because it creates pressure to make profits as soon as possible. "If the fridge is empty, the pressure is greater to make money than when you have millions from an investor to spend," he said, adding that he does not want the company to get listed on the stock exchange anytime soon. "We don't want to sell. It's better when the profits go into your own pocket."

Corresponding statements are heard from one of the founders of Nano-X GmbH (B.2), Reimund Krechan. "Permanently, I have been addressed by venture capital firms, but so far I have rejected their offers," Krechan said. "Money makes only lazy" (in German "Das Geld macht nur bequem").

"In hindsight, this scarce funding had also a positive impact as we had to focus on the essentials," said Niels Fertig (founder of German Nanion Technologies GmbH), "we have learned how important it is to orient ourselves toward the market and not toward the investors."
("Im Nachhinein hatte diese äußerst knappe Finanzierung auch positive Folgen, weil wir uns auf das Wesentliche fokussieren mussten," erläutert Fertig. „Dabei haben wir gelernt, wie wichtig es ist, sich am Markt zu orientieren und nicht an den Investoren.")

And ChemCon GmbH (B.2) follows a strict strategy that growth (in terms of employees and office/laboratory space) follows cash/profit development; positive growth *allowed* expansion. (in German: Das sehr positive Wachstum erlaubte eine Erweiterung des

Teams um 10 Mitarbeiter bis zum Ende des Jahres). ChemCon's principle from the beginning: continue to invest only if a new order was caught ("...weil immer erst dann investiert wurde, wenn auch ein neuer Auftrag vorlag.)

A graphical reflection of this attitude is also seen for German Nano-X GmbH (B.2) after having bought a building in 2001. After having caught a big order for its products, Nano-X additionally bought the adjacent area for a new production hall and office. At first Nano-X increased the number of employees to then later execute the request (Figure I.137).

Looking at growth of young firms is associated with perspective (Table I.64). For instance, the policy perspective (and expectation) is related to job creation (and growth). We shall take primarily a perspective of the firm and its founder(s) (and probably stakeholders).

Indicators of Growth

Development of a new firm can be tackled along several dimensions which characterize "input," internal structures, events and processes in a (new) firm or relate to performance and output, for instance, growth in terms of revenues or number of employees over time or interrelate both of these aspects (productivity, Figure I.130).

Garnsey et al. [2006] have critically discussed the various indicators that can be used for observing and measuring growth according to relation to input, value and output. The most common ones are absolute figures of revenues and numbers of employees (ch. 4.1). Each is subjected to limitations. Employment figures are the most commonly used measure because they seem to offer standardized, comparable data on the rate and direction in which a firm has been expanding. But, developments of revenues (disregarding inflation effects over time) and numbers of employees are not necessarily correlated.

Absolute figures of revenues and numbers of employees, particularly with regard to NTBFs, may have special issues. To a certain degree it has become common practice of NTBFs to incorporate money (funds or grants) from public sources for R&D into reported revenues. But, reliance on public grants or funds must be viewed as time-delimited projects. If this money is used for temporarily hiring personnel and without knowing about this may easily lead to misinterpretations of an effect of the decrease in the number of employees.

Depending on the situation the following discussions of NTBF developments will usually focus on "*absolute years*" for a time period or a sequence of individual years (for instance, 1997-2006 or 2003, 2004, 2006, 2008) rather than years of existence. Furthermore, we shall usually consider

- Absolute figures of revenues *and* numbers of employees *and*
- Productivity (revenues per employee),
- For production-oriented firms also output-oriented indicators, such as pieces of items (devices, instruments) sold or produced, volume of material (pounds or kg, barrels or liters) or overall performance of items sold, such as producible power (cf. Figure I.150, Figure I.154).

Getting data on revenues and numbers of employees is rather difficult, particularly for privately held firms. Except for public firms which are required to publish such data by law and according to standard accounting rules one has to rely on data collected by special for-profit firms, data reported to and published by the media or published by firm founders (in interviews).

Hence, one is plagued with issues of missing data for timelines and reliability and quality of the data. And even official data, for instance, in annual reports sometimes vary – and it is well known that such data for a year to year may be distorted in the framework of accounting rules to fit corporate policy.

Concerning growth rates we shall sometimes refer to the "Compound Annual Growth Rate." **Compound Annual Growth Rate (**CAGR) is a widely used mathematical formula that provides a "smoothed" growth rate. CAGR is often used to describe the growth over a period of time of some element of the business, for example revenues.

The compound annual growth rate is calculated by taking the nth root of the total percentage growth rate, where n is the number of years in the period being considered (Equation I.10).

Equation I.10:

$$\mathbf{CAGR}(t_0, t_n) = \left(\frac{V(t_n)}{V(t_0)}\right)^{\frac{1}{t_n - t_0}} - 1$$

$V(t_0)$: start value, $V(t_n)$: finish value, $t_n - t_0$: number of years

In any case, specifically for a GST approach, understanding a new firm's development requires taking firm-internal and external factors into account. Following the axiom that at any point in time, metrics of *firm size change* will show the firm undergoing growth, stability or decline and transition points between these expressions of development will indicate change of the firm from one state to another one, one may encounter problems of interpretation over a time period if economic recession is not taken into account. Its effect may show up in one or the other indicator or, worst, it may show up as growth in one (in revenues), but decline in the other (employees) as is observed, for instance, for Novaled (Figure I.148).

Growth discussions are especially critical if a sample of firms is considered using growth indicators based on "year of existence" (year = 1, 2, 3, ...) of NTBFs. Any observation and related comparisons referring to "year of existence" may cut off any external effect with a relevant impact for growth a startup may be subjected to and compares or group together "apples and pears" and other fruit.

Referring again to recession this could mean to group together startups with lacking necessary (external) financing during foundation, startups in their third year and almost no chance to catch customers due to the recessions and struggle for survival and "stable" NTBFs in their sixth year which just encounter a manageable reduction of revenue due to reduced orders of customer with conviction to overcome the recession and will be back to growth.

Generally, recessions may lead to difficulties in interpreting just growth curves (revenues or employees) which show setbacks (back to growth, plateauing; Figure I.107) in terms of firm internal problems or recessions or both. Additionally, revealing firms with delayed growth require careful examination of whether this is a reflection of slow firm growth in the early development phase or whether the situation reflects structurally effects of the firm, a characteristic of good or high growth hindered by recession and a rather long time for catching up to structurally appropriate characteristics – or just a favorable event that leads from a state with a medium or low growth to an "explosion of growth."

In addition to recession, there is a further environmental factor for (technology) entrepreneurship one must be aware of, the situation of industry genesis which may show up in waves of segments, such as the chemical industry and particularly the synthetic dye industry or automotive industry [Runge 2006:266-269, 274-275]. Currently such waves occur in the I&CT industry and here particularly the hard- and software industry and the "Internet industry" – and also in CleanTech with the biofuels segment (A.1.1)

Entrepreneurship in a segment of industry genesis usually has a special constellation concerning attracting entrepreneurs and financial backers as well as the government. This may induce a "push" for founding and funding affecting the early state of new firms compared to startups in other industry segments.

Finally, to understand growth of new technology venture it is important to be aware of factors and their interrelations (Figure I.114, Figure I.115) which led to decline and ultimately closure of the firm.

To proceed selected aspects of common growth models and their key fundamentals shall be discussed to reveal how to integrate them or, at least, put them into the context of an embracing GST framework.

4.3.1 Life-Cycle Models and Stage-Based Views

Life-cycle models [Bhidé 2000:244-259; Dorf and Byers 2007:459-465] similar to those of technologies and industries (Figure I.32) reflect stage-based views and suggest that organizations evolve in a consistent and "predictable" manner.

One widely-cited conceptual life-cycle work was published by Larry Greiner in 1972 and 1998. His emphasis is on five *continuous phases of growth* ("stages of evolution") interrupted by "stages of revolution." A growth phase (evolution) is characterized by a *particular style of management* (Figure I.118). Each "revolution" is specified by a *management crisis* and the solutions that lead to the next phase of growth. *The same organizational practices are not maintained throughout a long life span* (Table I.68).

The model's phases cover explicitly the development of firms till maturity. However, in the context of entrepreneurship the main emphasis will be only on foundation, early and later growth phases which are the "S-part" of a development curve (Figure I.32). Greiner tackles key issues that are relevant for entrepreneurship as *effects and organizational processes.*

Table I.68: Larry Greiner's Five Phases of Business Growth [Greiner 1998].

Size: From Small to Large ↓	**Life-Cycle (Age):** From Start Up to Maturity	
Revolution: stages of crisis	**Growth through**	**Crisis of**
Practices become outdated.	Creativity	Leadership
Companies that do not change will fold or cease to grow.	Direction	Autonomy
Often solutions for one crisis become a major problem in the next crisis.	Delegation	Control
	Coordination	Red Tape
	Collaboration	

Generally, a stage-based view may be viewed as a *"stimulus-response" mechanism.* Stage-based models presume that development is achieved by firm-internal change initiated by controllable or even uncontrollable events or processes or intentionally initiated change by entrepreneurs/managers or responding to external effects (cf. US Osmonics [Runge 2006:91]; German Zweibrüder Optoelectronics (B.2); Google (Figure I.160, Figure I.163; Box I.24).

An internal "event" could be the detection that the productivity of employees has decreased or grasping an opportunity and pushing a firm into a new direction. Fundamentally, in staged-based models new firm growth will be described in terms of some *prototypical triggers* affecting a stage and responses to initiate a new stage.

Mostly related to entrepreneurs starting in niche businesses who initially do not face obvious growth opportunities an emphasis can be that by intention – and dissatisfaction with the status quo – entrepreneurs decide consciously to go for larger opportunities or change development direction and invest in broadening the offering line or become active in another area of the value system [Bhidé 2000:252].

Drucker's pitfall number three – outgrow your production (deliverables) capabilities, management capabilities (Figure I.118) – seems to provide a good rule of thumb for entrepreneurship for growing private firms and associated effects on the organization of the startup. Furthermore, its empirically founded occurrence after four to five years shows up when the early period of risk of failure of the startup has passed the maximum and decreases (ch. 4.2).

But the new firm enters a first growth crisis. We look at the period of the new firm's ca. four years of existence simultaneously as the "*startup thrust phase*" for "lift-off" (ch. 4.3.2, Figure I.125) into potential growth of the startup. This shifts the focus to special *organizational life-cycle* (OLC) models for NTBFs.

The "startup thrust phase" coincides temporarily largely with a definition of "*early-stage entrepreneurship activities*" (Figure I.15) that covers a time span which begins with the initial communication of startup intentions, continues with the transition into active business as defined by the actual start of business activities (first sales revenues), and includes the ensuing survival or failure of the new venture. Thus the period covers the subject of nascent entrepreneurship as well as that of new venture survival/failure matching recent work [Keßler et al. 2010].

The paper of Keßler et al. [2010] based on a model (not specifically for NTBFs) consisting of the *person, resource/environment and founding process* forwards the basic assumption that *characteristics of nascent entrepreneurship* have significant explanatory power for founding success and new venture survival.

Growth of a firm is associated with *development of organization*. **Organization** refers to *structure* (order of components/parts) and *function* (order of processes/activities) and *coordination* for the pursuit of a goal.

Organization in firms is seen essentially as being determined by lines of *communication* and *influence*, *authority or power* as well as data and information flows through these lines of communication. Lines of influence/authority/power to structural components of the entity will be superimposed by *coordination* among the structural and functional components.

Corporate culture is usually seen as the basement of communication and coordination. Treating organizational development is often reduced to a special "one-dimensional" aspect in terms of the organizational structure and management complexity.

Malone [1988] defines **coordination** operationally "as the *additional information processing performed* when multiple, *connected actors* pursue goals that a single actor pursuing the same goals would not perform." (Emphases added) Coordination means extra organizing activities related to achieving a goal.

Boiled down to entrepreneurship and micro- or small firms this definition of coordination implies the following components: 1) a set of (two or more) actors, 2) who perform tasks, 3) in order to achieve goals.

The common problems that have to do with coordination are [Malone 1988]:

- How can overall goals be subdivided into tasks?
- How can tasks be assigned to groups or to individual actors?
- How can resources be allocated among different actors?
- How can information be shared among different actors to help achieve the overall goals?
- How can the different knowledge, education and professionalism and professional (special) languages be overcome and conflicting preferences of different actors be combined to arrive at overall goals?
- How can we track and measure the level of overall goal achievement referring to a system of activities and sub-processes? (added by the author)

These problems have been tackled partly in the discussion of the single entrepreneur (A.1.2) versus team constellations (ch. 2.1.2.5) and specifically entrepreneurial teams (Table I.41; Figure I.72, Figure I.73), but shall get more attention for entrepreneurship.

A paper by Hanks et al. [1993] presents an effort toward an empirical taxonomy of life-cycle organization-related stages in young high-technology organizations. In GST language, however, one would speak about navigation through a series of (system) "*states*" rather than "stages" each of which can be induced by a significant change or development challenge.

Two aspects of particular importance for organizational change (the internal dynamics of growth) are "specialization" and "centralization." *Organizational specialization* means how many functional areas exist in the firm so that at least one full-time employee fills the function. It should be noted, however, that specialization may be present already by functional roles at the start and in the very early phase of a young firm (Table I.41, Figure I.72). It may appear as a path from personal operational competencies to corporate functions (and establishing the value chain).

Formalization has been introduced as a further division of functional areas according to particular activities and professionalization, such as finances as financing versus controlling versus accounting.

Organizational centralization was measured by Hanks et al. [1993] referring to a list of five decision issues and finding out the level of management that must approve the decision (authority/power) before legitimate actions may be taken.

Cluster analysis has been used by Hanks et al. [1993] to derive empirically a taxonomy of growth stage/state architectures (originally called configurations) in a 1988 sample of high-technology organizations (R&D intensity on average 3.1 percent) reflecting a situation in 1987/1986. The derived architectures were interpreted as a sequence of four growth stages.

The notion architecture in the context of a firm may represent various expressions of organizational structures (Table I.69) [79]. It was proposed in this study that life-cycle stages could be defined and operationalized as unique architectures of organization context and strategy.

However, one has to note that in the sample firms which are seven years in existence could be affected by two recessions and those of sixteen years even three recessions in the US (Nov 1973–Mar 1975; Jan–July 1980; July 1981–Nov 1982), if sales and growth rates are considered. Suffering from a severe recession may require up to two years to catch up to the growth level before the recession.

Our interpretation of State I in Table I.69 is that it represents early growth until the "management crunch" after four to five years of existence (Figure I.118) reaching 6-7 employees with organizational settings probably close to the startup architecture. The NTBF is still a "micro enterprise" (Table I.4). State II has emerged after having crossed the "management crunch" and having adopted a functional basis of the organization and further differentiation toward less centralization but more formalization (a small enterprise; Table I.4).

State III having an average age of seven years similar to State II seems to represent a quite rapidly growing firm in terms of number of employees and sales with an operationally functioning *manufacturing* orientation and related necessary functional specialization, such as customer/product service, production planning and scheduling and quality control.

State II and State III show comparable original productivity which is on average $59,200 sales per employee for State II and $59,000/employee for State III. And the proportion of number of employees per special functions is 7.0 (State II) versus 6.2 (State III). However, the clusters differ markedly by level of centralization.

Hanks et al. [1993] view their State III as a fast expansion stage of development and this may tentatively assumed to be associated with a transition from the stage of Direction in State II to Delegation in III in the sense of Greiner (Table I.68).

The increase of number of employees above "25" would reflect a change of the firm's organization in terms of communication including reporting lines and coordination following the "10 - 25 - 150" rule of thumb (Table I.72).

Table I.69: Growth states and related organizational features for young technology-based firms [Hanks et al. 1993].

State I – Average growth in sales				
Relatively young, small, highly centralized and informal firm, focusing on the development and early commercialization of their technology-based product(s)				
Age (Years)	**Average Sales ($)**	**Number of Employees**	**Employee Growth**	**Special Functions**
4+x	271,000 (404,000) [1)]	6.46	29%	1.5 (mainly R&D)
	Organization Structure	**Organization Level**	**Centralization**	**Formalization**
	Simple	Little (2.2)	High	Quite informal
State II – Average to low growth in sales				
Appears to represent an expansion state of development; in addition to research and development, specialized functions present in at least 50 percent of firms; include sales and accounting, indicating that these firms are actively involved in the commercialization				
Age (Years)	**Average Sales ($)**	**Number of Employees**	**Employee Growth**	**Special Function**
7.36	1.4 million (2.0 million) [1)]	23.64	94% [2)]	3.4
	Organization Structure	**Organization Level**	**Centralization**	**Formalization**
	Adopted generally a functional basis of organization	Compared to State I firms an additional organization level (3.18)	Still very centralized, but less so than in State I	Little more formal than in State I

Table I.69, continued.

State III – Good-growth in sales				
Firms are still growing quite rapidly. In addition to research and development specialized functions are present in at least 50 percent of the firms and include sales and accounting; include functionally also shipping and receiving, finance, purchasing, customer and product service, production planning and scheduling, quality control and payroll. This appears to indicate expansion and increased professionalization, particularly in the *manufacturing* arm.				
Age (Years)	**Average Sales ($)**	**Number of Employees**	**Employee Growth**	**Specialized Function**
6.6	3.7 million (5.5 million)[1)]	62.76	28%	10.17
	Organization Structure	**Organization Level**	**Centralization**	**Formalization**
	Employ a functional organization structure	Average four levels of management	Have the lowest centralization mean of all the clusters	The second highest level of formalization
State IV				
Specialized functions present in this architecture, over and above those present in the other states; include personnel, building maintenance, advertising, market research and inventory control. Presence of these specialists may suggest greater formalization of human resource programs and policies, cost control, and market expansion; a divisional structure has emerged in several firms. They have overcome obstacles, such as a recession, and via healthy growth have achieved the status of a medium-sized firm. They have achieved the characteristics of Drucker's forth pitfall (ca. $60 million with 300-500 employees after 8–14 years of existence (ch. 4.2.3; Table I.67; Figure I.143, Figure I.144).				
Age (Years)	**Average Sales ($)**	**Number of Employees**	**Employee Growth**	**Specialized Function**
16.2	46 millions (69 million) [1)]	495	57% [2)]	15.3
	Organization Structure	**Organization Level**	**Centralization**	**Formalization**
			Centralization is low	Formalization is the highest of all the clusters

1) To make monetary values from 1986/1987 comparable with those observed for 2006/2007 we assumed an average inflation rate of 2 percent, which is multiplying the original figures by 1.49. 2) Not clear; due to the way of measuring growth, our emphasis will be more on absolute employee numbers.

State IV is characterized by a productivity (sales per employee) of $93,000. Hanks et al. [1993] put forward several limitations which hinders clear interpretation of State IV. But it may be tentatively associated with a further development of the high growth State III. The nearest comparable example to the situation of State IV could be the German SAP AG which after fifteen year of growth in 1987 had 468 employees and sales of €77.7 million (Table I.67).

Other clusters revealed by Hanks et al. [1993] are notable characterizing essentially *non-growth firms*:

> While similar in size to Clusters A (State I) and B (State II), respectively, they are significantly older than their counterparts. Companies average 18.7 years of age, yet employ a mean of only seven employees. Employee growth is non-existent, actually declining slightly. These firms have virtually no specialization, less than two organization levels (1.71) and employ a simple organization structure. Centralization is the highest of all the cluster groups and formalization is lowest. These are old small firms. They are presently not growing, and appear to have their product(s) fairly well developed.

This is in line with the discussion of "non-growth" NTBFs (ch. 4, Table I.63). For these firms entrepreneurs or owners, respectively, have consciously chosen to keep their firms small with a size of five to nine employees. The phenotype growth pattern corresponds to that of the "Growth Setback – Plateauing" phenotype or "Continuous Growth – Asymptotic" (Figure I.107).

Specialization in terms of developing functions of startups may appear by more or less fast hiring of appropriately targeted personnel of an otherwise unsystematic hiring process during the startup's development so far.

For instance, Cisco System's founders relied on improvised staffing for its first four years, but obviously ran into a "management crunch" (Figure I.118). John Morgridge "built the management structure" after he was recruited as President and CEO of Cisco in 1988 (Figure I.145, Figure I.158). He concentrated on hiring of "professional and experienced people in all the main functional areas – a Chief Financial Officer, a Vice President of Engineering, a Vice President of Manufacturing, and a marketing person" and recruited a professional sales staff. [Bhidé 2000:285].

Organizational development of NTBFs concerning specialization may follow a path from personal competencies and roles and activities of founders' functions outlined in Figure I.72 and Table I.41.

An illustrative development of internal and external organizational structures through specialization, leadership structure and networking during ca. nine years is given for the German NTBF IoLiTec GmbH (A.1.5; B.2). Furthermore, by the end of 2011 IoLiTec achieved a status concerning the level of progression from R&D via piloting to commercialization of offerings. One can also see how IoLiTec's *technology strategy* is implemented in the firm's state reflecting *executing the strategy*.

Disregarding the environment, for instance, that business growth is determined by technology and market environment of its industry, as a summary of internal effects for the firms' early growth one finds:

- Problems tend to change with increased number of employees and sales revenue.
- New functions emerge (specialization).
- Structural hierarchy increases.
- Formalized processes are established – for control (formalization).
- Jobs become more interrelated.
- Coordination and communication become more difficult.
- Organizations that do not grow can maintain the same structure for longer periods of time.
- Prior to change being possible the owner-managers need to develop skills and competencies in leadership, coaching and management before effective delegation and team building could take place.

A growing firm needs "feeding" (ch. 5.2); it must draw in new resources to support growth, but it may face coordination and delivery problems and planning delays as it is very difficult to synchronize resources to requirements in a dynamic system.The need for internal coordination and resource allocation sets a brake on the rate at which business opportunities can be pursued.

Start-up and early growth of NTBFs encounter several organizational challenges for growth. Parallel to increasing the number of hierarchies (organizational levels) relevant changes include

- The way of communication and the flow of information,
- Individual working processes and their coordination,
- The scope of tasks for employees,
- Employee development (Figure I.121),
- The felt extent of responsibility of the individual employee for the firm (commitment).

However, revisiting development of specialization in new firms requires lifting the assumption that the new firm builds functions or functional activities only internally to get results of particular activities. In contrast to the notion "outsourcing" meaning transfer of existing activities or whole functions of a firm to external service providers we look into "*outcontracting.*" This will be understood here as an NTBF buying in "contract services" for functional activities or a "whole function" that are needed for the business, but do not exist so far in the NTBF and will not be established in due course. Concerning the necessary activities of the startup a decision must be made between "do-it-yourself" or "let others do it." "Outcontracting functional activities" means additional and other kinds of coordination efforts.

Gottschalk et al. [2007] found that ca. 87 percent of young firm of research-intense industry sectors "outcontract" functional activities in Germany. They found the following levels of totally or partially utilizing contract services:

- Accounting (74 percent)
- Payroll (57 percent)
- IT Infrastructure (26 percent)
- Production (24 percent)
- PR/Marketing (18 percent)
- R&D (15 percent)
- Distribution/Sales (12 percent).

The vast majority of these contract services are done by domestic firms. The highest costs for outcontracting have been observed for outside manufacturing.

Young German firms have various reasons for outcontracting functional activities [Gottschalk et al. 2007]. Only 47 percent of the firms put forward cost as an argument. The major line of arguments followed a longer term strategic decision to outcontract functional activities to concentrate on core competencies and activities.

The insufficient endowment with personnel, technology and laboratory or pilot plant facilities, but also lack of experience in the context of application-oriented R&D projects imply that necessary scientific resources and research services, such as chemical or technical testing or engineering activities, must be purchased from external firms or research institutes to push the own R&D projects or to produce own products.

For young high-tech firms it is important to gain or complement, respectively, the lacking or insufficient technical and administrative infrastructure by external providers. Reasons for outcontracting are as follows [Gottschalk et al. 2007].

- Data security (16 percent)
- Reduce risk (28 percent)
- Lack of special personnel (33 percent)
- Lack of personnel (44 percent)
- Reduce cost (47 percent)
- Get better quality (55 percent)
- Gain flexibility (61 percent)
- Access to technology and know-how (65 percent)
- Focus on core competencies (97 percent).

A particular interesting case of outcontracting is the German firm Xing AG (ch. 3.4.2; B.2). For starting his Internet company the founder who perceived the opportunity implemented his business idea essentially by outcontracting a very large proportion of needed functionality from the beginning, in particular, software development.

Effects similar to directions of outcontracting in the US have been observed by Ardichvili et al. [1998] who looked into timing and sequence of startup teams' delegation of business functions in growing entrepreneurial ventures.

In 1986/87 their related study surveyed startups founded in the period 1979-84; and they repeated the survey in 1992/93. They found that there was no difference in the patterns of delegation of functions between manufacturing and service firms. Irrespective of the sales level and timing top level delegated functions included Accounting, Warehousing and Shipping, Production, Computer Systems, Purchasing and Personnel.

Currently, for NTBFs with larger production facilities also "*Quality Management*" has become a very important formalization (cf. IoLiTec (A.1.5, B.2), Nano-X GmbH (Figure I.137; B.2), Novaled AG (Figure I.148; B.2), Heppe Medical Chitosan GmbH (B.2)).

For NTBFs, also other kinds of specialization show up rather early – according to applications (IoLiTec (A.1.5, B.2), products or product groups and business lines (for instance (all described in B.2), WITec GmbH, Novaled AG; Zynga, Inc. (online games as products; ch. 3.4.1.1).

Finally some facts with regard to resources and organizational development of startups and NTBFs are to be noted.

- Startups, particularly RBSUs, often need internal and also a *distinct external organization* in terms of networking and cooperation with universities and public research institutes (as exemplified by the IoLiTec GmbH case (B.2)).
- VC-based startups and NTBFs usually get organizational structures of medium to large firms imposed by the investors which may occur rather early (Novaled AG (Figure I.148), biofuels NTBFs – A.1.1);
  NTBFs with management and organizational structures closely related to those of large firms are characterized by Bhidé [2000:4, 21] as "*transitional firms*" to indicate their way to become large firms.
- Founders with industry experience in (very) large firms tend to set up early an organizational structure for the NTBFs that mirrors business and organizational processes they encountered with their previous employer (ATMgroup AG, Polymaterials AG; B.2).

To inquire further into the details of NTBF growth complementing the sales and number of employees figures from Table I.69 one can refer to the study of Gottschalk et al. [2007] which provides averages of related indicators for German NTBFs (Table I.70). The data differentiate the high-tech area according to Table I.1, but do not differentiate initial funding and ownership (Table I.74). The data cover the 2001 dot-com recession.

According to the study, only ca. 7 percent of the high-tech firms in Germany did not achieve any revenues in the first year after foundation. In the first year of business all firms of the sample had average revenue of €190,000 which increased to €840,000 after ca. six years in 2006. The extents of growth of revenues and employees depend on the industry segment of the NTBF. On average the firms started with three to four persons.

Concerning employee growth after ca. seven years the firms had on average nine employees with five employees having a university degree; after ca. five years the firms had seven employees (see next page) roughly comparable with State I in Table I.69.

Young technology firms' growth during the first few years is rather strong with the growth rate decreasing significantly after four years. However, one must be aware of the fact that for startups high growth rates are often due to the low level they have started from. More realistic impressions would consider growth rate calculations to start with year four of the firm's existence. This would be after the "*startup thrust*" phase (Figure I.125).

The data of Cowling et al. [2007] reported for an NTBF sample from Germany and the UK collected 1997-2003 fit these findings: in the median firm in its 12th year there were 12 persons in Germany and 10 in the UK.

Furthermore, for the NTBFs growth in sales is consistently larger than growth by employee numbers. That indicates the productivity (sales per employee) has continuously increased over the time period under consideration, probably due to learning effects.

The data in Table I.70 show that after foundation on average revenues for NTBFs from the industrial high-tech areas TVT and HVT are considerably larger than those from the software and TBS areas. The yearly averages across the Dot-Com Recession shed some light on an recession's impact and the issue of sampling for statistical descriptions of NTBFs, such as that of Hanks et al. [1993] [79] (Table I.69).

Table I.70: Sales, number of employees and average growth rates of German NTBFs [Gottschalk et al. 2007].

Averages	**Sales Year 1 1); 1,000 €**	**Sales 2006; 1,000 €**	**People Year 1 2)**	**People 2006 3)**	**Growth, Employees**	**Growth, Sales 4)**
All 5)	190	840	3	7 (3)	24%	34%
TVT	260	1,130	4	8 (3)	25%	34%
HVT	350	1,530	4	10 (2)	29%	39%
Software	140	480	3	6 (3)	25%	37%
Other TBS	160	670	3	7 (4)	22%	32%

Table I.70, continued.

Year Started						
1998-2000	210	1,270	4	9 (5)	13%	28%
2001	170	740	3	7 (4)	17%	34%
2002	190	690.	3	6 (4)	18%	40%
2003	220	760	3	6 (3)	23%	50%
2004	170	650	3	6 (3)	39%	–
2005-2006	150	260	3	5 (2)	60%	–

Growth rates are actually calculated in terms of a CAGR formula (Equation I.10).
1) Sales in first year of business; 2) Number of full-time employees;
3) Total employee number and employees with university degree in parentheses;
4) Foundation years 1998 - 2003; 5). For acronyms cf. Table I.1.

In addition to the above outline on NTBFs' growths the following observations were made for German NTBFs. With regard to the foundation dynamics *NTBFs differ already during their early years by their development paths* [Creditreform - KfW - ZEW 2009]:

> *Most young technology ventures exhibit only little economic activities.* Except for the founding persons there are rarely additional employees. Except for the year of foundation there are no further investments and neither a product nor process innovation is launched.
> They probably represent largely RBSUs characterized (for the UK) as a "spin-out becoming one of the 'living deads' with little prospect of success." They have limited growth and hence, become the 'living deads' rather than disappear completely [Fyfe and Townsend 2005] – They are "*non-growth NTBFs*."
>
> For a *second* group of NTBFs' foundation indicators for development reflect *a build-up process which, after some time, comes to an end* (cf. Clusters A (State I) and B (State II) text below Table I.69): Growth in terms of number of employees and volume of investments slows down and the NTBF does not launch additional new products or processes.
> Technical developments concentrate on updates or improvements (phenotype: growth setback – plateauing, back to (low) growth or asymptotic growth (Figure I.107) – they are "*marginal NTBFs*."

The third group of NTBFs comprises firms which in their core domain exhibit high activities. Over several subsequent years they show high growth rates in terms of number of employees, they launch new products or processes or issue high investments (innovation and investment persistence; ch. 4.2.3, Figure I.117) in expansions or renewal of their capacities. The related phenotype could be assigned to "continuous growth or growth setback – back to growth" or even non-linear exponential growth (Figure I.107) – they are "*promising NTBFs.*"

This grouping is in line with data in Table I.63 and associated text, which is: most NTBFs start small and stay small, having roughly between twelve and fifty employees (Table I.69). A general analytical description of the developments is displayed in Figure I.155.

The above grouping is also reflected by a study of German *NTBFs founded in incubators with entrepreneurial teams*, the survey's population being from 2007 [Zolin et al. 2008]. Accordingly, all firms with an entrepreneurial team are rather small. On average they have twelve employees, 80 percent of them achieved less than €2 million in revenues and 82 percent report a balance sheet total of less than €2 million. Since no firm in the sample exceeds €50 million in sales or a balance sheet total of €43 million, according to Table I.4 the sample can be classified as reflecing a sample of small and medium-sized enterprises mainly consisting of micro firms.

Though not strictly related to technology ventures, a rough "80:20" rule seems to be also applicable for mid-sized firms. An investigation of 1,300 German mid-sized, usually family-controlled firms [Fröndhoff 2008] and comparisons with the 180 firms (14 percent) showing the strongest growth revealed that the top firms have growth rates of 10 – 39 percent, far exceeding the others (ch. 4.3.6).

Concerning the proportion of NTBFs with growth or fast growth characteristics, in line with general results from GEM (ch. 4.1), a recent US study [Stangler 2011] found for new (technical and non-technical) firms, for instance, that about two-thirds of job creation came from young firms, many of which were small and never got much bigger. Only a small number of firms creates a disproportionate share of such additional jobs; these are the top-performing firms. According to the study

- Measured by employment growth, in any given year, the top-performing 1 percent of firms generates roughly 40 percent of new jobs.
- The top 5 percent of companies (measured by employment growth) creates two-thirds of new jobs in any given year.
- The fastest-growing young firms (between the ages of three and five years) account for less than 1 percent of all companies in the economy, yet generate 10 percent of new jobs each year.

The "average" firm in the top 1 percent contributes eighty-eight jobs per year. The large majority of these companies end up with somewhere between twenty and 249 employees. On average, fast-growing young companies create about twenty-seven jobs per year, with most growing to a size of about twenty to ninety-nine employees. The average firm in the economy as a whole adds two or three net new jobs per year [Stangler 2011]. This has also been described for NTBFs in Germany (ch. 4.3.1).

Importantly to note, however, many of the jobs created by these fast-growing firms will disappear. Most of the companies in the top 5 percent and top 1 percent are young and so susceptible to failure even if they have been creating jobs. That means promoting fast developing, high-growth companies will not guarantee the creation of sustainable firms. Three or four years later the business may fail and the jobs disappear.

Specifically, controlling for industry sector and ranking on the *Inc.* 500 lists, it was found [Markman and Gartner 2002] that extraordinary *high growth – in terms of sales and number of employees – was not related to firm profitability* (cf. also LinkedIn Corp., ch. 3.4.2.1, B.2).

Venture capital and large private investment are viewed in the US to be important for the development and growth of certain firms. However, it does not appear to be universally important for creating high-growth companies.

Of several hundred fast-growing companies on the *Inc.* list over the last decade, only sixteen percent ever received a venture capital investment (cf. also the German situation, Table I.24, Table I.25, Figure I.54, Figure I.55). *Venture capital, moreover, is highly concentrated in just a few sectors*, but high growth companies can be found in nearly every sector of the economy [Stangler 2011].

We assume that foundations of technology-based firms are roughly a quarter of all foundations (Table I.2). But young, *high tech firms create a greater percentage increase of net new jobs than do the other categories of young firms* [Kirchhoff and Spencer 2008]. Hence, we propose that job creation by NTBFs can be approximately related to a kind of "80:20" pyramid displayed in Figure I.119.

From a point of view targeting practice providing *rules of thumb*, similar to Drucker's pitfall No. 3 (ch. 4.2.3), and in a context of a stage-based view of firms' growth one is led to search for approximate time lines with development-related "markers" for NTBFs, keeping Bhidé's [2000:247] basic approach into account that "does not try to force empirical regularities into a recurring temporal sequence."

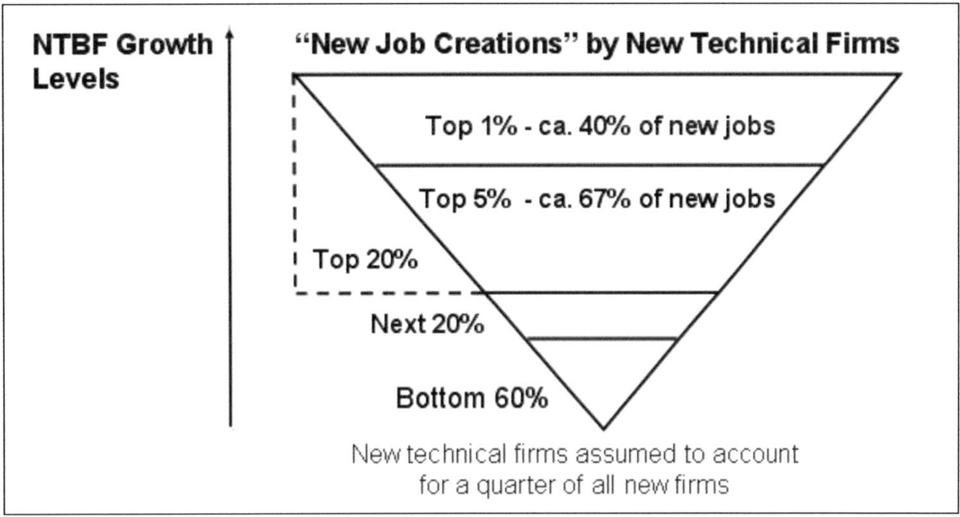

Figure I.119: The inverted pyramid of growing NTBFs related to employment contributions according to a rough 80:20 rule.

For instance, all the previously described observations suggest the following tentative time-related markers (Table I.71) concerning revenues for NTBFs with founders having full or majority ownership and full control of the venture (Table I.74) – unless significant firm-internal or external events interfere strongly with the development. The attribution also takes productivity (revenues per employee) into account.

Table I.71: Approximate time-markers for NTBFs with above average growth (non-VC-controlled startups).

Years of Existence	Revenue	Number of Employees	Average Growth Rates *)
4 - 5	$400,000; €300,000	7 - 8	Year 2: 60%
5 - 6	$1.6 mio.; €1.2 mio.	9	Year 3: 50%
7	$5.5 mio.; €4.1 mio.	30-60	Year 4: 40 %
15 - 17	$50 – 70 mio. €37 -52 mio.	500	Year 6: 30%

*) Estimations using data from Table I.70.

Other time markers for such kinds of NTBFs can refer to three rules of thumb reflecting organizational issues of young firms (Table I.72).

Table I.72: Approximate time-markers for development phases of NTBFs without VC participation and control via an implemented management.

- Drucker's third pitfall, the management crunch after 4-5 years (Figure I.118) for startups without a professional management team – and the NTBF having 9-11 employees;
- Fourth pitfall, the "Let go" issue (Figure I.118) after 10-15 years and ca. $60 million revenues (Table I.67, Table I.71);
- The "10 - 25 - 150" rule of thumb relating the number of employees to organizational issues, particularly related to leadership/management, specialization, communication, coordination and delegation.

The "10 - 25 - 150" rule of thumb, covering essentially the ranges 8-12 or 20-30, relates to the approximate number of employees of a growing firm or an overall range of 8 to 30 employees without discrimination. It has emerged from various discussions with entrepreneurs and confirmed by several of those who contributed to the author's Technology Entrepreneurship curriculum. For instance, in the time table of Attocube AG (B.2) one reads: "2005 >10 employees, introducing new organizational structures."

The situation inducing new firm structures was characterized by P. Klausmann [2011] by relating to communication and coordination issues: "You can gather up to ten people at one table to have a productive meeting." And N. Fertig, founder of Nanion Technologies, explained: "Sticking points for organizational problems occur with ten to fifteen employees. Then, internal communications becomes the decisive factor for the future development." (in German "Knackpunkte für Organizationsprobleme treten bei zehn bis fünfzehn Mitarbeitern auf. Dann wird die interne Kommunikation zum entscheidenden Faktor für die weitere Entwicklung." [Fertig 2012:Slide 42].

The rule relates to typical thresholds associated with necessary adaptation of the organization and its infrastructure to changed conditions of growth. The "10-threshold" is in line with Drucker's "management crunch" in year four to five of a new firm having on average seven to nine employees and Greiner's "revolutions" (Table I.68).

Overcoming these thresholds requires time, energy and money (including conflicts with employees and partners or stakeholders). According to Weber [2002] the *ability of large groups or organizations to coordinate* 12 people successfully (ch. 4.3.2) represents a certain threshold of efficient coordination *per se*.

The "10 - 25 - 150" rule of thumb is corroborated also considering growth of a startup in terms of the number of employees and associated expansion of building facilities. This means relating "communication intensity" (probability of communicating at least once per week) of the firm's employees in relation to separation distance in meters between the employees. Accordingly, communication intensity is ca. 30 percent, decreasing to 15 percent for a distance of 5 meters and converging to 12 – 10 percent if

10 meters are exceeded. The last measured distance was 80 meters [Ullrich and Eppinger 2000:Exhibit 14-12].

Concerning the "25 threshold" we hear from Niels Fertig (Nanion Technologies GmbH, B.2) when the firm founded in 2002 after seven years had 30 employees: "We now have reached a size which requires new structures (in German "Wir haben jetzt eine Größe erreicht, bei der man neue Strukturen braucht."; cf. also Table I.90).

Lars Hinrichs, the founder of German Xing AG (ch. 3.4.2.1; B.2), pointed to another relevance of the number 30 for firm development. During the early phase of Xing he held regularly meetings with his employees every Friday. Every employee was motivated to speak out, what went well and what went wrong – and to make suggestions for betterments. This worked until the group reached the number of thirty as a threshold because with more people no longer everyone could or would speak up. For more people in the firm corresponding meetings were organized by department in a face-to-face manner or for distributed departments via video conferences.

> According to the above view one can assume that the "30 marker" more or less ends a startup's state of *organizational success as an outcome of a dynamic organization-wide and all employees embracing process.*

It is notable that, for instance, Autio [2005] interconnects the "25 threshold" to "high-expectation activity" which is attributed to "all startups and newly formed businesses which expect to employ at least 20 employees within five years' time." His rationale: achieving the size of 20 employees is not simple (cf. also Solazyme – Table I.90). Firms of this size, typically, will have a developed internal specialization, identifiable management roles and some separation of ownership and employees, in the sense that not all employees are also owners of the company.

Several entrepreneurs the author spoke to mentioned that the "twenty-five employees threshold" has shown up as a dip in the firm's productivity chart (revenue per employee). Crossing successfully the "twenty-five employees threshold" has been identified as inducing a period with a rather long lasting organizational stability – unless other factors like recessions interfere.

> An economic recession may show up every 5-10 years (with predictable occurrence, but ignorance when it will occur and with what intensity). Furthermore, *a recession requires usually other qualities of leadership/management than a "normal" growth phase.* In particular, it may turn an NTBF's state into "survival mode."
>
> Hence, a recession does not only affect the financial state of the organization, but also the organization in terms of management, control and execution.

The "150 rule" has been put forward, for instance, by Sonnete [2011] through a social hypothesis, which covers also the "10 - 25" part: The brain size constrains the size of social networking (by group size) which requires memory on relationships and social skills. And the human brain can only remember 150 meaningful relationships.

Indeed, evolutionary anthropology suggests a hierarchy of social networking according to the number of sub-group members [Sonnete 2011]: Support Clique (ca. 3-5 members), Sympathy Group (ca. 10-15), Camp (ca. 25-30), Village (ca. 130-180) and Tribes (ca. 500-1,000). This fits also with the organizational hierarchy of the army with Sections (squads) (10-12 soldiers), Platoons (of 3 sections, ≈ 35 soldiers), Companies (3-4 platoons, ≈ 120-150 soldiers) and Battalions (3-4 companies plus support units, ≈ 550-800).

The "150 threshold" is also seen in Microsoft's (Table I.67) and Xing's (B.2; Figure 8) productivity data.

Gladwell [2000] also remarks upon the unusual properties tied to the size of social groups. Accordingly, groups of less than 150 members usually display a level of intimacy, interdependency and efficiency that begins to dissipate markedly as soon as the group's size increases over 150. This concept has been exploited by a number of corporations that use it as the foundation of their organizational structures and marketing campaigns.

Staged models mostly emphasize the fit between the design of the organization and growth stage. Growth is considered to distort the balance between the design of the organization and the stage of growth, and the task of leadership/management is to restore a balance. In this sense, the models are also *metamorphosis* models, as the organizational architecture of the firm needs to be changed for each stage.

The above patterns do not generally apply to VC-based NTBFs and NTBFs striving for large scale productions. Here, the following rules of thumb may be useful (Table I.73).

Table I.73: Approximate time-markers for developments of VC-based and production-oriented NTBFs.

- Venture capitalists assume that "lemons" ripen in two or three years, but the "pearls" take seven or eight.
  A new business rarely establishes itself in less than three or four years.
- Developing a company of sufficient size out of a new business development (NBD) initiative of a large firm will take five to eight years [Runge 2006:464].
- A seven-eight years requirements is also often observed in non-technical areas, e.g. it took FedEx eight years from idea to become operational [Bhidé 2000:168-185].
- For (existing) companies observed for a period of almost a century it took 6-11 years to transform laboratory results into *mass production* [Runge 2006:654-655].
  This seems also to be the case for NTBFs (cf. Nanophase Technology (Figure I.140) and biofuels startups – A.1.1). The related phenotype of growth corresponds *often* to that of "delayed growth" (Figure I.107).
  However, delayed growth may also be associated with an "unexpected event," such as a political interference as in the case of US firm First Solar (Figure I.154) and the German firm Enercon GmbH (Figure I.150; B.2) or due to firm-specific development issues, as observed for US Closure Medical [Runge 2006:98-103].

As a summary, staged approaches do have the merit of making observations of firms' dynamics with regard to revenues and number of employees as growth indicators – and combining both into productivity. However, without observations at the firm level, the mechanisms and processes of growth remain obscure.

Disregarding VC-based startups with their stage ("series") approach (Table I.27), "life-cycle and staged models" fail to adequately account for the great variety in the manner in which new technology-based firms grow. Moreover, comparing developments of "Ford versus General Motors" there appeared "great differences in how the two competitors developed their assets and routines, and in the role of their founders." [Bhidé 2000:245-247]

VC-based NTBFs follow often the VC-staged process (Table I.27) for (anticipated) growth which is structurally related to the Stage-Gate® (PhaseGate) process for innovation in large firms (ch. 1.2.7.2 Figure I.76, Figure I.180). In large firms intrapreneurship is often following strict phased processes with decision points and a restricted number of progression paths for innovation, NPD or NBD processes. Here, the prototype is the Stage-Gate® process (Figure I.79, Figure I.180).

Moreover, the early phases of a new product development project (which is concept generation and product planning) are commonly acknowledged to play a central role in the success of product innovation of large firms. Early decisions are unlikely to be changed during downstream phases, unless high costs and time are experienced. They have therefore the highest influence on project performance.

However, early analysis and problem-solving is also a difficult task, because the necessary information and insights are not available until one gets into detailed design. Most companies are locked in this dilemma between anticipation (anticipating decisions in the early phases of product development, where influence on performance is substantial) and reaction (delaying decisions to downstream phases, where information and opportunities are manifest).

A structurally similar situation is also found for startups. Though for a large part of new technology ventures developments seem to evolve in a phased manner following "stimulus-response" mechanisms and representing particular states of the firm the absence of prototypical *initial startup constellations and conditions* in stage-based models and the *absence of predetermined development paths* precludes so far some common features in the growth of NTBFs.

4.3.2 The Initial Architecture and Initial Configuration

The startup constellation represents a key for entrepreneurship as the basis for further development of the firm and its assessment by others ("observers," particularly financial backers). It will generate expectations concerning the development or even generate expected values for the firm (,for instance, financial projections of VCs).

In line with the GEM model (Figure I.15) we assume this constellation to be preceded often by a "pre-startup" phase comprising linking idea generation and revealing opportunity and conceptualizing the firm (Figure I.125). In the GEM-approach (Figure I.15) the pre-startup phase covers activities of the "potential entrepreneur" and the "nascent entrepreneur."

The diaries of the German firm Suncoal Industries GmbH (B.2) provides a lucid insight into the pre-start phase of an NTBF.

Basically, one must differentiate the *founding success* and the *survival* of the new firm. We shall denote the *one to four years period* after firm's foundation as the "*startup thrust phase*" (Figure I.125). Our discussion will follow an approach focusing on the founder person(s), resources and environment as well as the founding process. Here the vision and mission of the founder(s) (ch. 2.1.2.7) are important for building the firm and expectations.

Previous research suggests that the *initial choices of the entrepreneur or the team have a lasting impact on the way the company evolves*. In particular, they may facilitate *self-reinforcing* mechanisms (ch. 2.1.2.5), positive feedback mechanisms, by which a system's conversion processes and output (Figure I.5) or states are enhanced or brought into a more favorable situation.

For instance, group formation is often associated with processes of system's development by self-reinforcement and self-enforcement (ch. 2.1.2.5). Early decisions referring to intangible and tangible resources (Table I.8) will be important (ch. 4.2.2).

The discussions so far emphasized the fact that startup and NTBF development will be affected not only just by financing, but also by associated issues of ownership and control of the new firm. Correspondingly, initial architectures of NTBFs must be differentiated – by one of several classes of *taxonomy* given later in Table I.74 and Figure I.128 – to discuss the various NTBF developments adequately (cf. also A.1.1.6).

Initial *resource endowments* – the stocks of resources including the founders' experiences (ch. 2.1.2.4) that entrepreneurs contribute to their new firms at the time of founding – may explain the different life chances of new firms during start-up (cf. [Bhidé 2000; Klepper 2001]). In that way certain firms, such as spin-outs of industrial firms or serial entrepreneurs already control a relatively large productive base and some financial reserves at start.

Initial financing and very early development will often rely on own resources and resources of family and friends (3F) and significant cash flow (Table I.23, Table I.24, Table I.25).

And there is a number of NTBFs which were started because the founders had already customers (notable cases cited in this book are ChemCon GmbH (B.2) and PURPLAN GmbH (Box I.21) whose founders started as consultants or engineering planners, IoLiTec GmbH and Solvent Innovations GmbH (A.1.5), WITec GmbH (Table

I.41, Figure I.123; B.2) and US Cambridge Nanotech ([Yang and Kiron 2010], Table I.80; B.2) , SAP AG (A.1.4) and Concept Sciences, Inc. (CSI; Box I.11).

Furthermore, family and family members may not only provide financial resources (3F, ch. 1.2.7) or buildings/land to the entrepreneur, but also other business-related support and advice or other commitments as observed for William Henry Perkin in Britain, the "father" of the synthetic organic dye industry (A.1.2.), ChemCon GmbH (B.2) and Nanopool GmbH [Runge 2010].

Additionally, the Family & Friends System (ch. 1.2.2, Figure I.16, Figure I.17) may provide emotional support and role modeling. For instance, spouses with similar or complementary talents, skills or education (for instance, commercial rather than technical orientation) may enter the startup of the entrepreneur full- or part-time, as described for the role of spouses of the US firms Osmonics, Inc. and Avery Dennison [Runge 2006:91-94, 474-477]. Spouses also founded Cisco Systems in the US (both with IT competence; Figure I.145) or in Germany ATM Group AG (B.2), CeGaT GmbH ([CeGat 2011], B.2) and OHB AG (end of ch. 2.1.2.4).

Hence, many entrepreneurs owe much of their success to parental education or inherited or won family contacts. WITec GmbH could take advantage from two factors. One of the founders, J. Koenen, got a loan from his father-in-law who also served as a role for being self-employed (Table I.41).

Special family-related cases occur if entrepreneurial activities sprout out of existing firms of the family. One case concerns business succession. Another case is "branching" or "specialization" of a family-owned firm. For instance, the German startup "Heppe Medical Chitosan GmbH" providing nano-chitosan as well as high-quality and high-purity chitosan and chitosan derivatives for medical and cosmetics applications is run by Katja Heppe (now Katja Richter), whose parents run the company Heppe GmbH. The parents' firm produces among other things chitosan for industrial paper and textile applications.

Chitosan is a natural product, a linear polysaccharide, which is produced for commercial use from the shells of shrimps and other sea crustaceans. It is one of the most important renewable raw materials of the world, cellulosics being the most important one.

At the startup phase the founders' aspirations, ambitions, experiences, competencies and skills are crucial to the company's growth, as are financial and business resources. But, potentially interconnected to financing aspects, the initial startup setting is influenced by the fact whether the startup will strive for *large-scale production* or whether the characteristics of the opportunity requires *fast* development for entering the market.

Large-scale production requires tremendous amounts of capital. To be successful a rule-of thumb says: For markets with short product cycles the time till market entry is important, for longer product cycles cost of production is critical.

For the startup the environment of the NTBF is relevant, for instance, statuses and developments of technology, existing or emerging functionally equivalent technologies and industry (segments, markets and competition, and other relevant systemic influences). Therefore, the entity of inquiry will be the *initial configuration* which covers the startup's *initial architecture* (ch. 1.1.2; Table I.5).

> Technology entrepreneurship may be compared with an across country auto racing with a starting point and various initial conditions, but with a given finish. At the start there are cars with various technical and design features having a driver or driver team with various capabilities and experiences. But to reach the finish (goal) there are several courses with various hindrances, barriers, ups and downs as well as bumpy or smooth roads and an environment affecting the course, for instance, wind and rain making the course slippery.
>
> And there will be gas stations to get needed power and cross-roads requiring a decision to follow the planned route or changing direction. The "starting constellation" of the entity determines the momentum for "lift-off" and then decisions will have to be made which route to follow initially (the "expected path to success") and in which manner ("action and execution and speed").

A complementary metaphor can focus on the start of an aircraft: A runway is used for preparation and then there is a needed thrust for the departure of the aircraft (lift-off). All that is in the aircraft's "potential"; and sometimes there may be an additional momentum, for instance, wind from behind (for instance, reflecting a startup having already customers on foundation, as described above).

The metaphor concerns the "*startup thrust phase*" (Figure I.125) which may cover a *one to four years period* with the startup and the founder(s) having their new firm steam up (Figure I.122).

For those NTBFs which started with having customers the thrust phase will be very short, probably one or two years. The startup thrust phase of NTBFs will be associated mostly with characteristic "average" growth rate patterns (Table I.70, Table I.71) reflecting "acceleration" to lift-off. The startup thrust phase must be differentiated from the "*scale-up period*" of new firms with large-scale production which usually takes six to eleven years until product launch into the relevant market (Table I.73).

> If vision is where the entrepreneur wants to arrive at (ch. 2.1.2.7), culture is the foundation he/she can get there (cf. also Table I.43).

In this context of direction, execution and control Reid Hoffman, co-founder of LinkedIn (ch. 3.4.2; B.2), made some recommendations for entrepreneurship [Kaiser 2004]:

"Smart people tend to think that they can execute on a complex plan. Executing on a complex plan is generally a recipe for failure. If you can't make a startup work on a simple plan then your chances of success are very low.

Another lesson is that as an entrepreneur you want to choose a project that is far enough away from what others are doing. You want as much distance from other players so that you have the opportunity to create something.
The last thing I would add is that you want to measure your endeavor as soon as possible. You want to be able to gauge the viability at the earliest possible moment, so that you can change and adapt your model as needed.
Entrepreneurs tend to want to launch only when their product or service is perfect. The problem is that waiting undermines the ability to evaluate whether the idea works as quickly as possible, so that you can correct course. Correcting course frequently is key to success." (cf. overshooting, Figure I.88)

Creating Firm Culture and Developing Employees

In Silicon Valley, it is often said that the founder is the startup.

One of the components of initial resource endowments of NTBFs, being also a differentiator, is corporate culture which plays an important – though generally hidden – role. It emerges in the first few years of the startup. Particularly as an *intangible asset* a unique culture offers an advantage over competitors. It will affect product design, prototyping and realization, hiring practices and the values empowering employees to live the mission [Lowry 2011].

Usually it is assumed that the entrepreneurs play the key role in shaping corporate culture which is "intrinsic" to the initial architecture. Focusing on the founders we put forward a model of corporate culture to be formed essentially by a *self-replicating* process. Figure I.120 illustrates this for "*behavioral patterns*" as the important and observable expression of culture.

Self-replication is any process by which an entity will make a copy of itself – usually by distinct steps increasing the number of entities by one or more. The important ingredients to such a step comprise entrepreneurial *leadership by example*, *influence and employee development* (Figure I.141). Establishing behavior and "how things are done here" through "leading by example" ("walk as you talk") is a characteristic of many CEOs of Hidden Champions (ch. 4.1.1).

Metaphorically, assuming the founder(s) to determine the DNA for behavioral patterns can be compared with *gene expression*, which is the process by which information from a gene is used in the synthesis of a functional gene product.

An NTBF usually starts with up to five persons. That means shaping culture starts almost from scratch with the entrepreneurs/owners driving its direction. There is mostly one "leader" who has the greatest influence – usually the founder or the "primus inter pares" of a team. Disregarding situations were only two persons form the new firm, the model of a self-replicating system can start with a three-person constellation.

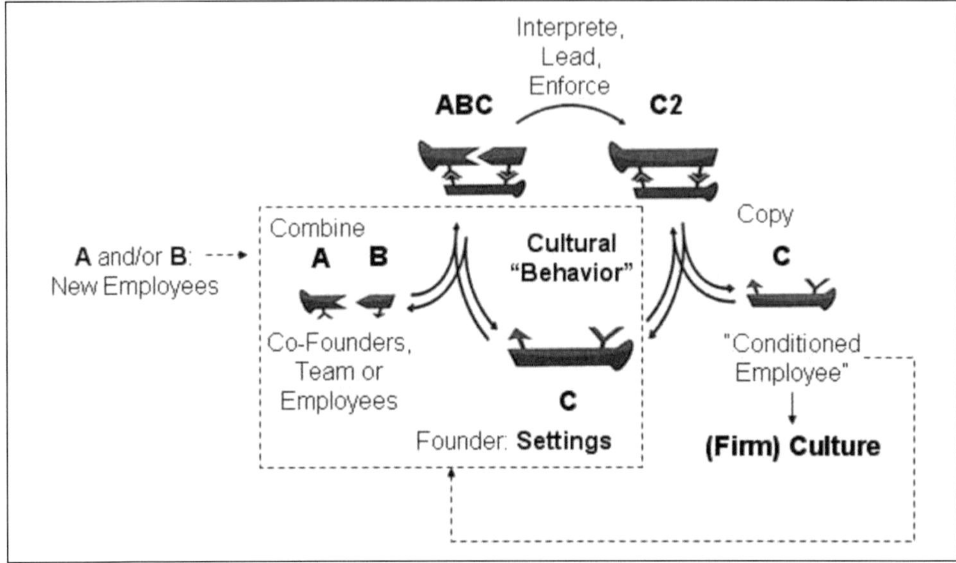

Figure I.120: Development of firm culture expressed by behavior of a firm's leader or founder by a self-replicating system.

Whereas Figure I.120 refers to a pattern-related process of firm culture evolution an interpretation of this process takes the following route. A small group, such as the founding members of a firm, does not face substantial difficulties in coordinating efficiently. Once they have done so, they can *establish a set of self-reinforcing rules or norms either tacit or formalized governing what actions and behavior are appropriate*. These norms allow the group or organization to continue to successfully coordinate activities.

As the group grows, new entrants' exposure to these norms allows the entrants to be aware of the appropriate behavior and creating an expectation for everyone in the group of what everyone else (including the new entrants) will do. Therefore, relatively slow growth and exposure of new entrants to the group's previously established norms can overcome large coordination failure.

The missing part of this interpretation is the "new employee" ("entrant") of the firm, which is associated with the firm's hiring process. A complement to the interpretation could tentatively refer to a "*targeted selection*" process which is ubiquitously applied by large firms for hiring activities focusing on a primary facility (a technical, commercial or administrative/managing specification) and a secondary facility of importance and additionally "needed to have" or "nice to have" competencies or personal traits and overall fitting with the corporate culture. [80]

Creating NTBF culture is the leader's responsibility to drive *employee development* by a directing or supporting style according to the employee's status of competence and commitment (Figure I.121). Leadership focusing on people in startups will emphasize influence rather than power as a means of assertiveness and execution (influence > attitudes > behavior; ch. 1.1.2).

Delegation and transfer of responsibility to the individual employee increases the level of identification with the firm. The leader's goal must be:

Keep the corporate culture across growth processes.

The human resources philosophy of 3M made in 1948 can serve as prerequisite of the interconnection of employee development and firm growth: "*As our business grows, it becomes increasingly necessary to delegate responsibility and to encourage men and women to exercise their initiative. This requires considerable tolerance. Those men and women to whom we delegate authority and responsibility, if they are good people, are going to want to do their jobs in their own way.*" [McLeod and Winsor 2003]

It is interesting to note that Greiner's staged model of firm growth and associated style of management (Table I.68) for growth of the firm also emphasizes directing and delegating as emphasized for leadership style for (individual) employee development in Figure I.121.

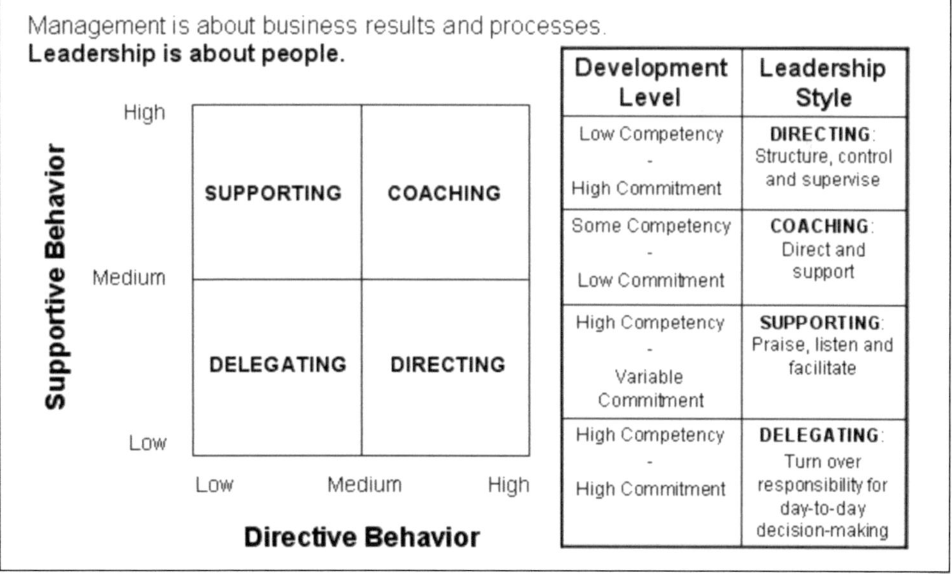

Figure I.121: Leadership styles for employee development referring to employees' development levels.

Self-reinforcing mechanisms as described above occur specifically in the context of *coordination* (ch. 1.2.1, 2.1.2.5; Table I.8, Figure I.73). There is ample evidence in experiments on coordination games that what a team did previously is likely to become a self-reinforcing norm about what to expect in the future. Game theorists have also recognized that organizational culture is one way in which a precedent helps select an equilibrium by reinforcement.

> Based on game-theoretical experimental settings with group sizes of 2 – 16 persons it was shown [Weber 2002] that the *ability of large groups or organizations to coordinate* successfully is *critically affected by the group's growth process itself*. Consistent with previous experimental research, coordination is much easier in small groups. It was shown that, even though coordination does not occur in groups that start off large, efficiently coordinated groups can be "grown." This corroborates also the interpretation of self-replication.

The experiments, indeed, indicated that efficient coordination in large groups is possible when groups start small and then grow slowly (coupled with the exposure of new entrants to the group's history). Moreover, the early failure of groups in the growth sessions of Weber's experiments appeared to produce an instance of the common view in the business world that firms can "grow too fast."

Following Kaplan [2003b] skills related to building the firm's culture comprise

- Leadership (ch. 1.1.2; Equation I.1): ability to build consensus in the face of uncertainty
- Communication: ability to keep a clear and consistent message
- Being a good team player: knowing when to trust and when to delegate
- Decision-making: knowing when to make a decision.

Trust, competencies and commitment will always be the foundation of success through leading decision-making and related courses of actions and execution (Figure I.111, Figure I.117). However, the starting point is always beginning with *communication* to ensure understanding of the vision and mission (ch. 2.1.2.7) and agreeing on common priorities.

Hence, 1) the leader must share his/her goals and objectives with the staff. And, after having a concept of activities and operations and specification of tasks in place he or she must 2) clarify coordinating instructions and 3) explain control and how to measure level of achieving objectives – having success (ch. 4.1).

Building consensus means *achieving agreements among a set of possibly very diverse people*. Building consensus is in the context of risk and uncertainty (ch. 4.2.1) and *making decisions* (ch. 4.2.2) at the right time. The expected result is that *everyone of the new firm is aware what his/her role and job is and what he/she contributes to success*. This implies to continuously check whether the firm's operations (Figure

I.5) have generated value and, moreover, whether value creation has been achieved efficiently.

Communication based on explaining *what is* and *why what to do* is a key capacity to *influence* someone else's behavior as compared with one's own behavior. In the context of entrepreneurship and building consensus among all the people of a startup it is advantageous to refer to operational facets related to *each individual's decision problem in the context of objectives* introduced by R. Ackoff.

Ackoff's suggestion as cited by Runge [2006:341] is concerned with the valuation of the objectives, the possible courses of action and the efficiency of each course of action in achieving each objective and the individual's probability of choice for each course of action. Accordingly, there are three *effects of communication* with a related basis, options and results referring to individual behavior and actions:

Motivation: Values of the objectives

Information: Possibilities of choice of the available courses of action

Instruction: Efficiencies of the available courses of action.

Information usually affects a person's mental state of cognition. Instruction refers to a finite number of actions to achieve a result which, as the simplest case, may proceed sequentially, but will become complicated through conditional branching options or loops. An everyday example of an instruction is the "cooking recipe." Motivation is the major effect leaders should focus on.

Concerning operations, behavior and social interactions corporate culture in large firms develops for new employees usually quite differently. Here one often observes three paths:

- Official and Codified:
  "Manuals"; for instance, Operations Manual, Office Manual etc. (on paper or the firm's Intranet); spelled out by supervisors
- Official and Not-Codified:
  Expected behavior; "tacit," rarely spelled out
- Unofficial & Not Codified:
  Do's and Don'ts; individual leaders' approaches to coaching and supporting.

This means, it takes rather long (and is costy) before a new employee has learned not only about the firm's offerings and strategies, but also the culture.

Initial Architecture and Initial Configuration with Corporate Culture Given

The notions *architecture* and *particularly configuration* for (technology) entrepreneurship (ch. 1.1.2; Figure I.16, Figure I.72, Figure I.73) are fundamental if we are looking for that "*footprint of a firm*" which may be sufficient to understand, explain and make statements about what can be expected from a firm's development.

According to GST the architecture is always related teleologically to the founder's goals interconnected with his/her aspirations and expectations. But there are various routes for the firm founder to reach the goals (Figure I.122). Accepting bounded rationality (ch. 4.2.2) the founder may not necessarily be aware of all the paths that are possible. As a reflection of competitive situations for startups striving to reach the "same goal" the systems view implies that there may be, for instance,

- (Almost) identical architectures having different initial configurations (think of startups in different countries, think of different input material as is observed for the biofuels industry – A.1.1).
- Initial architectures that may differ, for instance, just using generic technologies.
- Different architectures which will usually follow different paths (such as various conversion processes);
  however, at a particular "cross-roads" (Figure I.122), one architecture may switch to a path of another one; think of the many processes in biofuels when an initially thermochemical route switches over to a bioengineering process, becoming a hybrid process (A.1.1.3; Figure I.175).
- But, different architectures can generate the same value for firms and thus would not provide competitive advantage.

The initial configuration (Figure I.122), structurally initial architecture plus environment, depends essentially

- on the industry's or economy's situations like a recession or a "bubble" or the start of a new industry and
- on the market level and type, whether foundation is oriented toward a megatrend, a niche, requires building a market (disruptive innovation strived for) or start in a policy-driven market or mediatorial market (Table I.15).

Entrepreneurship in an area of *industry genesis* represents a special initial configuration for an NTBF (ch. 4.3.5; Figure I.143, Figure I.144, Figure I.145) and is often associated with the phenotype of "exponential" growth (Figure I.107).

Financing is a key component of a startup's architecture. Initial financing of technology-based startups often begins before the formal incorporation of the new firm. It begins during the *pre-startup phase* (Figure I.52, Figure I.125).

The "pre-startup phase" covers a time span which begins with the initial communication and discussions of startup intentions, driving forward related specific scientific or technical inquiries and continues until the formal foundation as a legal entity that has been authorized to operate by a state or other political authority ("incorporation").

After firm's foundation there is the startup thrust phase which coincides essentially with the phase of "early stage entrepreneurial activities" covering the first 3.5 years after formal firm's foundation according to the GEM model (Figure I.15).

Incubation of RBSUs (ch. 1.2.6.2; Figure I.) represents a particularly illustrative example of financing toward formal firm's foundation. Concerning availability or gathering of resources the pre-startup phase may overlap partially or totally with an incubation phase, which, on the other hand, may also extend into the startup thrust phase (Figure I.125; ch. 1.2.6.2; Figure I.).

Independently from sources of financing (Figure I.52, Figure I.59, Table I.30) one should take the statement given by a "bioentrepreneur" as a general recommendation:

> "The key thing I have learned over the past six-to-seven years is the importance of having enough cash in the bank. ... I have learned that it is a good strategy to raise money, even when it is not needed, so that there is always a sufficient cash cushion for when the market is uncertain." [DeFrancesco 2004]

Initial financing as one basis of pursuing the business opportunity is a key component of the startup's (initial) architecture and configuration (ch. 1.2.7). As founders' attitudes (and goals) toward financing in relation to control/ownership of the new firm differ the initial financing structure is an important factor and issue when addressing a firm's development and growth.

For proper discussion we shall introduce a taxonomy for financing of NTBFs (including initial financing) according to ownership and control by founders as given in Table I.74 (cf. also Table I.30). Control affects essentially leadership/management and strategy or strategy logics, respectively, and, hence, decision-making concerning the firm's development or growth path and pace. Initial NTBF configurations may show initial financing according to all these types. For instance, a large number of biofuel startups started with venture capital (A.1.1). They are "*VC-based startups*."

Obtaining massive amounts of venture capital or capital from an IPO does not necessarily mean loss of control. For instance, due to a dual class structure of common stock, a Class A share may be accompanied by five voting rights, while a Class B share may be accompanied by only one right to vote, or *vice versa*. Also dissection into three classes is observed.

A detailed description of a company's different classes of stock is included in the company's by-laws and charter. In this way the founders of the US firms Google (Box I.24), Groupon, Zynga, LinkedIn and Facebook (ch. 3.4) retained control over their firms.

Table I.74: Taxonomies of financial structures of technical startups' initial architectures in terms of ownership and control by founders and related typical sources for financing.

(Almost) Full Ownership, Full Control	Majority Ownership, Almost Full Control	Minority Ownership, Little (almost no) Control
Examples (any combinations or left out of one or more of the given components)		
Bootstrap; Cash flow; 3F + Bank loans; 3F + bank loans + "angels" + public R&D projects	3F, cash flow; bank loans + public R&D organizations (e.g. universities', research institutes' ownership in exchange for IP) + CVC + angels + silent partnerships of (public) investors or even VC	3F + VC +CVC + private large-scale investors ("*VC-backed startups*")

The vast majority of NTBFs fall into the two groups in which the founders' control over the firm is retained (over the first ten to twelve years of their existence). *VC-based NTBFs* account for no more than 5 percent of all the NTBFs (Table I.23, Table I.24, Table I.25). Their economic significance results particularly from the number of jobs they generate for the national economies (ch. 1.2.7).

We shall consider *VC-based startups* always as a class of NTBFs where venture capital firms or corporate venturing companies exert control over the firm. Sometimes there are firms which may have venture capital from investors or companies, but control remains with the founder(s), such as the above mentioned companies (Google etc.) or in Germany CeGaT GmbH (B.2) or Gameforge AG (ch. 3.4.1.1, B.2).

The transition from of a fledgling business or NTBF to a well-established medium-sized or large company is associated with fundamental transformations. If a startup becomes a VC-based NTBF the associated change from leadership and management roles of the founders does not only mean a change to professional management (Figure I.118), but usually also a switch to approaches of Management Science and Technology and Innovation Management (TIM, Figure I.1) with formalized structures and work processes, such as New Product Development (NPD), New Business Development, PhaseGate innovation processes (Figure I.79, Figure I.134, Figure I.180) etc.

Particularly, approaches, visions and leadership/management style of a CEO established by VCs – usually persons with one or two decades of experience in relevant positions in large firms – may clash with that of a founder taking a CTO or CSO position as observed for Amyris Technologies (Table I.99; A.1.1.5).

In line with phased models we assume that the paths of the startup to reach its goal are interrupted by "transitions of one into another state" with a certain period of duration (ch. 4.3.4), the "*transition states*." These are depicted as squares in Figure I.122

and occur as "crises" in Greiner's model (Table I.68) or "adaptations" in Bhidé's [2000:22] view. Transition states will "relax" to new states which will be "dynamically stable" characterized by a new configuration until the next "interruption" occurs. Additionally, it is assumed that

> the "transitions" and associated development phases are distributed uneven and *usually not foreseeable* for the whole path to sustainable growth.

"Interruptions" of a firm's development are initiated by events, happenings or "perturbations." And we regard a "firm event" (or "firm happening") as any effect from inside or outside of a purposeful system with a positive or negative impact on reaching the organization's goal(s) or keeping the organization's current state.

If these events or happenings become aware to the organization's control system (for instance, the firm's leadership or management team) they will require to make decisions how to proceed to reach the goal(s). Here, it is to be noted that for purposeful systems the principle "one cannot not decide" applies. This means, even no decision will have an effect on the firm's further development (ch. 4.3.5.1).

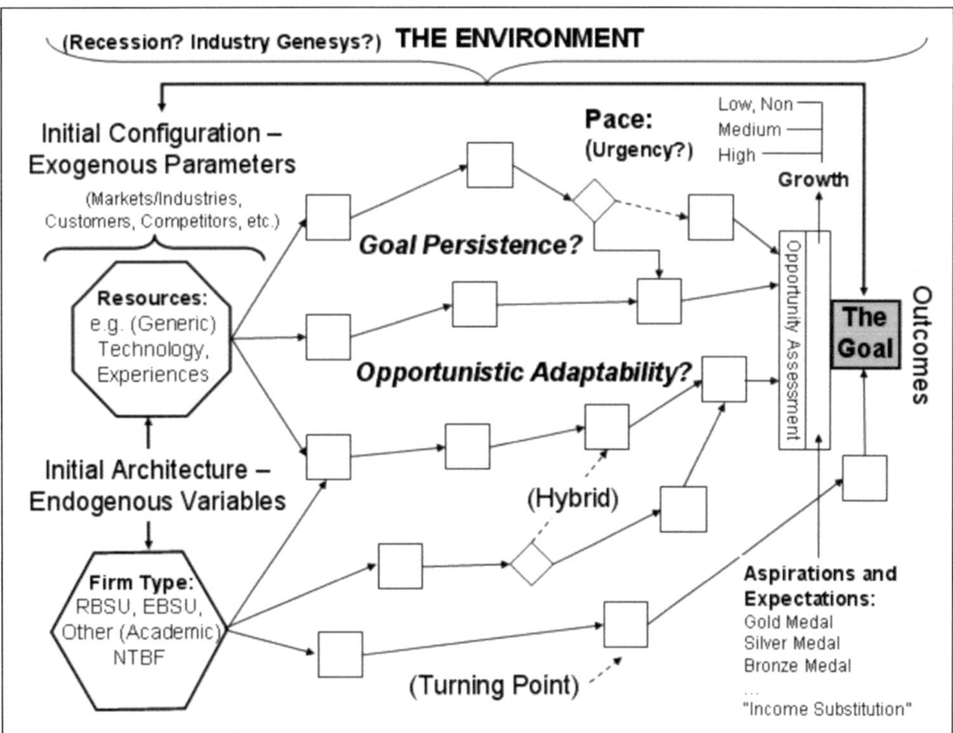

Figure I.122: Relating initial configuration with associated architecture and variability of paths to goal achievement for NTBFs.

Usually *unpredictability of specific growth paths* is assumed to be due to the increasing complexity of the firm as it grows. But, moreover, a purposive system like a firm must always exhibit choice of alternative courses of action (Table I.6), a choice of paths. The mix of tangible and intangible resources required for growth is rather precise, but shortages of any one resource, for instance, by a recession, can create bottlenecks with knock-on effects.

Apart from consideration of goals there may be requirements of *pace* for the growth of a startup. Pace can be defined by the entrepreneurs themselves (Box I.20). But in case of external financings there may be pushes for pace (or urgency) by investors or lenders. Another criterion for pace and speed up will be a conscious decision to compete against another large or small firm on pace if the entrepreneur knows that the particular competitor has a slowly moving organization, for instance, a bank [Hoffman 2007b].

Hence, pace or urgency as a driver of development may have an influence on the decision concerning the path to be followed to reach the goal.

On the other hand, Bhidé [2000:61] pointed out that there may be a number of startups which have no clearly defined goals, but follow an approach of adaptation to emerging or revealed opportunity: One third of the *Inc.* 500 sample entrepreneurs changed their initial concept.

Additionally (ca twenty years ago) 41 percent of the *Inc.* 500 entrepreneurs had *no business plan* at all and 26 percent had just a rudimentary plan [Bhidé 2000:54].

For technology ventures, however, one can assume that the situation has drastically changed. One can assume that currently the vast majority of the related technical startups have clear goals and plans (and will use a business plan). But there continue to be successful foundations without explicit and formal planning documents and are based essentially only on purposeful action (German PURPLAN GmbH, Box I.21).

Box I.21: Jump starting a company by finding, attracting and binding customers rather than writing an explicit business plan or doing detailed planning [Wintzenburg 2009; PresseBox 2012].

The German NTBF PURPLAN GmbH was founded in 2003. It had revenue of €14.2 million, 114 employees and 15 apprentices in 2011 and had revenue of ca. €18 million with ca. 120 employees in 2010 and an export rate of ca 50 percent (in 2009 €10.6 million and 120 employees; 2008 data: ca. €16 million and 103 employees; 2007 data: 80 employees, 2006 data: ca. €7 million and 65 employees). Since the year of foundation revenues have increased by a factor of ca. eight.

PURPLAN plans and constructs complex plants for storing and refining liquid substances – many of which are water-contaminating or inflammable, such as MDI and TDI and polyols for polyurethanes, hardeners, binding agents, pentane, solvents or glues. PURPLAN with a distinct focus and targeting consequently a niche in established markets has been recognized by the German National Founders' Award as one of the Top-Climbers (Box I.17).

PURPLAN carries out demanding tasks combining high-tech-planning and precise workmanship and executing its contract jobs through project teams in which engineering competence and practical project-handling skills complement each other successfully. Running successfully the interfaces between engineers, master craftsmen and craftsmen is PURPLAN's *core competency*.

Usually, there are firms which are good with planning or good with fast and cost-efficient construction, but only few can combine both areas. The interface is very difficult to manage. Characteristic features of all PURPLAN plants are holistic system solutions, high user comfort, quality and pronounced safety systems. In the area of automation programming is performed and software-solutions are compiled which meet the most sophisticated requirements.

PURPLAN has two mid-aged (ca. 45 years old) engineers as founders and firm's foundation resulted from frustration. One of the founders, Andreas Sandmann, was an employee and engaged considerably for fifteen years in building a machinery firm. However, in the end his boss rejected to delegate more responsibility to him. Sandmann wanted to shape more, wanted more decision-making authority and more reward.

Fully upset by this situation he decided to leave the firm and founded together with a colleague, Oliver Schawe, the firm PURPLAN – without a business plan, credit and starting capital. Their major "capital" was experience and contacts and networking. The team just called potential customers by phone ("cold call") and caught the first orders for selling a plant for production of polyurethane foams – via sub-contractors.

At the beginning they just wanted to plan the plants. But with an increasing number of orders growth started and customers suggested that they themselves build the plants – and they started PURPLAN. In retrospect Sandmann's conclusion was: "I now reap the fruits of several years of employment. This would not have been possible immediately after leaving the university."

This path of experience-based entrepreneurship starting with not much planning to change over to production compares with starting with consulting (ChemCon GmbH, B.2) and focusing directly on customers without much explicit planning; and it is similar to the path the founders of SAP AG (A.1.4) followed.

PURPLAN followed "*opportunistic adaptation*" [Bhidé 2000:18] which means without much effort to prior planning and research or only sketchy planning and rather high risk entrepreneurs adapt to unexpected circumstances in an "opportunistic" fashion.

PURPLAN GmbH is one of the above mentioned NTBFs which almost from the start are involved in commercial activities and production-oriented firm's foundation occurred through initiatives of customers. Some key components of the initial architecture include the founders having industrial experience in the startups activities, the firm type being "other (academic) NTBF" (Figure I.122; Table I.2) and relying on unique core competencies (Box I.21). Firm developments followed an exponential growth until 2008, when the Great Recession began.

As market entry by catching customers represent a key step and event for a startup and having already customers at the foundation provides simultaneously an emergence of an initial configuration out of the initial architecture.

> Having orders and/or customers already at firm's foundation is a special *favorable initial configuration* as it does not only provides revenues (and positive cash flow) from the beginning, but makes further planning of penetration into a largely known market rather reliable.
> Correspondingly, it will be much easier to attract external financing, if needed.

The below Figure I.123 illustrates this situation for the case of WITec GmbH, which was founded as a university spin-out (firm type RBSU) by an entrepreneurial triple (Table I.41). Similar to PURPLAN growth of WITec in terms of revenues exhibits a dip when a recession (here Dot-Com Recession) occurred. Otherwise growth follows roughly a linear pattern. Though revenue data are still missing, WITec does not seem to exhibit a decrease of revenue in 2010 as most firms do during the Great Recession. A "full" description of WITec's entrepreneurial configuration is given in Table I.80.

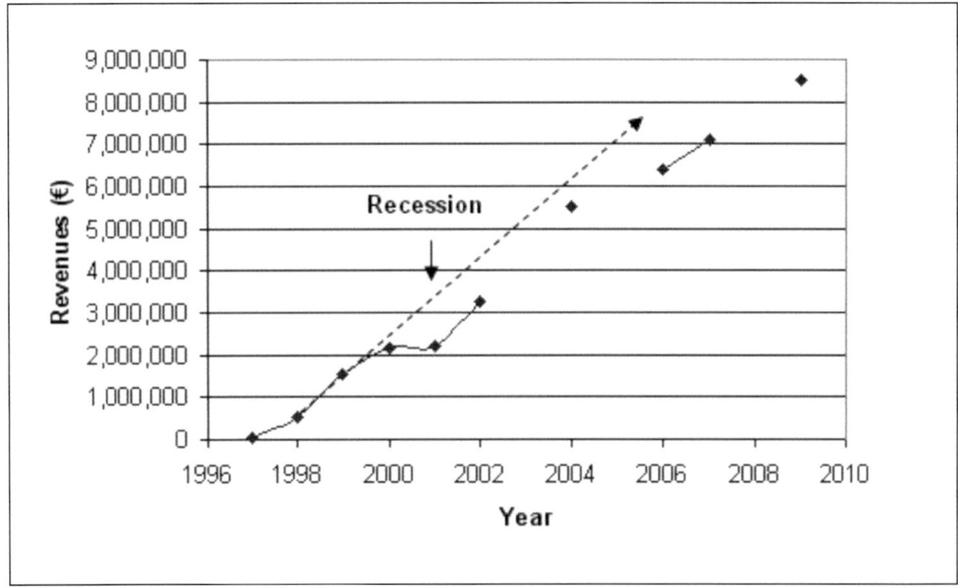

Figure I.123, continued.

German WITec GmbH (from Ulm in Germany; B.2) started 1997 with a real customer and some more customers with purchasing intentions and provided reliable expected values of profit data for 1998 [Koenen 2010]. This opened options to choose among three banks which offered needed loans for expansion (financial resource base). WITec's initial architecture is additionally characterized by a favorable leadership team (Table I.41, Figure I.73) and resource building.

Since 1998 revenues of WITec exhibited strong linear growth (squares represent real data, the dashed arrow extrapolates early revenue data). CAGR (Equation I.10) between 1998 and 2009 is ca. 29 percent. Since 2000 productivity of WITec is extraordinary and consistently around €271,000 per employee (configuration in Table I.80). WITec shows both innovation and investment persistence.

WITec could finance its growth essentially from generated cash-flow ("organic growth"; Figure I.127). In 2010, however, it added non-organic growth to its strategy. It acquired a majority stake of the Ulm company omt optische messtechnik GmbH which focuses on industrial process control.

Figure I.123: Revenue development of German WITec GmbH.

Similar situations of initial configurations (Figure I.124) when entrepreneurs start with customers with foundation of the new firm are the management buyout (MBO) and business succession (as described for Aluplast GmbH and KWO, B.2). A key differentiator here is the architecture with regard to education and/or experience of the entrepreneur(s). Admittedly, business succession is rarely a "new firm." It could be one, if the successor(s) reduces activities in the old business considerably and adds a new one or changes over to a totally new business model for the firm.

If there is an *a priori* strong interconnection of the founders' intentions and an explicit goal, there may be situations happening inside or outside the company that require changing the path. Furthermore, if the founders evaluate particular paths to be equivalent so that they cannot decide which path to follow they find themselves in a situation of "experimenting" – probably by trial and error.

The theme remains, but the way of implementation and execution changes. Then, it is of extreme importance to stick persistently to the vision and the essence of the goal – there must be *goal persistence* (Figure I.122). Take, for instance, the US company PayPal (A.1.7) that provides people the opportunity to pay online. Recently, payment by PayPal became also available for smartphones and PC-tablets.

According to Wikipedia, PayPal, founded in 2000 as a merger of two startups, is an e-commerce business allowing payments and money transfers to be made through the Internet ("e-payment"). Online money transfers serve as electronic alternatives to traditional paper methods such as checks and money orders. A PayPal account can be funded with an electronic debit from a bank account or by a credit card. The recipient of a PayPal transfer can either request a check from PayPal, establish his/her own

PayPal deposit account or request a transfer to their bank account. In October 2002, PayPal was acquired by eBay for $1.5 billion (ch. A.1.7). PayPal dominates e-payment globally.

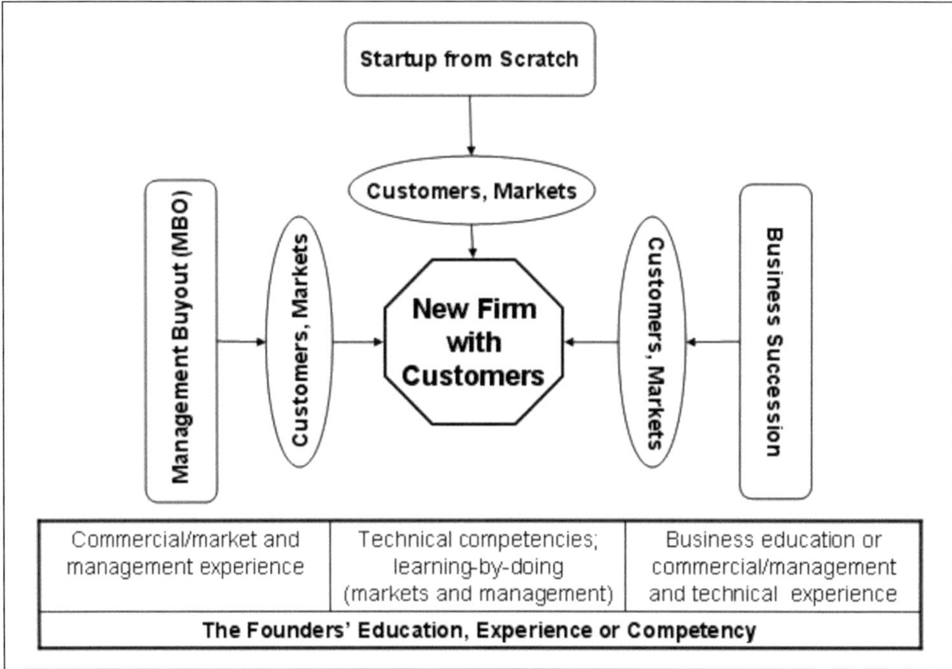

Figure I.124: Initial configurations for new firms having customers with foundation extended to management buyout (MBO) and business succession as an aspect of entrepreneurship.

According to Reid Hoffman who is also a co-founder of LinkedIn (ch. 3.4.2.1; B.2), over the years PayPal has made *multiple significant pivots*. The company started as a mobile encryption platform. Then it was a mobile payments company. Next PayPal was a combination of mobile and Web site payments company. Finally PayPal became an email payments company.

Each pivot over the life of the company was the result of *rethinking the business but maintaining the vision.* The focus was always to become a *payments operating system*; but the nature of the operating system changed multiple times [Hoffman 2010]. Such *goal persistence* can show up when it is operationally possible to accept long-term duration of developments to reach the goal (versus short-termism).

Goal persistence will be intimately related to the reason why to become an entrepreneur (Table I.39, Table I.40, Figure I.66), for instance, if the entrepreneur follows a vision or dream or wants to transfer science or technology into applications. In this case there is a directional relation to addressees, customers and markets.

On the other hand, if the driver of *firm's foundation is reflexive to the entrepreneur*, such as wealth generation, having self-determined work or seeking challenges, rather than goal persistence entrepreneurship may follow a path of *opportunistic adaptability* (ch. 2.1.2.4) indicated as an option in Figure I.122. However, focusing on self-determined work and independence without a clear direction following *opportunistic adaptability* may end up in a situation to be one's boss for a business one is able to manage, but does not like or is not convinced of.

Furthermore, goal persistence does not make sense, if the founders of a new technology venture recognize during the very early ("thrust") phase of their startup that their business idea will not materialize. Such a *false start* requires an immediate change of business direction and redefinition of the business goal ("failing fast" principle [Runge 2006:787]).

A corresponding situation is described for the 3M Corporation (ch. 1.2) or more recently observed for the German startup NanoScape AG which appears currently as a provider of porous, nanocrystalline materials (particularly nanocrystalline zeolites) and a developer of tailored application solutions for the CleanTech markets. Customer orientation of NanoScape means modification of NanoScape's materials to suit the needs of each individual application or adaptation of NanoScape technology to fit its customers' processes.

The road to the current orientation, however, was rather illustrating. NanoScape AG, founded in 2001, is a spin-out of Ludwig-Maximilians University in Munich (Germany) and the Fritz-Haber Institute of Max-Planck Society in Berlin. It was highly awarded in a business plan competition and got much publicity in the media through its catalytic (nanotubes-based) process of producing styrene, a fundamental raw material for the chemical industry (with the prospect that the technology could halve the cost of making styrene, targeting bulk chemical producers such as BASF, Dow and RoyalDutch Shell as customers) [NanoScape 2002; NanoScape 2005; Marsh 2003].

However, the original business orientation on catalysis and high-throughput technologies did not materialize which the founders realized soon. Due to the (dot-com) recession the founders encountered additional problems of financing. Hence, they ceased activities in the area of carbon nanotube catalyst development and some other activeties and repositioned their business model. NanoScape has so far survived on equity investments totaling less than €1 million. With 13 employees, it has built sales up to about €1 million in 2008 [Marsh 2008].

The NanoScape founders characterized their false start in the following way:

> "After an exciting jump start in the wrong direction NanoScape has now the right position to start toward a successful future."
> (in German: "Nach einem begeisternden Frühstart in die falsche Richtung ist nun die NanoScape in der richtigen Position, um einer erfolgreichen Zukunft entgegenzustarten.") [NanoScape 2005]

The additional lesson learned here is: Even being perceived as a high potential start-up on the basis of a business plan assessment and gaining general recognition and much attention in the public are no secure basis and guide to establish a successful NTBF.

The same can appear even to large firms. For instance, Dow Chemical won an R&D 100 Award regarded by The Chicago Tribune as "the Oscars of Invention" for developing a new plastic material, ethylene styrene interpolymers introduced as INDEX® Interpolymers, which was a novel family of plastics. However, this "new-to-the-world plastics flopped" terribly [Runge 2006:280].

Expanding the Resource Base by Networking

Finally, the role of *networking providing resources for startups*, not necessarily only for their initial architecture or configuration, has been described for the national innovation systems in terms of the concept of a "networked economy" (ch. 1.2.6, A.1.3; Figure I.39, Figure I.51). In its function to provide tangible (financial) and intangible resources to startups, such as access to analytical, testing or information services, corporate venturing (CV) arms of large firms (ch. 1.2.7.2; Figure I.125) are special. Corporate venturing, for instance, is not amenable to all kinds of NTBFs as CV follows either a strategic intent or specific interest in a particular technology and market(s).

CV may appear as the lead investor who then increases credibility of a startup to catch also other investors. Additionally, access to resources like those of the investing firm, for instance, to analytical or testing services and consulting, is also possible by networking with universities or public research institutes, as described for IoLiTec GmbH (A.1.5) and in many other case studies (B.2).

Variability of resources in a networked environment promises that there is scope for NTBFs to obtain leverage from financial resources ("smart money") affecting access to other resources (ch. 5.2) and tackling different aspects of resource requirements of the NTBF. But there are also pitfalls. The startup's leaders may not be aware of the gain of knowledge and experience of their (human) resources cooperatively involved with partners of the investing firm and, hence, may not fully leverage the (CV) resource from a strategic point of view.

The other situation concerns the situation if the firm investing in the startup will be simultaneously a customer, which may look for customized products. This apparent advantage may turn to serious problems when the initially favorable configuration of having a customer (and also money) becomes flawed.

For instance, as part of the business model of the US NTBF Nanophase Technologies (Figure I.140), the company expected its market partners to fund equipment that is primarily dedicated to produce their partners' products – not just only providing equity and consulting and services with the characters of an alliance.

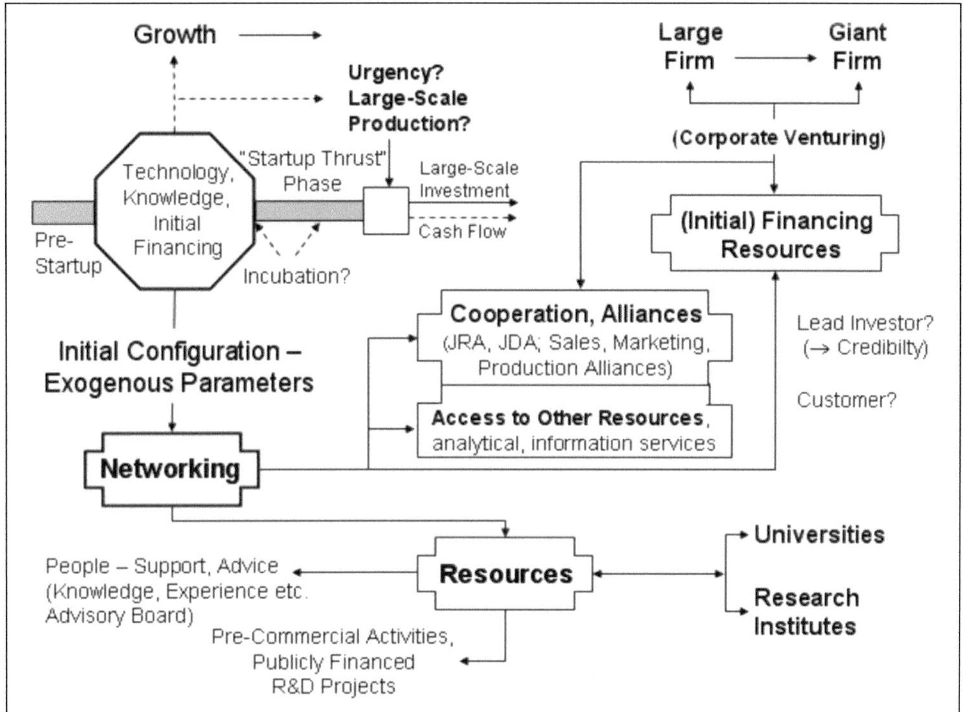

Figure I.125: Example of leveraged startup resources by networking with corporate venturing and academia.

German specialty chemicals firm Altana uses Nanophase's nanomaterial products as ingredients and additives for paints, coatings, polymers, plastics, inks and sealants under its NanoBYK brand. Altana made a $10 million equity investment in Nanophase during 2004. Altana also lent Nanophase $1.6 million to purchase and install nanomaterials production equipment during 2006 to support capacity requirements related to volume growth.

Early, during 2000, also German chemical giant BASF loaned Nanophase $1.3 million to purchase and install production equipment to produce nanomaterial products for its Z-Cote brand (zinc oxide into sunscreen applications). In particular, the role of BASF as a customer of Nanophase turned out to be highly critical (Figure I.140).

Apart from acquisition of an NTBF after corporate venturing into an NTBF (for instance, hte AG, B.2) also a marketing and sales agreement with a large firm may ultimately lead to acquisition, if the alliance was viewed as a strategic partnership with the NTBF (Closure Medical [Runge 2006:99-101] or Solvent Innovation GmbH; A.1.5).

More "Interruptions" for NTBFs' Developments

What was widely observed in the past and also in the present for a startup's initial architecture (for instance, in biofuels; A.1.1) is that its need of large investments often means a change of ownership to venture capital or similar investment firms with additional significant changes in leadership/management, fast increasing number of employees, organization and imposing urgency of development.

Figure I.126 illustrates these situations, often for startups aiming at large-scale production, referring explicitly to a VC case (cf. also Equation I.11). In this case not just the strategy of the firm to reach the firm's goal may be changed, but even the original goal may be changed.

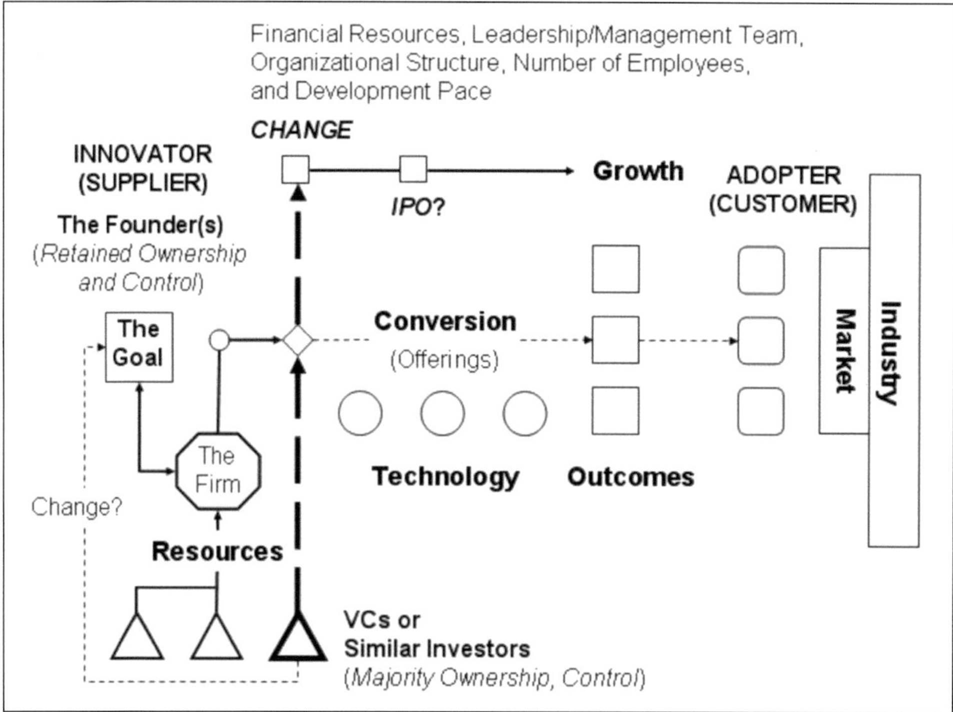

Figure I.126: Leadership/management change and equity ownership/control changes inducing organizational implications.

On the other hand, also changes of a startup's leadership/management to professional management or persons with profound leadership/management experience in industry or large research institutes without affecting ownership and control may be associated with a notable change of organizational (and probably strategic) settings.

Both effects can be expected to significantly influence development (and growth) of young firms, as does the "management crunch" (Figure I.118) for privately held

NTBFs (with full or majority ownership and control). We regard all such effects as relevant "interruptions" for new firm development.

A further related notable effect of cash influx for the young firm's development is a rather early IPO issuing public stocks which may or may not affect the majority stake situation. Going public of startups sometimes occur very fast, three to five years after foundation. Examples are the US Internet firms Groupon (founded 2008, IPO 2011; ch. 3.4), Zynga, Inc. (foundation 2007, IPO 2011; ch. 3.4.1.1; B.2) and Google (founded 1998, IPO 2004; Box I.24, Figure I.159) and German Xing AG (founded 2003, IPO 2006; ch. 3.4.2.1; B.2).

Basically, after initial financing and early commercial activities growth of NTBF may proceed along two different processes (Figure I.127):

- Organic growth:
  Growth essentially by own resources of the firm (cash flow or ownership and control keeping financial options)
- Non-organic growth:
  Growth through investment in other (micro or small) firms or acquisitions of or mergers with other (micro or small) firms.

While developing organic growth resembles the same entity to wind up by persistent innovation and investment cycles to reach the goal(s). Non-organic growth resembles an independent loop initiated by investment in a different entity – a new asset. The value results from financial contributions and – hopefully – also particular synergies of resources. Both kinds of processes may contribute to the overall observable growth. In case of a merger or acquisition to generate a new firm issues relate essentially to leadership structure, creating a generally accepted (new?) mission and fit versus clashes of firms' cultures.

> Due to the amount of money involved investment at any level in another company may represent an "interruption" of a firm's development path in the sense of Figure I.122.

Non-organic growth by acquisitions is often associated with a change to a group ("holding") structure of several firms at different locations. Legally, the "holding" covers controlling the individual "subsidiaries." The other option is integrating the acquired firm at the location of the acquiring firm. Very fast growth of startups is often achieved by non-organic growth (LinkedIn and Zynga; ch. 3.4).

Basically, both modes of growth will (have to) be linked to innovation and investment persistence.

In the context of entrepreneurship non-organic growth has notable consequences. When observing numbers of employees (or revenues) a jump in the figures may lead to erroneous conclusions concerning the firm. And concerning the embracing economic system, the total number of jobs created by firm merger or acquisition may not have increased. On the contrary, firms' mergers and acquisitions are often associated with job losses.

A central issue of investment decisions and activities (illustrated by Figure I.154) is associated with the firm's deciding on

- invest following demand or
- invest anticipating (before) demand.

For instance, private firms relying much on generated cash flow for development tend to invest after demand (Box I.20).

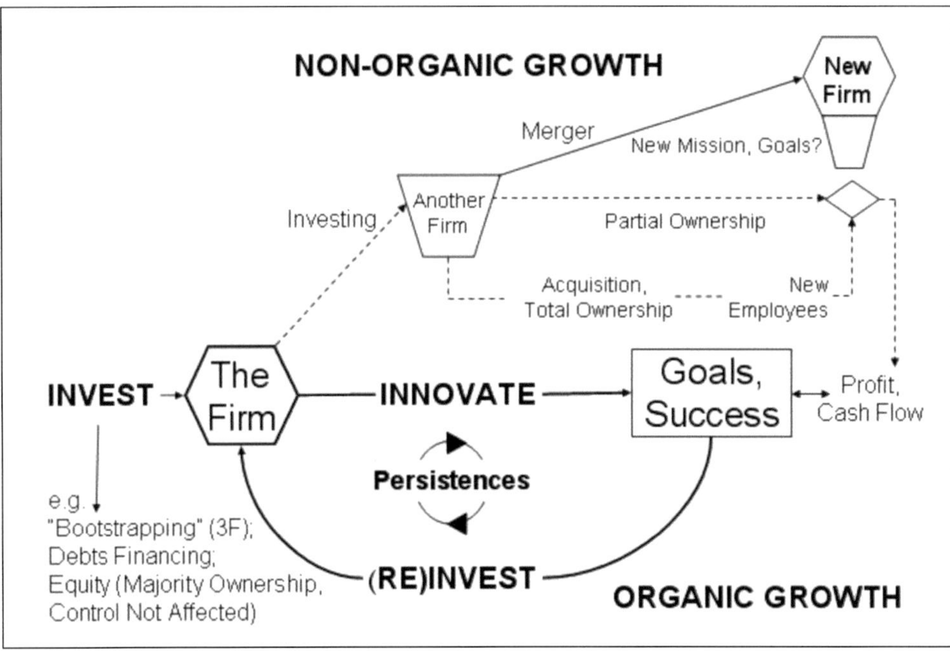

Figure I.127: Differentiating organic and non-organic growth of (new) firms.

The most prominent example of organic growth versus non-organic growth of NTBFs is the "Ford versus General Motors" case [Bhidé 2000:246]. The Ford Motor Company (FMC; started in 1903) evolved in line with Henry Ford's engineering and manufacturing interest and was build essentially using initially raised capital from friends. He turned his vision of large-scale production to manufacture cars that everyone could afford through raising capital without losing majority control.

General Motors (GM), founded by William Durant, followed a different route. In its early years it grew through acquisitions of smaller companies then in the automobile production rather than building its own plants, for instance, in 1904 the Buick Motor Company etc. GM as a holding company for further acquisitions was formed in 1908.

Organic growth and non-organic growth may occur very early during the development of the NTBFs, already during the startup thrust phase. And there is no way to differentiate these processes referring to the typical growth indicators – revenues or number of employees.

A striking example is the Nanogate AG [81] / Nano-X GmbH (B.2) cases in Germany. Here, both NTBFs where spun out at almost the same time from the same research institute, both having a person with high leadership/management experiences in the foundation team and both targeted markets for surface coatings on the basis of chemical nanotechnology. Both firms had different initial financing sources and very different growth strategies, in particular, non-organic growth versus organic (Figure I.137, Figure I.141).

Furthermore, it seems that firm's foundation by persons with deep and broad experience in technology, markets, management and simultaneous responsibilities for several subsidiaries of a large firm are more ready to build their grow strategy on non-organic growth (ATMgroup AG, B.2).

Finally, on their road to SMEs based on organic growth, ten or more years after start-up, NTBFs may also select non-organic growth as an additional path to growth. This is observed, for instance, for the German optical instruments ("nano-tools") firm WITec GmbH (Figure I.123). On the other hand, the German optical instruments firm JPK Instruments AG, which addresses similar markets as WITec with a related technology followed non-organic growth rather soon after foundation (Figure I.141). Also Google followed a non-organic growth path (Box I.24).

Features of new firms' development described so far in terms of a general systems view, using teleology and specific effects from inside and outside the firm influencing further development, suggest evolutionary models to contribute to additional insights into growth of new firms.

Evolutionary models [Bhidé 2000:249-254; Runge 2006:7-8] essentially map startup and firm developments to Darwinian Theory of hereditary factors as the origins of change variations and "natural selection." Accordingly, history (of the initial architecture) matters – and firm development would be largely "path-dependent" unless a "mutation" induced by external factors gives rise to spontaneous and permanent change in the DNA of a "gene" of the firm.

In Darwinian Theory hereditary factors are determined by the principle of homology. Hereditary factors will only be transformed in small steps and the degrees of freedom of these hereditary factors will be retained (under negligible or slowly changing external conditions) and will only change according to certain patterns. In social systems, for instance, one has processes of homologation for values and behavior of persons, layers of groups and larger organizational units.

The related Darwinian natural selection according to reproductive fitness, however, cannot explain all the shaping. Functionality has to be differentiated from "pattern formation." *Structural evolutionary theory* is concerned with pattern formation and works to explain why Nature favors relatively few structural archetypes. It limits the choices available for selection and explains formation of similar patterns as "*Variations on a Theme*." Some aspects of this theory can be used for a GST approach to development of firms (A.1.6).

4.3.3 Resource-Based Views

Entrepreneurs who are involved in the early stages of business creation are unlikely to become preoccupied with life-cycle issues of decline and dissolution. Inspection of major factors for decline of young firms refers often to lack of resources (Figure I.114, Figure I.115) – often already at the start of a new venture.

"What an organization knows at its birth will determine what it searches for, what it experiences, and how it interprets what it encounters" [Huber 1991:91]. One implication is that a new firm's learning and capability accumulation may influence its decision-making and paths of development markedly (cf. Figure I.129 and below text).

When dealing with startup architectures and configurations and their significance for the development of young firms the previous discussions have emphasized a number of *tangible and intangible resources* (Table I.8). Some of them are sufficient, some being necessary and some fulfilling both requirements to expect viability and growth of a technical startup of a particular type (RBSU, Other academic NTBF, EBSU).

Apart from country-specific factors, we dealt, for instance, with a variety of resources of the entrepreneur(s) or the startups, respectively. Personality (ch. 2.1), cognitive abilities, creativity and revealing opportunities, education and experience (ch. 2.1.2; 3.2, 3.3), personal competencies (Figure I.72), a team for firm's foundation (ch. 2.1.2.4; Table I.41, Figure I.72, Figure I.73) and team extensions (including an Advisory Board), communication and coordination (ch. 2.1.2.4), networking and cooperation issues (ch. 4.3.2), financial endowment (ch. 1.2.7) etc. were emphasized.

We also dealt with macro-economic events like recessions, markets, industries, competitors and national legislations etc. and interconnected endogenous factors of a firm's internalities with exogenous factors from external analysis (Figure I.114, Figure I.115). Finally we observed a "strategic group" in Germany with very similar architectures to exhibit from the start good to high growth patterns to become mid-sized or large firms (Hidden Champions, ch. 4.1.1).

Concerning viability and growth of startups all this directs special attention toward *resources* (ch. 1.2.5.2) *and sustainable competitive advantage*. Hence, it is important to inquire into the role the economics-oriented resource-based theory or the resource-based view (RBV) [82] of firms [Bhidé 2000:214-216; Alvarez and Busenitz 2001] and which role it should or can play for dealing with developments of new technology ventures in the context of technology entrepreneurship.

Resource-based theory was extended to include the cognitive ability of individual entrepreneurs [Alvarez and Busenitz 2001], but having corresponding attitudes, aspirations and intentions to act and pursue resources are also important. *Resource heterogeneity* is the most basic condition of RBV and it assumes at least some resource bundles and capabilities underlying offering generation are heterogeneously distributed within the firms.

Competency and skill heterogeneity of the founder team has been discussed (ch. 2.1.2.5; Figure I.68, Figure I.72, Table I.41). Skill heterogeneity plus a constraint is emphasized, for instance, by a founder's experience: "Early on, one of the most inspiring lessons we learned was the value of bringing people together with different backgrounds and skills" but, furthermore, "almost all {interviewed founders} would agree on the importance of getting the *right people*, based not just on their expertise, but on their attitude as well." [DeFrancesco 2004] This attitudinal aspect was also a reason to dissolve the founding team (in agreement) of the German NTBF Attocube AG (B.2).

Driven essentially by the currently dominant view of corporate strategy RBV regards a *company as a collection of resources and "capabilities."* In RBV a **capability** is usually considered a "*bundle*" of assets or resources to perform a business process (but cf. Equation I.2). The business is composed of individual activities or, in the sense of GST, a "*system of activities*." For instance, the product development process involves conceptualization, product design, pilot testing, new product launch in production, process debugging, marketing etc. across the firm's functions – for NTBFs usually tasks and roles for activities along a value chain (Figure I.7).

Note that in RBV it is generally assumed that resources are tradable and non-specific to the firm, while *capabilities are firm-specific* and are used to engage the resources within the firm. [82]

However, a firm will usually focus on certain capabilities consistent with its strategy or, in case of entrepreneurship, with strategy logics to reach the founders' goal. The resource-based perspective fits the GST framework if it highlights the need for a fit between the whole external context in which a company operates and its internal capabilities and potentiality (Equation I.2).

That is, *dynamic capabilities* (ch. 1.2.1) are required, the adaptability to the environment to renew and re-configure the competencies in response to key factors and conditions of the environment.

The essence of the resource-based perspective is a *knowledge-based view* and taking GST this will include foreknowledge which means *intelligence* (ch. 1.2.3). This is associated with a particular constraint. Tacit knowledge (or technology) which can be learned only through personal experience is an example of know-how as a resource that is difficult to have or transfer *ex ante* (before or on foundation). Furthermore, assuming bounded rationality (ch. 4.2.2) cognitive limitations imply that no two individuals possess identical stocks of knowledge.

That all means, entrepreneurship and RBV adopt precisely one common unit of analysis – the resource. These resources may manifest themselves in several different ways and different phases of NTBFs' developments.

One kind of difference occurs in the ways entrepreneurial leaders and managers of small versus established (large) firms utilize the resources to exploit (business) opportunities. Another important difference is that for NTBFs founders often learn on the job how to utilize and build up resources, for established (large) firms there are educated and trained "professional managers." And importantly, the relevance of resources emerges in the face of competition when the NTBF must answer four questions about its competitors' resources:

- How do they compare in terms of size and components?
- How efficiently are they used?
- How effectively can we learn from their experience and practice?
- How do we maintain our own competitive advantage?

According to RBV each organization is a collection of *unique* resources and capabilities that provides the basis for its strategy (logics) and that is the primary source of its *performance* and, hence, returns. According to common views [Autio et al. 1997] to provide a competitive advantage resources and capabilities must exhibit the attributes listed in Table I.75.

Table I.75: Key attributes to provide competitive advantage.

Attribute (VRIO)	Resources, Effects
Value	Allow the firm to exploit opportunities or neutralize threats in its external environment with the resource/capability
Rarity	Control of the resource/capability, possessed by few, if any, current and potential competitors
Imitability (Non-substitutable)	When other firms cannot obtain them; the significant cost disadvantage to a firm trying to obtain, develop or duplicate the resource/capability
Organization	The firm is organized appropriately to obtain the full benefits of the resources/capabilities in order to realize a competitive advantage (cf. also Equation I.2)

For NTBFs value and imitability in the sense of VRIO is related to *technical value* concerning the producer/supplier (ch. 1.2.5.2). To the supplier or producer technical value is measured by how *protectable from the competition* the product is or how *exploitable* the product is as a basis for further offerings, for instance, based on a platform technology (Table I.12). A discussion of applying the VRIO-approach to the competitive advantage for a startup in the new technical field of "ionic liquids" is discussed for the case of German IoLiTec GmbH (A.1.5, B.2).

The resource-based view suggests that *heterogeneity* is necessary but not sufficient for a sustainable competitive advantage. For instance, it has been shown that often heterogeneity requires interfaces or gatekeepers or management of these, respectively, reducing inherent barriers imposed by the differences of particular (human) resources (ch. 1.2.3; Figure I.20, Figure I.73).

As Conner and Prahalad [1996] argue, any (strategy-oriented) theory that seeks to understand performance differences between firms must incorporate a theory that addresses the question of why firms exist. This key question is also fundamental for GST and its emphasis on purposeful behavior and the strive to reach goals. It is tackled with regard to the entrepreneurs' reasons and motivations to found a firm (ch. 2.1.2.5; Table I.39, Table I.40), the ideas and opportunities they pursue (ch. 3) and the risk they take and decisions and actions they make (ch. 4.2).

Often it is the founder (or founding team) who possesses much of the technical and managerial knowledge that make-up the assets of the firm. Hence, an entrepreneur's expanding knowledge base and absorptive capacity may become the entrepreneurial firm's competitive advantage. Furthermore, concerning intangible assets these are (almost) inimitable because they have a strong tacit dimension and are socially complex.

In the entrepreneurship domain, tacit socially complex assets are often also directly generated by the founder(s) and spread across the organization, such as firm's culture (Figure I.120) and hiring and developing human capital (Figure I.121). These are *idiosyncratic assets*, distinctive, even unusual features of individuals that are more valuable when used in the particular firm than outside of the firm. Such assets tend to be difficult to observe, describe and value but have a significant impact on a firm's competitive advantage.

Related, in an interview [Hof 2008], one hears from Eric Schmidt, then Google's CEO, how Google manages the tricky process of innovation and its relation to corporate culture which is not transferable:

> "Why aren't many other companies doing this, too? I think it's cultural. You have to have the culture, and you have to get it right.
>
> So we're likely to see even more acquisitions by Google? I would think so. But small. The likelihood of us doing big things is pretty low because we'd have to assimilate the culture. Nobody works the way we do. The Google culture makes sense if you're in it, and no sense if you're not in it."

RBV sees companies as different collections of tangible and intangible assets and capabilities, which determine how effectively and how efficiently a company performs its functional activities. To apply RBV it is essential to *identify the firm's potential "key" resources.*

> According to GST, however, the perspective of RBV has to be complemented. Resources are dependent on interactions and combinations with other resources and therefore *no single resource or a set of individual resources* – tangible or intangible – *can become the most important one for firm's performance. Resources and capabilities form open systems* which require time to achieve a quasi-equilibrium but, due to the open system paradigm, they will evolve over time – even in a phase of stable growth (cf. ch. 5.2).

Open system means: the system under consideration is adding or destroying, increasing or decreasing, exchanging or sharing mass-based or power-based assets, information, people and values (including money) with other systems.

Systemic effects in small or large firms do not only emphasize interactions of resources, but also feedback and reinforcement mechanisms, *largely out of the control of leaders/managers, which affect the firms' development (growth).*

The processes of *combining, organizing and leveraging resources let "new" systemic resources emerge.* For instance, sharing knowledge resources within a network of partner firms may add to the originally shared knowledge resource of the individual partner; close customer relationships and common projects may add to the innovativeness of the "supplier."

For technology entrepreneurship Autio and Garnsey [1997] introduced, for instance, an extended RBV model emphasizing a firm to be an open system interacting with others in its environment to identify incentives and constraints which originate from the environment and those which form through the internal dynamics of growth. This RBV extension concerns the influence of network relationships (Figure I.51, Figure I.125) on growth processes and competence building.

The authors take growth-reinforcing and growth-offsetting effects into account that determine the firm's capacity for resource accumulation. They present a systemic evolution model which, however, is a specialization to NTBFs founded either as a spin-out firm or as a private venture. The technology, application, and the capabilities of the management team determine the potential of the firm to reach stand-alone growth.

The development path follows a link of the NTBF to an innovation network or manufacturing chain. Within these structures there is scope to obtain leverage from resources and to pursue external opportunities. The internal pressures will be reinforced by external pressures in the growing firm, as funders, customers and distributors call for expansion so that as a next step a cluster is formed.

In the approach of Autio and Garnsey [1997], for individual firms there is the opportunity to *grow with the network.* But, at this point, the NTBF is often very dependent on a network's "locomotive firm" which drives the development. As an example that has been observed in practice the many of Nokia Corporation's small supplier companies in the Finnish telecommunications industry are hinted at.

Therefore, apart from the business idea and opportunity, according to RBV *assessing startups* will have to put a strong emphasis on *people and organization*:

- Most of the value of the venture is attributed to people – the entrepreneur and his/her "team" – a key resource.
  People build the startup; leadership, corporate culture and execution associated with a sense of market urgency are important.
- The other large part of value of the venture relates to a successful and *sound structure* and *processes right at the start*.

Without a structured and validated offering development, operation, market, sales, and financial *plan* (reasonable financings) *and team for execution* the entrepreneur(s) will constantly reacting to competitive or other forces and will be often in "fire fighting mode." Entrepreneurs cannot expect, in uncertain businesses, to gather reliable data on potential demand and competition. In many niche businesses the specific information generated by *doing* is therefore more valuable than excessive search for relevant information [Bhidé 2000:59].

One can believe that a technical startup merits a positive "premoney valuation" derived from intellectual properties, human capital and other intangible resources (Table I.8). The capacity of *Inc.* companies in the US [Bhidé 2000:29] and NTBFs in Germany to finance high growth rates through internally generated "funds" (cash flow) suggest that their profit margin were significant.

The founders' capacities to *differentiate their offerings* through their personal efforts seem to be an important reason for profitability (cf. CEOs' knowledge of customers with Hidden Champions; ch. 4.1.1). The entrepreneur(s), rather than a product or technology, represent the source of the startup's profits [Bhidé 2000:47] – and the firm's productivity and performance.

As a summary, the question is whether a firm's model *defining a firm as a distinctive bundle of assets, resources and capabilities* (a firm's "business system") and *characterized by a related portfolio* allows explanation or even reliable expectations and has prescriptive implications for new technology ventures. Here, the following restrictions are notable.

The resource-based model refers to a constellation at a particular point in time, usually in the past to describe and explain the development of a firm to that point in time and often uses that for "predicting" future developments – mostly without considering developments of the firm's human resources and the role played by developments of social interactions of employees, such as the formation of "Communities of Practice" (CoPs) [Runge 2006:372] and reinforcement of behavior (Figure I.129), information resources for decision-making of the firm's leaders etc. (cf. Ashby Memory [84]).

In contrast, for instance, to an Input/Output (I/O) Model, the resource-based view is grounded in the perspective that a firm's internal constitution, in terms of its resources and capabilities, is more critical to the determination of strategic action than is the external environment. "Instead of focusing on the accumulation of resources necessary to implement the strategy dictated by conditions and constraints in the external environment (I/O model), the resource-based view suggests that a firm's unique resources and capabilities provide the basis for a strategy. The business strategy chosen should allow the firm to best exploit its core competencies relative to opportunities in the external environment." [Hitt et al. 2005]

RBV often assumes that resources are relatively immobile. However, both people (with tacit knowledge) and information are transient, both can disappear immediately. And this may be highly critical for young firms (with only few employees). Furthermore, RBV does not consider serendipity or luck as a factor of firm's foundation and growth.

Technology entrepreneurs can use market forms of governance and political/public forms of governance to coordinate many resources necessary to realize an economic opportunity (Figure I.59, Table I.30).

RBV refers essentially to (only) economic markets; the role of other types of markets (Table I.15) is non-existent or underdeveloped.

In particular, the massive interference of policy with technology entrepreneurship and support of (technology-based) SMEs and the direct and indirect resource provisions in terms of financial support, legislation (Box I.1, Figure I.34 ; A.1.1), research grants and networking in terms of competence networks, funding R&D projects, science and technology parks and clusters (ch. 1.2.6.2) restrict applicability of RBV to firm's foundation and early development to only selected aspects.

Finally, strategic orientations of technology ventures often follow several aspects which are complementary rather than exclusive:

- *Resource-oriented* - Resources of the firm push
- *Market-oriented* - The market drives
- *Interrelation-oriented* - Alliances and networking are key
- *Opportunity-oriented* - Fast opportunity identification and exploitation drives.

4.3.3.1 Bootstrapping a Technology Startup

Initial and growth financing of NTBFs relies very often on the founder's (founders') own funds and 3F funding as resources and subsequently on cash flow from business operations (Figure I.52; Table I.23 (for *Inc.* firms), Table I.24, Table I.25, Table I.26).

That means initial funding of a technology venture occurs to a large extent by "**bootstrapping**" (bootstrap financing). However, there is no generally agreed upon definition of the notion "bootstrapping." Often, it means "to start a firm by one's own efforts and to rely solely on the resources available from oneself, family and friends." [Dorf and Dyers 2007:111-113] However, for most cases of German NTBF foundation dealt with in this book the startups did not rely only on resources available from the founder(s), family and friends.

A more elaborate definition is given by Eckmann [2008]: “Bootstrapping is a means of financing a small firm through highly creative acquisition and use of resources without raising equity from traditional sources or borrowing money from a bank. In short, ‘bootstrapping’ means starting a new business without external start-up capital.”

Here, on the one hand, emphasizing “creative acquisition and use of resources” opens the spectrum of accessible resources and methods of financing. On the other hand, it remains unspecified what actually “traditional” would embrace. Eckman adds a specification: “It is characterized by high reliance on any internally generated retained earnings, credit cards, second mortgages, and customer advances, to name but a few sources.”

Wikipedia [83] emphasizes that “financial bootstrapping is a term used to cover different methods for *avoiding using the financial resources of external investors*.” And, moreover, bootstrapping can be defined as “a collection of *methods used to minimize the amount of outside debt and equity financing needed from banks and investors*.” (Emphases added)

Emphasizing the notions “avoiding” and “minimize” the author thinks that last definition provides the necessary scope to discuss bootstrapping in the context of technology entrepreneurship where a wide variety of financial and other resources and methods of financing are available for entrepreneurs (ch. 1.2.7.3; Figure I.59, Table I.30).

The financial requirements of the startup and the availability of capital in the market will determine if bootstrapping is an appropriate means (ch. 1.2.7.1). But choosing that way is also related to attitudes toward the various sources of capital for technology entrepreneurs (Box I.20). Software-based ventures typically require less start-up capital than, for instance, either electronics or biotechnology ventures, thus is more likely to rely solely on personal funding (ch. 3.4). Furthermore, growth orientation (Table I.63, ch. 4.1) will also influence the decision for bootstrap financing.

Bootstrapping is often associated with the opportunistic adaptability approach of technology entrepreneurship when time and effort trade-off are considered: Many months spent trying to raise money (with no guarantees!) versus same time spent starting business, establishing proof of customer and product and building traction.

The interconnections of bootstrapping and opportunistic adaptability are lucidly described by Klaas Kersting [2012], the co-founder of German Gameforge AG and Flaregames GmbH (B.2). He put it into several steps for “Building a Startup” with the premise “fail early, fail often – and learn” (ch. 5.1).

- Find some money. (Hint: friends, family or fools are a good starting point.)
- Focus and prioritize (Hint: Just do the important things.)

- Get to market fast. (Hint: you don't know the market until the market knows you.)
- Know your numbers. (Hint: you cannot know too much.)
- Avoid overhead. (Hint: you might not need to hire your cousin as a consultant just yet.)
- Cash flow is everything. (Hint: buy low, sell high; collect early and pay late.)

Key questions for bootstrapping are:

- How much cash do you need, and when (Figure I.57)?
- If nothing changes, when will you run out of money?

For the early phase of NTBFs it is often difficult to separate acquisition of financial resources and methods of financing by business operations. Bootstrapping in the broad sense is characterized largely by high reliance on any internally generated monetary reserves. For understanding the role of business operations for bootstrapping basics of accounting and financing one can refer to Dorf and Byers [2007:403-436] and the author's Course Material (Handout Lectures 10-13; pp. 1-18) of the Technology Entrepreneurship Web.

Bootstrapping offers many advantages for technology entrepreneurs and is a good method to get a startup operating and well positioned to seek equity capital from outside investors at a later time – if needed. In particular, a business that makes money builds its credibility – with suppliers, employees and customers. Keeping costs *consciously* below revenues will position the company to survive in lean times which will always come!

> Fundamentally, NTBFs and RBSUs have the possibilities to go for capital focusing on research and development grants, scholarships, financial contributions or subsidies of federal or state governments, NGOs and national science organizations (such as, NSF or DFG) or grants for technology projects, which are sometimes cooperative projects (ch. 1.2.6, 1.2.7).

Learning the nuts and bolts of running a business takes time. Start learning from the birth of the firm.

The portfolio of bootstrapping targeting sources, methods and activities minimizing external financing (debt and equity) focuses simultaneously on expenses versus profits and cash flow. *Superior execution is the key* for the components of such a portfolio.

Basic operating expenses comprise ("buy low"):

- Location selection (cost of renting offices and laboratories, etc.; in an incubator, science or technology park – ch. 1.2.6) and networking including utilization of infrastructure of the parent organization, if the startup is a spin-out (RBSU) or the founder of an NTBF has strong ties to research institutes or academia.

- Renting (or leasing) sophisticated technical instruments or devices rather than purchasing them.
- Outcontracting selected activities of the NTBF's value chain through external contract services (ch. 4.3.1) or strategic alliances.

Profit orientation means

- Go fast to market (customers); have the ability to adjust to a rapidly changing industry or environment; build experience and know-how as you go.
- Focus on cash-generating activities (Hint: Just do the important things.).
- Look for quick breakeven (ch. 1.2.7.1, Figure I.53).

Offer high-value products or services that can sustain direct personal selling:

- (Bootstrapping) entrepreneurs should pick high-value products and services where personal salesmanship can replace an expensive marketing scheme.
- Meet customers' specifications; do not overshoot (Figure I.88).
- Provide high-value service and support to customers.
- Focus on one offering (of probably few more) which represents a "cash cow."
- Learn from the customer(s) and adjust the business model, if needed.

An issue of going fast to market is the question of *how to bring the product to market.* Is it going to require a change of behavior on the part of intended customers? Most startups *underestimate the difficulty, not to mention the time and money required, to get a product launched and established in the marketplace.*

Overcoming customer inertia is easier and cheaper if a product offers some tangible advantage over the alternatives. Concrete product attributes – with data to support – can lead to sales. Make the risk of dealing with the startup small for the customer as compared with the risks associated with not solving his/her problem (ch. 4.2.1.1).

Cash flow management ("is everything"):

- Keep cost to a minimum and have positive cash flow (ch. 4.2.3).
- Adjust the revenue (income) and expenses (loss) curves, the profit curve.
- Carefully track currency exchange rates, if the startup has international orientation.

In particular, "*working capital*" (ch. 1.2.7.1) is primarily concerned with the day-to-day operations rather than long-term business decisions. Managing working capital has to ensure a company has sufficient cash flow in order to meet its short-term debt obligations and operating expenses.

Following Investopedia "*working capital management*" is a managerial accounting strategy focusing on maintaining efficient levels of both components of working capital (Figure I.130), current assets (essentially cash, accounts receivable, inventories) and current liabilities (such as accounts payable), in respect to each other ("collect early and pay late"; defer your payments as long as possible).

The accounting entry accounts receivable (giving credit or allowing late payment by customers) are assets of the customer and must be financed by the startup. Accounts payable are a way of financing the startup's assets.

Accounts Receivable (collection of what a business/startup is owed)	The receivable conversion period (RCP) is the time between the sale of the final product on credit and cash receipts for the accounts receivable (cf. DSO).
Inventories	The inventory conversion period (ICP) refers to the length of time between purchase of raw material or input for production of the goods or service, and the sale of the finished product.
Accounts Payable (Payment of what a business/startup owes)	The payable deferral period (PDP) is the time between the purchase of raw material or input on credit and cash payments for the resulting accounts payable.

"*Days sales outstanding*" (DSO) is a measure of the average number of days that a company takes to collect revenue after a sale has been made. A low DSO number means that it takes a company fewer days to collect its accounts receivable. A high DSO number shows that a company is selling its product to customers on credit and taking longer to collect money (Investopedia).

Be careful with discretionary expenses, such as

- Going for a highly professional representation on the Web.
- Sales and marketing programs.
- Growth initiatives.

Keep growth in check:

- Start expanding, once the new venture starts growing while keeping the cost curve below the revenue curve (Box I.20).
- Expand at a rate that you can afford and control. This enables you to develop management skills slowly and to iron out problems under less pressure.
- Re-invest profit for growth; target investment and innovation persistence (Figure I.117, Figure I.127).
- Invest in new people if there is no other alternative, not in advance of needs (Box I.20).
- Hire workers the business needs (but only pay what you can afford).
- Develop people internally (Figure I.121).

Cultivate banks before the business becomes creditworthy:

- Keep good financial records, sound balance sheets.
- Look for bank overdrafts and line of credits.
- Prepare early for the next step of financing (ch. 1.2.7.3).

4.3.4 Cybernetic Principles and Concepts for Technology Entrepreneurship

Building on preceding and previous discussions and metaphors (ch. 2.1.2.9, ch. 4.1; Equation I.2) concerning observation and expectation we shall approach understanding and explaining growth of new (technology-based) firms by achieving appropriate qualitative or even quantitative descriptions.

At best, *expectation*, not prediction, will appear as being comparable with weather forecasts which are made by collecting quantitative *data about the current state of the atmosphere* and using scientific *understanding of atmospheric processes* to project how the atmosphere will evolve. However, incomplete understanding of atmospheric processes mean that forecasts become less accurate as the difference in time between the present moment and the time for which the forecast is being made (the range of the forecast) increases – being also one of the issues of RBV.

Observation will refer to *indicators* of tangible output, outcomes or benefits of a conversion process (Figure I.5). But input and the conversion process in terms of variables and parameters (ch. 1.2.1) will address often intangibles which will refer to *intervening variables* – interpretations of observed facts.

An intervening variable reflects theoretical processes that are assumed to take place between what is observed as the "before" conditions and the "after" conditions. The situation is displayed for learning as an intervening variable in Figure I.3. Consequently we shall often follow a "reasons for thinking that" approach rather than a "reasons why" rationale (Figure I.2).

Information on intangibles (firm's culture, leadership, entrepreneurial commitment, team interactions, interactions with the Advisory Board, networking, etc.) can, at least partially, be grasped by direct observation in the firm or by telling of the founders or the firm's employees or reporting by others. But we shall not know how founders will decide to respond to serious events or crises, unless observation takes place during such effects (cf. volition; ch 2.1.2, 2.1.2.9).

It has turned out that fundamental cybernetic principles and concepts can provide a basis for progressing. With regard to measurement of *human-activity systems* one often refers to a difference scale related to a concept of "change."

Following one of the fathers of cybernetics, in line with Ashby [1957], we shall assume in all cases that the *changes occur by finite steps in time and that any difference is also finite*. The change will occur by a measurable jump, a discontinuity or a "turning point" (in German Wendepunkt) of a measured curve reflecting growth or decline, if the change is tracked on the basis of an appropriate, measurable indicator.

In the sense of Equation I.6 we shall follow a differentiation between a *state* and its related characteristic in terms of an observable attribute, a value of the related *indica-*

tor. Within GST we concentrate on a system's states: A **state of a system** ψ is any well-defined representation of the conditions of its existence and an associated property that can be recognized if it occurs again.

Every system will naturally have many possible states. That which is acted on will be called the **operand**; the factor inducing a change will be called the **operator** (given in Script font similar to the Hamilton operator in Equation I.6), and *what the operand is changed to* will be called the **transform**.

The change that occurs, which one can represent by a relation in terms of a mono-directional graph, A → B, is the **transition**. A set of transitions, on a set of operands, is a **transformation.** The series of positions taken by the system in time defines a *trajectory* or *line of behavior* [Ashby 1957]. The transition is specified by the two states ψ_1 and ψ_2 and the indication of which changed to which.

$$\Delta\psi = \psi_i \rightarrow \psi_{i+1}$$

A priori, the transformation is defined in the sense of cybernetics, "not by any reference to what it 'really' is, nor by reference to any physical cause of the change, but by the giving of a set of operands and a statement of what each is changed to. *The transformation is concerned with what happens, not with why it happens.*" [Ashby 1957].

> Cybernetics does not treat things but ways of behaving. It does not ask "what is this thing?" but "*what does it do*?" and "*what can it do*?" (for which purpose) [Ashby 1957].
>
> However, with GST as the overall framework, to make these concepts *applicable* to the field of entrepreneurship, one must consider the related fields (psychology, sociology, economics, business administration, etc.; Figure I.1) and any kinds of their relevant observations and their ways of measurements, principles and concepts. This will provide an *abstract system* of combined theories, empirical basements, principles, concepts, etc. from the various (scientific) disciplines as the basis.
>
> The resulting abstract theoretical system will rely on "borrowed" knowledge, approaches and methods from the various involved fields. This means, a particular approach provided in one abstract system may be "switched" to another (usually higher) system to find an "appropriate" description or explanation or causal interrelationship for presenting expectations (or probably forecasts) for a company system under consideration as characterized in Figure I.128.

<table>
<tr><td rowspan="3">A transformation will be called closed if all the transforms involve only the elements of the original basis set (all transforms restricted to A, B, F, D) [Ashby 1957]. There is no “inflow” and no “outflow as in open systems.</td><td>{A B F D}</td></tr>
<tr><td>↓</td></tr>
<tr><td>{D B F A}</td></tr>
</table>

This set of transforms obtained contains no element that is not already present in the set of operands. A closed transformation creates no new element, the “domain” and the “range” being identical. And for non-interacting elements the closed transformation corresponds to a permutation. A large capital influx into a firm by investors would represent a typical “open transformation” which is common for open systems.

A test for closure is made by reference to the details of the transformation itself. It can therefore be applied even when one knows nothing of the cause responsible for the changes.

Furthermore, a transformation increasing the number of entities it acts on is an “*exact transformation*” in the sense of *self-replication* if the original is retained (copy of A B F D as above), otherwise it is a “*similarity transformation*” like (D B F A). Hence, in reality forming firm culture (Figure I.120) in this sense should be considered a similarity transformation rather than a copy.

A special transformation is the *identity transformation*, in which no change occurs, in which each transform the same as its operand is.

Which effect of a transformation we observe (or disregard or do not detect) enters essentially into our “*reasons for thinking that.*” For instance, the change from a square with four corners to a four-pointed star, one with a fourfold and the other with a twofold rotation axis with regard to the plane in Figure I.2 can be achieved by similarity transformations such that the ratio of the two diagonals in the square and the stars is kept, leaving them invariant.

Hence, investigating or observing just ratios of diagonals and not additionally the shapes of the objects and the lengths of the individual diagonals “make both objects identical,” the result of an identity transformation. Furthermore, there may be more attributes associated with change. Including also colors (“colored symmetries” [Shubnikov and Koptsik 1974]) for representing objects the right hand side of Figure I.2 exhibits three different objects (or systems).

A transformation is *single-valued* if it converts each operand to only one transform. If it is not single-valued and not one-to-one it will be *open* and correspond to a “one-to-many” situation.

A transformation of the kind

A	B	C	D
B or D	A	B or C	D

is not single-valued.

We have just seen that after a transformation $\mathcal{T}$ has been applied to an operand α, the transform $\mathcal{T}(\alpha)$ can be treated as an operand for $\mathcal{T}$ again, getting $\mathcal{T}(\mathcal{T}(\alpha))$, which is written $\mathcal{T}^2(\alpha)$. In exactly the same way $\mathcal{T}(\alpha)$ may perhaps become an operand to a transformation $\mathcal{P}$, which will give a transform $\mathcal{P}(\mathcal{T}(\alpha))$. Generally, operators are not commutative: $\mathcal{P}(\mathcal{T}(\alpha)) \neq \mathcal{T}(\mathcal{P}(\alpha))$.

In the current context of time developments a transformation of a firm's state emphasizing a change by an event ("interruption") will always refer to a current state. This means, for instance, an event that a venture capital firm (VC) participates in an NTBF (Figure I.126). Empirically founded, the transformation will induce a change by the combined or interrelated, respectively, effects of several factors which are sufficiently strong to be observable in totality after a certain period of time and described by the transformation Equation I.11. Here $\mathcal{VC}$ is the corresponding operator changing, for instance, equity, ownership, control, management and number of employees and organization of the firm (Figure I.126).

Similar to Equation I.6 (ch. 2.1.2.9) the following notation will be used to describe the *intrinsic relation between the two components, operator and operand.* The result will be a changed state of the entity associated with an observable and measurable value *VC* induced by several presumed effects given in braces.

Equation I.11:

$\mathcal{VC}\ |\psi_i> \rightarrow$ *VC* {add equity – (other, more) owners – other firm control – install management – add employees – re-organize firm} $|\psi_j>$

or, in short, $\mathcal{VC} \rightarrow$ *VC* {add equity – (other, more) owners – other firm control – install management – add employees – re-organize firm}

The above notion separating contribution by dashes represents a "*systemic transformation*" were the various changes represent an overall change by the combined action of the given transforms. The dashes bare the relation to a representation by an "array." Viewed in this sense for an array to be unchanged, each component must be unchanged.

The effects of the individual transforms can rarely be observed in isolation. Moreover, a selected measurement by a selected indicator may not be meaningful due to the "impact time" periods of the individual "(non-systemic") transforms. These may be rather different. Compare adding equity being observable for a particular day or week versus number of employees observed at the end of the year or as a yearly average.

The particular types of measurement have different "time resolutions." In soft sciences the minimum or even common time span between two subsequent measurements that can be meaningfully interpreted (as a change) is the *resolution* of the measurement. For instance, tracking numbers of employees year after year will make most associated transforms with shorter impact time unobservable and will induce the transformation to appear as an overall systemic effect.

On the other hand, if there is a possibility to (largely) separate the individual transforms of a transformation ("weak coupling") we will continue to use the common set notation ("listing"), separating elements by commas {A, B, C, ...}.

In our context, a transform with an observable effect would be, for instance, if an NTBF catches a very huge order of the NTBF's product from a major customer. On the other hand, changing the legal form of a limited liability company (LLC, GmbH) to a non-public stock company (AG in Germany) may represent a transform retaining largely everything else of the firm and may be a transform without any measurable effect on the other sub-states of the firm.

If we associate a transform with a measurable quantity of interest, the transform may turn out to be a "positive" ("growth-inducing") or a negative ("decline-inducing") change of the related particular measurable quantity. Consequently, existing *transformations are not necessarily observable*, positive and negative effects may level off, if they occur (almost) simultaneously.

In terms of cybernetics **control** was previously defined as the *purposive influence* toward a predetermined goal involving continuous comparison of current states to future goals ("is" versus "shall" assessment). The above transformation "$v\ e \rightarrow$ " with attributed transforms has an inherent shortcoming. It does not take into account any possible changes that concern the pace (time) to reach the goal or even modifying the original goal (Figure I.122, Figure I.126).

A *teleological relation* (start to end; Figure I.78) which may be associated with significant changes (observable through appropriate variables, parameters and indicators) can be viewed as a set of transformations. Specifically, according to GST, a *firm's foundation (the "birth") is the first transform* which causes the founder or founding team with their ideas and perceived business opportunities, motivation, aspirations and expectations to strive for their particular goal(s) by means of a firm. And this changes, for instance, their states of personalities (Figure I.16, Figure I.122).

Connected to the goal expectation can be viewed as a transformation of the personality. The entrepreneur as a firm founder, owner and leader with control over the firm has explicit qualitative and quantitative goals which induce expectations. Basically, there are founders who tend to emphasize keeping control over the firm (autonomy and vision) or making money (wealth) accepting venture capital, losing control as seen in Table I.39, Table I.40, Figure I.65 and Figure I.66. The corresponding "*entrepreneurial expectations*" $\mathcal{E}\,\mathcal{E}$ (*type*) can be represented, for instance, by Equation I.12. The second example is not single-valued.

Equation I.12:

$\mathcal{E}\,\mathcal{E}$ (*aspiration, motivation*) → (*autonomy, vision*) {confidence in own business idea – revealed opportunity – internal locus of control – perseverance – risk taking - tolerance for ambiguity – self efficacy – ownership/control – organic growth}
$\mathcal{E}\,\mathcal{E}$ (*aspiration, motivation*)→ (*wealth*) {confidence in own business idea – revealed opportunity – internal locus of control – perseverance – risk taking – tolerance for ambiguity – self efficacy – accept venture capital – lose OR keep control – organic growth OR non-organic growth – selling firm is an option}

Following Ashby [1957], although the system may be passing through a series of changes, there is (often) some aspect that is unchanging. Hence, some statement can be made that, in spite of the incessant changing, is true unchangingly. The simplest case occurs when a state α and a transformation are so related that the transformation does not cause the state to change. Algebraically it occurs when $\mathcal{T}(\alpha) = \alpha$. That means, the state α is *a state of equilibrium* under $\mathcal{T}$.

The same phenomenon may occur with a set of states. Take $S(\downarrow)$ to be a non-systemic transformation or one in which the operands of a state are only slightly coupled, that is *open* ("unclosed") and has *no state of equilibrium*, but exhibits a domain that generates no new state. Such a state is **stable** with respect to S.

S(↓):=	*a b c d e f g h*	S(↓):=	*b g*
	p g b f a a b m		*g b*

Using transformations and the stability concept are associated with an issue of reducibility of a complex situation that avoids dealing with a situation that every factor (variable or parameter) had an effect, immediate or delayed, on every other factor. When a dynamic system can vary continuously, disturbances are, in practice, usually acting on it incessantly. For this reason the only states of equilibrium that can, in practice, persist are those that are "stable" in the above sense.

The (last b g) transformation is closed, so something persists, and the observer who looks only at this level of discrimination can say of the sub-system: "it persists," and can say no more [Ashby 1957]. Classification and taxonomy according to given criteria is related to such a persistence (Figure I.128).

> In a new firm's development there is *no invariant overall state* because different problems arise, affect particular sub-states and are addressed correspondingly in different ways leaving the "rest" stable. Without observations at the firm level and identifying the prototypical problems, the mechanisms and processes of growth as linked to expectation or goals remain obscure.

"That something is 'predictable' implies that there exists a constraint." If an aircraft, for instance, were able to move, second by second, from any one point in the sky to any other point, then the best anti-aircraft prediction would be helpless and useless. The latter can give useful information only because an aircraft cannot so move, but must move subject to several constraints. There is that due to continuity – an aircraft cannot suddenly jump, either in position or speed or direction. There is the constraint due to the aircraft's individuality of design" [Ashby 1957:132] and its "resources" (engine and fuel) determine the distance it can cross.

This reference to a process is essentially that of using a mapping – using a convenient (for instance, mathematical or graphical) representation rather than the inconvenient reality.

Figure I.128 summarizes the landscape of elaborated constraints for technology entrepreneurship which provides a "navigator" for the entrepreneur where he/she wants to be or be active in and the advisor or consultant to properly advise, propose and guide the entrepreneur and the entrepreneurship researcher to properly select criteria to create samples and interpret measurements and findings.

Leaving out a constraint reduces the strength of "prediction" and, in case of complexity, requires being very conscious about the limits of the domain of interpretations and even more "predictions."

The many seemingly different, controversial and even contradictional results and findings concerning growth of young firms [Garnsey et al. 2006] means that the people selected different constraints for inquiry and often are talking about different systems of investigation. The related issues of statistics are often associated with selecting a sample which is assumed to provide class properties and, furthermore, how response rates of questionnaires distort the originally selected sample structure.

For the taxonomies of industries related to characteristics of technologies (Table I.1; TVT, HVT) in terms of "*research intensity*" (RI = R&D expenses / total revenues) the differentiation is based on the proportion of financial quantification of research and development expenditures which cuts across industry taxonomies according to business or offerings.

Figure I.128 illustrates a fundamental dilemma of technology entrepreneurship, the complexity of constellations and the question whether and how results of macro-approaches have relevance for practice and, in particular, for individual entrepreneurs and those providing advice and consulting to them.

Industry taxonomy as given in Figure I.128 combined with firm type (RBSU versus other academic NTBFs) and ownership/control and financing (VC-based versus non VC-based) seems to be the bare minimum. How just one industry – biofuels – for understanding technology entrepreneurship has to be boiled down is illustrated in Figure I.183, Figure I.184, Figure I.185 (A.1.1.5) and Table I.17.

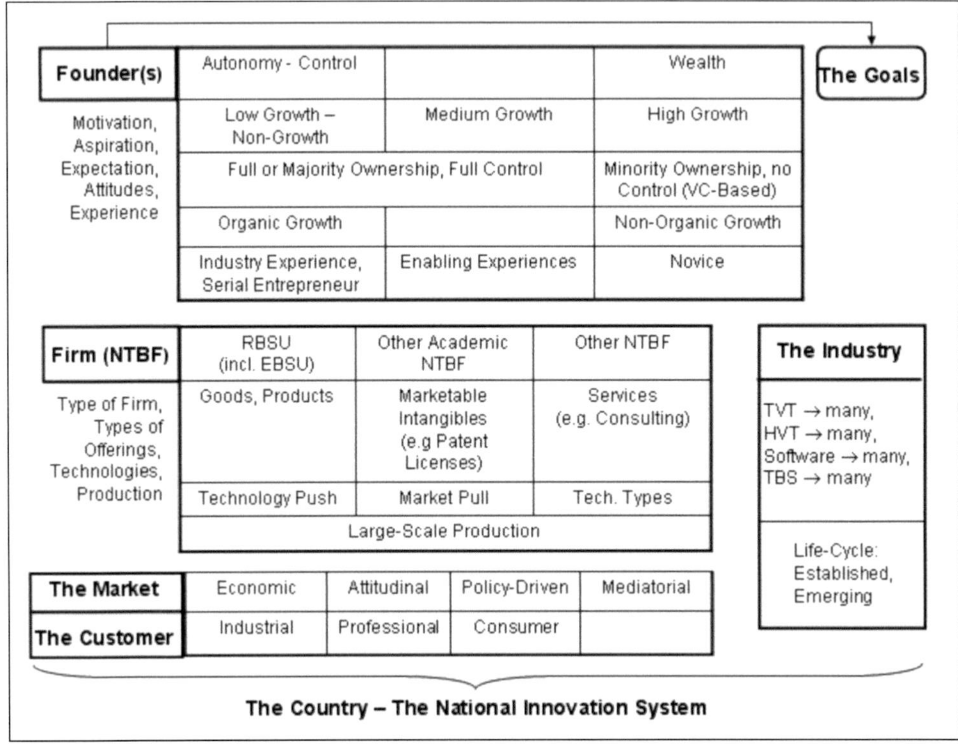

Figure I.128: Constraints as a basis of taxonomies for technology entrepreneurship to characterize configurations of NTBFs (read sub-tables from left to right).

And there seems to be even notable differences in industry segments in attracting entrepreneurial personalities. Entrepreneurs may take big risks to bring the latest scientific tools to market. For instance, the people who take personal risks to bring *new scientific instruments* to market are a special breed. *Many of these entrepreneurs are well-educated scientists* who could make a fine living working as consultants or as employees in high-technology companies. Yet they risk their livelihoods and their own money for the chance to start up their own firms [Reisch 2011a].

Stability is commonly thought of as desirable, for its presence enables the system to combine of flexibility and activity in performance with something of permanence, something "generic" which, for instance, is the focus when dealing with entrepreneurship over time (history) and space (regional culture; comparing Germany and the US).

Goal-seeking behavior is an example that stability around a state of equilibrium is advantageous. Nevertheless, stability is not always good, for a system may persist in returning to some state that, for other reasons, is considered undesirable or proceeding to some new state that is highly necessary, due to a changed environment.

In this way, these concepts may be used to explain and illustrate the transition from core competencies to core rigidities (ch. 2.2.1, Box I.8) as a combined effect of persistence and self-reinforcement ("success breeds success") resulting in a firm's "routines" and "routinized decisions" – how things are done here or how things are decided here (Figure I.129). This kind of persistence is, of course, a special property of the whole system focusing on just one aspect.

Self-reinforcement is essentially determined by decision-making self-reinforcement (that is, past acceptances make future acceptances more likely) and adaptive expectations (further belief in prevalence; Box I.17) (ch. 2.1.2.5).

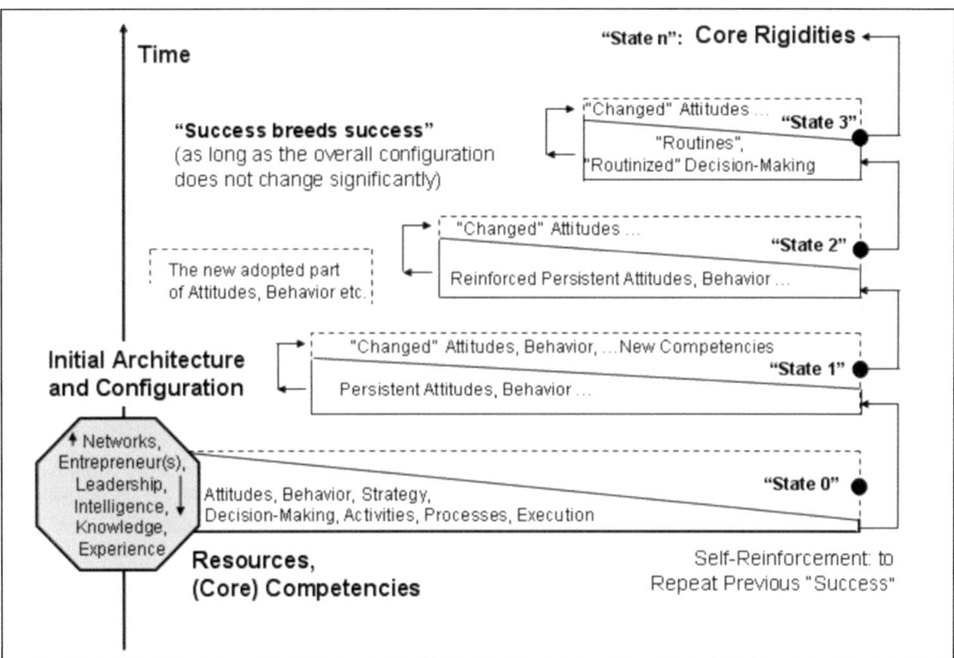

Figure I.129: The progression of core competencies toward core rigidities.

The descriptive path to "core rigidities" can help understand the well-known fact that the founding configuration (including firm culture) and the early development of a start-up influences further development of the firm. This means it can account for *path-dependency of NTBF development*, if observed or searched for.

If the founder has industrial experience (or is a serial entrepreneur) corresponding path-dependency (decision-making, behavior) may already enter the starting configuration of a new firm. The same is true if the founder(s) has hired early on an "experienced manager."

The Special Focus on the Single NTBF

The following approach to technology entrepreneurship and NTBF developments (growth) will have to make more use of concepts and principles of economics and business administration and findings (including situations in large firms) for NTBFs as given in this book so far and the author's previous book [Runge 2006]. The emphasis will be the entrepreneur(s) and the firm and *tracking changes of states of the firm through reference to mainly the finances related indicators*. The corresponding situation and framework is depicted in Figure I.130.

The use of an abstract system of concepts – principles etc. from various scientific disciplines for a GST- and cybernetics-based theoretical framework for (technology) entrepreneurship – refers also to the levels of describing phenomena. When dealing with growth of startups reference will be made, for instance, to how physics approaches the phenomenon of light. For light effects physics refers to a "particle model" (light as composed of discrete quanta called photons) and use this to explain the photoelectric effect and the "model of continuous waves" to explain light interference.

This approach is a reflection not just of the particular subject of inquiry, light, but for explaining experimental settings including interactions with a particular substrate or with itself. Hence, for a given experimental setting "behavior of light" can be explained for the particular context "as if ..." According to the Copenhagen Interpretation of Quantum Theory, the wave and particle pictures, or the *visual* and *causal* representations, are "complementary" to each other. That is, *they are mutually exclusive, yet jointly essential for a complete description* of quantum *events*.

Switching between knowledge of different disciplines or using metaphors and analogies also means switching between epistemology. Epistemology is the investigation into the grounds and nature of knowledge itself. It is important because it is fundamental to how we think. Without some means of understanding how we acquire knowledge, and how we develop concepts, describe and explain in the various disciplines, we have no coherent path for our thinking.

Sound, though basic epistemologies are necessary for sound thinking and reasoning to deal with the interdisciplinary phenomenon of entrepreneurship. In particular, we shall switch always between a "why-thinking" (cause-effect thinking) and a "how-thinking" (how an effect of an operator on an operand leads through a process to a result).

It is clear that growth indicators reflect the outcomes of many different interacting causes that influence new firms' growth paths. Figure I.130 does not only indicate that the numbers of employees (as a resource) and revenues (indicative of the financial state) are taken as *indicators for the whole state of the growing firm*.

Furthermore, due to non-accessibility to other data, *performance* during development (growth) and particularly *productivity* as defined in Equation I.2 and referred to in Figure I.130 will be taken as an indicator how efficient the goal is achieved which may be *related to leadership and/or management and organization (specialization, communication, coordination etc.* (Table I.69) of the firm and, hence, an "organizational state."

In doing so we are always aware that even reported revenues in official documents, for instance, in income and loss statements, may contain "Extraordinary Items," one time expenditures for a given year which may "deform" the revenue indicator by an unexplained effect.

Reformulating the productivity, a capacity, defined in Figure I.130 generates an expression for (financial) **strength** or "**energy**" (Equation I.13).This is the typical form used in physics and chemistry for various types of energy and performance, some of them being also listed (in normal font).

Equation I.13:

Revenue = Productivity x Number of Employees
→ (Financial) Strength or "Energy"
[$,€/Employee];

Energy-Related Phenomenon → Capacity ⊗ Intensity

Volume Energy → Volume ⊗ Pressure

Thermal Energy → Entropy ⊗ Temperature

Shaping Energy (Power)→ Form ("Gestalt") ⊗ Elasticity

Note that *capacity* in this sense is different from *capability* as understood by RBV: According to Merriam-Webster capability is *the quality or state of being capable*; the facility or potential for *an indicated use* or deployment; *a feature or faculty capable of development* ("potentiality").

Figure I.130, Figure I.87 and Figure I.5 allow a further metaphor to be established between the factors for development of the state of "total energy" E of a new firm mapping what the firm "has, can, gets and owns" to potential energy E_{pot} and "does" (decision, implementation, execution) to kinetic energy E_{kin}:

$$E \rightarrow E_{pot} + E_{kin}$$

These notions clarify differences between performance and productivity which are often used as synonyms. According to Equation I.2 performance is related to the *comparison of current achievement* on the basis of existing resources and constraints (Actuality, A) and what could be achieved by developing resources and removing constraints (Potentiality, P) which is A / P.

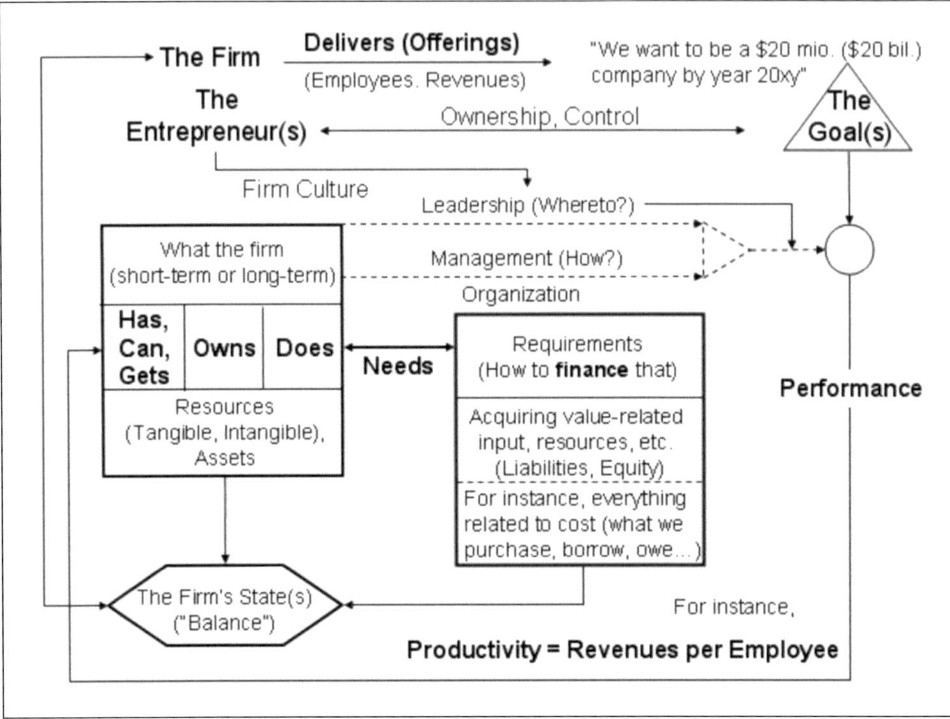

Figure I.130: The constellations and interrelationships that characterize states and performance of a growing NTBF.

On the other hand, productivity was related to Capability C, the possible achievement with existing resources and within existing constraints, as A / C. Actually, as defined in Figure I.130, *productivity is an indicator* derived from other, for NTBFs more or less readily available indicators for NTBF growth. Correspondingly, we view:

> *Change of productivity* is an indicator of performance
> (as illustrated in Figure I.132).

Conceptually, productivity is often related to how well an organization converts input, resources (labor, materials, machines etc.) into output and outcomes (goods and services). This is expressed by ratios of outputs to inputs. That is, for instance, (input) cost per (output) good / service. It is not on its own a measure of how efficient the conversion process is. But this definition is not practical for systematic investigations (of NTBFs) as relevant data are rarely accessible.

Performance is a relation between “what is” and “what could be,” or verbalized “more with the same or even more with less.” In essence, in the current context *performance is related to the first derivative of productivity* and “*performance persistence*” is expressed by “constant” productivities (cf. also Figure I.10, Figure I.132).

According to these definitions we associate *leadership* with an emphasis on performance and intangible resources (developing firm's culture, employees, etc. to get more out of these resources), but *management* with an emphasis on productivity and tangible resources.

Figure I.130 provides an outline of a model for the description of NTBFs' developments (growths) in terms of sub-states which will be characterized by relevant, accessible indicators. The "has, can, gets" block (essentially non-financial tangible and intangible resources and assets), the "resource state," is interconnected to what the firm needs in terms of financing ("financial state") and how effective and efficient the goal is achieved in terms of leadership and management and organization ("organizational state") related to productivity.

A mismatch between available input and resources (has, can, gets) and required input and resources (needs; finances) to reach the goal(s) may constrain the amount of development that can be undertaken at any given time.

Figure I.132 illustrates performance increase with a "more with the same" or "more with less" constellation, thus, for a firm, achieving higher profitability. Notably, market share as an antecedent of organizational performance is consistent with models proposed in numerous empirical studies [Den Hartigh et al. 2002].

"Full" information concerning relevant indicators about the states of a firm is readily available only for public stock companies in terms of annual reports and documents for the stock exchange. For the overwhelming proportion of NTBFs information about the firms' states is often rudimentary.

Furthermore, the input-conversion-output cycle in Figure I.130 and the extended one of Figure I.5 generate an impression of linearity of the involved processes. However, the GST/cybernetics approach emphasized that for new venture growth a number of systemic interactions generates self-reinforcing processes, such as

- Founder team formation (Figure I.70);
- Building company culture (Figure I.120);
- Organizational learning (the learning curve, ch. 2.1.2.5);
- Scale effects, often referred to as" economies of scale" which imply that the average total cost will decline with growing production volumes (Figure I.154).
- Decision-making self-reinforcement (that is, past acceptances make future acceptances more likely; ch. 2.1.2.5);
- Spiraling innovation persistence and investment persistence (Figure I.127);
- The transformation of core competencies into core rigidities (Figure I.129);
- Adaptive expectations (further belief in prevalence; Box I.17).
- Downfall of the young firm's financial state by interaction of internal factors and processes and external effects and processes bound to customers, markets, factor markets and financing (Figure I.114).

The last aspect represent interaction effects with the environment, *market-bound* self-reinforcing mechanisms whereas the other ones are *firm-bound* self-reinforcing mechanisms. Adaptive expectations are also important for the NTBF's interactions with customers. This occurs when a customer's preference for a product is dependent on the opinions or expectations of other (potential) customers (ch. 4.2.1.1). The interdependence of opinions is based on information sharing and "escalation" (ch. 4.3.5.2). Startups, hence, try to utilize this effect by publishing a list of "reference customers" on their Web sites.

A further market-bound self-reinforcing mechanism refers to "utility of an offering," when the economic utility of using a product becomes larger as its network grows in size. Network size is determined by the number of suppliers and users of products based on a common technology standard. Network size is important in many markets, but most visible in the markets like telecommunications, computer equipment and software [Den Hartigh et al. 2002].

4.3.5 A Bracket Model of New Technology Venture Development

It is the theory that decides what we can observe.
Albert Einstein

The basis of the following "business bracket model" of development (growth) of new technology ventures follows largely the above described lines incorporating aspects of the stage-based and resource-based views and reliance on a number of inferences from empirical observations outlined so far. It concentrates on the social and economic context. It establishes a relation between the firm's development due to internal and external factors and appropriately selected observable indicators.

Focusing on NTBFs it is important to re-emphasizing that we are dealing with relatively low level *organized complexity*. This is we are dealing usually with human-activity organizations of two to forty or rarely one hundred persons, at the highest.

Furthermore, we often encounter young firms with very few (1-3) products and few customers (1-5) whose number does not exceed a dozen – even if their revenues exceed one billion dollars (or euros)(cf. First Solar Inc. in ch. 4.3.5.2 and Figure I.154).

The central constellation for growth (cf. also Figure I.5) is in terms of business processes, functions, resources and capabilities and organization (Figure I.131). The bracket model emphasizes growth, but includes not intended growth, and may also cover firm's failure.

> HP co-founders David Packard and Bill Hewlett "did not believe that growth was important for its own sake," but concluded that "continuous growth was essential" for the company to remain competitive and they continued, "Growth: to emphasize growth as a measure of strength and requirement for survival." (citation [Bhidé 2000:230]).

> In this regard, for NTBFs "growth persistence" is in line with goal persistence (ch. 4.3.2), innovation and investment persistence (ch. 4.2.3).

Pressures for growth and capacity for growth are discussed by Bhidé [2000:230-233]. Apart from d*riven to grow by the entrepreneur's aspiration and intention* pressures for growth result partially from self-reinforcing processes. Some of these are the following.

- A firm cannot remain small if its rivals increase market share by exploiting economies of scale or if customers believe that size is a precondition for long-run survival. Longevity goes hand in hand with growth.
- The external labor market may initiate pressure for NTBFs to grow. A dependency on highly talented people and researchers means the firm has to offer opportunities for personal development and progress. A stagnant firm risks losing its talented employees.
- The accumulation and experiences and the development of decision-making routines may lead to an increase in the capacity of the firm's managerial and supervisory personnel, creating more efficient capabilities, which increases the potential for growth.

The bracket model assumes that t*here are no "invariant states of activity" in young firms* the way a stage-based view often suggests. In the sense of open systems with continuous input/resources and output/outcomes (Figure I.5) there is continuous conversion activity including continuous learning of leading and managing, resource and asset re-grouping and resource allocation characterizing a firm's states which will be interrupted by changes through (*unbalanced* internal and/or external) events and decisions and actions. And, furthermore,

> "A firm cannot easily stop growing after it has reached some fixed critical mass." [Bhidé 2000:230]

For technology entrepreneurship a critical mass for non-growth NTBFs could be ca. a dozen employees (ch. 4.3.1; Table I.69, Table I.71).

In this regard one can create certain associations with Newton's First Law of Motion – sometimes referred to as the Law of Inertia. The First Law is often stated as:

An object at rest stays at rest and an object in motion *stays in motion* with the *same speed and in the same direction* unless *acted upon by an unbalanced force* (italics added).

Relating the notion "object staying in motion" to a *firm with continuous growth* one can speak of **dynamic stability of sub-states** of a growing (open) human-activity conversion system or firm (Figure I.5) if it keeps "*growth regularity*" among sub-states. This would be described by a (mathematically) strictly monotonic increasing function between state variables S, like $S_i(t+1) > S_i(t)$ or their respective indicators.

The description would be valid over a certain period of time, the "*dynamically stable interval*" $<t_0, t_n.>$ (*uninterrupted* progression in one direction). Though not of particular interest in the current context, the model will allow also monotonic decreasing relations for dynamically stable sub-states, with no stop leading to firm failure.

NTBF growth for a period of observation, hence, will be represented by dynamic stability of subsequent sub-states of the firm in terms of their intervals of existence that appear with unchanged growth regularity (including constancy) of the appropriate indicators separated by a variety of to-be defined interruptions.

This shows structural and functional similarity with what Abrahamson [2000] regards in the context of "change management" for firms as "dynamic stability: a process that alternates major change efforts with carefully paced periods of smaller, organic change."

Dynamic stability in our context of open systems is not what usually is understood as dynamic stability in engineering. Here this is the property of a body, such as a rocket or plane, which, when disturbed from an original state of steady flight or motion, dampens the oscillations set up by restoring movements and thus gradually returns the body to its original state.

> It is important to be noted that dealing with an open system a transition into another stable state does not mean the system may recover the initial stable situation (state). After overcoming the perturbation it will proceed in another, new state targeting the system's goal. Every new dynamics drives the system's states toward the final state of goal achievement. In so far, there is a link to dynamic capabilities.

Inquiring into dynamic stability investigates primarily what happens during the time after a disturbance. To an outside observer a dynamically stable state may appear rather insensitive to certain "small" perturbations as, if necessary, a system may adapt to such effects in a non-observable way.

Conversely, strength and stability means that, for observation, only the severe perturbations may disrupt dynamically stable states to make these distinguishable from "underlying noise" of measurements. A summary of such usually observable perturbations of dynamically stable (sub)-states based on the outlines in this book will be given later (Table I.76).

The notion of the "small perturbation" is associated with another issue, that changes are relevant due to *unbalanced* internal and/or external events and related decisions and actions.

Observing a particular *indicator* may reflect the combined, interaction of oppositely acting factors which may result in rarely observable small perturbation, though the overall state of the system may be affected significantly. For this reason *the current approach will usually consider not just only one indicator, but usually several ones* which may reflect the perturbation differently. Reference to the details of the firm's internalities, a switch to the micro level, can provide necessary insights.

When accepting that an economic recession can usually be viewed as a severe perturbation of the growth of an NTBF Figure I.123 shows this to be the case for the German WITec GmbH for the Dot-Com Recession (2000/2001). However, the same figure suggests that the even more pronounced Great Recession is not reflected by the growth curve. Understanding and explaining this effect is only possible focusing explicitly on internalities of the firm and the markets it serves.

Internalities of the firm will emphasize a balance of basic business processes of the value chain (Figure I.7), indicated, in the sense of GST, by circular arrows in Figure I.131. It is the responsibility of leadership/management to keep innovation and investment persistence and own continuous learning and build related business experiences. The innovation process is interwoven with all other processes.

For NTBFs with an early on export orientation it is very important to organize distribution, usually by local distributors as these know the specifics of the local markets. And for complex technical products there must be user training, consulting and technical service in the countries.

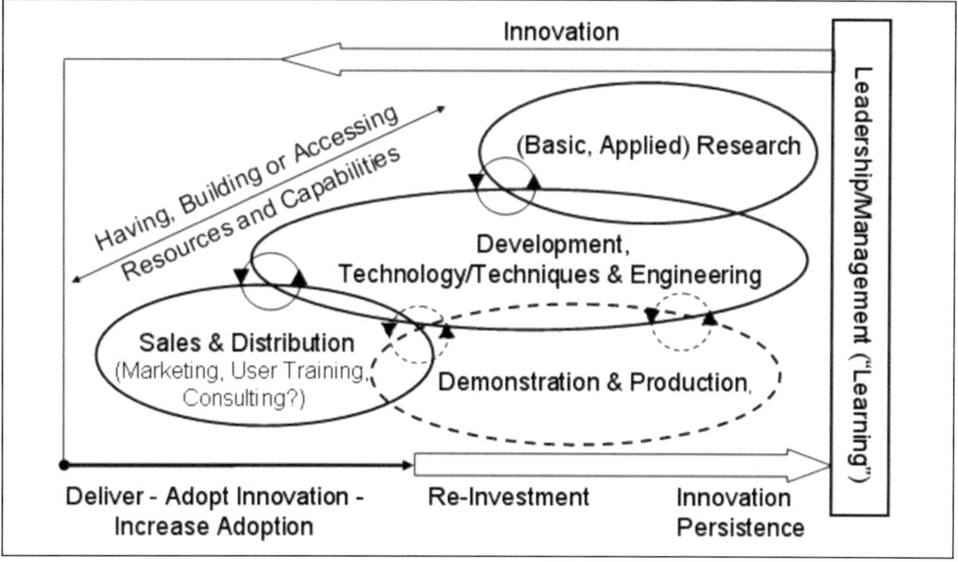

Figure I.131: Key activities to be balanced for a growing NTBF.

Figure I.131 exhibits implicitly the *customer-facing* process when researchers (scientists and engineers) are actually interacting or communicating directly with customers (face-to-face, visits) – which is not a business service feature experienced or seen by the customer. This combination of development, demonstration and distribution by researchers is not only relevant for disruptive innovations, but also for getting ideas.

Formalism, Constraints and Special Cases of the Growth Model

Following the formalism of quantum theory for treating energy levels of molecules which separates electronic, (nuclear) vibrational and (total molecule) rotation states by factoring,

$\Psi = \Psi_{electronic} \cdot \Psi_{vibrational} \cdot \Psi_{rotational}$,

we shall use the "bra(c)ket notation" (Equation I.6, Equation I.12) to describe the overall system state of an NTBF Ψ by *weakly coupled sub-states* (Equation I.14).

This factoring of states means that there is only weak coupling between the states and that their related observables can be measured (almost) independently from each other. Weak coupling, however, may be observable, for instance, for molecular excited states. The dominating bands of electronically excited states measured by UV or CD spectroscopy [85] will often be superimposed by a weak structure of vibrational states.

Equation I.14:

$|\Psi$ (Overall System State)> $\rightarrow$ $|\psi_{SF}$ (Financial State)>·$|\psi_{RS}$ (Resource State)>·
$|\psi_{OS}$ (Organizational State)·$|\psi_{NS}$ (Networking and Coop State)>·…

For NTBF development (during the first ten-twelve years) we assume that *for certain time intervals weak coupling between sub-states exists*, such as the financial (FS), resource (RS) and organizational (OS) states. This means, in terms of indicators ("expectation values" for given states in the sense of Equation I.6), for instance, due to only loose correlation increasing revenue is not necessarily associated with increasing productivity or strictly monotonically increased revenues ($S_i(t+1) > S_i(t)$) may not mean strictly monotonically increased number of employees (Figure I.149, but cf. Figure I.145).

And increasing the number of employees very fast may result in organizational coordination problems (ch. 4.3.2) leading to decrease of productivity, as David Packard put it (cited Bhidé [2000:233]), "more businesses die from ingestion than starvation." It is to be noted that some of the entrepreneurs the author had discussion with mentioned to track productivity and used decreasing productivity as a trigger to initiate organizational change.

We associate the organizational state with only the internal situation of the firm, whereas the firm's interconnections to outside entities will reflect "network organization." If we would assume strong coupling between OS and NS for an exceptional case we shall note it and use just one indicator to cover both and continue to call it an "organizational state." Generally NS may contribute to FS and RS ("add-ons").

Concerning the combined organizational and networking states leadership and management capabilities and "coordination capabilities" are important which refer to the maximum heterogeneity that the firm's coordination routines allow to derive a positive net benefit from.

Strong coupling, not separable sub-states, may occur when large amounts of capital infusion by investments or venture capital (financial state) result in adding large numbers of employees (resource state) and adding a management team (organizational state). Strong coupling may also show up for corporate venturing which combines finances, other resources and networking and cooperation (Figure I.125).

Based on *available* observable data and *simplifying complexity* we associate the overall state of an NTBF Ψ to the firm's indicators of sub-states, namely revenues, number of employees and productivity, plus an empirical bulk "multiplier" N which may account for remaining influences on the sub-states' including firm-internal and external effects (Equation I.15).

Equation I.15:

S(Overall System State) $\rightarrow$ N $\otimes$ FS("Financial State"; Revenues) $\otimes$ RS("Resource State"; Employees, R&D) $\otimes$ OS("Organizational State"; Productivity)

We regard revenues (R) as an indicator of the financial state allowing to finance a firm's development – for instance, organic growth via own cash flow or capital infusion via external investments or non-organic growth via acquisition of another firm. The financial state FS is susceptible to self-reinforcing processes with positive or negative effects (Figure I.114). FS may reflect investment persistence and RS innovation persistence (Figure I.127).

As an interesting aspect of new firm development a study by Chandler and Hanks [1993] suggests that the great majority of entrepreneurs have growth concerns that far outweigh their concerns about profitability.

The resource state is assumed to cover non-financial resources, in particular, employees. Specifically, in Equation I.15 R&D means the number of employees in a research and development function (department). R&D as a resource is usually measured by R&D Intensity (ch. 1.1.1; Table I.1; cf. also TVT, HVT in Figure I.128).

R&D is a "potential" and is a resource for innovation by a company-wide process which, when strongly coupled with innovation persistence as an organizational competence and strength, builds sustainable competitive advantage. The role of R&D activities of the R&D function of a firm becomes more obvious in terms of revenues, if the firm's offerings refer to inventions, for instance, by selling IP (licenses) or providing contract research (Table I.3).

As the approach to dynamically stable sub-states allows equality ($S_i(t+1) \geq S_i(t)$) a growing firm with *constant productivity* (P) is a special situation of growth with a stable organizational situation for a particular "*dynamically stable interval*" $<t_0, t_{n'}>$. For this interval, with revenue (R) and number of employees (EN), one can write the formula

$R_1 \cdot (EN_1)^{-1} = R_2 \cdot (EN_2)^{-1}$ = constant, resembling the well-known relation of ideal gases which is a good approximation to the behavior of many gases under many conditions and leads us to consider the notion equation of state. Classical thermodynamics is much concerned with equations of state.

The *equation of state* for an "ideal gas" is $p_1 \cdot V_1 = p_2 \cdot V_2$ = constant for a given temperature T with pressure (p) and volume (V) as state variables or $p \cdot V = n \cdot R \cdot T$ (R is a universal gas constant and n is equal to number of moles, the mass (m) divided by the molecular mass (M)).

In cybernetics there is a special condition of systems in growth states called the *steady state condition*. This special condition is possible after some time, when all input and output quantities are and remain constant. A relation between input and output quantities for a system in a steady state condition has been called "Static Transfer Response of the Dynamic System" [Ruhm 2008].

In our context, we shall not be so strict when speaking of a steady state condition. For dynamic stability we relax the steady state condition emphasizing OS(Organizational State; Productivity) (Equation I.15) to require only "negligible" change of productivity over a certain period. We shall denote these as "**dynamically stable states**" – of the whole system.

This situation of NTBFs with "constant" productivities shows a metaphorically comparable situation with the notion of a "steady state" as a model in cosmology (an alternative to the Big Bang theory). Accordingly, a steady state has numerous properties that are unchanging in time; the universe is always expanding but, by observation, is maintaining a constant average density (mass in relation to volume).

Therefore, one could speak of a *"steady state firm growth"* for the special case when growing through states of relatively stable organization, when expanding but maintaining a constant average "density" which here would be related to (inverse) productivity. This is illustrated in Figure I.132. Admittedly, such a relation between counts of components and size (firm revenues) does not differentiate components and their interactions in terms of organization including structures of leadership, coordination and communication.

In *purposeful and organized social (human-activity) systems* like firms change is usually associated with resistance by those who are or perceive to be affected by the change. Hence, it requires structured efforts ("energy") of leadership/management of firms to implement necessary changes of activites and processes and organization ("change management"). Such a period of implementing change proceeds via an *unstable "transition state"* (Figure I.133). Due to a lack of observing details such a transition state is regarded as a "black box". However, to a certain degree, a transition state may be compared with the startup thrust phase directly after firm's foundation which is often an unsteady process with considerable risks of failure, searching for structure and organization dealing with changes for all the persons involved in it.

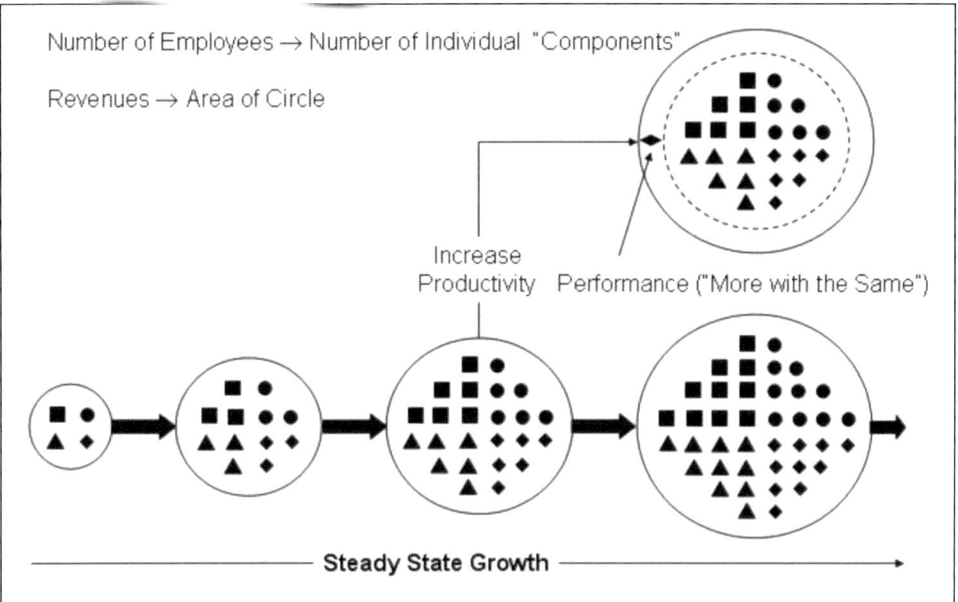

Figure I.132: Illustrated steady state growth and performance versus productivity of an organized entity.

After firm's foundation with a particular purpose and goal in mind and formalized by a transformation (Equation I.12), the startup thrust phase (ch. 4.3.2, Figure I.125) represents the first transition state of more to come for a new firm's development. Due to proportionately high risk of failure a new firm may operate here in "survival mode."

For firms, unless balanced by a simultaneously occurring second change, the transition state will exhibit a change of the relevant indicator function (revenue or employees) in the positive *or* negative direction. It may also be observed by a temporary increase/decrease of productivity and/or performance. Transitions states can be conceptually related to Greiner's "crises" (Table I.68). Internal change of the firm can be initiated by external effects. But there may also be intentional change of a firm's organization by the founder as described for Osmonics by Runge [2006:91-94].

Metaphorically a new firm's development via transition states can be compared with chemical reactions where an energetically stable initial state of educts (A + B) requires energy to bring the (closed) system via a transition state (admittedly with a very short life-time) into a more stable "product state" (C + D).

The non-stable transition state will have certain duration between two dynamically stable (sub-)states. Apart from referring to the startup thrust phase an idea of the duration of such an "intermediate" state can be got from revenue and employee developments of the German Nano-X GmbH, when after catching a big order, it increased first the number of employees, then increased production capacity and finally turned to delivery of the products. (Figure I.137, Table I.77).

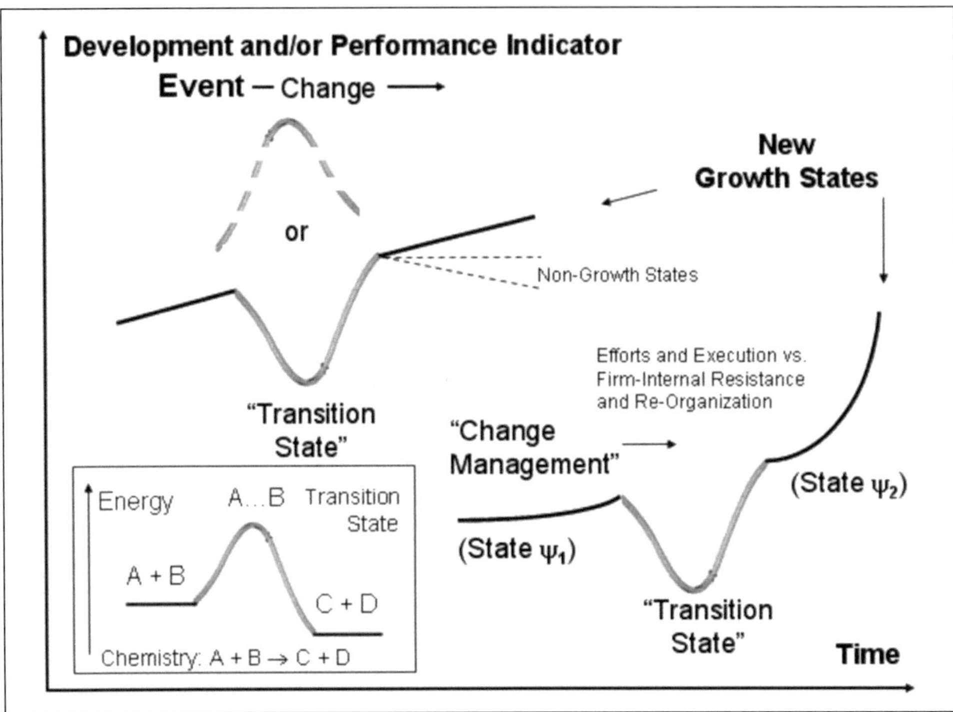

Figure I.133: Interruptions of NTBF developments by "transition states."

A transition state requires usually another style of leadership/management and execution, sometimes even a totally different style if the new firm has to enter "survival mode."

As described previously in various contexts a transition state initiated by a switch to (or start as) a VC-based NTBF or one getting huge private investments follows usually a path into a state that corresponds largely to intrapreneurial states and initiatives of innovation management of existing large firms (Figure I.134). Most notably, change to a VC-based NTBF is usually associated with a change of the leadership/management team, organizational structure and organizational processes (Figure I.126) and expressed by a transformation described formally, for instance, by Equation I.11.

Figure I.134 and Figure I.130 contrast implicitly the challenges of building a firm with those of managing ("running") an established firm. Building requires the entrepreneur(s) to develop tangible and intangible assets, organization and coordination mechanisms more or less from scratch. The focus of executive managers of established companies revolves around existing assets, their strategic uses and related mechanisms.

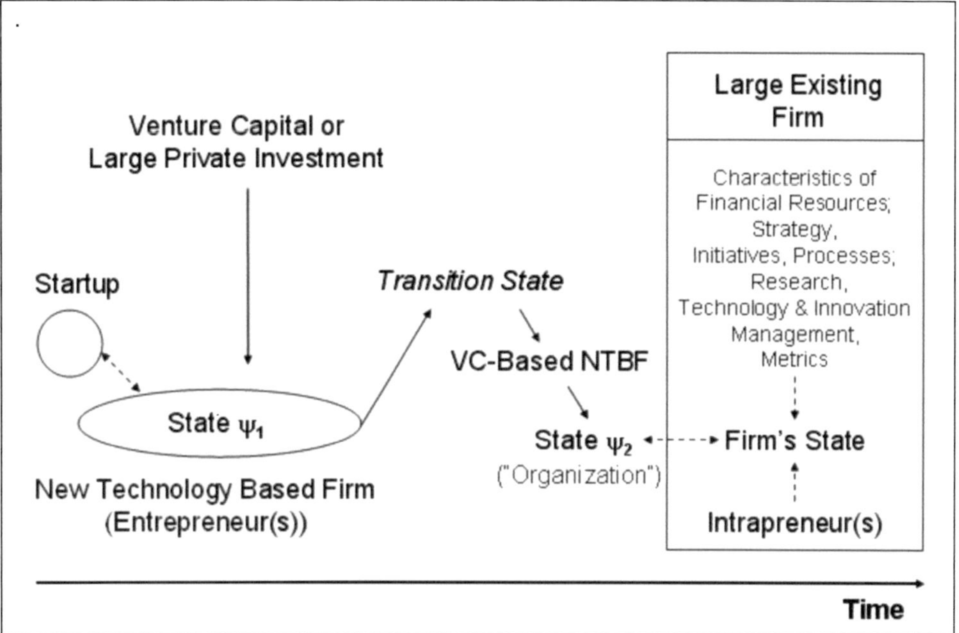

Figure I.134: The transition of an NTBF into a new state after becoming a VC-based NTBF characterized to be similar to those of large existing firms.

VC-Based NTBFs represent an entity intermediate between a privately held or privately controlled firm and a large corporation. Key differences refer firstly to

Opportunities

- Uncertainty is less – VCs tend to go into rapidly growing, large markets with proven teams (or replace founder entrepreneurs by experienced management if things do not develop as anticipated)
- Potential profits are higher than typical privately held or controlled, respectively, firms.

There is more planning than in startups. VC-based startups work from a business plan and run "standard" financial management and technology/offering development processes as in large firms. In particular, fund raising by VC-based startups is synchronized with development phases and achievements of milestones (Figure I.52, Table I.27) similar to the Stage-Gate process of large firms (ch. 1.2.7.2; Figure I.79, Figure I.180). The second aspect concerns resources:

- There are more resources available than in bootstrapped (ch. 4.3.3.1) or private firms, but less than in big corporations (for intrapreneurship).

CleanTech (Table I.52) and particularly biofuels and biobased chemicals (A.1.1) are areas where technology entrepreneurship is focused largely on VC-based NTBFs.

Considering all the previous outlines there are bridges to fundamental concepts used so far in treating new firms' dynamics according to:

- Stage-based views (ch. 4.3.1),
- Resource-Based Views (RBV, ch. 4.3.2)
- Input/output (I/O) models (4.3.2).

The current *phenomenological approach* seeks essentially for qualitative and quantitative descriptions in terms of equations of state rather than explanation. That part relates to a basic cybernetic model (ch. 4.3.4) and relates metaphorically to quantum theoretical and physical fundamentals. The bridge between description and explanation is achieved focusing on "interruptions" or "perturbations of growth, which means, "the transformations concerned with what happens, not with why it happens."

Explanation is approached by providing information on the nature of the "interrupttions," "why it may happen," in terms of a list of assumptions and common ways of perceiving forces in firms and forces in markets and industries and their interactions dealt with by business administration and economics (Table I.76).

4.3.5.1 The Bracket Model

The following approach will focus on NTBFs, young technology ventures in no more than the first twelve years of their existence (ch. 1.1.1.1). It takes the position of an observer and identifies and utilizes mechanisms and drivers of *changes* of *states* of a new firm entrepreneurs being a part of.

> Phased firm growth in the sense of GST is a result of a *relation-oriented purposeful process of development* in which a linear or circular interacting series of *changes from inside or outside the firm* lead irregularly to new development states associated with increases (or decreases) of appropriately selected observable indicators of the developing entity.

That means, the approach focuses on the firm and its relevant super-systems, such as the economic and political systems, and consequently *switches between macro- and micro-levels*.

The model follows a *teleological approach* emphasizing often *not "why things happen"* but emphasizing ways of behaving and asking

- What can it do? (the resources)
- What does it do? (the phased process)
- How does it proceed to reach the goal?
- What can be expected with regard to reaching the goal – often providing "reasons for thinking that" (Figure I.2)?

Necessary resources during the pursuit of the entrepreneurial goal(s) are specifically given in Figure I.120 and Figure I.121, Figure I.125, Figure I.126 and Figure I.130.

"That something is 'predictable' {"*expectable*" in our context (ch. 4.1)} implies that there exists a constraint." [Ashby 1957:132] (Braces added). The below presented "bracket model" of development of NTBFs takes time as an implicit variable (states at given times, usually on a year-by-year basis) and focuses on some fundamental constraints for a hierarchy of taxonomies (Figure I.128):

- Firm type RBSU ("spin-outs from academia") versus "other academic NTBF" and "other NTBFs" (Table I.2)
- Full and majority ownership and control versus minority ownership, little (almost no) control (*VC-based NTNFs*; Table I.74)
  VC-based NTBF do not only affect ownership and control, but also endowment with large financial resources and change in terms of management and organizational structure and probably strategy as well as execution following closely approaches used by large firms (ch. 1.2.7.2, Figure I.126, Figure I.134; A.1.1).
- Differentiation of organic and non-organic growth (Figure I.127).

It is to be noted that concerning VC-based NTBFs the emphasis will be on combined ownership, control and financial endowment. Other types of (large) investments, for instance, for large-scale *production-oriented NTBFs*, may decouple these categories, for instance, through association with "silent partnerships" in the NTBF.

Furthermore, there may be always cross-roads when NTBFs with full and majority ownership and control change over to VC-based firms (Figure I.52). Prototypes of VC-based NTBFs will catch capital often during or at the end of the startup thrust phase (Figure I.125) or within the first five to eight years of existence. Several new biofuels firms were started immediately with venture capital (A.1.1).

The differentiation between organic and non-organic growth may be a "moving" typology. For its growth and a restricted phase of its development an NTBF may switch between both modes (Figure I.123). Non-organic growth exposes a firm usually to integration risk. While having a stake in another firm mostly improves revenue opportunities. Full acquisition improves often revenues, but iintegration risks means issues of management, coordination and performance.

Development of new firms is irregular. It proceeds via states of different life-times and modes of development. These exhibit three basic observable types of development (growth) as displayed in Figure I.107 depending on the founders' goals and characteristics of the firm which can be measured by appropriate indicators, for instance,

- *Linear* growth which may be "high" (steep) or "low (cf. Figure I.123, Figure I.137, Figure I.141, Figure I.149);
- *Exponential* (or hyperbolic) growth, which is typically "super high" and proceeds often very fast (EGCs, "gazelles"; for instance, Figure I.145, Figure I.159) or occurs after a delay (Figure I.143, Figure I.144, Figure I.154);
- *Asymptotic* growth, which will express "*non-growth" situations*.

Furthermore, as growth will be "perturbed" or interrupted, respectively, by firm-internal or external factors or both and associated with "transition states" (Figure I.133) change between different phenotypes of growth may occur (cf. delayed growth in Figure I.107). NTBFs with large-scale production will usually show significant growth patterns only after five to eight years of development and scale-up.

Basically, we assume to be sufficient that for growth dynamically stable states exhibit only (continuously and monotonously increasing) linear, slightly curved or exponential behavior for an observable interval $<t_1, t_2>$ (Figure I.133).

If we can associate two subsequent dynamically stable states to the intervals $<t_1, t_2>$ and $<t_5, t_6>$, then the life-time of the in-between transition state (or states) will be attributable to $<t_3, t_4>$. For instance, for the German WITec GmbH (Figure I.123) the interval <2000,2002> would relate to a transition state (cf. also Figure I.156). This indicates that *transition states of new firms* may have life-times which are comparable in duration with dynamically stable states.

Formally, transition states may show up as an overlay on the overall growth pattern which let a gross observable indicator appear its indicator function to keep continuity, but not necessarily with increasing character.

If strong change is associated with a very short "life-time" of the transition state, the growth function will exhibit a "jump" as illustrated in Figure I.21 and seen for the revenue curves in Figure I.137, Figure I.139 as well as Figure I.140, where a basically exponential growth pattern of Nanophase Technologies between 1994 and 2000 is interrupted at 1997.

The bracket model is resource-oriented as well as activity- and process-oriented where "events" trigger a corresponding change of the state of the firm including the state(s) of the involved persons or groups of persons, respectively (Equation I.14, Equation I.15).

> We define a **(business) bracket** as a *generally observable, but also expected* (Figure I.136) *impact* of a transform or a transformation into a *new state* of a firm in a specific business area. The associated observable quantity, reflecting the *time development of the impact*, will start with a usually difficult to detect onset, then there will be a peak-like or wider, skewed bell-shaped center representing the essence of the impact, and finally a very long-tail characterizing formally a slowly but continuously reducing effect of the transform or transformation, respectively (Figure I.135).
>
> A business bracket will be represented by a *graphical map of an event*, the "perturbation" of the time-development of a properly selected observable quantity characterizing a change into a firm's transition state whose life-time ends when a following new dynamically stable state can be detected by an observable regular shape of the indicator function (Figure I.135, bottom).

That means one may identify a "*front bracket*" rather well, but principally never an "*end bracket*." For practical purposes and special cases one could "declare" an end bracket with regard to the point when the impact of the transform/transformation is assumed to become so small that it can be viewed as having no further significant influence on the measurement result – and this may be the end bracket to coincide with the emergence of a new bracket.

Business brackets originate

- externally from *non-controllable* effects or internally (controllable or uncontrollable) ones or
- from *unintentionally or intentionally effects* – initiated by *firm-internal* decisions and actions.

Structurally, a bracket refers to a *relation* associated with a *change of a systemic state* of the entity under consideration. A *long tail of a bracket map* simulates artificially that a "bracket event" leads to a new state of the affected entity which will also influence future states, for instance, in terms of decision-making, actions and behavior. In this regard it simulates "organizational memory" (Figure I.136, Figure I.129) which actually corresponds to an "*Ashby memory*" [Ashby 1957:115-117], a theoretical construct evoked to explain behavior of an incompletely observable system by reference to an event in the past.

> "If a determinate system is only partly observable, and thereby becomes (for that observer) not predictable, the observer may be able to restore predictability by taking the system's past history into account, i.e. by assuming the existence within it of some form of 'memory'." [Ashby 1957:115] Ashby provides a very illustrative example of reference to "memory" for a "living system." [84]

A business bracket reflects a business event which may affect a company's *competitive position* in a positive or negative direction. It reflects influences on the development of a (new) firm as it affects measurably what a firm has, can, gets, owns, does (also as a response to external effects) or delivers (Figure I.130) in relation to the firm's mission and goals.

Business bracketing emphasizes the micro (entrepreneur and firm) level of entrepreneurship with the issue of connection to the macro level.

- It relies on observations at the firm level to reveal mechanisms and processes of growth to be operative and focus on dynamics – through sequencing of events and their interactions and mutual reinforcements (Figure I.114, Figure I.115).
- It provides *explanatory guidance* to make sense of the mechanisms and processes that give rise to new firms' developments.

The business bracket concept has to be differentiated from the concept of "bracketing" of social theory and sociology, in particular, temporal brackets (ch. 1.2.2). Here, bracketing would be a process related to the *entrepreneurial person(s)* in the context of the venture's development, specifically a focus on people.

Temporal brackets, for instance, refer to cognition when a company-internal process or event will start (the *front bracket*) and an expectation how long an event or process will take to complete (*end bracket*) or, after a period of time, the cognition of an event that will start a subsequent event or an intentional start of a new bracket in anticipation of an event to occur. Their identification requires observation and direct inquiry into entrepreneurs' intentions, decisions and behavior. Defining front and end brackets means the entrepreneur is largely in control of the bracket.

Temporal brackets can, for instance, be related to strategic actions or decisions taken by actors or to planning. That is, each temporal phase started either by cognition of a significant exogenous or endogenous event and a related or an unrelated endogenous decision and action taken by organizational decision-makers. Some of these brackets get formalized into timetables such as business plans. Some critical brackets associated with pacing may be made explicit as sub-goals or milestones. Hence, temporal brackets may structure venture development, which is at least partially controlled by the entrepreneur [Bird 1992].

For ventures to emerge as recognized and reliable entities with a competitive advantage, the timing must be right. For this to happen, the entrepreneur needs to be aware of and understand the time requirements of the different events and processes.

Entrepreneurial growth of NTBFs will be described by periods of *dynamic stability of states* interrupted by transition states (Figure I.122, Figure I.133) initiated by business brackets.

With regard to *observable indicators* maping a business bracket and its time-dependency to *a "wave" with positive or negative amplitude* (the event's impact) has the consequence that, for a curve, overlapping "waves" may exhibit, for instance, a dip or may even "extinguish" each other ("balanced forces"), as is seen for the interference effect of light. This is illustrated for the case of a bracket by an economic recession with a negative impact in Figure I.135 (for an example cf. also Figure I.123).

Depending on the subject under observation it is important to select an adequate indicator as a bracket may show up for one indicator, but not for another. To reveal an indicator Figure I.89 demonstrates this effect referring to the example of choosing the right indicator to demonstrate a relevant macro-trend. For this reason the current *bracket approach* to NTBF growth *will usually consider both, revenues and numbers of employees* (cf. Figure I.149) – and productivity.

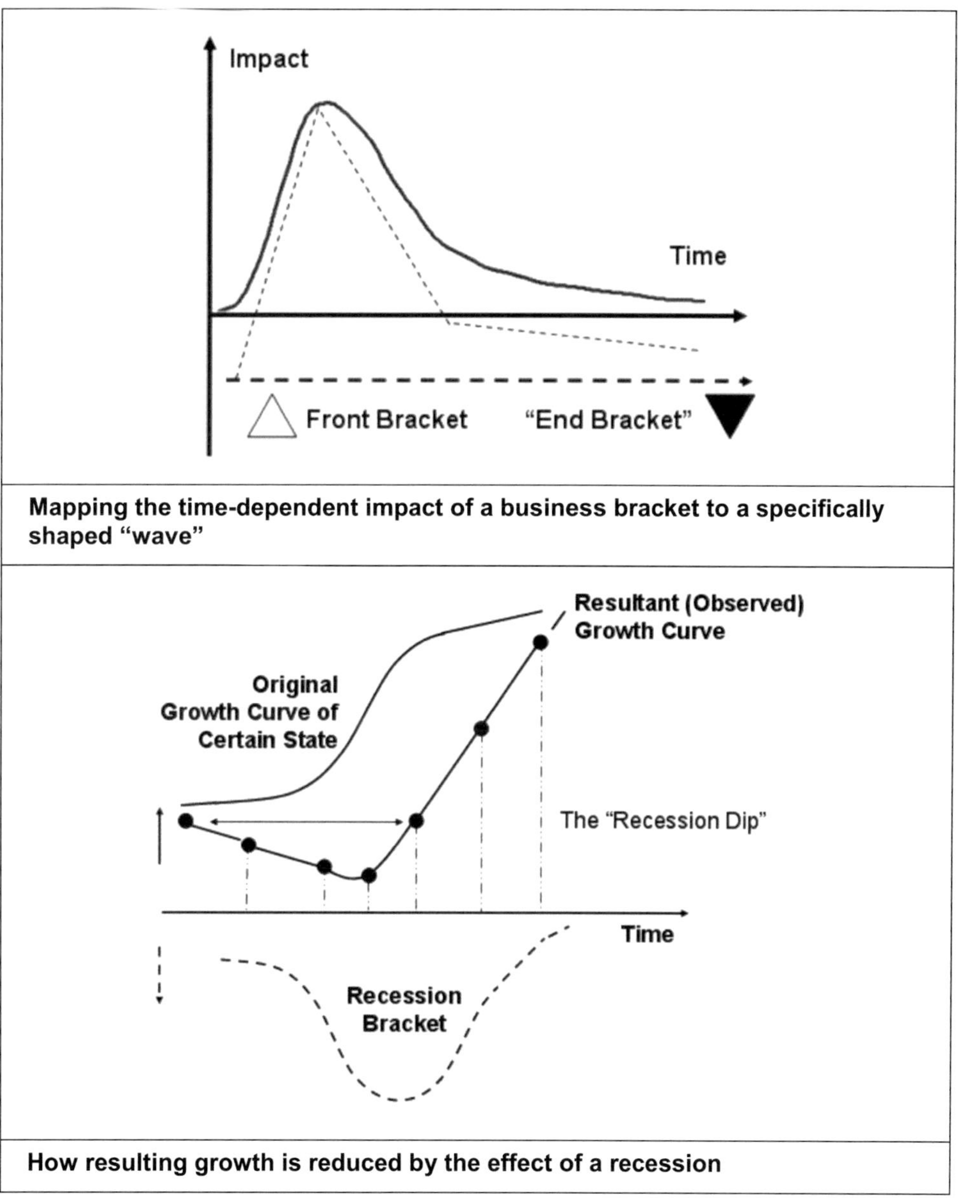

Figure I.135: Defining a bracket's time-dependent development concerning its impact and how brackets with negative impacts add up through overlap resulting in an observable dip in an observable growth curve.

For instance, recession stems primarily from the (enlarged) economic system and may affect (almost) any kind of an entrepreneurial firm. It is a *general bracket*. If only a particular kind of NTBF is affected, for instance, of a specific industry or market, we have a *special bracket*.

> Concerning observability and meaning the bracket approach depends on an (almost) complete data set for the relevant effects (transforms, transformations) whose collection for NTBFs is generally a challenging exercise.

> Brackets may, or usually will, affect the whole system relevant for (technology) entrepreneurship (Figure I.13), but will often be associated with only a particular sub-system's states due to limitations of access to relevant needed data (viewing "pars pro toto"; Equation I.15).

According to the bracket model entrepreneurship appears as a series of transformations accompanying a new firm's development into a viable small or medium-sized firm. Each transformation corresponds to a bracket generating a transition state. The transition state is a critical period of the firm's development into a new dynamically stable state involving decision, actions and activities of change to respond adequately to the initiating bracket or brackets.

> As brackets represent *relations*, between an operator and an operand (or a firm and a market/customers or a firm and the economic system in a recession) they will exhibit *regional dependencies* concerning their impacts.

Differences, for instance, in sales effects to customers may be due to different cultures, attitudes and preferences in different countries. An example for the global level would be: An internationally operating food company introduces a new product based on genetically modified objects (GMOs) and increases distinctly its revenues specifically through this product in the US. On the other hand, in Germany the product may be rejected largely by the public – and, moreover, the firm's reputation may decline generally in Germany so that revenues for all its other products also decrease.

And there may even be country-specific regional differences. A case of introducing one kind of innovative self-cleaning roof tiles in Germany by the firm Erlus Baustoffwerke AG from the south of Germany [Runge 2006:237-239] is an intentional bracket generation. If considered as one bracket for whole Germany, it would have a distinct regional impact – fewer sales than potentially possible. The reason is, there are clear regional differences for color preferences for roof tiles in Germany: In the north preferences are for grey/black, whereas in the south red/brown is preferred.

Focusing on the above growth and bracket indicators (measured revenues, number of employees, probably profit and productivity) sometimes does not allow to detect a bracket by an observer – though the firm "feels" all its impact. This is illustrated in Figure I.136 for a growth situation where two brackets occur within a rather short period of time, and the latter having at least the same impact as the previous one.

inspecting simultaneously several indicators may be a clue that may reveal the impact of the latter one. This applies to two positive, but also negative brackets.

An example that several brackets occur which will not be resolved by our common indicators is shown in Figure I.144. The brackets may occur so fast after one another that the in-between phases cannot be observed. If the time period between the two bracket events is larger, the corresponding growth curve may exhibit a "*shoulder*" of the clearly emerged latter bracket.

A final remark concerning brackets and their time dependencies should be made. A bracket and its time development is a representation of an individual, single event which will not change sign of the amplitude (impact) over time (Figure I.135). It reflects *a change of a firm's state* in one direction, by a positive or negative impact.

However, brackets may sometimes apparently appear to be reflexive. Envision a pharma firm having launched a new "blockbuster" drug in the market that made $700 million in sales over the first two years – product sales associated with a positive bracket. But, in the third year, after critical lethal side-effects were detected, the firm was subjected to a number of lawsuits, the cost of related litigations amounting to say $600 million (in the US; cf. the Vioxx and Lipobay cases, Table I.66). This seems to be a conversion of the original bracket, a change to a "negative wave" induced in the firm and by the same set of customers that generated the positive impact.

However, we encounter here decoupled brackets! The impact of the first positive bracket, being *an intrinsic part of a relation (to customers)*, means the firm has changed its state and when the firm encounters the negative effect it meets the impact (of the customers) in a different state: these are *separate, different brackets by concept*.

> The result of a bracket in the current understanding and GST approach to initiate a firm's new state has the consequence that its development (and growth) corresponds to *an irreversible process*.
>
> There is no "fall-back position" to the previous state (*status quo ante*) as a strategic option for decision-making. What can principally be re-established is an "empty shell," a constellation which will be used by persons and agents of the system who, however, will have changed knowledge, attitudes, decision-making potential etc. for initiating actions.

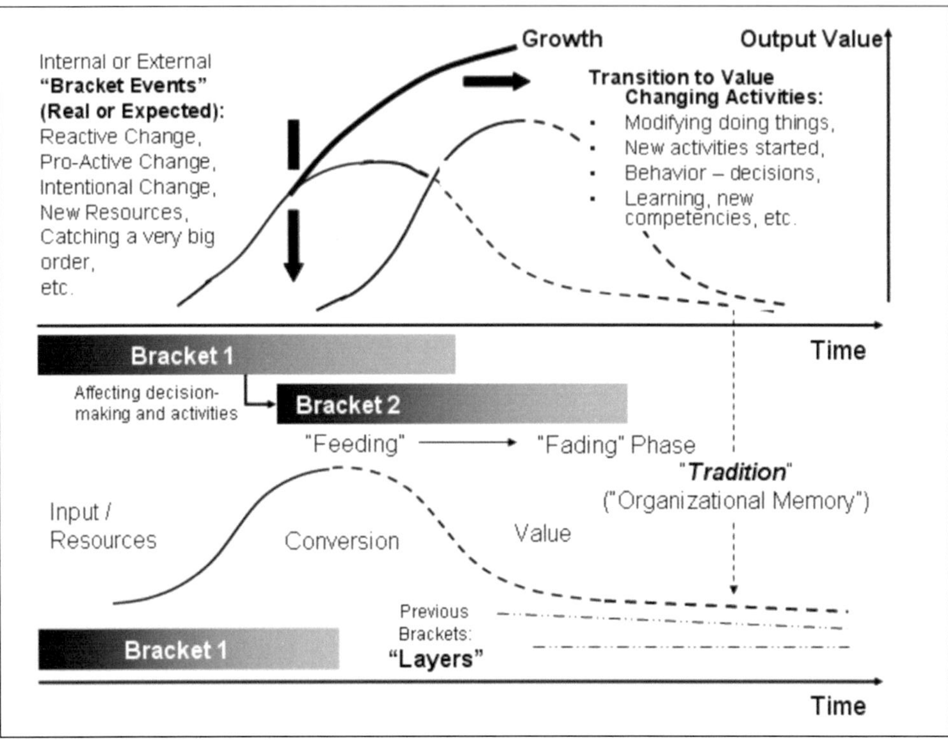

Figure I.136: Burying a bracket from observation such as revenues resulting from impacts within a short time difference.

The bracket approach is a "*perturbation theory*" treating firms' developments and associated states as essentially dynamically stable growth, for instance, with innovation and investment and productivity persistence and appropriate sets of input, resources, activities and decisions, etc. to be perturbed by *company-internal or external effects*.

In this way the notion "back on track" has a particular meaning: After a perturbation of a dynamically stable state the firm will proceed to a new dynamically stable state which, however, may or may not be structurally comparable to the previous one (linear – linear; Figure I.133, left) or (almost linear – exponential; Figure I.133, right).

The bracket model is *an irregular phase model of new firm development* contrasting the common stage-based views, such as that of Greiner (Table I.68) which associates growth with a set of structurally and operationally predefined phases initiated by a very small set of given initiators and related generic activities to proceed through the related phase.

We shall proceed assigning a variation in the measured/observable curve to *changes of the particular state of the firm* (Equation I.11; Figure I.135, Figure I.136) by one or more *unbalanced forces* and providing the *reasons for thinking that the observed* changes are due to particularly significant effects out of a set of conceivable effects.

These possible effects are derived either from empirical facts, cases, experience and/or rules of thumb established for NTBFs or existing large firms or from general theories like Porter's Five Forces Model and the Encapsulated Six Forces Model (ch. 1.2.5.3; Figure I.33, Table I.16, Table I.18) and reference to firms' failures (Figure I.114, Figure I.115) and finally hazards and risks for NTBFs (Table I.65, Table I.66).

A key role will be played by *generally known effects* (of super-systems) affecting NTBF growth like an *economic recession* or, for instance, the "*silicon cycle*" affecting specifically the semiconductor and photovoltaic industries as well as legislation for *policy-driven markets* (CleanTech like wind and solar power or biofuels). In this line, the recent (March 2011) Earthquake/Tsunami catastrophe of Japan affected, at least in certain countries like Germany, the nuclear power industry and attitudinal markets.

It should be noted, however, that for new firms endowed usually with only little resources brackets may emerge which will affect medium-size and large firms to only a little extent.

> The *first business bracket of entrepreneurship has been attributed to firm's foundation* with aspects of initiating the legal form of the firm and ownership and control (Table I.74). This event induces a startup state which can be *viewed as a transition state* (Figure I.133) which including its duration (lifetime) corresponds to the *startup thrust phase* of the initial configuration (ch. 4.3.2, Figure I.125).

Firm's foundation is particularly also a fundamental temporal bracket for the founder person(s) implicitly represented in Figure I.15 (firm's birth) and Figure I.16, the transition from intention to deed.

Instability of that very early state may be due to the formation process of agreement concerning the new firm's mission, roles and responsibilities of the leadership team and, if present, functions of employees and formation of the firm's culture. This state may also require, for instance, to relief initial group tensions as described specifically for a team in Figure I.70.

The startup thrust phase may also be associated with financing issues and establishing network connections, such as building the firm's advisory board or setting up cooperative connections with other firms. And there may be intentional changes of the original business idea or goal – becoming aware of a false start (ch. 4.3.2; 3M, NanoScape) – or revealing a new opportunity to be pursued.

A probably incomplete list of corresponding perturbations of NTBF growth will be summarized in Table I.76 in a structured manner referring partially also to risk classes for technology entrepreneurship (Table I.65).

The suggested growth models will be framed by classifying bracketing along several, often interconnected dimensions, sometimes associated with rough time markers for brackets to occur (Table I.71 - Table I.73). In particular, environment-related brackets reflect that we are dealing with open systems (Table I.76). There is

- Risk taking (ch. 4.2.1; Table I.65, Table I.66)
- Financing steps (ch. 1.2.7.2, Figure I.52; ch. 1.2.6.2)
- Firm growth and internal organizational factors, such as specialization, integration, coordination, communication etc. (Table I.69)
- Scale-up (for production related NTBFs) (Figure I.8, Figure I.9)
- External: economic and industry cycles or legislation (Figure I.34).

Bracket events show up often with "polar" effects, such as the economic boom – bust occurrence, getting a (singular) big order or making big sales versus losing the biggest customer (of only a small customer base), getting a key researcher or leader or manager, the actual company architect (Figure I.126), versus losing one. Loosing key researcher is often observed when technical firms merge.

The unexpected bracketing is often associated with opportunistic adaptation as described by Bhidé [2000:53,61,63].

In this way the bracket model has become a form of *exploratory data analysis*. It seeks to find patterns in data that are of theoretical, empirical and conceptual relevance. It is related to a *sequence analysis* involving the temporal ordering of events that mark the transitions of one stable state into another one. The issue is: getting sufficient data to find out whether hypothesized brackets will show up in the figures. [85]

Table I.76: Selected expected bracket events for new technology-based firms during their first ten years of existence.

Bracket Events	Comments, Examples
General – Global or Local; the Overall Economy or a Special Industry	
General economic recession	Usually reduction in demand; change of a firm's organization; change of management style (control, execution); some sectors are little affected by recessions, e.g. health and pharmaceuticals (life science"), the same is true for industrial research or public research as a customer
Special industry boom/bust cycle	"Silicon Cycle" of the semiconductor industry (Figure I.152)

Life cycle stage of the industry, particularly "birth of the industry"	Examples through, SAP, Microsoft, Cisco
Societal changes of attitudes or behavior	
Natural disaster, hurricane, overflooding, earthquake etc.	Hurricane Katrina for the US Golf Coast in 2005; the March 2011 Tsunami and earthquake in Japan
Firm-Related Brackets	
Firm's foundation as the first bracket	This bracket may coincide with selling an offering to the first customer(s) – e.g. WITec, Cambridge Nanotech (Table I.80), etc.
Both growth inducing and growth limiting factors can create organizational problems.	Hitting a critical mass of customers or employees is a "tipping point" which upon crossing leads to significant growth (decline of productivity?)
Issues within the founding team; change of organizational structure (integration, coordination, communication issues); change of the leadership/management team, in particular, a simultaneous change of ownership (Figure I.126) and VCs establishing "professional management" and organizational changes; change of management, for instance, "firing" by VCs, uncertainty, concern and anxiety of employees (effects on productivity); one key person/researcher leaves the NTBF or a key person appears in the leadership or management team	Ca. 14 percent of young German closed firms went bankrupt due to disagreements and tensions within the leadership team without any economic reason (ch. 4.2.3). It is always a negative signal if a CEO and co-founder leaves the NTBF. After firm's foundation a CEO appears from outside as a leader and ultimately the "company architect" (3M, Avery Dennison, Bayer AG; ch. 4.2.3)
Introduction of "quality management" process for NTBFs with production is often seen to be associated with organizational frictions.	This is particularly pronounced for the research function.

Table I.76, continued.

Intentional change of the NTBF's development and growth direction; "business model innovation," intentionally changing the business model, stepping up in the value system	False starts (3M, NanoScape); Business Self-Assessment (BSA): Tired of buying reverse osmosis membranes from other companies for use in equipment, the founder of Osmonics, Inc. Dean Spatz wanted to accelerate growth by directly manufacturing membranes himself [Runge 2006:92]; change intentionally firm's direction totally (Zweibrüder Optoelectronics GmbH, B.2); in 2008 MetroSpec Technology, Inc. was migrating from a service engineering/design company to a manufacturing company (Figure I.167) German Xing AG (ch. 3.4.2.1, B.2) introducing new sources of revenue
Intentional change of organization	Osmonics, Inc.
Breakeven	Reaching profitability, Figure I.53
"Sudden" changes of the NTBF's financial states	"Sudden" large capital inflows affecting operations; successful IPO; "sudden" stop of capital inflow, for instance, if agreed upon payments do not show up due to problems of the investor (seen during the Great Recession, MnemoScience, B.2) or investors/backers stop financial contributions (Zoxy Energy Systems AG, B.2)

Technology/development steps; a technological milestone (Novaled AG, Zweibrüder Optoelectronics GmbH) or breakthrough; achieving a dominant design, achieving an "industry standard"; achieving the "Holy Grail" of an industry	Affecting a technical process like scale-up or a general break-through effect (large processing cost reduction) or overcoming a "critical mass" effect
Successful launch of a new product, instrument, de-vice or service; disruptive or breakthrough innovation	Henkel (Figure I.138)
Patent, trademark (brand) infringement or other lawsuits	Especially if associated with a big loss
Non-organic growth; getting a stake in or acquiring another firm	Google, founded in 1998; bought Applied Semantics in 2003 that had a little piece of software called AdSense (Box I.24) or merging of new firm's (PayPal, ch. 4.3.2; A.1.7), ATMgroup AG (B.2)
Environment-Related Brackets	
First commercial activities (selling goods/products, services, licenses); specifically start already with a customer or customer base or startup backed by a supply agreement with a customer	Found a firm after having a cus-tomer or having already sold something forming the basis of the startup; IoLiTec and Solvent Innovations (A.1.5), WITec, ChemCon, Attocube, PURPLAN, Cambridge Nanotech Concept Sciences, Inc. (CSI) - Box I.11
Establishing sales and distribution	Own organization of sales and distribution ("sales forces," e-business, franchising) versus "out-contracting" sales and distribution (and probably marketing) includ-ing cooperation and networking with large firms (Figure I.51, Figure I.125). internationalization for an NTBF's offerings (including issues of currency exchange rates; Figure I.116)

Table I.76. continued.

Market-related brackets: catching a very big order from a customer (including military) or special customer segment	Examples: Nano-X GmbH (Figure I.137), Perkin & Sons (Table I.100)
Losing a major customer (from a small customer base) or a large reduction of orders of customers;	Nanophase Technologies (Figure I.140)
"Sudden" appearance of a competitive (substitutive) offering on the market or disappreance of a competitor	
Adverse event due to supplier-customer relationship;	German Nanopool's issues with a customer/distributor [Runge 2010]
Infrastructural effects (enlarging the basis of offerings' usage for technical or other reasons);	The role of IBM for SAP's and Microsoft's growth (Figure I.143, Figure I.144)
Exchange rate issues;	ChemCon, First Solar
Election of a new government	May influence S&T policy, regulations, subsidies etc.
Blocked export by changes of national tariffs or industry standards;	
Changes of national laws, regulations or innovation policy, phasing out materials (chemicals) or technical components	Policy-driven markets; renewable energy and energy efficiency
The Unexpected Bracket, Chance-Related Bracket	
Windfall gains: can occur due to unforeseen circumstances in a product's market, such as unexpected demand or government regulation; luck or serendipitous technical or commercial events	Tsunami/earthquake in Japan: German Heyl GmbH & Co. KG (Berlin) [86] offers e.g. Prussian (Berlin) Blue for treating radioactive cesium contamination by binding and washing out radioactive cesium spilled by the Fukushima nuclear plant. There was not only tremendous demand in Japan, but also in other countries [Dankbar 2011].

Though playing a central role in the models of industry forces (Figure I.33), in Table I.76 a bracket related to competition (rivalry) occurs only once. Competition basically may adversely affect revenues and profitability (decrease in profits or market share).

On the other hand, strong competition has been listed as one of the reasons for firm failure (Figure I.114). Furthermore, the previous outlines strongly emphasized competition or competitive advantage, respectively (Figure I.94, Figure I.114, Figure I.117, Figure I.122, Table I.75).

One rationale for this situation is that the bracket model emphasizes effects which show up observably for a relatively short period markedly in the development indicators of an individual NTBF. This requires competition to induce a "sudden" change of the particular development indicator as given in Table I.76. Or competition in a market corresponds to a permanent activity and represents a "ground noise" which cannot be separated from the development indicator's curve.

Disregarding new Internet and consumer service firms or software firms, most technology startups are not late entrants into a crowded space. They are often innovators entering new fields or new ways of doing business. They need to get to a significant size before established companies take notice. Furthermore, the sizes of the markets startups are operating in are usually too small to satisfy the needs of large firms (ch. 2.2.2).

Referring to 1989 *Inc.* 500 *privately held startups* Bhidé [2000:40,41] found that concerning competition fewer than 5 percent competed against large (*Fortune* 500-type) companies, 5 percent against midsize companies and 73 percent compete against small companies or other startups.

According to Saras Sarasvathy [Buchanan 2011] technology entrepreneurs see themselves not in the thick of a market but on the fringe of one, or as creating a new market entirely. "They are like farmers, planting a seed and nurturing it," she says. "What they care about is their own little patch of ground." However, things may become pronounced, if they get involved in patent issues (Nanion Technologies GmbH; B.2).

Analyzing the cases of competitive groups in nanocoatings (Nanogate AG [87], Nano-X GmbH (B.2), Nanopool GmbH [Runge 2010], Nanofilm LLC – B.2) and ionic liquids (four cases; A.1.5) as well as discussing the issue of competition with other founders of NTBFs who gave presentations at the author's course [1] suggests that technology entrepreneurs worry not so much about competitors.

Not just those following competitive antagonism (Table I.32), but all entrepreneurial firms whose founders address directly the known competition operate consciously and continuously vis-à-vis the competition, but being confident and convinced that their offerings are and must be always better, lighter, higher quality, easier to use etc. (cf. "what if" questions of ideation, ch. 3.3; innovation and investment persistence).

A different situation may occur for VC-based or other startups or NTBFs which have become very large in few years in a densely occupied market. Here adverse effects of competition could result in a sudden reduction of market share – particularly, when there is an "explosion" of startups, many new entrants into a market or industry in a boom period or in policy-driven markets, such as the biofuels field (A.1.1) or other areas of CleanTech.

Here "super-competition" between a very large number of small and large firms may emerge fast based on technologies, types of offerings, processes and business models (Figure I.181, Figure I.184, Figure I.185, Figure I.183, Table I.17).

In this case startup entrants often hope that the market size is so large that even capturing only a tiny market share will satisfy the entrepreneur(s) ambitions and expectations. However, the situation will become soon more complicated when, for instance, for wind energy (wind turbines) and photovoltaic, companies from China and India enter the scene competing fundamentally on price (ch. 4.3.5.2).

Usually, for new market or industry genesis, there are very many market niches to be occupied by startups to avoid a highly competitive field (cf. ionic liquids applications, A.1.5; WITec GmbH versus JPK Instruments AG, Figure I.141).

The faster the startup can define its own market and secure customers the more likely it is to end up being the dominant player in the (new) segment. The time to start looking at competition is later in the life cycle. That does not mean startups do not need to do their homework though. Knowing who else is out there that may be a threat is important, but execution at early stage pays way more dividends than deep competitive analysis and defensive strategy.

The concept of dynamically stable states and transition states and bracketing for new firm development implies that an entrepreneur (leadership/management team) will undertake several *different types of entrepreneurial activities at different times* in relation to the goal of the firm and to "perturbations" from inside or outside the firm. Perturbations require *different competencies* and *different kinds of decision-making* to lead growth, but also lead and execute through a "crisis."

Different types of entrepreneurial activity require different resources and skills of the entrepreneur, with different types of risks and the rewards. These different types of activity should not all be lumped together in empirical research. But one can hypothesize that viable NTBFs that face similar initial configurations and sensitivities to external effects and similar developmental problems in sequence and access comparable markets will go through similar phases of activity and exhibit similar growth patterns (Table I.80, Figure I.163; A.1.6).

So far, the emphasis to systematize technology entrepreneurship was on types of technology ventures (Figure I.128) and architectures and configurations (ch. 4.3.2, A.1.6). On the other hand, Kunkel [2001] suggested focusing on an alternative approach, classifying types of "entrepreneurial activities" which would be attractive when dealing with NTBF growth. However, we think that the current approach integrating structure, the order of componens, and function, the order of processes, is appropriate for the subject.

4.3.5.2 The Bracket Model for Framing Empirical Observations and Explaining NTBF Development

Observation of actual and continuous development (growth) among NTBFs requires a long period (ten to twelve years) of investigation. But to deal with survival and growth of new firms the startup thrust and early development phases of NTBFs, particularly the initial configuration (ch. 4.3.2), will provide appropriate quantitative variables and parameters and criteria.

The bracket approach provides a way to a systematic analysis of a firm's ability to resist external shocks: "If there is a high probability of any negative event occurring and the hardship it imposes are generic, then one can incorporate the effect of random events through the venture's capacity for withstanding a common set of probable difficulties" [Woo et al. 1994:520]. In this sense the menu of (positive and negative) events in Table I.76 can be seen as a conceptual basis for the following discussion.

> The bracket approach boils down to "*curve analysis*" ("curve resolution"), *origins of change* and *discussions of dynamic stability intervals of growth*.

For instance, when encountering a significant (positive) jump in an early linearly increasing growth curve, as is observed in Figure I.137 for the German nanocoatings firm Nano-X GmbH (founded in 1999), the most straightforward assumption would be to associate the bracket with a remarkable *sales* activity. The earlier increase of the number of employees to prepare the related sales *success* (a further "reason why we think that") indicates such an effect. And, indeed, this is the case. However, the *actual cause* of the jump, what the bracket represents, can only be grasped by switching to inspection on the firm level.

In 2004 Nano-X (B.2) captured a very big order for a new *innovative product* including this order to provide continued demand in the near future. The firm's first reaction was expanding production capacity and manpower (2003/2004 bracket of the employees' curve). The second (2004/2005) bracket is the reflection of the revenues' jump, doubling revenues. The product concerned surface coating (scale protection of metal sheets avoiding high-temperature oxidation) with the automobile industry as the end-user. The product is sold to German steel producer Thyssen-Krupp delivering the coils nano-coated by Nano-X technology for several types of metal sheets for the car body of the Volkswagen Passat model.

The Nano-X case shows that future revenues and profits of young firms cannot be foreseen reliably and single big orders may exhibit tremendous influence – in the positive or negative sense if the firm depends on big orders of one of very few customers (Figure I.115).

A jump in revenues like that of Nano-X is also observed for German ATMgroup AG from 2008 to 2009. Without detailed knowledge about the firm's internalities there cannot be a straightforward explanation – as in the case of Nano-X. The individual details for an explanation are as follows: The firm was founded as ATMvision AG in 2005 and in 2008 it acquired two other firms to become ATMgroup (B.2) which expresses the jump as a "resulting explosion" through non-organic growth (Figure I.127). Other examples of non-organic growth (Figure I.141) do not show such a marked effect.

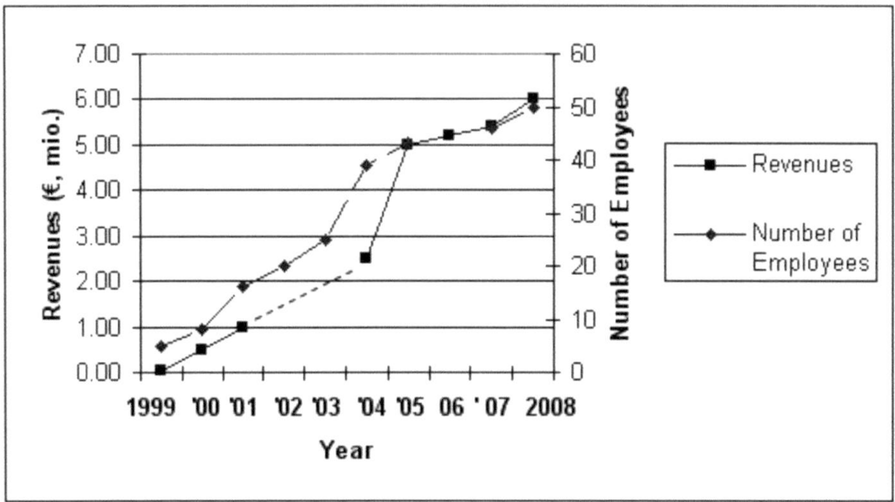

Figure I.137: German Nano-X GmbH with a jump of otherwise linear growth periods (dotted line replaces missing data points).

A linear growth curve changing over into much steeper "linear" shape after a *disruptive innovation* (Figure I.138) can be seen in the historical example of intrapreneurship of the German firm Henkel AG & Co. KGaA (Henkel & Cie. at that time). Founded in 1876 in 1878 the firm's first branded washing material "Henkel's Bleich-Soda" (Henkel's Bleaching Soda) entered the market. Almost linear growth of production from 1902 to 1907 (Figure I.138) is observed. Sales of the "Bleaching Soda" from 1884 (239,000 German Reichsmark (RM) to 1900 (1,155,000 RM) [Feldenkirchen and Hilger 2009] showed roughly exponential growth over the period. For comparison in 1914 the value of the US dollar had a factor of ca. 4.2 to the German Reichsmark [Runge 2006:473,474].

The total production increase after 1907 is due to the Persil innovation [Feldenkirchen and Hilger. 2009]. A rather steep exponential growth of Henkel's revenues due to sales of Persil, the *first "self-acting detergent,"* was prevented by a very long lasting, terrible global event, World War I (WWI; Figure I.139). On the other hand, the development in Figure I.139 reveals that even an extremenly strong force can be largely balanced by an extremenly successful innovation assumed to exhibit exponential development (indicated by the dotted curve progression): Henkel's revenue curve plateaued during WWI, rather than showing a marked dip.

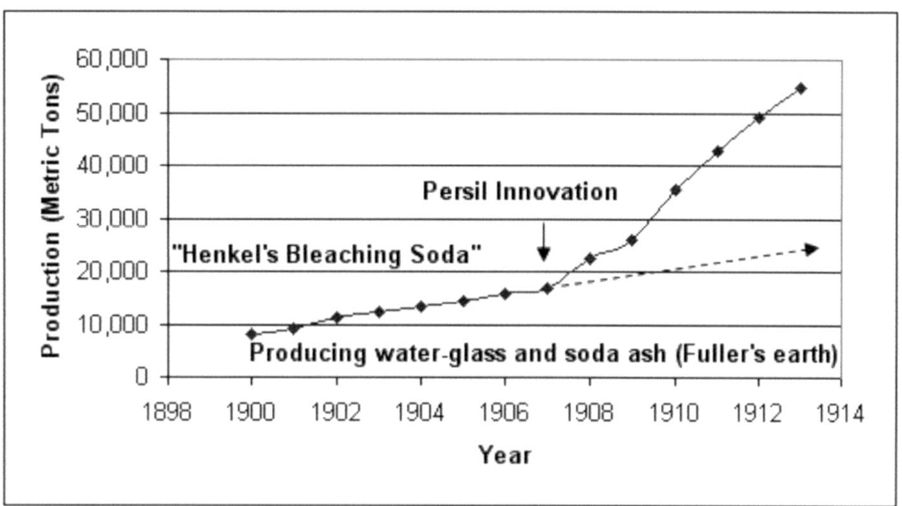

Figure I.138: The bracket of Henkel's Persil innovation in Germany (production of glycerin ca. 1% of total neglected).

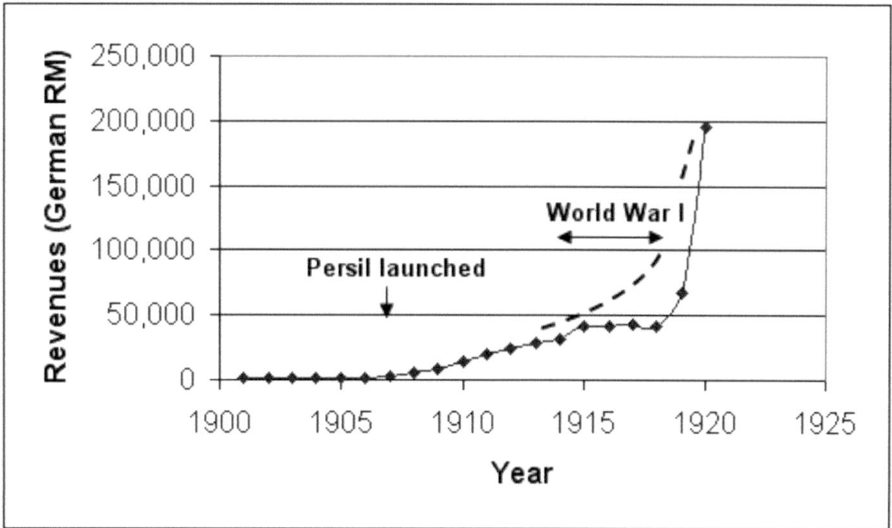

Figure I.139: Bracketing observed for Henkel's revenues by an extremely severe global factor.

It is not uncommon that some few (fast) growing firms exhibit steep exponential growth keeping yearly revenue increase of fifty percent to one hundred percent from startup. Such companies achieve one billion dollar (euro) in revenues in just a few years, usually less than ten years, as achieved by Cisco (Figure I.145), Q-Cells (Figure I.152, Figure I.153) or Google (Figure I.159, Figure I.160; ch. 4.3.6).

More than often assignment of brackets to entrepreneurship-related effects, however, may become tough as the below Figure I.140 shows referring to Nanophase Technologies Corp. Nanophase was founded in 1989 as a spin-out of Argonne National Laboratories. In the early years, Nanophase ramped up manufacturing technologies to commercial scale and established a viable manufacturing facility. In 1996 revenues were still quite small (approximately $600,000).

From its inception in November 1989 through December 31, 1996, the firm was in a development stage of scale-up. Since January 1, 1997, the Company has been engaged in commercial production and sales of its nanocrystalline materials, and the firm no longer considered itself in the development stage. It went public at the end of 1997 and the common stock traded on the NASDAQ.

The example of Nanophase Technologies shows the fundamental risk of new firms, namely dependency of revenues on very few customers. Nanophase shows both associated effects – much gain and much loss.

Furthermore, the example stresses an important issue of the current and almost all approaches to growth of new firms relying on revenues and/or number of employees – the longer-term viability of a firm without positive cash flow and, moreover, rather than profit generating having net losses over years. Here patterns of decreased net losses follow essentially increased revenues.

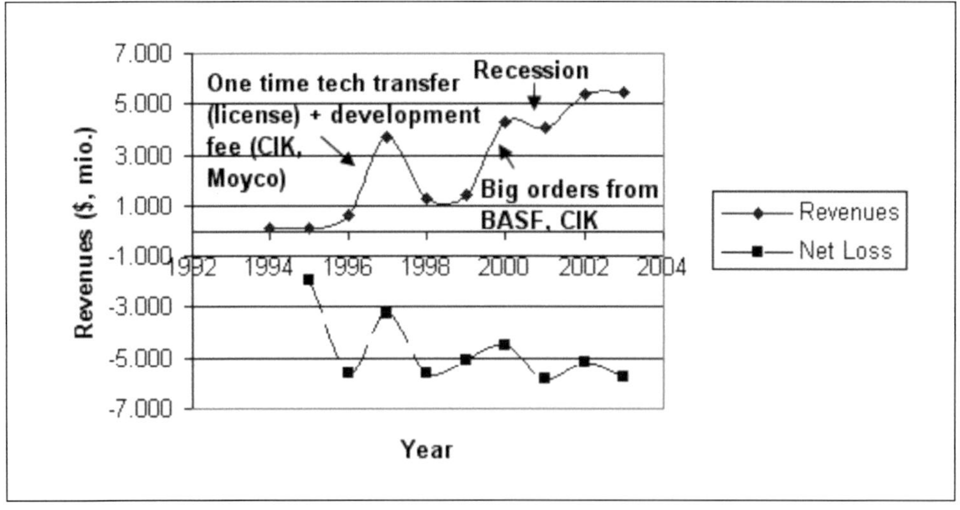

Figure I.140, continued.

The dip at 2001 of the curve is obvious – the Dot-Com Recession. It is not clear, however, whether there is a (negative) dip between 1997 and 2000 and how to assign it. There is no straightforward explanation for one bracket of such a markedly negative effect which stops an exponential growth till 1997.

Comparisons with other firms concerning the period 1996-1998 does not provide a clue. Hence, either there is a large one-time negative influence or the curve represents a one-time large addition to revenues which, for the 1996 – 1998 – 2000 period, would show exponential growth. The last option is indeed is the case.

Revenues from BASF and CIK constituted approximately 68.5 percent and 10.0 percent, respectively, of the company's 2000 revenues.

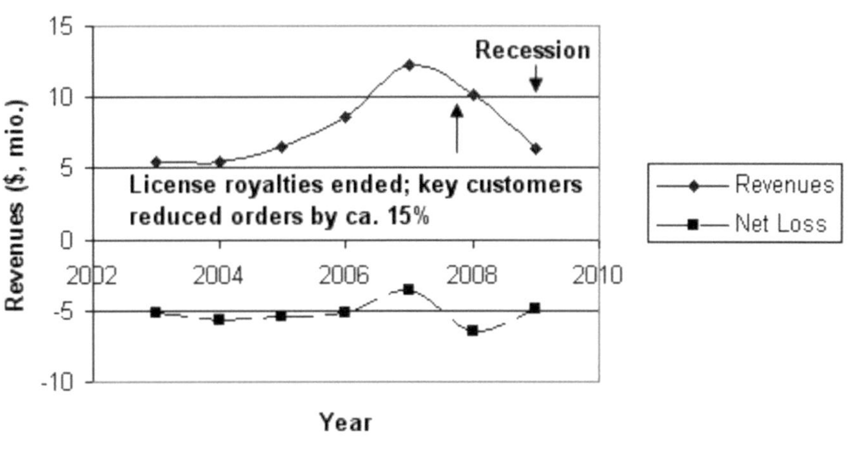

The reduction of 2009 revenues to the level of 2005 could be a huge effect of the 2009 Great Recession. The decrease of the revenue curve starting already after 2007 and its tremendous magnitude in comparison to revenue reductions of other firms during the recession suggests the existence of an additional (negative) effect. The cause is revealed by inquiry on the firm level.

Since 1996 Nanophase's number of employees was hovering around fifty five (with a peak of 61 employee in 1997 and currently around 55).

Figure I.140: Revenues and net losses of Nanophase Technologies.

The *actual financial state of a firm* and its overall "health" is not generally reflected by revenues. The assumption of revenues to be an appropriate indicator for the new firm's financial state or number of employees to reflect *viability of growth* is generally associated with some implicit premises – and the non-accessibility to more appropriate data unless dealing with the few NTBFs which are already publicly traded at a stock exchange.

Usually, one takes the "income state" (revenues) in the sense of "pars pro toto" as an indicator of a firm's financial state to be sufficient for further growth. But, ideally at least the "loss state" (net loss) and "financial reserves" (cash, equity, or number of basic and diluted shares outstanding) should be taken into account.

Patterns of *non-organic growth* (Figure I.127), which is essentially growing by having stakes in or acquiring other firms or establishing joint ventures (JVs), provide complex cases. But, for the few cases under consideration, related NTBFs do not show specifically significant patterns. An illustration is presented by the two German NTBFs Nanogate AG and JPK Instruments AG (Figure I.141). These do not only belong to the *same competitive group* as Nano-X GmbH (Figure I.137) or WITec GmbH (Figure I.123), respectively, but are structurally very close – also with regard to years of firms' foundation, being subjected to the Dot-Com Recession.

In particular, WITec GmbH and JPK Instruments AG show comparable education and leadership structures of the entrepreneurial team (Table I.41; Figure I.73) and for both the founding team is the owner or majority owner, respectively, with full control of their firms. They both are university spin-outs (of physics departments) and offer nanotools ("nano-analytics") on the basis of optical microscopy for research purposes addressing essentially "hard" substrates versus "soft" substrates like cells and provide "enabling technologies" for use in industrial R&D departments and private or public research organizations.

Furthermore, WITec and JPK have a large proportion of customers from life science (pharmaceuticals) which make them rather resistant against economic crises like the Dot-Com Recession or the Great Recession. Both exhibit dedicated innovation persistence and linear growth curve but have different CAGRs (JPG Instruments has 41 percent between 2002 and 2010, WITec 29 percent between 1998 and 2009), and show different productivities (WITec ca. €270,000 per employee, JPK Instruments ca. €125,000 per employee).

Both firms show marked innovation persistence. Contrary to WITec (Figure I.123). JPK Instruments (founded in October 1999) had it startup thrust phase during the Dot-Com Recession. It launched its first product into the market only in the fourth year after foundation (in 2002). It is not clear whether the strong increase in revenues of JPK between 2002 and 2004 (from €650,000 to €4.5 mio.) is the result of increasing sales of instruments or taking a stake in the firm nAmbition or both.

Furthermore, it is to be noted that, in terms of revenues, WITec and JPK do not show any effect due to the Great Recession in 2008/2009. The most likely reason is that both address to a large extent customers from public or industrial research areas which mostly did not encounter any significant shortages of their budgets during the crisis.

On the other hand, Nano-X GmbH (founded 1000) and Nanogate AG [87] (founded formally 1998), both having experienced leaders or managers, respectively (B.2), are active in nanotechnology for materials and materials' surfaces ("chemical nanotechnology") and are based essentially on "enhancing technologies" and less on "enabling technologies" for industrial and consumer applications.

Moreover, both are spin-outs from the same public research institute at almost the same time and both are located in close proximity. However, as Nano-X is a private company, Nanogate is VC-based, essentially a spin-out that resulted from a cooperation of the German Bayer AG and the public Institute for New Materials (INM) with immediate involvement of the VC firm 3i. Furthermore, it is to be noted that two of the founders from INM left Nanogate relatively soon after foundation.

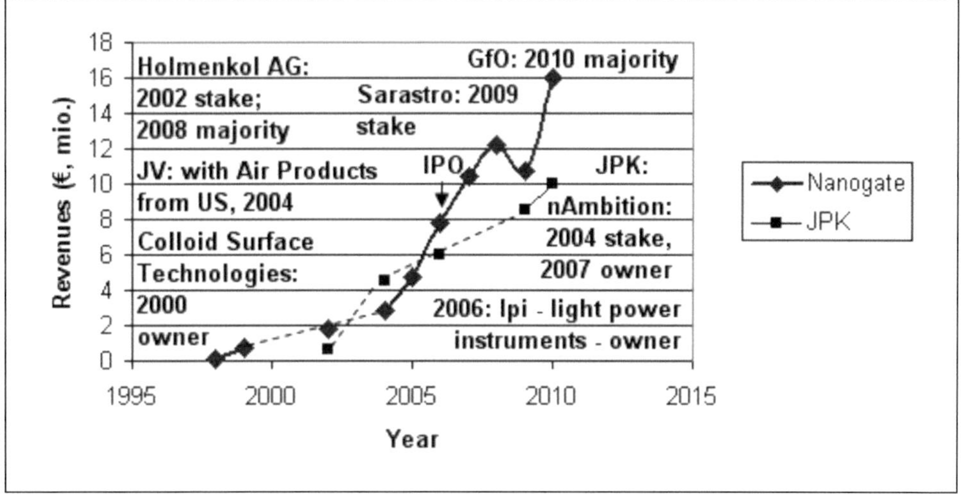

Figure I.141: Revenue developments of German firms Nanogate AG (founded 1998) and JPK Instruments AG (founded 1999).

The linear growth of Nanogate until 2004 and then the steeper linear growth until 2008 do not exhibit any special remarkable effect driving revenue growth. However, the big jump out of the 2009 Great Recession is to be noted. Only knowledge of the non-organic growth strategy and results of Nanogate would provide the clue to assign the jump in revenues to acquiring stakes in other firms rather than acquiring a big customer as observed for Nano-X (Figure I.137).

Productivities of both firms differ markedly: Nano-X (ca. €120,000 per employee), Nanogate (ca. €180,000 per employee). On the other hand, JPK Instruments with a non-organic growth approach follows linear growth, similar to the technologically directly comparable firm WITec (Figure I.123) without any indication about the role of its acquisitions or partial ownership of other firms.

For Nanogate the JV with Air Products and Chemicals from the US in 2004 opened up international markets the company might not have reached alone. But it is unlikely that the steeper increasing revenue after 2004 to be due to the JV event.

One may assume that building of different kinds of resource base involves different kinds of activity. And this may affect different development states and paths. Concerning financial resources for development, like cash flow, loans or equity provision without losing control versus equity from large, private investors, venture capitalists or IPOs, differences do not seem to be mirrored observably by the underlying growth indicators for WITec versus JPK but for Nano-X and Nanogate.

Also US NTBF Osmonics developed via non-organic growth [Runge 2006:91-94]. Considering also that Google in its early years took over another firm with no observable macro-effects (Box I.24; Figure I.159, Figure I.160) it seems that getting a stake in another firm with the possibility of revenues' development to behave like a big order does or massive addition of employees or both do not necessarily show up as a bracket in a revenue curve. Furthermore, the discussion of PayPal indicates that also mergers must be considered.

Related to the relatively small sample of NTBFs discussed in this book, we feel that it is not clear whether NTBF development by non-organic growth should be taken into account as a development path with its own observable specifics if revenues or numbers of employees are considered.

Nanogate exhibits a continuous pursuit of growth by both an internal and external focus. If, however, there is a marked increase in number of employees it can be expected to increase organizational pressure to integrate and consolidate, irrespective of whether it results from organic or non-organic growth.

As observed for NTBFs with organic growth also Nanogate shows a marked decrease in productivity when the number of employees increased significantly (here from 41 (2006) to 68 (2008) employees; Figure I.142). So far, a negative bracket has been associated with management and/or organizational issues. And this seems also to apply here. Nanogate's annual report 2007 tells us:

"As part of the implementation of the strategic growth programme NEXT, a new organisational structure was created, a management stratum was established and the central management strengthened with the addition of a commercial director and a management position for the buildings/interiors business segment."

Furthermore, in 2008 Nanogate Advanced Materials GmbH, the JV with Air Products & Chemicals, was fully acquired by Nanogate and in 2009 the related dip is additionally enforced which should be an effect of the 2009 Great Recession.

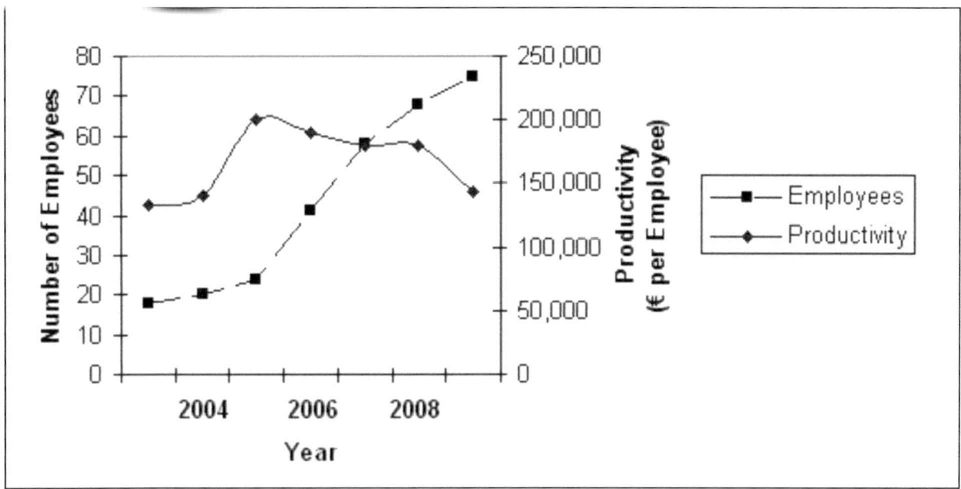

Figure I.142: Increase of number or employees and associated effects on productivity for German Nanogate AG.

After having taken a general economic and other external effects, recession and WWI, into consideration we shall consider the industrial system, essentially *early growth of particular new industries or markets* (providing high growth opportunities). Macro effects for the synthetic dye and automobile industries including the fast rise of gigantic firms like the German BASF, Bayer and Hoechst were cited by Runge [2006: 274-275] in terms of entry and exit firms of the industry or given by the Ford and General Motors cases in the US [Bhidé 2000:245-247].

The birth of an industry (segment) usually induces a "boom" of firms' foundation (cf. the biofuels situation; A.1.1). It represents a special opportunity for entrepreneurship. However, the particular "growth pushing events" for an individual new firm can rarely be foreseen or expected.

Basically, the growth curve of corresponding (very) successful firms has the phenotype of "continuous – exponential" (Figure I.107). Birth of an industry (or a segment) means everybody is new in the business. And in such markets buyers have to deal with a new company or forgo the product or service altogether.

A first more recent example refers to the "mainframe computer age," here the emergence of Enterprise Resource Planning (ERP) software and the German SAP AG (A.1.4) "grasping" the opportunity (Figure I.143). SAP had to overcome the recession of the first oil-price crisis around 1974, but actually could take advantage from a market-related infrastructural effect. Mainframe dominating IBM on which SAP's software ran introduced a faster and cheaper model which enlarged the customer basis for SAP tremendously.

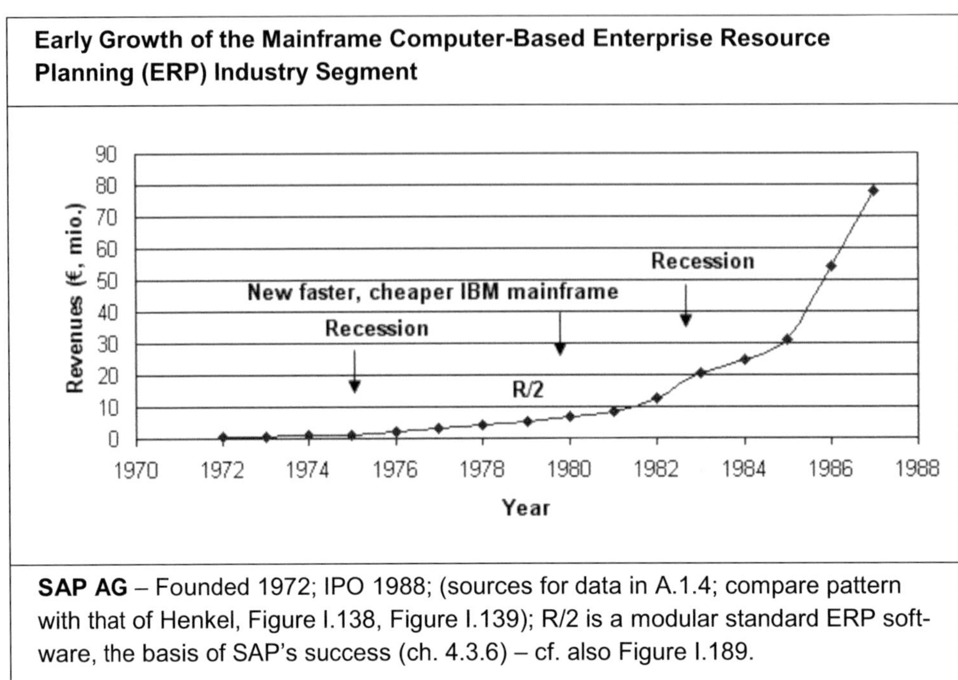

Figure I.143: Brackets for the early growth of SAP AG.

Before being able to fully take advantage from the favorable market effect, SAP had to overcome a further recession. This explanation interprets the middle growth phase (1981-1985) as an overlap of the 1980 bracket and the ca. 1983 recession. This means an interpretation overcoming the recession rather than introducing two brackets at around 1981 and 1985. Further revenues developments of SAP are given in the Appendix (A.1.4). The quite linear revenue increase between 1976 and 1981 is associated with productivity data of SAP (Figure I.147) which are almost constant for the interval <1976,1979> and, hence, represent a dynamically stable state.

In a similar way, increased growth of Microsoft after 1986 is seen as a sharply exponential growth induced by a whole series of brackets between 1980 and 1985, some of them being listed in Figure I.144. Any revenues reducing bracket would be leveled off by these tremendous factors. Early revenue growth data are given in Table I.67.

Microsoft has been very responsive to the demands of IBM to secure the crucial order for the PC operating system MS DOS from IBM in 1980. After the success of the IBM PC Microsoft could drive hard bargains with PC manufacturers to consolidate its position in the existing PC operating system market to ultimately conquer the market.

As for SAP IBM with its new mainframe to run SAP's application also for Microsoft IBM played a key role of "accelerating" the developments of the firm.

Early Growth of the PC-Based Industry

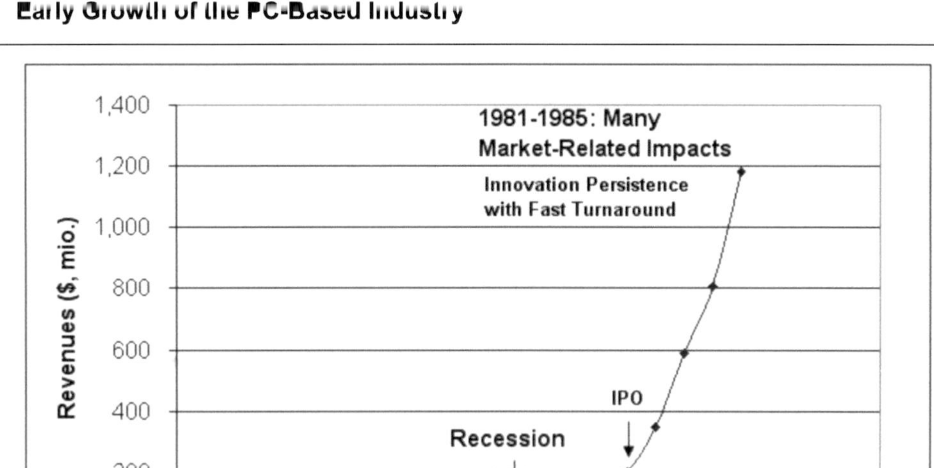

Microsoft Corp – Founded 1975; has been in *the Inc.* 500 List;
Many immediately following brackets and also the IPO in 1986, getting big orders and showing innovation persistence:

1980/1981: Huge contract with IBM to develop languages for their first PC including the operation system (DOS); arrival of the 16-bit IBM PCs.
Becomes the first major company to develop products for the Apple Macintosh.
1983: MS-DOS open to run on non-IBM machines; memory stretch beyond the original 640K limits of the Intel 8086 chip; MS Word for DOS 1.0.
1984: Leading role in developing software for the Apple Macintosh computer, for instance, the Microsoft Excel spreadsheet for Macintosh; provided the essential technology basis for the Windows versions of Microsoft Excel a few years later.
1985: Graphical user interface (GUI) introduced (Windows).

Figure I.144: Bracketing during Micosoft's early development (Sources of data from Note 88).

An example of early growth of a firm which did not suffer from any recession during its early growth phase is Cisco Systems, Inc. (Figure I.145). Furthermore, the Cisco example shows that productivity in addition to other growth indicators to be appropriate to track company internal effects. Cisco exhibits a bracket which correspond actually to two related company-internal effects, getting venture capital, installation of a "professional" manager and adding a huge number of employees (Figure I.126, Equation I.11) – definitely associated with serious organizational problems of integration and coordination seen in the productivity curve.

Cisco Systems represents an NTBF, which, just having passed the startup thrust phase, after three years, became a VC-based NTBF.

Early Growth of the Computer-Based Networking/Telecommunication Industry

Cisco Systems:

Husband and wife Len Bosack and Sandy Lerner, both working for Stanford University, wanted to email each other from their respective offices located in different buildings but were unable due to technological shortcomings.

A technology had to be invented to deal with disparate local area protocols. And as a result of solving their challenge – the multi-protocol router was born. Since then Cisco has shaped the future of the Internet by transforming how people connect, communicate and collaborate.

From the birth of Cisco (founded 1984, IPO 1990) until year seven of its existence no recession showed up and Cisco exceeded one billion in sales in 1993.

Developments of Revenues and Numbers of Employees:

Revenues and number of employees developed almost synchronously.

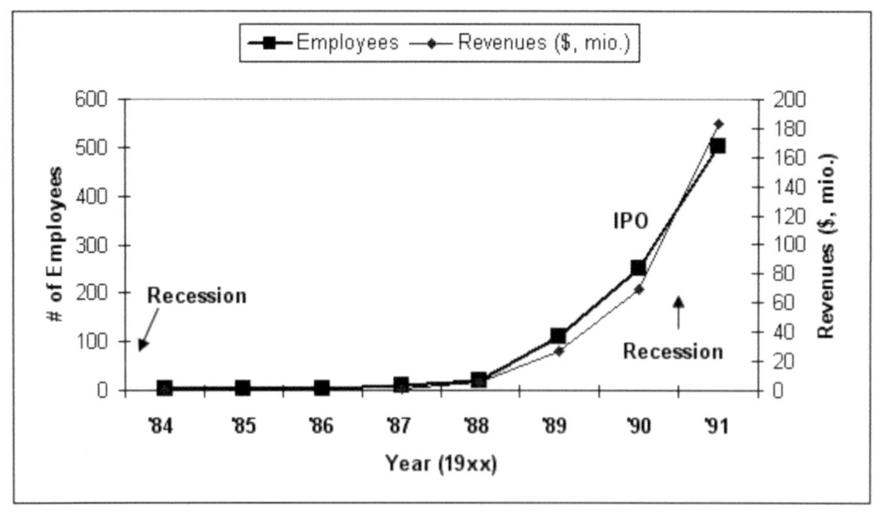

Figure I.145, continued.

Productivity and employees' numbers of Cisco:

There is a marked negative dip. Employees' numbers suggest that the large increase in employees reflects organizational frictions affecting productivity.

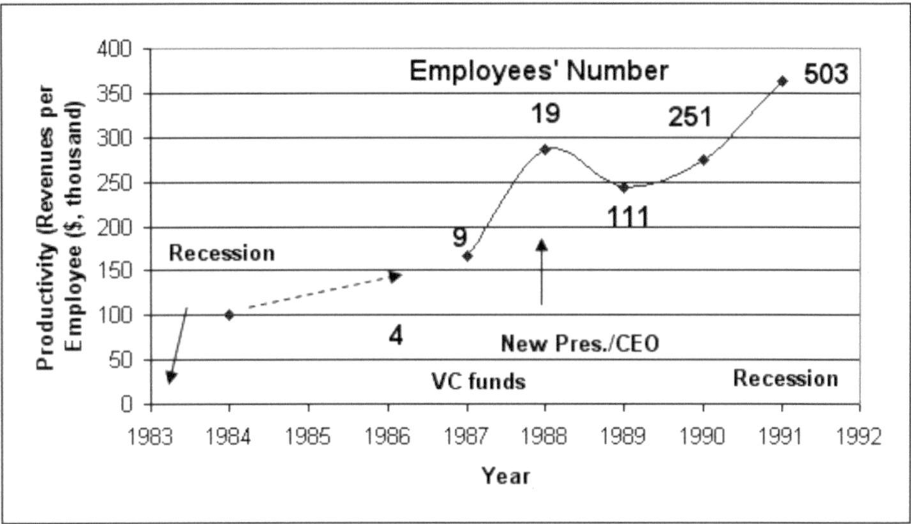

John Morgridge joined Cisco Systems in 1988 as President and CEO, and grew the company from \$5 million to more than \$1 billion in sales.

After 7-8 years of existence Cisco showed a remarkable productivity of ca. \$350,000 per employee.

Figure I.145: Characteristics of Cisco Systems (Sources of data from Note 89).

The management crunch (Figure I.118., Table I.72) and the "10 - 25 - 150" rule of thumb (ch. 4.3.1, Table I.72), for instance, reflects essentially organizational issues resulting in decreasing outcomes and results, especially productivity of the firm. This means, this rule is associated with a bracket having a negative impact. Associations to a time-scale of NTBF development (Table I.71 - Table I.73) may facilitate reasoning when attributing observed brackets to related effects as does the knowledge about the occurrences of recessions.

In a similar way to Cisco one can reveal organizational issues of Microsoft while showing strong revenue growth (Figure I.146) and of SAP (Figure I.147; cf. also Table I.67). As Cisco does, also Microsoft shows significant reductions of productivity when employees' numbers were increased by around ninety people (1980/1981). And, furthermore, for Microsoft obviously the dramatic increase in employees' numbers after 1981 induces further organizational issues – and subsequent brackets with *escalating* effects. Additionally, Microsoft data are in line with the "management crunch" and the "10 - 25 -150 rule" (Table I.72).

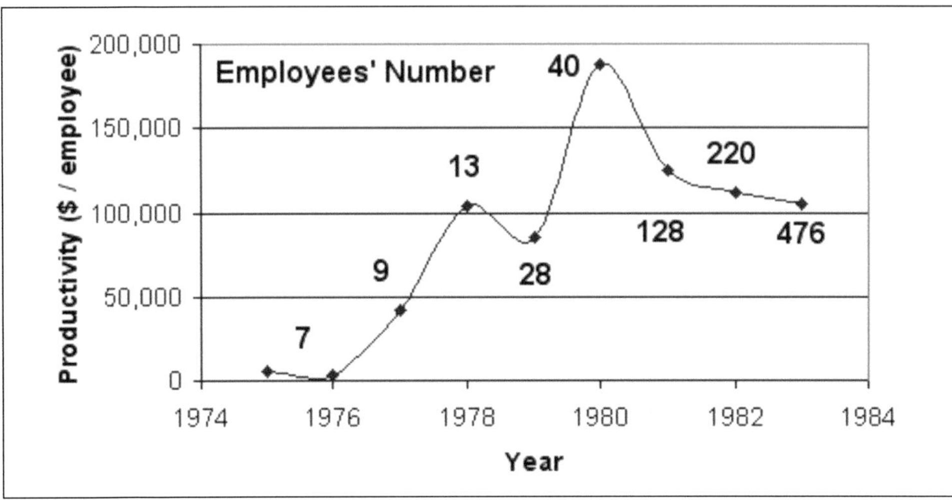

Figure I.146: Early organizational issues of Microsoft comparing productivity and number of employees.

We characterized the tremendous decrease of productivity as an escalating effect when Microsoft's number of employees tripled or doubled over a short period (40, 128, 220, 476, Figure I.146).

In general, we regard **escalation** as an expression of a *binary relation* originating with a special phenomenon or event and showing up either as an increase or rise, lifting something's extent, volume, number, intensity or scope *fast stepwise to a higher level* or a corresponding decrease. Escalations can be described in terms of transformations. Escalation is also possible among different kinds of brackets (effects), in particular, by mutually reinforcing interactions.

SAP exhibits two dips (eighteen and sixty-one employees) of the productivity curve paralleled by increasing growth in terms of number of employees and revenues (Figure I.147). One could be tempted to attribute the 1975-dip to effects of the 1973/1974 recession. 1979-data represent definitely pre-recession effects. We attribute the 13-18 employee dip to organizational issues (the "management crunch"; Figure I.118) without ruling out additional recession effects. SAP and Microsoft both have reductions of productivity after increasing the number of employees above thirteen.

After seven to eight years of existence, SAP and Microsoft exhibit productivity of ca. €85,000 or $100,000, respectively – similar to that of Nano-X (Figure I.137).

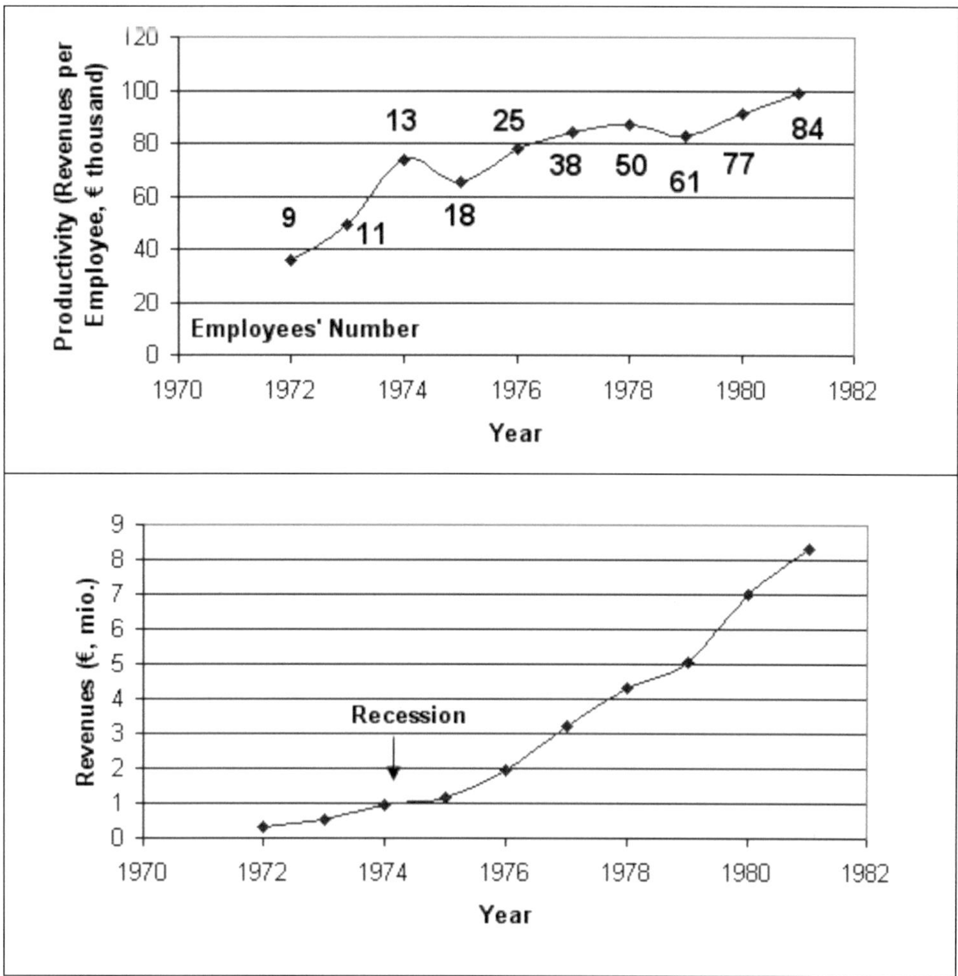

Figure I.147: Early organizational issues of SAP derived from comparing revenues and productivity.

Organic growth of NTBFs in terms of innovation and investment persistence (Figure I.127) can also be viewed as escalations.

For instance, the disappearance of an NTBF due to financial problems (Figure I.114, Figure I.115) is a decreasing escalation. It represents a circular situation of development in which a trigger in a system (a "cause") leads to an effect in the sub-ordinate system, this effect simultaneously becomes a cause affecting the super-ordinate system and, hence, the original cause initiates a reaction onto itself leading ultimately to an end state.

According to the cybernetic formalism (ch. 4.3.4) one can, for instance, describe such a development (decreasing revenues) triggered by "negative market effects" via the operator $\mathcal{NM}$ as a stepwise process affecting the financial state of the firm and resulting in a disastrous (numerical) financial situation DFS of the firm and, finally, bankruptcy:

$\mathcal{NM} \rightarrow$ {recession – reduced orders/demand – less customers} $\rightarrow n_S(M)$

$\mathfrak{FS} \rightarrow$ {revenues/cost ratio – rejected application/denunciation of loans – bad debts (accounts receivables) – poor financial management} $\rightarrow$ FS

Furthermore, we attribute the results of the transformations to corresponding (numerical) variables $n_S(M)$ and FS quantifying their effects. Disregarding the "hen or egg" problem, the start of the escalation (Figure I.114) is taken as the firm development by the transformation $\mathcal{NM}$, such as losing the most important of only few customers affecting FS.

For the escalation one can take a *logistic difference equation* (Equation I.16; as seen in Figure I.70, Equation I.5) and start from FS = medium = 4 (, then acceptable = 3, bad = 2, very bad = 1) that simulates the path to the disaster of a non-sustainable financial state DFS of the firm. The market influence is introduced by the factor $n_S(M)$ (assuming again -0.25):

Equation I.16:

$$FS_{j+1} = n_S(M) \cdot FS_j \cdot (1 - FS_j) + DFS \rightarrow DFS$$

A summary of many of the brackets and related attributions to effects discussed so far for new technical firm development is provided by the example of German (VC-based) Novaled AG (B.2) in Figure I.148. Novaled is active in and a key leader in the organic LED (OLED) business for flat panel displays (FPD) and lighting. It expected to break even in 2011.

Figure I.148 exhibits the origin of the firm (RBSU, as a GmbH – LLC) and its thrust phase (2001 – 2003), revenues and employee data, financing by venture capital in line with re-organization of the firm and the start of commercial activities, information for breakthrough technology developments and securing IRP by patents as well as further details for re-organizations associated with the first and second financing round.

The first financing round of Novaled is important with regard also to establishing a very experienced CEO, who, with an engineering education, was in high management positions of two firms, each representing one of Novaled's business targets. The related bracket is reflected in data for revenues and employees. The tendency of venture capitalists to establish management (and CEOs) in terms of "*veteran management*" having long experience in managing and in the markets and industries the NTBF is active in can be observed ubiquitously in biofuels startups (A.1.1).

On the other hand, the effect of the Great Recession 2009 is only seen in the number of employees (Figure I.149). The capture of further rounds of financing was definitely facilitated not only by previous progress in revenues but also the tremendous technological breakthroughs in terms of world records, increase of trust and credibility. This means, technological advancements may have also created a bracket which, however, cannot be detected by the used indicators.

Negative effects of re-organizations associated with the first (2004/05) and second round (2007/08) as seen in the productivity curve are probably more than leveled off by income (selling licenses, then also material) and overall productivity increase and probably less by the additional financial resources. Re-organizations associated with significantly increasing numbers of employees refers to building the Chemistry Group (in a more or less physics and engineering environment) and setting up a matrix structure for Novaled.

Notably the financing rounds of Novaled concerning amount of equity and duration until the next round follow roughly the outline of Table I.27. The listed world records and the particular financing rounds occurred (accidentally?) in the same time window.

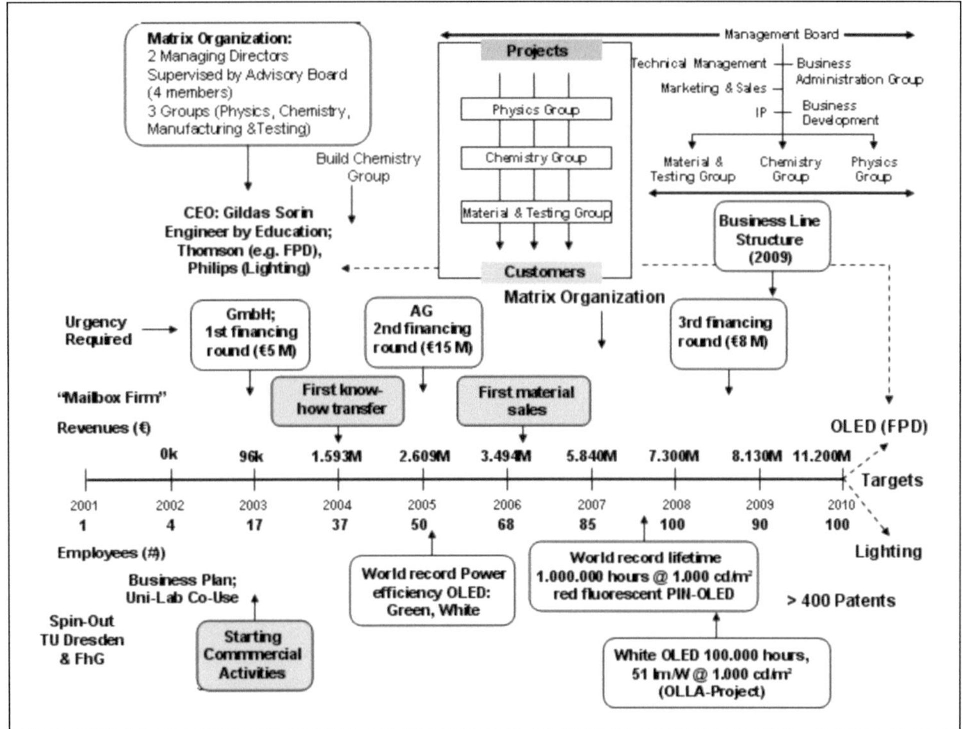

Figure I.148: Developments and characterizations of various states of Novaled AG (Extracted from Blochwitz-Nimoth [2011]).

Figure I.148 exhibits several states contributing to the overall firm state characterized by certain features in different layers. From bottom to top one sees the R&D and technology state ("world records"), the income of the financial state linked with the state of personnel (resource state), the commercial state (selling different kinds of offerings), the financial state (VC financing rounds) and the firm's organization including management state. For further illustration Figure I.149 displays development curves of Novaled regarding productivity and number of employees.

Re-organization of Novaled (B.2) concerning a business line approach in 2009 is outlined by Muth [2010], profit and loss development until 2008 by Böhme [2008].

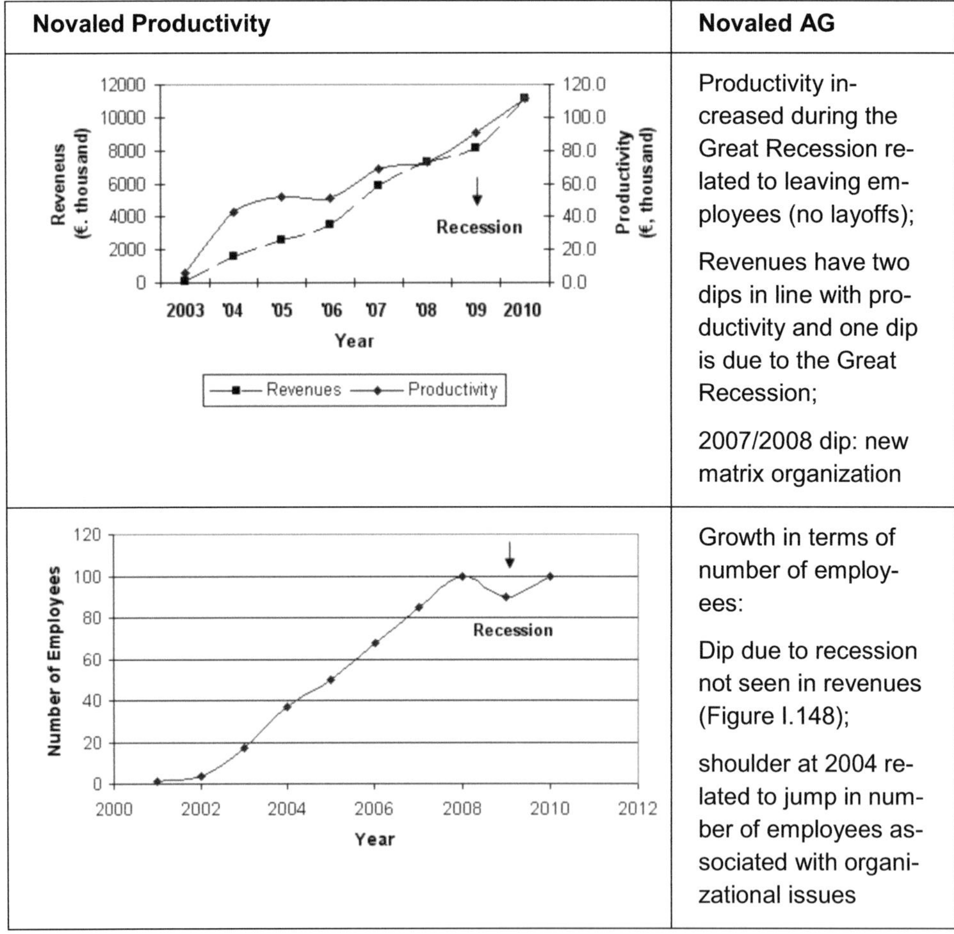

Figure I.149: Growth patterns of German Novaled AG.

A key founder of Novaled is the entrepreneurial professor Karl Leo (B.2). Key characteristics of an entrepreneurial professor for technology entrepreneurship are not just a business mindset, but also revealing, supporting and encouraging entrepreneurial talents.

Entrepreneurship in Wind Power and Solar Energy

So far, the discussion has focused on (largely) *economic markets*. In Figure I.34 the constellation of the *policy-driven* biofuels segment is outlined and in the Appendix (A.1.1) a rather detailed industry analysis for the *policy-driven* biofuels segment in the US and Germany is presented. However, many other segments of "renewable energy" are also policy-driven, such as wind power (wind turbines) and photovoltaic (PV, solar cells).

The most prominent beneficiaries of the German laws which induced a boom in entrepreneurship in solar cells and modules and windpower are the German Enercon GmbH (wind power) and, for photovoltaic, Q-Cells AG as well as First Solar from the US with its vast majority of revenues from Germany. Ultimately, all showed strong exponential growth in line with the advantages to be taken from the German Renewable Energy Act (EEG – Erneuerbare-Energien-Gesetz, Box I.22.).

Box I.22: Outlining the German Renewable Energy Act (EEG – Erneuerbare-Energien-Gesetz) referring to solar and wind energy.

Renewable energy is supported by governments of developed nations in various ways. In the US the solar incentive regime, for instance, is mainly on the basis of tax credits. Tax incentive programs exist at both the federal and state level and can take the form of investment tax credits, accelerated depreciation and property tax exemptions.

In Germany the Renewable Energy Act (EEG) [90] provides a favorable legal framework based on German power grids; not just solar power (Figure I.152, Figure I.153, Figure I.154), but also wind power (Figure I.150, Figure I.151). Since 2003 the focus has been on grid-connected ground or large roof mounted solar power plants in Germany and other European Union countries with feed-in tariff subsidies that enable solar power plant owners to earn a reasonable rate of return on their capital.

Governmental subsidies, economic incentives and other support for solar electricity generation generally include feed-in tariffs, net metering programs, renewable portfolio standards, rebates, tax incentives and low interest loans.

Under a feed-in tariff subsidy, the government sets prices that regulated utilities are required to pay for renewable electricity generated by end-users. The prices were set above market rates and may differ based on system size or application. Net metering programs enable end-users to sell excess solar electricity to their local utility in exchange for a credit against their utility bills. The policies governing net metering vary by state and utility. Some utilities pay the end-user upfront, while others credit the end-user's bill.

Under a renewable portfolio standard, the government requires regulated utilities to supply a portion of their total electricity in the form of renewable electricity. Some programs further specify that a portion of the renewable energy quota must be from solar electricity, while others provide no specific technology requirement for renewable electricity generation.

The EEG was designed to level the playing field by taking into account external costs of conventional electricity generation, provided investment subsidies guaranteeing fixed income of suppliers of renewable energy fed into the grid and usually had a time table for cuts of these kinds of benefits. The German EEG became a model for other European countries.

Already in the 1990s Germany required utilities to connect generators of electricity from renewable energy technology to the grid and to buy the electricity at a rate which for wind and solar cells amounted to 90 percent of the average tariff for ultimate customers. There was support of private and professional customers of PV through special interest rates and additionally a guaranteed feed-in subsidy to be *pari* by the big energy suppliers for two decades (reduced every year by 5 percent). This induced the photovoltaic boom in Germany – and corresponding technology entrepreneurship.

The 1991 Power Feed-In Law (Stromeinspeisegesetz – StEG) required "big" owners of the electrical grid in Germany to accept feed-in of power by suppliers of renewable energy. Furthermore, suppliers were guaranteed a minimum royalty linked to the average earnings through the big electrical power producers. For the producers of wind power these tariffs kept cost of the suppliers of wind power and, hence, led to a wind power boom in Germany – and associated firms' foundation. On the other hand, for suppliers of solar electricity the tariffs were far from cost recovery.

The 2004 Renewable Energy Act, on the one hand, reduced support of electricity by wind turbines. On the other hand, there was a re-orientation of the Federal Government toward solar energy after a significant impact of the so-called "100,000 Roofs Program" concerning solar energy.

A drive in addition to issues of lifting oil dependencies and climate change and change to renewable energy occurred after the Chernobyl disaster in 1986 in the Ukraine [91] focusing on replacing nuclear power. The related societal attitude was strengthened particularly in Germany. And the nuclear Fukushima disaster in Japan in 2011 again led to a societal and also political drive to phase out nuclear power by renewable energy. This means again opportunities for technology entrepreneurship.

The privately held German ENERCON GmbH (B.2) is a production-oriented wind power firm that illustrates how a policy-driven market in Germany influences firm development (Figure I.150) – taking advantage from the Renewable Energy Act (EEG) and its precursors. On the other hand, from its start ENERCON showed innovation persistence with regard to building wind turbines and wind farms and occupying step by step the whole value system for constructing and distributing wind turbines.

Already in 1986 ENERCON built the first ENERCON wind farm with ten E-16 / 55 kW wind turbines. In 1988 there was construction of the company's first own production facility and development and installation of E-17 / 80 kW as well as E-32 / 300 kW models. The year 1991 was occupied with development of the gearless ENERCON concept and the first prototype which was the fundamental technical breakthrough.

In the below Figure I.150 growth of ENERCON, a German Hidden Champion (ch. 4.1.1), is shown in terms of installed (electric) power by various types of wind turbines (usually for wind farms). Innovation persistence of the firm as the basis of growth in general and further growth strategies are displayed and the two growth "accelerating effects of legislation in 1997 and 2000. On the other hand, the two changes of the German EEG to the disadvantage of the supplier show up. These changes following reductions of feed-in tariffs weakened demand in key European markets. Notably, the 2000 EEG more than outweighed the Dot-Com Recession at that time.

Revenue data (with some specific ones missing) given in Figure I.151 do not indicate the effects as markedly as the data in Figure I.150.

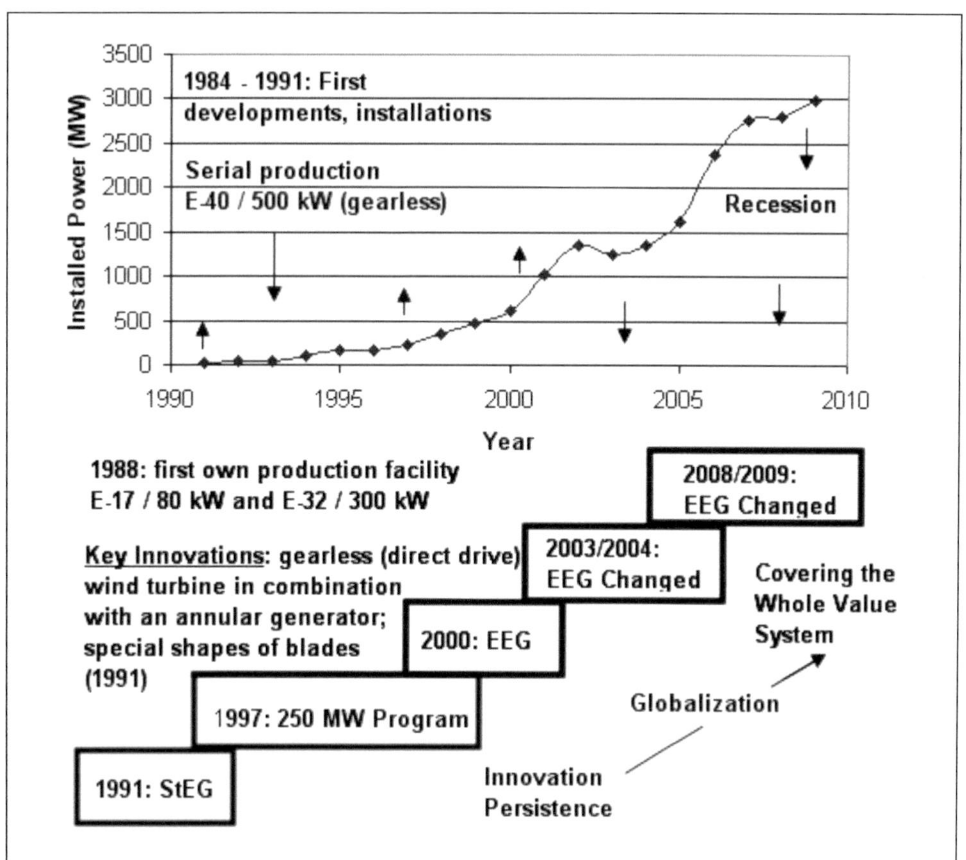

Figure I.150, continued (time line for Enercon's product development).

1995	2002	2004	2006	2007:	2010
E-66 / 1.5 MW	E-112 / 4.5/6 MW	E-48 / 800 kW, E-70 / 2,000 kW	E-82 / 2 MW	E-126 / 6 MW	E-126 / 7.5 MW

Figure I.150: Installed (electric) power by German ENERCON GmbH through various types of wind turbines in relation to the German Renewable Energy Act (EEG) and its changes.

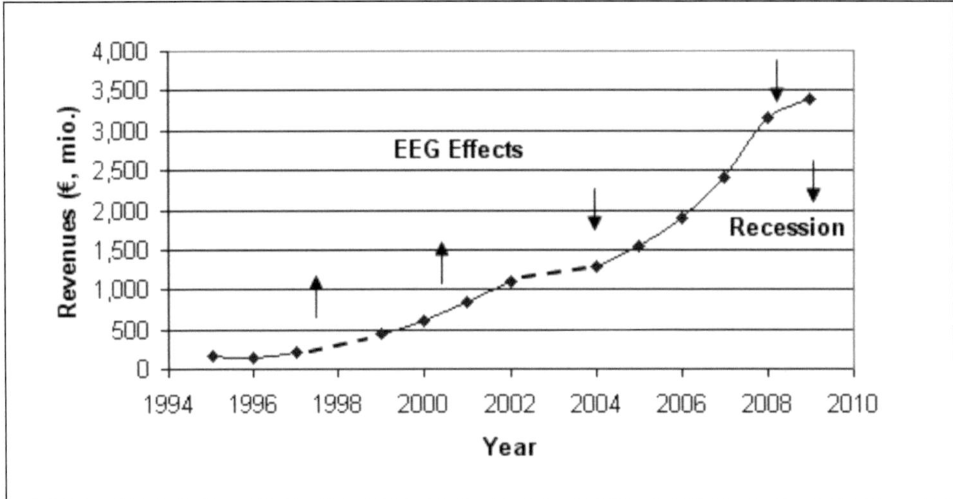

ENERCON GmbH: 2009 – revenues €3.4 bil., employees > 13,000;
founder Aloys Wobben (net worth €3.5 bil. according to the current Forbes World's Billionaires list);
in 2009 No. 4 (global market share 8.5%) of 10 firms covering ca. 80% or the global market, market share in Germany 60%;
holds ca. 40% of wind turbine related patents in the world (more than German Siemens and US General Electric together; cf. also Box I.7);
Business basics: gearless technology meaning less maintenance (> 50% of wind turbine failures are due to damages of the gear), innovation and investment persistence, high compatibility with electrical grids, longevity of windmills, quality and reliability, high level technical service.

Figure I.151: Revenue development of German ENERCON GmbH in relation to the Renewable Energy Act (EEG).

Growth of ENERCON can be viewed as the combined effects of the (German, European) Renewable Energy Acts, innovation persistence, global reach and integrated offering along the whole supply chain. Adverse effects of the EEG changes are probably levelled off by innovation efforts.

Growth accelerating effects of the German EEG for the photovoltaic solar cell areas show up for two marked representatives. German Q-Cells AG and First Solar from the US are production-oriented, but originally not VC-based firms. Both used publicly supported investment organizations and subsidized loan possibilities or wealthy private investors as their financial basis. Q-Cells and First Solar differ with regard to several aspects, most notably technology, position in the value system and (low) cost strategies (Figure I.11, Figure I.12).

Q-Cells (updated analysis of [Runge 2006:281-282]) was founded in 1999 and, after starting production in 2001, could take immediate advantage from legislation and exceeded one billion euros in sales already in 2008 (Figure I.152). In 2009 it suffered considerably from the Great Recession.

Q-Cells was founded in Berlin (Germany) by an entrepreneurial team of four experienced persons (engineers and physicist); two of them participated in the foundation of the German PV-NTBF Solon AG (also founded in Berlin) and one was a former employee of Solon. The foundation team was complemented by a (McKinsey) consultant Anton Milner with an engineering background who became CEO. Q-Cells co-founder Reiner Lemoine, also a co-founder of Solon, was the Director Production and Technology of Solon.

To establish production Q-Cells took advantage from more favorable financing conditions in the German state Saxony-Anhalt than those in Berlin. It had several financing rounds: The initial financing of ca. €15 million for the first factory and a further ca. € 20 million in a second round financing for the expansion of the Q-Cells 1 and Q-Cells 2 factories. For development in 2004 the company invested further € 20 million.

Investments were essentially by publicly supported investment organization, such as MBG and IGB, creating a private stock company. Investors had essentially silent partnerships getting a fixed rate for a given period plus a participation component of the firm's profit. In 2005 Q-Cells issued an IPO.

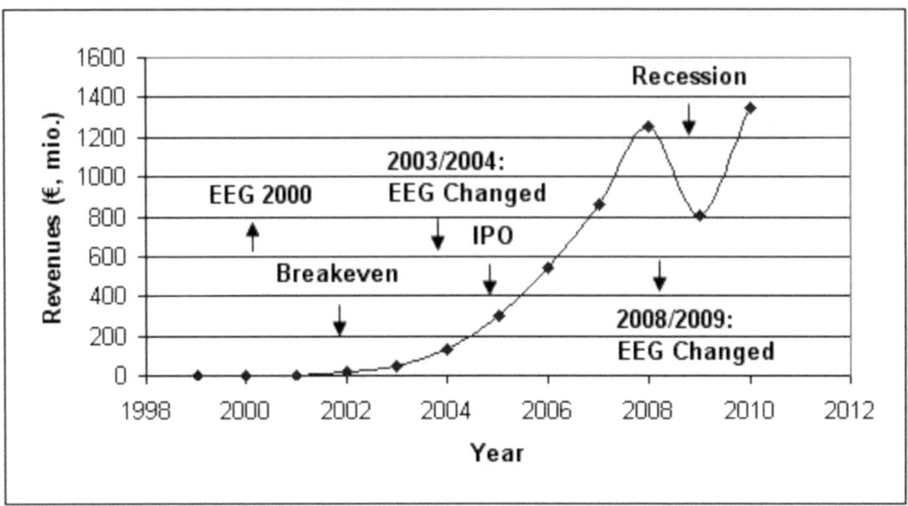

Figure I.152: Revenue development of German Q-Cells in relation to the German Renewable Energy Act (EEG).

Figure I.153 shows the innovation persistence of the firm. Concerning technology Q-Cells started being active essentially based on monocrystalline and polycrystalline silicon solar cells.

Between 2003 and 2005 in each year the number of employees almost doubled (from 207 to 767) which can be expected to reduce the productivity due to organizational and management issues. This is seen in Figure I.153 in the productivity chart. On the other hand, productivity provides another dip at 2006/2007. The time period of this dip coincides with the "silicon cycle" which is the economic upturns and downturns unique to the semiconductor market.

Usage of high purity silicon by the electronics and solar sectors (Figure I.11) had increased from approximately 31,000 tons in 2004 to approximately 35,000 tons in 2005 and 39,000 tons in 2006. The increase was driven by 5 percent year on year growth in the electronics sector usage and by a whooping 20 percent growth in solar sector usage. In the absence of an equivalent increase in capacity, the prices for high purity silicon increased sharply due to strong demand growth. [92]

Effects of the silicon cycle with higher cost of raw material could essentially affect profit of the firms. Or, if availability of silicon in the market is reduced, output would decrease and, keeping the firm's workforce, would affect productivity negatively.

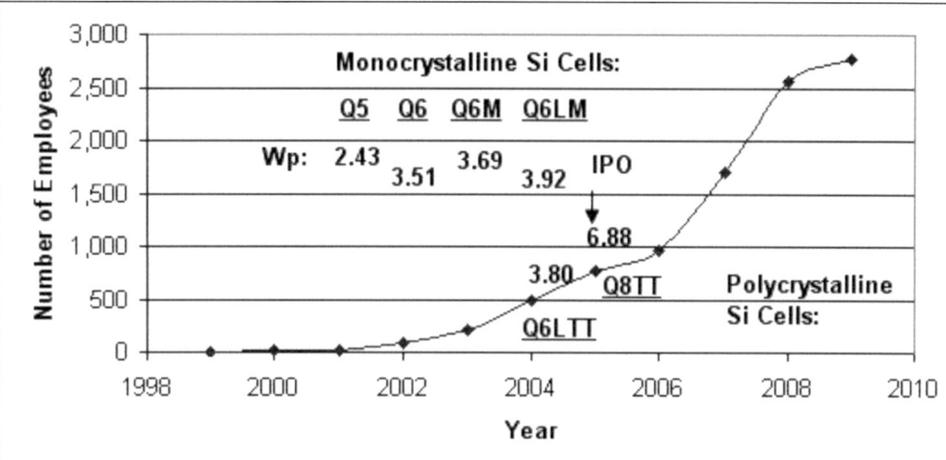

Large capital inflow (IPO) led to fast increase of employees' numbers and correspondingly can induce decrease in productivity – jump in Wp to polycrystalline Q8TT.

Photovoltaic power capacity is measured as maximum power output under standardized test conditions in "Wp" (Watts peak).

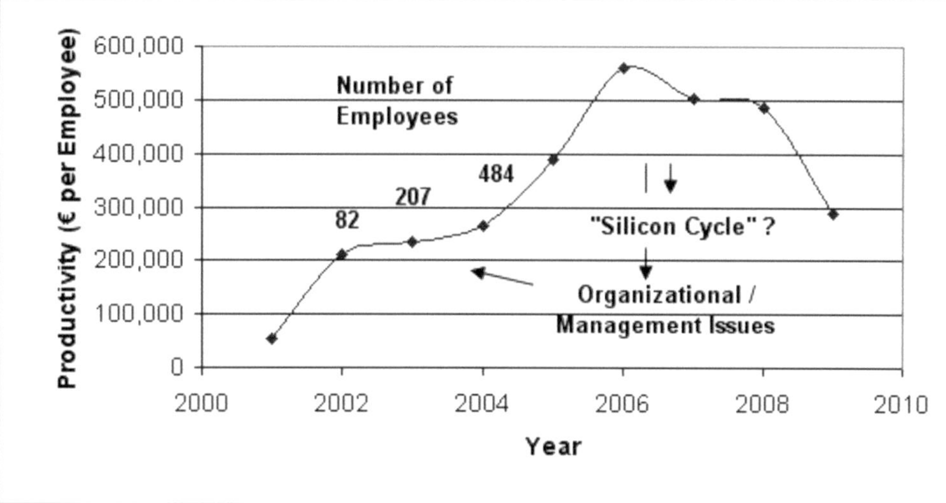

Figure I.153: Development of employee numbers of German Q-Cells AG in relation to innovation persistence and productivity.

However, the steady, continuous growth of Q-Cell's revenues (Figure I.152) makes it unlikely that the silicon cycle is significantly responsible for the 2006/2007 dip in productivity. Again it can be assumed that the tremendous rise in numbers of employees from 767 in 2005 to 2,568 in 2008 led to organizational problems.

The US company First Solar Inc. (as Closure Medical [Runge 2006:98-103]) represents a firm with a two decades history of entrepreneurial efforts (see below) and finally a breakthrough and big success based on the German Renewable Energy Act. For instance, its net sales in 2007 were obtained in Germany by 90.7 percent and 98.8 percent of net sales were generated from customers headquartered in the EU. Even after efforts to bring down this dependency in 2009 Germany still accounted for 65 percent of First Solar's net sales.

Furthermore, according to First Solar's annual report 2010: "During 2009, principal customers of our components business were Blitzstrom GmbH, EDF EN Development, Gehrlicher Solar AG, Juwi Solar GmbH, and Phoenix Solar AG. During 2009, each of these five customers individually accounted for between 10% and 19% of our component segment's net sales. All of our other customers individually accounted for less than 10% of our net sales during 2009. The loss of any of our major customers could have an adverse effect on our business."

Germany has become an aggressive subsidizer of solar power – well known in the world and especially also in China.

First Solar's production did not reach 25 megawatts until 2005 (in Perrysburg, Ohio in 2002 only 1.5 MW). The company built an additional line in Perrysburg, Ohio. But then First Solar took advantage form massive loans with favorable conditions from IKB Deutsche Industriebank AG, a long-term lender and service provider to medium-sized German companies ("Mittelstand").

When First Solar installed four 30 MW production lines in Germany cash outlays for the German plant were partially recovered through the receipt of $9.5 million and $16.8 million in 2007 and 2006, respectively, of economic development funding from various German governmental entities. The substantial German production incentives accounted for about 50 percent of capital costs. Additionally, it took advantage from feed-in tariffs though the 2003/2004 change of the EEG reduced these.

Furthermore, the Company has participated, or is currently participating, in laboratory and field tests in the US with the National Renewable Energy Laboratory (NREL), the Arizona State University Photovoltaic Testing Laboratory, and in Germany the Fraunhofer Institute for Solar Energy, TÜV Immissionsschutz und Energiesysteme GmbH and the Institut für Solar Energieversorgungstechnik.

First Solar had a predecessor Solar Cells Incorporated (SCI). SCI was founded in 1990 as an outgrowth of a prior company, Glasstech Solar (founded 1984) founded and led by the US inventor/entrepreneur Harold McMaster. McMaster was an expert at making glass plates. During the energy crisis in the 1970s, photovoltaic technology gained recognition as a source of power for non-space applications. In the 80s, McMaster became interested in solar technology and experimented with different ways to put photovoltaic materials on glass. He worked first with silicon and then cadmium-telluride (CdTe) as the company was called Solar Cells.

Viewed as "The Glass Genius" he developed a process for making high quality strengthened, or tempered, glass used for architectural glass (for instance, windows in skyscrapers) and automotive glass. By selling the machines to produce these glasses he made a fortune.

McMaster was aware that glass could easily be coated with thin layers of chemicals that change its color or ability to pass light. Glass is also an electrical insulator. Those two properties are essential for construction of photovoltaic cells – "solar cells" that change sunlight directly into electricity. Harold McMaster's treated the actual *solar cell* as simply *a different kind of coating on glass* and his vision was to use that technology in commercial-scale production of electricity that would ease America's dependence on imported oil.

McMaster revealed the essential cost element of large area solar arrays to be glass. At first, McMaster looked into amorphous silicon research but gave up. He then raised yet again money to create Solar Cells Inc., to work on a different thin-film technology, cadmium telluride. By 1997, Solar Cells had a prototype production machine. According to a former head of the Thin Film Partnership program at the Department of Energy's National Renewable Energy Laboratory (NREL), SCI was clearly the industry leader in thin-film photovoltaic technology.

In 1999, True North Partners LLC, an investment arm of the Walton family, purchased controlling interest, and renamed the company First Solar LLC. John T. Walton (heir of Sam Walton [Bhidé 2000], who founded Wal-Mart, the largest retail organization of the world) put ca. $150 million into the firm until his death in 2005. Simultaneously, he installed professional management.

The firm almost shut its doors. It is said that it was begging for government contracts with NREL. But it was lucky to have a generous supporter, John Walton. The ca. $150 million investment is not the amount of money that a single VC might fork over to a solar startup these days. First Solar did not set up a pilot line until 2002. It began production at its first commercial factory in the US in 2004, but as early as 2008 its revenues exceeded one billion dollar (Figure I.154).

Within the photovoltaic industry, First Solar faces competition from numerous crystalline silicon solar cell and module manufacturers, but also from efforts to commercialize silicon thin-firm solar cells/modules (Figure I.11).

The thin-film based solar cells or modules, respectively, are generally less efficient than silicon cells. However, thin-film systems can be utilized at various temperatures and light/radiation situations. Basically, production cost for thin-film systems are lower. The value proposition of First Solar is to deliver a high performance and high quality solar module at a more affordable price.

First Solar developed considerable competitive (cost) advantage over German solar cell producers which, like Q-Cells, relied on mono- or polycrystalline cells, by using

CdTe thin-film technology thus avoiding the silicon cycle and operating as a module supplier on a higher stage of the value system (Figure I.11). In contrast to the silicon batch process,

Si-Feedstock – Ingot – Wafer – Solar Cell – Module (Figure I.11),

First Solar's CdTe thin-film PV is a fully integrated and automated continuous process:

Glass in – Deposition – Cell Definition – Final Assemby & Test – Module Out.

Additionally, for the end of the modules' life-times it offers a module collection and disposal program in the PV industry.

Even when competing with falling polysilicon prices, First Solar still had a ca. "40 cents-per-watt installed cost-per-watt advantage," which delivers lower-levelized cost of electricity.

Figure I.154 shows that First Solar could take advantage from economy of scale reducing average manufacturing cost from roughly 3 $/W to 1$/W, and even to $0.74/W by the end of 2010. **Economies of scale** for a firm primarily refers to reductions in average cost (cost per unit) associated with increasing the scale of production for a single product type. Economies of scope are conceptually similar to economies of scale.

Economies of scope refer to lowering average cost for a firm in producing different products. Economies of scope make product diversification efficient if they are based on the common and recurrent use of proprietary know-how (cf. also platform technologies – Table I.12) and the learning effect (ch. 2.1.2.5). Further economies of scope occur when there are cost-savings arising from by-products in the production process (biofuel production with biobased chemicals as by-products; A.1.1).

Existing business firms usually grow by increasing their scale and scope. The scope of a firm then relates to the number of its products. The scale of a firm is given by the size of its products. A firm like Microsoft gets few big products while Amazon sells a huge variety of goods, each of small size in terms of sales. For startups, often only having one to three products, both these concepts rarely apply.

Overall competitive advantage of First Solar, expressed in terms of positive factors on revenue development, rests on the following number of combined factors compared, for instance, with Q-Cells. Compared with other PV firms which are not operating in Germany both have an additional support factor (n(EEG)) related to the German EEG.

- The value system advantage, modules versus cells – n(VS)
- Production cost reduction, advantage as a result of economy of scale – n(PCR)
- Being independent from the silicon cycle – n(non-Si).

Focusing on key components First Solar's overall growth driver n_{CA} can be expressed as:

$n_{oA} = n(EEG) \oplus n(\text{market demand}) \oplus n(VS) \oplus n(PCR) \odot n(\text{non Si}) \odot n(\text{internalities})$.

Finally, due to gaining most of its revenues in Germany favorable foreign exchange rate between the US dollar and euro could partially offset a price decline of First Solar's products in Germany. The other way round was observed for German ChemCon GmbH (B.2) which, after foundation, had ca. 80 percent of sales in the US (ch. 4.2.3; Figure I.116).

Considering these special effects as well as no indications of extreme sensitivity in the other considered NTBF cases in this book (B.2) one may suggest that dollar-euro exchange rates will not, or rarely, show up as an observable bracket effect in revenue data.

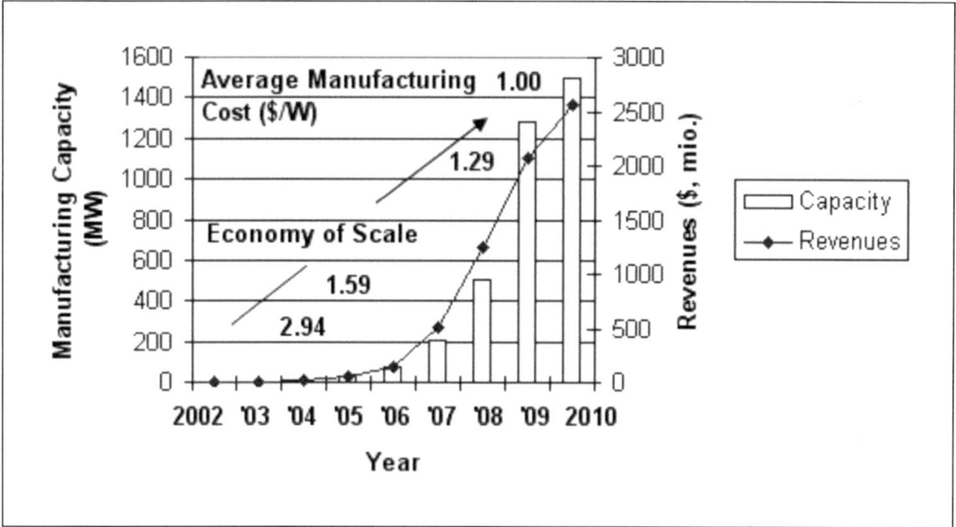

Figure I.154: Manufacturing capacity and revenues of US First Solar in relation to average manufacturing cost demonstrating economy of scale.

Figure I.154 reflects a fundamental dichotomy of firms for decision-making associated with production. It is the question whether (increasing) production capacity (or expand space of labs and offices) shall follow demand (indicated by revenues) – as seen for the 2006 - 2008 period – or whether increasing production capacity shall precede demand (2009/2010).

In the first case the firm can lose money that could be earned, in the second case you lose money due to "unnecessary" investments. And, in the particular 2009/2010 case, expected increase of manufacturing capacity to meet demand was hit by the Great Recession with distinctly reduced demand. NTBFs relying on cash flow and loans mostly follow the first approach – as described also for Nano-X GmbH (Figure I.137) or ChemCon GmbH (Box I.20).

The next foreign NTBFs taking advantage from the German and European policy-driven markets were the Chinese seeing firms Suntech and JA Solar emerging as "giants" based on their low cost positions.

Overall, the global PV market had entered into a phase of strong consolidation associated with disappearance of firms, mergers of firms and business model innovations of surviving firms and simultaneously capturing market share of thin film cell and module suppliers from suppliers of silicon-based offerings. Leading and surviving firms in the PV solar cells/modules area have left the class of NTBFs. They have achieved the situations of "normal" large to very large firms and operate correspondingly.

For the area of CdTe *thin-film* solar cells and modules First Solar had almost a monopoly or monopoly-like position. However, for First Solar there was a risk of disruption from higher efficiency technologies, and especially other thin-film technologies. For instance, there were some startups or other developing firms which were active in innovative CIS (copper-indium-disulphide) or CIGS (copper indium gallium diselenide) thin-film photovoltaic (Figure I.11), such as in Germany Würth Solar GmbH & Co. KG [Runge 2006:257-258], a subsidiary of the Würth Elektronik Group which belongs to the Würth Group (Box I.23).

The privately held Würth Group, with revenues of €8.6 billion in 2010 originally emerged as a Hidden Champion [Runge 2006:256-258]. It is a world market leader in its core business, the trade in *assembly and fastening material*. It currently consists of over 400 companies in 84 countries and more than 3 million customers all over the world. The group has been set up by (now) Professor Reinhold Würth who has become one of Europe's richest men. He has transformed a two-man business dealing with screws of his father since 1954 into a worldwide active group emphasizing the "*fastening business*."

While manufacturing can be scaled (Figure I.154) turning the modules into systems is mostly "variable" cost since the majority of the cost is materials (cabling, aluminum) and labor (BOS, Figure I.11).

Box I.23: German Würth Solar and its Approach to CIS Thin-Film Photovoltaic.

By 2009 Würth Solar saw itself as the world's leading manufacturer of innovative *CIS solar modules* and also *a provider of complete solar installations*. Its revenues were €116 million with 280 employees in 2009 and a manufacturing capacity of 30 MW (in 2005 €5.3 million and 103 employees; in 2004 €3.9 million with ca. 67 employees). In autumn 2006, Würth Solar started the first large-scale series production of its GeneCIS solar modules worldwide at its CISfab factory in Schwäbisch Hall, Germany which was especially built for this purpose.

This kind of technology, utilizing a fully integrated and automated continuous production process, put the focus on copper indium gallium selenium sulfur compounds (depending upon composition called copper indium gallium diselenide Cu(In,Ga)Se2

(CIGS) or $CuInS_2$ (CIS)). Würth's CIS technology and CIS modules can be used under the harshest environmental conditions, making them especially suitable for use in extreme climatic zones.

Würth Solar with the Würth (Elektronik) Group as the majority and controlling owner (of ca. 80 percent) had two partners, utilities (energy) supplier EnBW (Energie Baden-Württemberg) and ZSW (Stuttgarter Zentrum für Sonnenenergie- und Wasserstoff-Forschung) or Stuttgart Center for Solar Energy and Hydrogen Research. The *University* of Stuttgart (IPE; Institute for Physical Electronics) *developed the CIS cell technology* in the 1980s. *ZSW developed the CIS prototype modules* which continued to *carry out research and development work for Würth Solar.* Development and installation of a pilot plant cost ca. €40 million.

The timeline of Würth Solar's emergence is as follows:

- 1996 a predecessor firm, Würth Solergy, starts first photovoltaic activities as a subsidiary of the Würth Elektronik Group.
- 1999 foundation of Würth Solar GmbH & Co. KG
- 2000 Würth Solar and ZSW enter the commercialization phase running a pilot plant.
- 2006 Start of the first large-scale production of CIS PV-modules called GeneCIS in the world in the "CISfab plant" having a capacity of 15 MW
- 2008 Capacity of CISfab extended to 30 MW.

€55 million were invested in CISfab. Würth Elektronik Research provides support for and optimization of day-to-day production at the Würth Solar and CISfab production sites in Schwäbisch Hall (Germany).

Related to the electronics group and based on the extremely wide and deep distribution and sales force of the Würth Group Würth Solar occupied all positions of the whole PV value system, emphasizing not just electricity for private homes (roofs), but also integration of their modules into architectural designs and architectural components and module integrations into other customized products from a variety of applications. The firm showed up as a supplier of whole systems and solar power plants (utility-scale systems) as well as engineered, individual system solutions for solar energy.

First Solar and Würth Solar share a rather close "initial" configuration:

- Both relied on research results of public research institutions and relied for developments on public research institutes (NREL versus ZSW).
- Both are (or were) financed by wealthy persons related to large international firms.
- Both had to go a long way before finally launching their primary offerings, thin-film PV modules.
- Both could make use of subsidies or other financial support of (State or Federal) Government.
- Both make heavy use of the German Renewable Energy Act (EEG).

Concerning thin-film technology in the US Solyndra, Inc. founded in 2005 went for designs and manufacturing of solar photovoltaic (PV) systems, comprised of panels and mounts, for the commercial rooftop market. It used thin film CIGS technology as Würth Solar and was producing cylindrical modules depositing CIGS on the inside of glass tubes rather than using plates of glass. The company had approximately 1,050 employees around the world, revenues in 2010 were approximately $140 million.

The US Department of Energy (DOE) offered the Californian firm a $535 million loan guarantee of construction to expand its manufacturing capacity to 500 MW per year, expecting the new facility to employ some 1,000 people. After raising $1 billion (!), the company was forced to slash costs, close a factory, restructure the executive team and cancel an IPO. As a competitor for Würth Solar the Solyndra star faded in 2011 [Kuo 2011; RenewableEnergyWorld 2010].

Solyndra's CEO admitted to have made the twin mistakes of expecting too much growth and not putting enough focus on market development. Its $535 million loan guarantee from the Department of Energy was under investigation by the US Congress [GreenBeat 2010].

By the end of 2010 concerning the high cost of module production – the most troubling issue – the company said to have its average sales price over $3.20 per Watt, about 65 percent more than leading crystalline-silicon PV manufacturers and its cost of manufacturing to be over $6 per Watt. The major cost and price declines in the crystalline PV sector hurt the competitive chances of Solyndra's modules. The company's next step was planning on having over 600 MW of capacity online by 2013. But soon Solyndra said [RenewableEnergyWorld 2010] it will have around 300 MW of capacity.

In the second half of 2011 Solyndra filed for bankruptcy and was the subject of several federal investigations. Republicans on Capitol Hill have been critical of the Obama administration's handling of clean energy programs, saying that it gave out a $535 million federal loan in 2009 to a California solar energy company, Solyndra, without properly evaluating if its business plan made sense [Johnson 2011].

Worldwide solar photovoltaic (PV) market installations reached a record high of 18.2 gigawatts (GW) in 2010. This represented growth of 139 percent over 2009. The PV industry generated $82 billion in global revenues in 2010, up 105 percent Y/Y (year on year) from $40 billion in 2009.

In 2010, the top five countries by PV market size were Germany, Italy, Czech Republic, Japan, and the US – representing over 80 percent of global demand. European countries represented 14.7 GW, or 81 percent of world demand in 2010. The share of global PV demand in 2010 [93] were:

- Germany, 42 percent
- Italy, 21 percent
- Other Europe, 18 percent
- Japan, 5 percent
- United States, 5 percent.

The tremendous growth of photovoltaics let "giants" enter the photovoltaic scene largely determined so far by "grown up" NTBFs to build businesses in renewable energy – emphasizing the role of NTBFs for innovation approaches and new business development of large firms. These large firms usually waited until there is sufficient demand.

For instance, the German giant privately held Robert Bosch GmbH, the world's largest auto supplier (sales of €47.3 billion in 2010), but a global supplier of technology and services in the areas of automotive and industrial technology, consumer goods and building technology, took over Ersol Solar Energy AG which started as ErSol Solarstrom GmbH & Co. KG in 1997 with production of multi-crystalline silicon solar cell and modules. Bosch became majority owner of the firm in 2008 and in 2009 Bosch Solar Energy AG emerged.

From 2009 to 2012 Bosch invested around two billion euros in solar technology and bought with Ersol, Aleo and Voltwerk three companies [Weishaupt 2012].

In the US, General Electric (GE) took a bold step into solar PV with several announcements, mainly its acquisition of PrimeStar Solar (with whom it had been working on CdTe thin film R&D). The startup created an NREL-confirmed record 12.8 percent efficiency (aperture area) panel on its 30 MW line in Arvada, CO, the company said in a statement [ElectroIQ 2011].

GE's CdTe push was part of a planned $600M+ investment into solar technology and commercialization. The company projected global demand for solar PV will surge to 75 GW over the next five years, much of that in utility-scale plants, an attractive market to be in. But more importantly, there were clear opportunities to bring costs down.

While First Solar was leading the cost/W charge, GE seemed to think it can compete there too. Not only does GE planned to push efficiencies "much higher" than the current 13 percent, but "we probably can cut costs 50% over the next several years."

Further, GE said it will build a 400 MW thin-film solar panel in the US (employing 400 workers), what would be currently the largest in the country, reportedly ready by 2013. Monocrystalline-silicon firm German SolarWorld had the biggest US solar PV site, with its cell and module operation in Oregon scalable to a combined 500 MW.

French oil giant Total SA agreed to buy as much as 60 percent of SunPower Corp. for $1.38 billion, taking advantage of increased global interest in renewable energy. SunPower, with sales of $2.219 bil. in 2010 was the second-largest US solar panel maker. The deal was supposed to lead to more solar industry acquisitions as US and European suppliers sought for help competing against rival suppliers in Asia [Herndon et al. 2011].

But by early 2013 the PV market, particularly in Germany, had changed dramatically. Already by 2007 solar energy, the former niche market, had become a global business. The industry was growing rapidly – too fast [Zeller-Silva 2007].

Moreover, the order situation was good and positive long-term forecasts estimated the annual growth of the market by 2020 to about 20 percent annually. But this deceived the businesses around the world to increase their capacity strongly. And skeptics feared that the demand would not grow to the same extent as the range. Additionally, the emerging, lucrative markets attracted competitors. In the US and in Asia powerful companies emerged that could be a real competitor for the German solar industry [Zeller-Silva 2007].

It was envisioned that tougher competition, the threat of overcapacity and declining public subsidies would make it hard for the industry in the future, to make money. And Frank Asbeck, chairman and founder of German firm SolarWorld, even spoke of an impending shakeout, despite the growing global demand [Zeller-Silva 2007].

And, indeed, starting in late 2008, the solar market shifted from supply-related to demand-driven. This was due to the plunging price of crystalline silicon cells and modules caused mainly by 1) falling silicon cost, 2) constrained availability of credit, 3) demand decline in certain previously high volume regions such as Spain, 4) oversupply and 5) unfair competition of Chinese suppliers due to their extremely low prices bound to high export subsidies of the Chinese government. Furthermore, in Germany there were cuts of feed-in tariffs and a shift of emphasis from private rooftop installations to commercial utility-scale PV installations.

As a result, the state of the industry changed over to consolidation of weak competitors and NTBFs and bankruptcy of several PV companies, notably the above mentioned Solyndra in the US and Solon AG in Germany and even Q-Cells stumbled toward insolvency.

Q-Cells and others had responded to Chinese competition by outsourcing some of their own production to Asia to cut costs. But that was not enough to save them.

By 2010 the era of the founders was over. Si-based PV had become largely a mature technology which means it can be used and built anywhere in the world, particularly in China. The original PV startups had become large-size firms with different requirements for operation in a highly competitive environment with the Si-based PV technology having changed over into commodity products.

For too long, many solar companies have rested on the high subsidies. A concentration of the industry was imminent. And it was clear that from about 50 German companies only a handful would survive in the next two years probably as independent companies.

In February 2012 German installers and dealers could buy a Chinese module already for €0.77 per watt in wholesale. Modules from German production cost €1.03 per watt. Just the year before, the solar module prices had fallen by 30 to 40 percent. The share of German companies in 2011 in the total market was just 20 percent. The share of Chinese companies accounted for 54 percent [Weishaupt 2012]. Ca. 75 percent of the German solar cell market was governed by non-German firms.

Early in 2012 the largest solar cell manufacturers in the world were (DE = Germany, Ch = China) [Handelsblatt 2012]: Solarworld (DE, No. 20, 2007 No. 7), Q-Cells (DE, No. 13, once the world's largest producer of silicon-based solar cells), Trina (Ch, No. 5), Yingli (Ch, No. 4), JA Solar (Ch, No. 3), First Solar (US, No. 2), Suntech (Ch, No. 1).

Policy tried to help the endangered firms looking for tariff on imports of Chinese silicon photovoltaic panels – but in vain. A cascade of insolvencies showed up. Most prominent among the herd of PV startups, in 2012 Q-Cells with ca. 2,200 employees became insolvent and was taken over by Seoul-based Hanwha Chemical Corp.

And in 2013 Bosch decided that the company had no choice but to abandon solar energy, to shut down the Solar Energy division – along with its roughly 3,000 employees. According to Bosch, cuts to European renewable energy incentives and the drastic changes in the market, particularly the rapid increase in capacity in China, had put "unrelenting" pressure on Bosch's Solar Energy division, which sustained a loss of one billion euros. Consequently, Bosch announced to cease production of all wafers, ingots, solar cells and Bosch solar panels by the end of 2014. However, the company will hold on to Bosch Solar CISTech GmbH – its thin-film research facility in Brandenburg, Germany [Energy Matters 2013].

In 2012 Würth Solar (Box I.23) divested substantial parts of its solar business to the Agricultural Trade Group Baywa and sold the rest to a second partner in 2013.

And lately also Chinese solar cell and panel suppliers turned out not to be immune to insolvency. Chinese panel producer Suntech Power has put its largest subsidiary into bankruptcy [Riley, 2013] and a second Chinese solar company, LDK Solar Co., has defaulted on a debt payment to investors [Ma 2013].

4.3.5.3 Selected Quantitative Applications of the Bracket Model

It is the mark of an educated person to look for precision
only as far as the nature of the subject allows.
Aristotle (Stanford Encyclodia of Philosopy, Episteme and Techne)

For a quantitative discussion of NTBFs' growth we shall follow cybernetics (4.3.4) and use a *heuristic approach of reinforcement* for *periods of* growth of sub-states of a new firm. We use a phased process interrupted by brackets and associated transition states (Figure I.133, Figure I.134) and are guided by situations described by Equation I.16, Equation I.5 and Figure I.70. The heuristic will be verbally expressed by the well-known saying

Growth breeds growth.

In the same category of algorithm, but including growth or decline (Figure I.114 and Equation I.16) one can cite the "Matthew Effect." The Matthew Effect [Merton 1968] specifically refers to a statement in the Christian Bible (XXV:29):

> "For unto every one that hath shall be given, and he shall have abundance; but from him that hath not shall be taken away even that which he hath."

In this context of "growth breeds growth" and emphasizing VC-based NTBFs Gompers et al. [2008] present evidence of "*performance persistence"* (definition of Gompers et al. [2008])) in entrepreneurship and that entrepreneurs with a track record of success are much more likely to succeed ("success breeds success") than first-time entrepreneurs and those who have previously failed.

Similarly, related to *innovation persistence*, Flaig and Stadler [1994] revealed a *positive impact of previous innovative success to further innovations in the following years.* They studied product and process innovations of private German firms from the manufacturing sector (data between 1979 and 1986). And there is a combined internal and an external effect paralleling growth by innovation and investment persistence: As entrepreneurs build a track record, uncertainties about their ability and business propositions diminish; the entrepreneur gets more self-confidence, confidence in the firm, experience of raising money from external financial backers as well as getting suppliers and customers.

We focus on dynamic stability during firm development which will be expressed by "*equations of state*" for a given "*dynamically stable interval*" of time $<t_0, t_n>$ to describe time-dependent relations under a given set of conditions in terms of appropriate indicators, such as revenues. Concerning firm growth we look at its financial sub-states (Equation I.15) and regard these is indicative for the firm's growth (ch. 4.3.5.1).

Let R(t) be the value (of revenue) at time t for the financial sub-state taken also as a growth indicator for the whole system and R(t+1) the value of one unit of time later. Following conceptually and structurally "growth breeds grows," the innovation and investment persistence growth cycles (Figure I.127) and Equation I.13 for an equation of state for dynamically stable states of NTBFs, intensity ⊗ capacity, makes R(t+1) proportional to R(t): $R(t+1) \rightarrow n_R \cdot R(t)$.

The factor n_R is an intensity covering interacting firm-internal and external effects. In particular, it may include responsiveness of the startup toward (changes of) the market environment. To account for this effect and self-reinforcement of the capacity factor we complement R(t) by an additional factor [1 + (R(t+1) / R(t))] to finally suggest Equation I.17 as a fundamental systemic reflection of "growth breeds growth" for dynamically stable states of NTBFs. In particular, it shall cover *dynamically stable states* with a *steady state condition* as defined above (Figure I.132).

We interpret the capacity term $R(t) \cdot [1 + (R(t+1) / R(t))]$ as being related to capability in the sense of RBV (ch. 4.3.3). The factor n_R is assumed to be a constant for a specific bracket interval and treated as an empirically determined factor.

Equation I.17:

$$R(t+1) = n_R \cdot R(t) \cdot [1 + (R(t+1) / R(t))]$$

t = 0, 1, 2, ...n; n_R a state and environment characterizing (empirical) factor;

$R(t+1) \geq R(t)$ for all t; asymptotic behavior for $n_R = ½$ gives $R(t+1) = R(t)$

The equation of state describes the state over a period of time beginning after a front bracket at the time when the firm has adapted a new "dynamically stable" state (t = 0, R(0) until the occurrence of a new bracket ends the development of this dynamic state (cf. right part of Figure I.133)). This new bracket may be identifiable.

On the other hand, based on an equation of state like Equation I.17 we regard a strong deviation of calculated values of an otherwise good fit between calculated and observed values as evidence for the impact of a bracket that cannot be directly observed on the basis of indicator data.

Equation I.17 is defined "recursively" – resembling to a certain degree the Fibonacci numbers [WolframMathWorld].

For R_i ($i \geq 2$) we proceed taking R(0) and R(1) as the basis and then start with the mappings, e.g. $R(1) \rightarrow R_1$ and $(R(1)/R(0) \rightarrow R_2/R_1$.

Then the factors n_R and (R_2/R_1) are fitted to acceptably describe the whole sequence of data on the interval $<t_0, t_n>$:

$R_2 = n_R \cdot R_1 \cdot (1 + R(1)/R(0)) \rightarrow R_2 = n_R \cdot R_1 \cdot (1 + R_2/R_1) \rightarrow$ Numerical fit for all data

We are primarily interested in an *equation of state for monotonically increasing revenue growth R* on a given interval $<t_0, t_n>$ The numerical values of the state-characterizing factor n_R and the starting quotient R(t+1)/R(t) of Equation I.17 will shape the curve, usually after a bracket.

Curves as determined by different values of the factor n_R and fixed values for an initial growth rate (R(0) and R(1) may exhibit various appearances as are given in the below Figure I.155. Shapes range from exponential (hyperbolic) to asymptotic growth and non-growth shape (Figure I.107).

Equation I.17 reflects also a mechanism when early growth of revenues, due to an unfavorable situation and insufficient magnitude of the intensity factor n_R, can turn into decreasing revenues, as is illustrated in Figure I.114, and ultimately disappearance of a firm:

- Failure paths all begin with one or more fundamental problems;
- They all lead to a situation where the symptoms of the worsening situation become visible in the financial situation.

Furthermore, Equation I.17 can simulate the case that the “thrust capacity” for a startup’s development does not suffice for a successful “lift-off” for (revenue) growth.

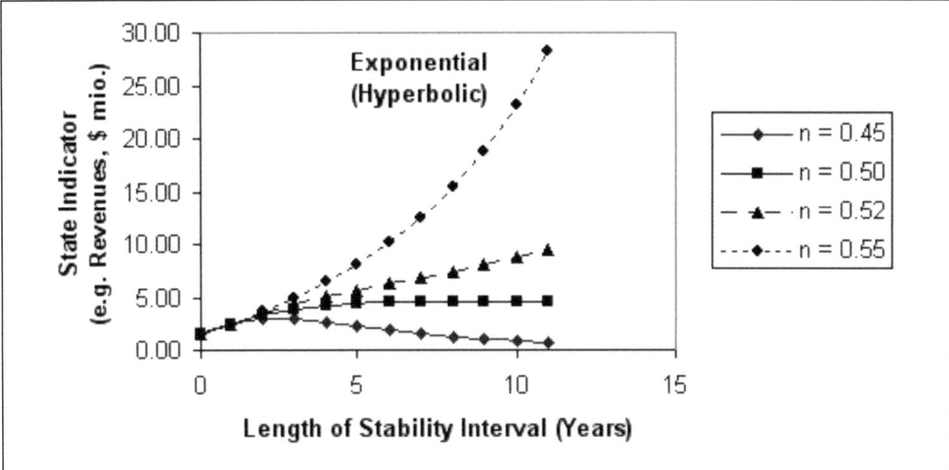

R(t+1) = n·R(t)·[1 + (R(t+1) / R(t))] (Equation I.17);

Start: R(0) = 1.5, R(1) = 2.5 units;
initial growth ratio: R(1)/R(0) = 1.67
(simulating very strong growth of the particular state)

An expansion of the curves for n = 0.50 and n = 0.52 is given on the next page.

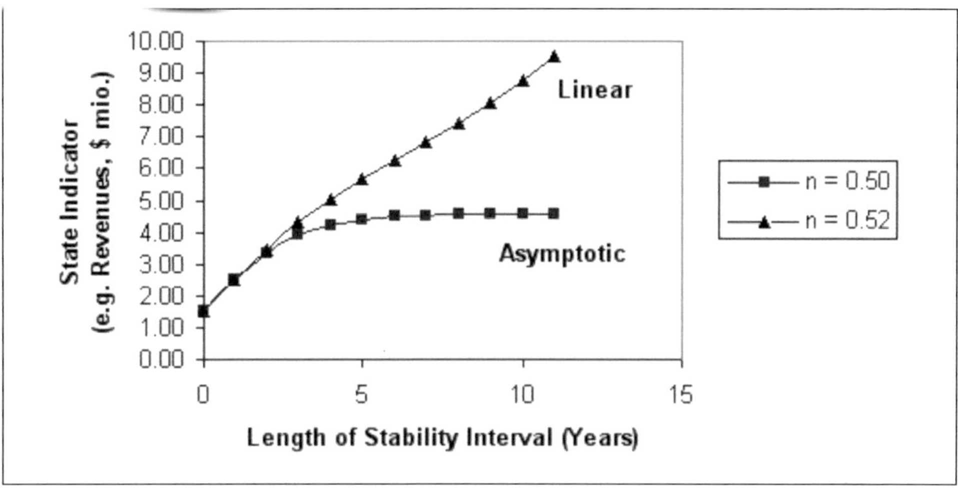

Figure I.155: Prototypical development curves as shaped by relevant parameters.

Equation I.17 becomes extremely simple if the quotient R(t+1)/R(t) is very small compared to 1 and, hence, can be neglected or can be viewed as a non-negligible constant g'. Then either we have

$R(t+1) = n_R \cdot R(t)$ or

$R(t+1) = n_R \cdot (1 + g') \cdot R(t)$

Therefore, at least for a particular interval $<t_0,t_n>$, we propose a growth period of an NTBF to be described by the very simple formula Equation I.18 ($n_R \cdot (1 + g') \rightarrow (1 + g)$). Though we shall denote g as a "growth factor" it should be noted that its numerical closeness to a common growth rate, such as CAGR (Equation I.10), must be viewed as accidental.

Through its relation to interwoven company-internal and external effects the growth factor g is related to n_R. If we interpret growth of a new firm to be determined essentially by growth of the market and firm-internal response (capacity) toward the market opportunity we can differentiate the *development categories for new firms* as:

- *Grow less than the market*
- *Grow with the market* and
- *Grow more than the market.*

Equation I.18:

$$R(t+1) = (1 + g) \cdot R(t) \text{ for } <t_0,t_n>$$

The following examples calculating revenue values for dynamically stable *financial sub-states* do not intend to provide numerically optimized solutions, but solutions with "simple" numerical relations that will *suffice to illustrate the essentials* of the bracket model.

The simplest case is provided by WITec's closely linear growth (Figure I.123) or the linear growth periods of Nano-X (Figure I.137).

The example of WITec (founded 1997) can be used to illustrate the essentials of the bracket approach. It is not about (statistical) curve fitting for an interval <1999,2009> or <2002,2009>. The bracket theory interconnects a time interval for development of states started by a front bracket as displayed in Figure I.123.

The first bracket is firm's foundation (ch. 4.3.5.1) with the startup thrust phase extending over the next three to four years being associated with rather unstable firm states and usually decreasing year on year growth rates (Table I.71).

In Figure I.156 Equation I.18 could be used as a numerically sufficient approximation for the interval <2002,2009>. But this interval is "perturbed" by the Dot-Com Recession. In Figure I.156 it is argued why <2004,2009> would be most appropriate interval characterizing a dynamically stable growth state of WITec. The interval would probably start at 2003, but there are no revenue data available for 2003.

Similar to WITec almost *linear,* parallel developments of revenues and number of employees for Nano-X GmbH (Figure I.137) express *almost constant productivities* for the intervals <2000,2003) (productivity ca. €63,000 per employee) and <2005,2008> (productivity ca. €118,000 per employee) and indicate these periods to be associated with dynamically stable organizational states.

Corresponding effects are observed for SAP productivity (Figure I.147) which show little changing for the interval <1976,1979>.

On the other hand, the Microsoft's revenues (Figure I.144) between 1977 and 1982 as well as 1980 and 1989 with starting growth proportions of ca. 3.6 or 2.1, respectively, exhibit *exponential* growth. The recession of 1981/1982 separates both phases.

Moreover, the period <1981,1986> is associated with many positive brackets and the IPO which together lead (probably) to accelerated growth expressed by a larger n_R-factor. The interval <1977,1982> of Microsoft's early growth is organizationally rather unstable (Figure I.146).

Equation I.17 is sufficient for the calculations (Figure I.157) of two dynamically stable financial sub-states of Microsoft separated by a recession bracket which, however, is not observable and obviously more than balanced by several positive brackets.

Formula: **R(t+1) = (1 + g)·R(t)**

The *tempting* approach:

Growth factor g = 0.13; R(0) = €3.29 mio. at 2002; CAGR (2002,2009) = 14.5%

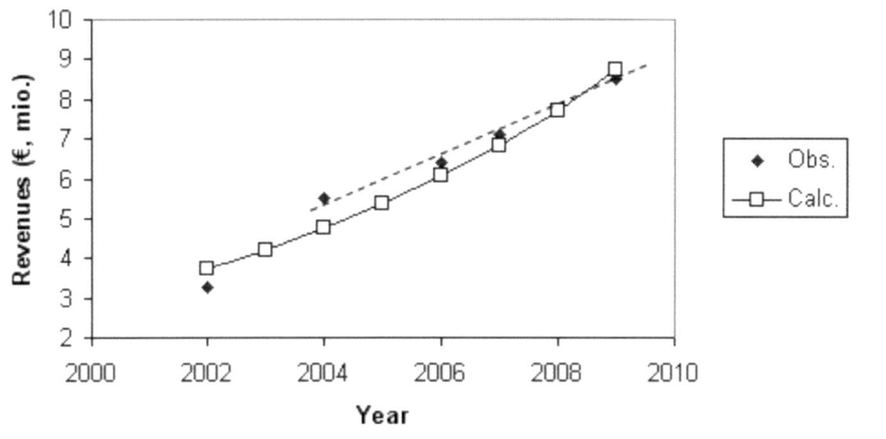

Comments: Actually, a very good fit is seen for the interval <2004,2009> (dashed line, g = 0.09) which is the most appropriate characterization of a *dynamically stable financial state* as the observed data for 2002 may still be affected by the Dot-Com Recession, data for 2009 by the Great Recession.
Furthermore, WITec exhibits *almost constant productivity* (Figure I.123). This means, for the related interval it has a financially dynamically stable state – and also a dynamically stable organizational sub-state showing a steady state condition with almost constant productivities (Figure I.123).

An approach with

growth factor g = 0.17; R(0) = €1.55 mio. at 1999; CAGR(1999,2009) = 18.6%

provides a *numerically acceptable approximation* for the 1999-2009 period. However, it covers also the Dot-Com Recession. Such *pure numerics* are not in line with the bracket theory and, hence, are lacking explanatory power.

Figure I.156: Calculated and observed revenues of WITec GmbH for a "perceived" dynamic stability interval <2002,2009> and contrasted with the theory-related <2004,2009> interval (dashed line).

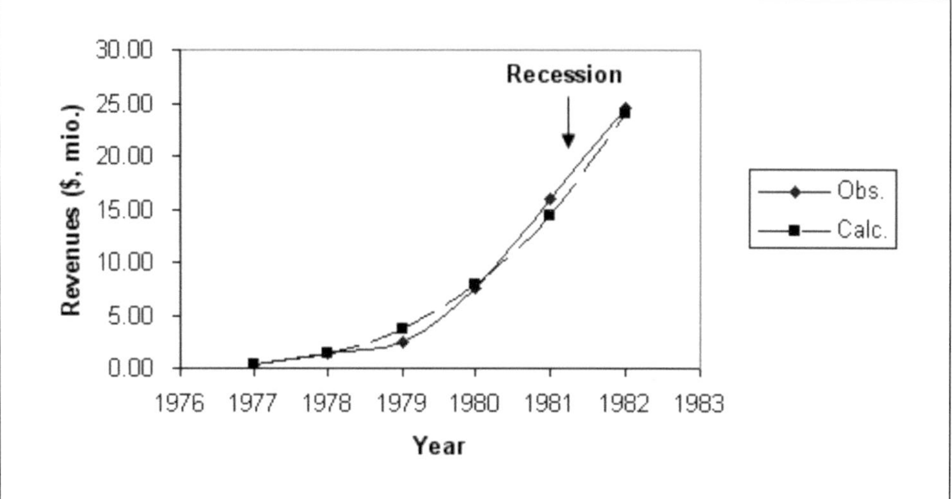

Microsoft Corp., founded 1975, IPO 1986; Equation I.17 used:

Above figure: R(0) = \$0.38 mio. (1977), R(1) = \$1.36 mio. (1978); n_R = 0.585; R(1)/R(0) = 3.57

After 1882 there are large deviations between calculated and observed revenues (calculated values much larger than observed ones) indicating a transition state.

Below figure: R(0) = \$7.52 mio. (1980), R(1) = \$16.06 mio. (1981); n_R = 0.605; R(1)/R(0) = 2.13

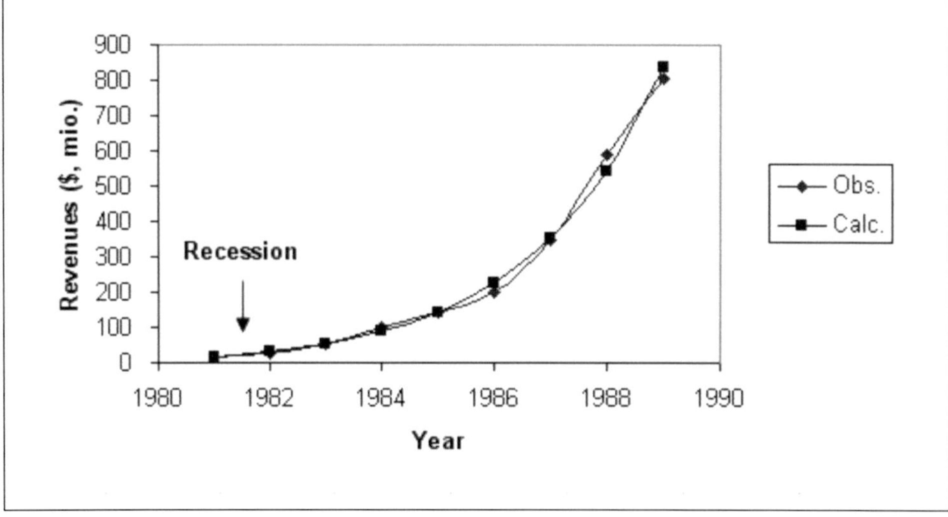

Figure I.157: Calculated and observed revenue growth of Microsoft.

When looking at the examples of Cisco and First Solar (Figure I.158.) Equation I.17 turned out to be inadequate and required modification. Cisco's starting growth proportion was 3.63 and that of First Solar is 4.21.

For both cases it has turned out that Equation I.19 different from Equation I.17 by the reinforcement factor is more appropriate for calculations. Its applicability is also shown for Google in Figure I.159.

"Similar" to Equation I.10 for CAGR growth relates to a fixed starting value R(0) in the series of R(t) across the dynamically stable state. Disregarding the construction of some sort of geometric mean through the t-eth root the emphasis on a fixed value of R(0) for the state seems to show that for the (financial) growth a particular initial constellation exerts a decisive influence.

For Cisco a good fit is observed and its revenues for the 1991 recession are obviously more than balanced by sales (and more employees). On the other hand, referring to productivity Figure I.145 shows that Cisco's organizational state is by no means stable across the early 1987-1994 period.

Due to the small number of cases treated by Equation I.19 one can only speculate about what is behind R(0) – a large capital injection at the start of the dynamically stable period adding to the "base line" development of each of the R(t)'s which will drive the growth over the whole period, a jump in other resources like employee number, or an extremely successful shift to another source of revenue.

Equation I.19:

$$R(t+1) = n_R \cdot R(t) \cdot [\, 1 + \{R(t) / R(0)\}^{1/t} \,]$$

t = 1, 2, …n; n_R a state-characterizing (empirical) factor

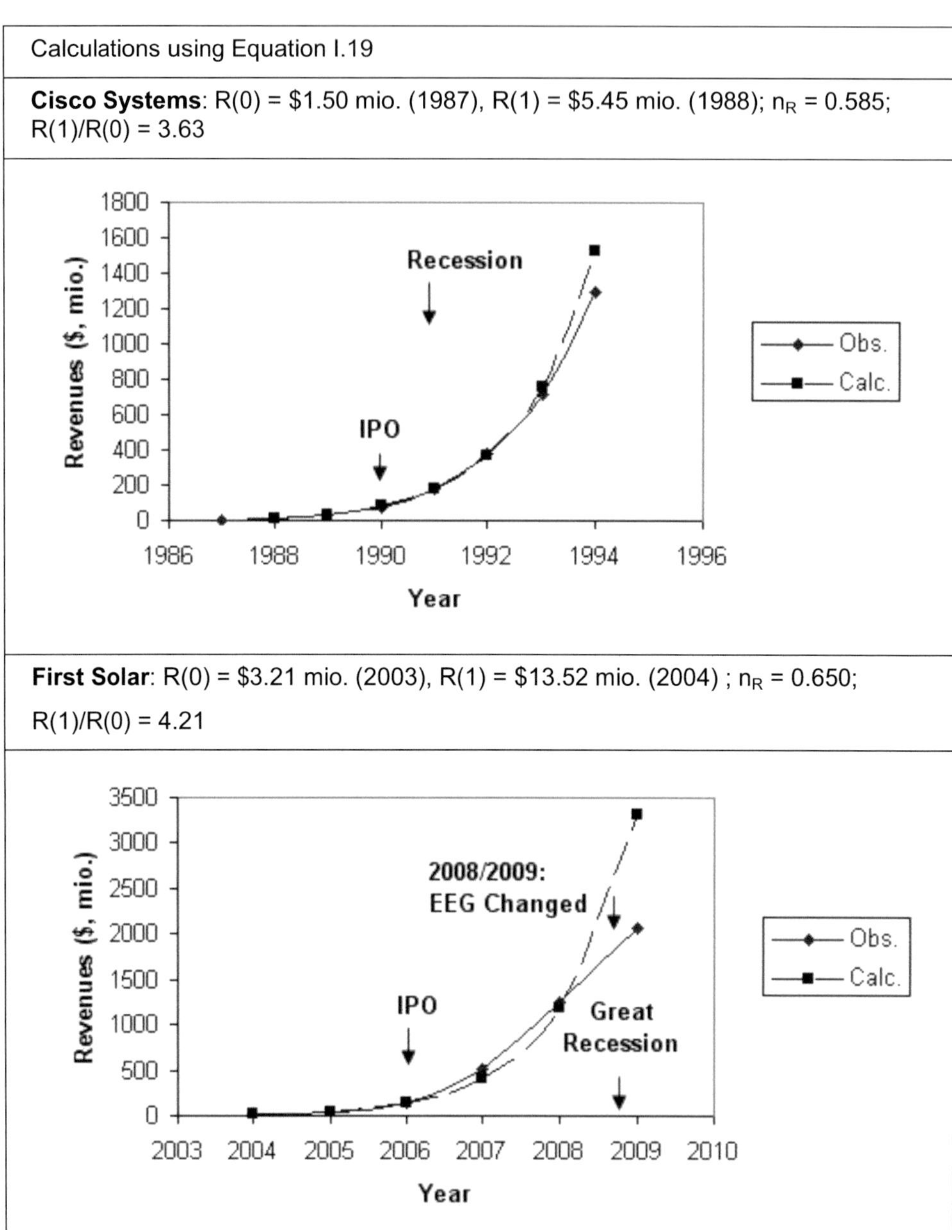

Calculations using Equation I.19

Cisco Systems: R(0) = \$1.50 mio. (1987), R(1) = \$5.45 mio. (1988); n_R = 0.585; R(1)/R(0) = 3.63

First Solar: R(0) = \$3.21 mio. (2003), R(1) = \$13.52 mio. (2004) ; n_R = 0.650; R(1)/R(0) = 4.21

Figure I.158: Calculated (Equation I.19) and observed revenues of Cisco Systems and First Solar.

The large deviation between observed and calculated revenues for First Solar after 2008 due to the Great Recession and the change of the EEG in Germany shows that the financial sub-state to grow regularly only between 2004 and 2008. This is different from Cisco whose regular growth was obviously not affected essentially by a recession.

As a summary, the preceding discussions have presented *growth categories* (Figure I.155) for early development states of young firms (NTBFs) in terms of equations of state which are essentially recursive relations (Equation I.17 - Equation I.19) and reflect self-reinforcement verbalized as "growth breeds growth." This means, a series of interacting internal sub-states' changes leads to increases in size (observed for revenues) accompanied by not necessarily synchronous changes of the characteristics of other sub-states of the growing entity over a certain period of time (Figure I.149).

> Based on the calculational results in this chapter it appears that irregularities ending an otherwise good fit when comparing calculated and observed revenues may provide a means to detect brackets which may not show up in observed curves.

The development patterns represent systemic features of different types of growth paths of new firms for a given time period starting at a specified point in time that may initiate the search for explanations on the firm level. In the above text it has been pointed out that after a perturbation, a bracket, the firm will not return to the original state, but rather develops further into a new state.

This shall be illustrated by different theoretical descriptions in terms of different formulas to be used before and after a bracket for the German NTBF Nano-X (Figure I.137; Table I.77) and Google (Figure I.159).

For Nano-X the discontinuity, the bracket R(2004) $\rightarrow$ R(2005) of +2.33 units for the interval <2002,2008>, is surrounded by two states differing by the g-factor of Equation I.18. Table I.77 provides a "full" theoretical description of the (financial) states of Nano-X. The jump suffices qualitatively to identify the different states. Admittedly, the quantitative approach of the early state with g = 1.52 is not satisfying.

The significantly more pronounced growth rate for the 2000 to 2004 period cannot be explained straightforwardly. This period covers the Dot-Com Recession with no observable negative impact for the revenues. However, Nano-X financed the first years essentially via R&D projects ("Verbundprojekte"; ch. 1.2.6) of the German federal/state governments, the EU and NGOs as well as cooperation with other SMEs. As project money is usually counted as revenue, it can be hypothesized that the significant revenue growth during the first years, from almost its start, are due to larger contributions from projects to revenues.

Therefore, the theoretical description of the pre-2005 dynamically stable state of Nano-X should be confined to the interval <2002,2004>, though it cannot be ruled out that it covers also the 2000 and 2001 range.

On the other hand, more successful applications of Equation I.18 are discussed later (Figure I.163, Figure I.164, Figure I.165, Figure I.166).

Table I.77: Theoretical descriptions (Equation I.18, <2002,2004>) of two developing financial states of the German Nano-X GmbH separated by a bracket with pronounced discontinuity, a "jump."

Year	Revenues (€, mio.)	Calculated Revenues (€, mio.)	
R(t+1) = (1 + g) * R(t)		g = 0.52	g = 0.05
2000	0.50	0.50	
2001	1.00	0.76	
2002	1.4 *)	1.16	
2003	N/A	1.76	
2004	2.50	2.67	
R(2005) = R(2004) + 2.33			
2005	5.00		5.00
2006	5.20		5.25
2007	5.40		5.51
2008	6.00		5.79
*) Estimated from average productivity (70,000)		Average Productivities (€ /employee):	
		70,000	118,000

It is interesting to note that in its early years also SAP's growth follows Equation I.18 (g = 1.70 for <1972,1974> and g = 1.63 for <1975,1978>; Figure I.147).

For Google (Box I.24) two growth periods according to different theoretical approaches can be identified with a transition state ("bracket") at 2002/2003 (Figure I.159) when revenues "jumped" from $86 million (2001) to $440 million (2002). Starting with the period 2000 – 2003 it turns out that after 2003 the gap between calculated and observed revenues widens drastically. Hence, by trial, it was found with which formula to describe the 2003 – 2008 period.

Early Growth of Contextual Advertisement on the Web.

After 2002 ca. 97 percent of Google's revenues were from advertisement! In 2001 ad revenue accounted for 77% and in 2002 it was already 92%. Advertising income is earned as Google operates its own Web sites, but it distributes ads also to partner Web sites. Its software AdWords (launched in 2000, major overhaul in 2002) is Google's unique method for selling online advertising. AdWords analyzes every Google search to determine which advertisers get each of up to 11 "sponsored links" on every results page. It is one component for the link between searching and advertising. The second part is AdSense which relates to semantics [Sullivan 2004].

In 2003 Google launched its AdSense contextual ad program and then greatly expanded AdSense, meaning ad serving application. AdSense placements are almost certainly the reason why Google has seen network-derived ad revenue rise so sharply.

There is another factor whose effects, however, cannot be assessed. Google bought Applied Semantics in 2003 and also three other companies [Sullivan 2004].

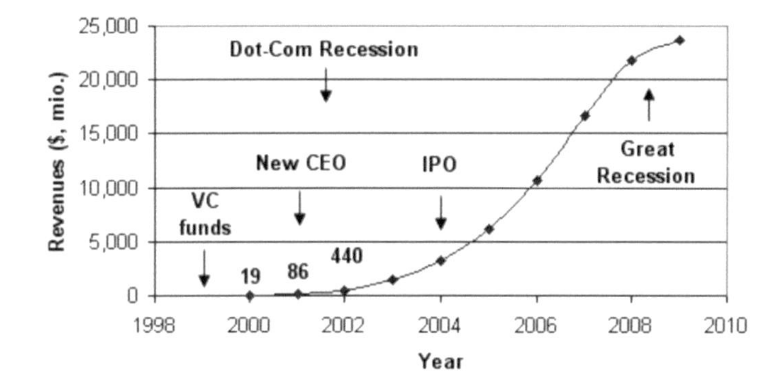

Google:
Founded 1998 by Larry Page and Sergey Brin

primary focus: a better search engine for the Web (Box I.24)

Calculated values according to

Equation I.18 (small chart), g = 3.75, R(0) = \$19 mio.);
Equation I.19 (large chart), R(0) = \$440 mio., R(1) = \$1,467 mio., n_R = 0.50

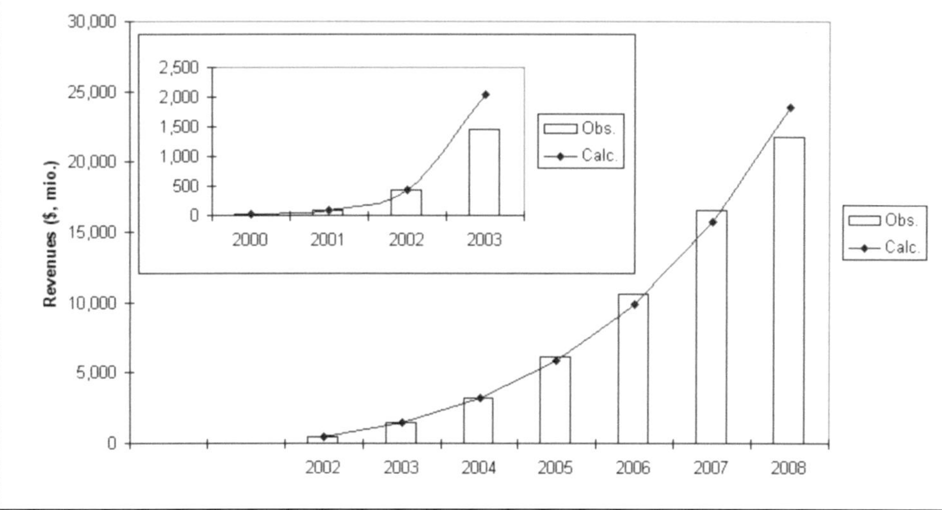

Figure I.159: Theoretical descriptions of two developing financial states of Google separated by a bracket with pronounced discontinuity, a “jump” (data from [Tech Crunchies 2010]).

So far, brackets were identified on the basis of revenues or employees/productivities. In case of Google revenue data do not reveal a 2002/2003 bracket (cf. Figure I.136). Fortunately, for Google the theoretical description is corroborated by other data: The pronounced "jump" characterizing the transition into the second state in clearly seen in the development of Google's profit (Figure I.160).

Based on the limited number of cases, it has turned out that IPO brackets could not be detected consistently though they may represent drastic changes of a firm's sub-states. Change will relate to ownership and potential change of control (change of leadership or management), respectively, to huge addition of capital (financial resources) which can be used, for instance, to pay back debts or increase of human resources in terms of R&D and marketing, sales and distribution (exploiting existing markets and enter new markets) or increase capacity of production.

IPO brackets are sometimes associated with increase of the number of employees which is observable by a decrease of productivity (Q-Cells, Figure I.153). But for Cisco (Figure I.145) in a steady state this does not show up.

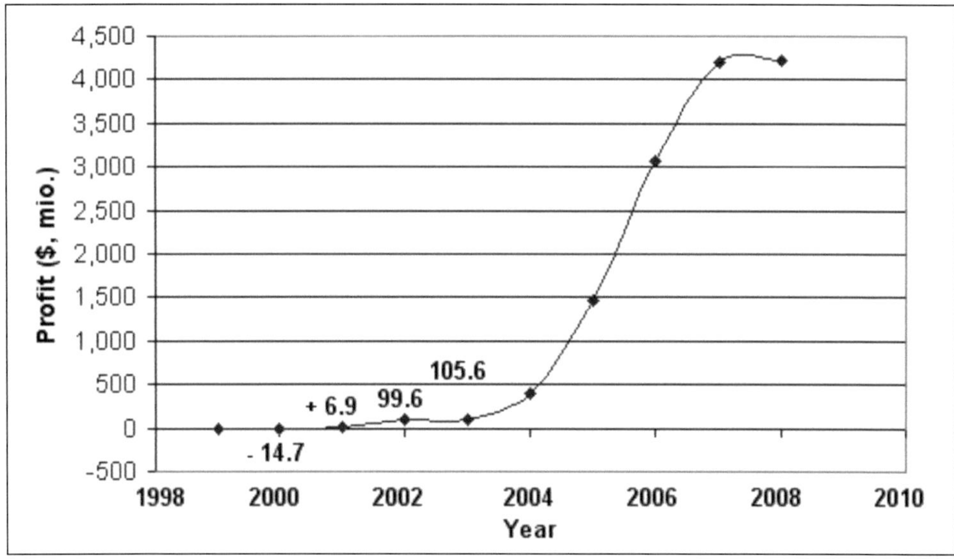

Figure I.160: Timeline for development of Google's profit (data from [Tech Crunchies 2010]).

Summarizing the presented aspects of the theoretical approaches to early growth of NTBFs one is led to suppose that for early unperturbed growth phases Equation I.17 and Equation I.18 are appropriate if the growth is essentially driven by cash flow from income (WITec, Nano-X, SAP, Microsoft and Cambride Nanotech and US LED given in ch. 4.3.6). This would be reflected by the cyclic process of innovation and investment persistence in Figure I.127. On the other, if large capital comes from outside sources, but not through an IPO (Cisco, First Solar, Google), Equation I.18 seems to be adequate. However, this is still a proposition.

The equations do not only provide numerically satisfying descriptions of dynamically developing growth states for NTBFs, but in certain cases various structures for different intervals may differentiate states without prior identification of brackets. In so far, for the developmental processes of NTBFs' patterns of growth are created that characterize growth states. However, a more detailed understanding of the observed effects must be bolstered by reference to the micro-level of the firm under consideration if common features for such states shall be revealed.

Specific initial startup configurations and random occurrences of largely conceivable bracket events and related proceedings concerning decision-making will lead to overall differences in the growth of new firms. But there are periods, time intervals of growth, which are structurally comparable for the dynamics of developments of different firms.

4.3.6 Expectations of Growth of Technology Ventures

Past performance is, as investors like to say, no guarantee of future results.

In the context of entrepreneurial success (ch. 4.1) various perspectives of success and aspirations were discussed, in particular, those of the entrepreneur(s), financial backers, such as venture capitalists, and policy with a special focus on job creation (Table I.64). Currently, particularly in the US and Germany, there is great interest of policy in *high growth* and *fast growth* young firms that become major employers. In the following discussions we shall put the emphasis on details of generating and "measuring" expectations (Box I.17).

The issue is the assessment of technical startups by third-parties and entrepreneurs to generate *expectations* of survival and growth or even growth levels which is often associated implicitly with mixed arguing along the lines of "reasons why" versus "reasons for thinking that" (Figure I.2).

Covering samples of technical and non-technical samples various studies have shown that *the activity pursued in the startup process (organizational effort) has a major impact on founding success*. Furthermore, both *the founding success and (short-term) new venture survival of nascent entrepreneurs improved when the nascent entrepreneur engaged in early and careful planning activities* [Keßler et al. 2010].

In particular, one can argue that, disregarding cases with issues of scale-up for production, a *talented team*, with a *large market in which to innovate* and *excellent execution* justifies expectations of survival and good growth of a startup.

Bhidé [2000:209] has argued generally "that the ambition and the capability of individual entrepreneurs have a significant impact on firm longevity and growth." High growth of NTBFs is essentially related to the aspirations of the founders if aiming to maximize long-term value of the business, or merely seeking an increase in income, wealth creation and independence, for instance, by selling a successful firm after a few years (cf. William Henry Perkin; A.1.2).

In the following discussion we shall focus on expectations of growth, in particular, high growth of startups/NTBFs from an observer's or venture capitalist's point of view rather than "high-expectation" startups which, according to Autio [2005], refer to "all start-ups and newly formed businesses which expect to employ at least 20 employees within five years' time." (ch. 4.1)

In the teleological environment of GST expectations make special connections between goals of the firm founders and their related strategy, their opportunities in terms of markets and industries including competition and accessible resources and should reflect *ex ante* statements (Box I.17).

In order to inquire into expectations of the evolutions of NTBFs we use *comparative approaches*. One of it is looking for features which make the NTBFs' next steps of development – become a medium- and large-size firms – survive, be successful and keep competitive advantage for their evolution. This relates also to Bhidé's aspect of firm longevity.

As this is a frame for the transition between two constellations looking at medium- and large-sized firms in a particular industry (Figure I.118) can be useful to create expectations of developments of NTBFs by searching for fits between relevant features of both these classes. Furthermore, we shall focus on top *decision-makers* who control the enterprise and who have a significant stake in its fortunes.

This is to a certain degree in line with Bhidé [2000:2009] who stated that "knowledge of the origins and destinations of the origins and destinations of the typical long-lived corporations will help us identify the important common elements of their evolution."

Another way to create even semi-quantitative expectations compares the initial configurations of startups (ch. 4.3.2, A.1.6). A given NTBF with a given initial configuration providing information of its modes of growth and achievements will be used to generate expectations. Hence, knowledge and tools derived from *ex post* analyses are used to generate *ex ante* expectations (Table I.80, Table I.81).

Success Factors of Mid-Sized Enterprises

So far, we have learned that developments of privately held or privately controlled technical SMEs (Table I.4, Table I.74, Figure I.128) in the US and Germany are rather similar and reference to the German situation will allow some generalizations. Notable differences may occur with regard to extent of globalization and internationalization, where generally German firms put much more emphasis on than American SMEs.

As will turn out small and large high growth firms put much emphasis on strategy and execution, summarized, for instance, by the characteristics 1-7 of the German Hidden Champions (ch. 4.3.5.2). Thus there will be a relation to **strategic groups**, defined by

M. Porter as "*a group of firms pursuing similar strategies along strategic dimensions*" [Runge 2006:221].

An investigation of 1,300 German mid-sized firms, usually family-controlled, and comparisons with the 180 firms showing the strongest growth revealed that the top firms have growth rates of 10 – 39 percent, far exceeding the others. And, furthermore, they also *invest significantly* more than others. This was attributed to the following factors [Fröndhoff 2008].

1. Taking advantage from megatrends.
   Adapting the business model very early to global market trends, in particular, mobility, health, energy and process and control technologies.
2. Strong internationalization
   The firms with the highest growth rates have an export rate of 50 percent and more. Even small firms have set up a distribution network including sales and service offices or even production facilities.
3. Premium products and services
   The firms with the highest growth rates have decade-long experiences in their segments and focus on premium products. They have high innovation rates and thus can keep their lead times and withstand competition. They specialize and focus on few businesses in global markets. Their lead in knowledge and experience allow them to launch tailored products in the markets. Furthermore, they offer additional services for their products.
4. Manufacturing and networking
   Bundling product and associated service is seen as a successful strategy for competing with young firms from developing countries. Firms with the highest growth rates set off themselves through difficult to copy knowledge of producing high-tech products. And they often cooperate with universities and public resource institutes and develop products together with customers or firms of other branches.
5. Local roots
   The high growth firm has local ties expressed by social engagement.

A study of Ernst & Young [2011] inquired into 68 mainly technically oriented (German) mid-sized and large firms, actually finalists of the "Entrepreneur of the Year" contest, which showed above average growth over a series of years. They had on average revenues of ca. €115 million per year and 815 employees. The majority of the firms generate their revenues in the home market.

The majority of these firms (ca. 60 percent) operated in *lucrative niches* or *promising and growing market segments.* A high proportion of sales – on average 14 percent – is attributed to the *research and development* departments. Sustainable growth relies on *permanent innovations* including internal processes and *high appreciation of the firm by employees and customers* [Ernst & Young 2011]. Main success factors include:

1. Look at the bigger picture ("Über den eigenen Tellerrand schauen").
   During an upturn 45 percent of the firms turn to new markets, ca. 30 percent turn also to new target groups. Apart from the well-known European markets growth markets in Asia and South America are seen as big opportunities.
2. Perceive competition as chance.
   Though three quarters of the firms complain about higher competition they accept the challenges to optimize their offerings and arouse new needs. Globalization is not seen only as a threat, but also an opportunity to access ideas, talents, customers and businesses.
3. Innovation and investment persistence (Figure I.127).
   cf. for instance, German Erlus AG [Runge 2006:237/238].and its cooperation with Nano-X GmbH (B.2).
4. Continuously improve not only offerings, but also organizational processes.
   During an upturn, if competitive pressure tends to decrease, 75 percent of the firms take the time to assess their organizational processes and to reconfigure them so that new ideas can spread and get promotion and support.
5. Inform and motivate employees.
   The majority of the firm sample has a style of cooperative leadership concerning firm orientation, strategy and goals of the firm. The targeted incentives, but also demands of employees are appreciated – 80 percent of the firms can trust that their employees are committed to these plans and new developments.
6. Plan ahead.
   The firms prepare for possible uncertainties and issues in their business, in particular on those, which they can influence even in the worst case.

All the above success factors correspond essentially also to those which are typical for the class of the German Hidden Champions (ch. 4.1.1).

In a comparative study of technical and non-technical high-growth companies from around the world in ca. 30 industries versus their less successful competitors Kim and Mauborgne [1997] presented a rationale for the differences. They found high-growth to be achieved by small and large organizations, in high-tech and low-tech industries and private and public firms. The origin of the differences was the companies' fundamental implicit assumptions about *strategy* which sought to make their competitors irrelevant through a strategic logic they called "*value innovation*."

They inquired into the five textbook dimensions of strategy given in Table I.78. For value innovation logic the first and last dimensions exhibit typical systemic features. They found that managers of less successful companies all thought along conventional strategic lines.

The high-growth companies used a value innovation approach, and it was consistently applied to business initiatives in the market place. This means, value innovation logic requires *execution* (Figure I.87).

Table I.78: Two types of strategic logic for value innovation [Kim and Mauborgne 1997:106]. Reprinted by permission of Harvard Business Review. Copyright ©1997 by Harvard Business Publishing; all rights reserved.

Five Dimensions of Strategy	Conventional Logic	Value Innovation Logic
Industry assumptions	Industry's conditions are given.	Industry's conditions can be shaped.
Strategic focus	A company should build competitive advantages. The aim is to beat the competition.	The competition is not the benchmark. A company should pursue a quantum leap in value to dominate the market.
Customers	A company should retain and expand its customer base through further segmentation and customization. It should focus on the differences in what customers value.	A value innovator targets the mass of buyers and willingly lets some existing customers go. It focuses on the key commonalities in what customers value.
Assets & capabilities	A company should leverage its existing assets and capabilities.	A company must not be constrained by what it already has. It must ask: What would we do if we were starting anew?
Product & service offerings	An industry's traditional boundaries determine the products and services a company offers. The goal is to maximize the value of those offerings.	A value innovator thinks in terms of the total solution customers seek, even if that takes the company beyond its industry's traditional offerings.

Value innovation is expressed by the departure from the conventional logic of the particular industry and can emerge from a so-called "value curve" which relates valuation of the customers (Table I.13) versus experiences of the customer (Figure I.161).

Value innovation provides inquiries into answering four key questions [Kim and Mauborgne 1997]:

1. What factors should be eliminated that our industry takes for granted?
2. What factor should be reduced well below the industry standard?
3. What factor should be raised well above the industry standard?
4. What factors should be created that the industry has never offered?

This often is a relation of the kind: increase utility drastically while reducing similtaneously own costs or, alternatively, give customers more of what customers need or appreciate most and much less of what they are willing to do without.

As an example in curve 2 in Figure I.161 the customer will get a device from three suppliers with above average advantages for elements C, D. E concerning functionality (C and E) and delivery, but the price (A) is rather high and it is not easy to use (B). Curve 1 represents a device with extremely favorable elements A and B (price and ease of use), but serious disadvantages concerning C, D. E.

The innovation displayed by curve 3 means combining advantages of D and E concerning delivery, functionality and B (ease of use) and disregarding C, which is optional features and meaning cost. The offering of curve 1 (maybe by an entrant) targets essentially customers who so far did not use the particular device due to a high price and learning efforts.

Value innovation is different to technological innovation, considered as innovation that integrates interfaces, marketing and operations. The emphasis is on product, service, and delivery. Competition is the key building block of strategy and positioning is by differentiation from the competitive pack by "breaking" the rules of the "standard game" to create fundamentally new and superior customer value.

In so far, value innovation can be useful as an instrument to assess young firms concerning growth in a competitive market by assessing in how far the they may provide appropriate answers to the above four questions.

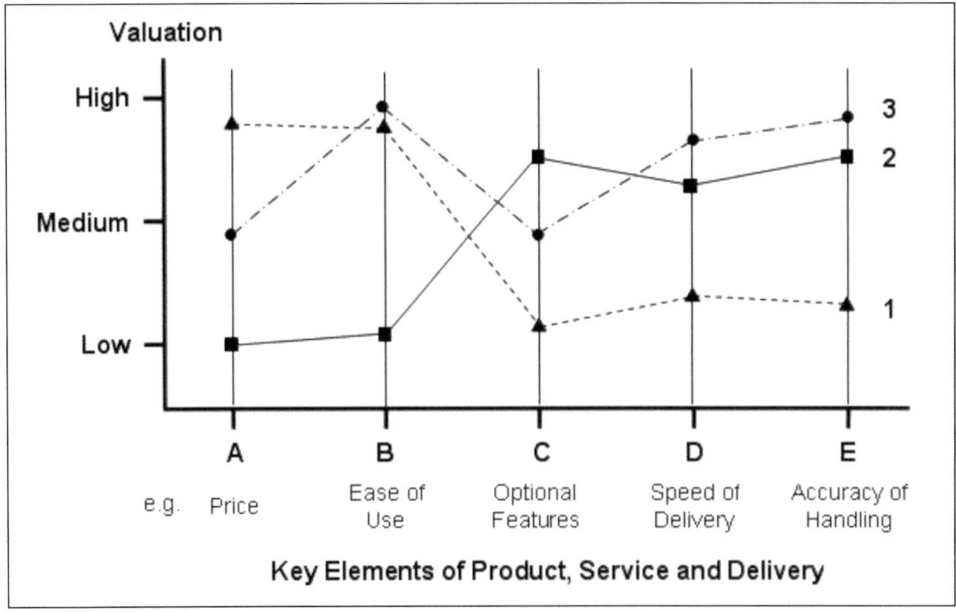

Figure I.161: A "value curve" for value innovation.

How the logic of value innovation translates into a company's offerings in the market has been described by Kim and Mauborgne [1997], for instance, for the German originally Hidden Champion SAP AG (ch. 4.1.1; Figure I.143; A.1.4) as follows.

Until the 1980s Enterprise Resource Planning (ERP) appeared as "business-application software." Providers focused on sub-segmenting the market and customizing their offerings to meet buyers' functional needs, such as payroll, human resources, production management and logistics. And the emphasis of the makers was focusing on improving the performance of particular software products.

Instead of competing on customers' differences, SAP sought out important *commonalities in what customers value*. SAP's founder/leaders correctly hypothesized that for most customers the performance advantages of highly customized, individual software modules had been overestimated. Such modules forfeited the efficiency and information advantages of an integrated system, which allows real-time data exchange across a company.

In 1979, SAP launched R/2 (Figure I.143), a *real time integrated business-application software* for mainframe computers. R/2 had no restriction on the platform of the host computer; buyers could capitalize of the best hardware available and reduce their maintenance cost dramatically. Most important, R/2 led to huge gains in accuracy and efficiency because a company needed to enter its data only once. And R/2 improved the flow of information. A sales manager, for instance, could find out when a product will be delivered and why it is late by cross-referencing the production database. SAP's growth and profits have exceeded its industry's.

Concerning technology entrepreneurship our interest in expectation of levels of success is the top ca. 20 percent of firms that create ca. 80 percent of jobs by high or fast growth (Figure I.119). The interest is to assess the fate of a startup from its initial configuration, its "birth" including the "startup thrust phase" (ch. 4.3.1, 4.3.2; Figure I.125) in a given environment.

We shall tackle the scope of *ex ante* expectations referring to three situations

- Statements about essentially survival and rough growth levels at the time of firm's foundation based on its initial configuration ("year 0" of existence and the first year at the highest)
- Statements about the situations after the startup thrust phase (year three or four of existence when the period of highest level for a firm's failure has successfully passed) or generally,
- Firms' development after a significant bracket in terms of dynamically stable states (as outlined in ch. 4.3.5.3).

In restricting to the time one must keep in mind that startups with (anticipated) production often need a period of four to eight years before they can commercialize their offerings ("delayed growth"). Such long "projection" into the future restricts expectations seriously, as not only the market and competitive landscape will have changed. The expectation would require assumptions about the scale-up process into large-scale production and a successful entry into the market.

The first case will be similar to that of natural science when usually the initial conditions and corresponding equations of motion (or situation-related differential equations) suffice to predict time-dependent trajectories. For the current discussion the initial configuration (ch. 4.3.2) would be the starting point.

We shall take care as far as possible of the caveat that the very dynamic developments of new firms in the first years in business (Table I.71) and short-term success does not allow inferences about sustainable success.

Serious issues for *ex ante* approaches are raised by the high complexity of new technology-based firms (Figure I.128) which makes statistical approaches largely questionable due to sample selection and keeping the structure in the answer sets. The multi-dimensionality of factors contributing to survival and growth of an NTBF and the much randomness of environmental changes impacting firm size result in situations which are often not expectable.

The Unexspectable

Intrinsic to a discussion of expectations is the *unexpected*, the "surprise" which may lead to:

- the unexpected success,
- the unexpected failure
- the unexpected outside "unique event," the unexpected development of a situation or an entity (ch. 1.2.1, 3.2.1).

Drucker [1995] illustrates the unexpected for systematic innovation which monitors the innovative opportunity from the point of view of innovators or management of firms.

The *unexpected success* is described as the area in which no other area offers richer opportunities for successful innovation. "In no other area are innovative opportunities less risky and their pursuit less arduous. Yet the unexpected success is almost totally neglected; worse, managements tend actively to reject it" (p. 37). The unexpected success is seen as an opportunity, but it does make demands (p.45). "In exploiting the opportunity for innovation offered by unexpected success requires analysis" (p. 41).

Unexpected success is a symptom; however the question remains; a symptom of what? Drucker describes the unexpected success as the underlying phenomenon which may be nothing more than a limitation on our own vision, knowledge and understanding (p. 41).

Then there is the *unexpected failure*. "Failures, unlike successes cannot be rejected and rarely go unmotivated. But they are seldom seen as symptoms of opportunity. A good many failures are of course, nothing but mistakes, the results of greed, stupidity, thoughtless bandwagon climbing, or incompetence whether in design or execution" (p. 46).

Finally, there is the *unexpected outside event* (cf. the German firm Heyl; Table I.76) This area indicates that "outside events that is events that are not recorded in the information and the figures by which a management steers its institution, are just as important. Indeed they often are more important" (p. 52).

Unexpected events or developments are often associated with the metaphor of the "Black Swan." And specifically (according to Nassim Nicholas Taleb) technological breakthroughs are referred to as "positive Black Swans" – *unexpected events or developments* with huge positive consequences *that in retrospect look inevitable*. Some of them, such as Google (Box I.24) in technology entrepreneurship, come seemingly out of nowhere to dominate within a short time. Others take years to mature and are surprising only as people forgot they were there.

In the GST context expectation of a startup's development by an outside observer is teleologically bound to known, explicitly measurable goals (objectives) or an explicit definition of what is viewed as "success" (Figure I.78, Figure I.122, Figure I.130) by the founders referring to the level of achieving the objectives – making intrinsic connections with varieties of strategy expressed in various forms.

Therefore, Google does not represent a "Black Swan." Its early success is incommensurable (impossible to measure or compare in value or size or excellence) with its early intentions and goals. It was "unexpectable" (Box I.24)!

Box I.24: Could one have expected extremely high growth on and soon after Google's foundation?

Google Inc. is an American multinational Internet and software corporation specialized in Internet search, cloud computing, and advertising technologies. It hosts and develops a number of Internet-based services and products and generates profit primarily from advertising through its AdWords/AdSense programs. The company was founded by Larry Page and Sergey Brin in 1998 while the two were attending Stanford University. [94]

For Google an exorbitant jump in revenues is observed very few years after foundation (Figure I.159, Figure I.160). Google achieved crossing explosion-like the marker of $10 billion just eight years after foundation. Google has grown into one of the world's biggest Web companies by market capitalization since its 2004 initial public offering.

According to Larry Page Google's rise was due to being an *innovator in both technology and business*. In order to be successful in technical innovation, said Page, you must understand the business and marketing side of the equation [Page and Schmidt 2002].

We shall consider the first five to six years of existence of Google, but shall focus, in particular, on the business orientation at and shortly after foundation (the first three years). This would provide the relevant information input into the expectation of further development (and growth) of the firm by an outside observer.

Google began in January 1996 as a *research project* by Larry Page and Sergey Brin when they were both PhD students at Stanford University in California. [94] The primary focus of "the "Google project" was to create a better search engine for the Web – better, for instance, than AltaVista or Excite. The project was called "The Anatomy of a Large-scale Hypertextual Web Search Engine" or simply, The Anatomy of a Search Engine.

The challenge was to crawl the Web efficiently and provide more relevant results than the search engines that were available at that time (better search and information handling). The project should provide a *solution of the following problem*: The dramatic growth of the Web presented problems for crawling the Web – keeping the crawled information up to date, storing the indices efficiently, and handling many queries quickly. The Google project relied on the *PageRank technology* that the pair developed. [Woopidoo].

The famous search algorithm of Larry Page and Sergey Brin is essentially applying the ranking method used for academic articles (more citations equals more influence) to the sprawl of the Internet.

PageRank is not *the* Google ranking algorithm. Instead, it is just one of many different factors. However, it is the most known and a key element of what Google does. Stanford University, where PageRank was developed by Google's co-founders, owns the patent on PageRank. However, Google's IPO filing revealed that Google has been granted a perpetual license and that in October 2003, it extended an agreement giving it exclusivity to PageRank through 2011 [Sullivan 2004].

Page and Brin decided to convert their research project in Stanford University's computer science graduate program into a formal company. Originally, Google was run from within the university under the Stanford University Website, with the domain google.stanford.edu. [94]

The founders started with their own funds and those of their friends and family, but the site quickly outgrew their own available resources. In its early year Google allowed no advertising in their search engine results. The search engine became profitable in 2000 with the introduction of unobtrusive text advertisements placed along side search results. They eventually received private investments [Woopidoo].

The early assumption was that although ads would be an important source of revenue, but licensing search technology and selling servers would be lucrative. However, Internet search was considered such a low priority at the time that Page and Brin could not find anyone willing to pay a couple of million dollars to buy their technology [Lietdke 2008].

The first external funding for Google was an August 1998 contribution of $100,000 from Andy Bechtolsheim, co-founder of Sun Microsystems, given before Google was even incorporated. They filed incorporation papers so they could cash a check made out to Google Inc. (incorporated on September 4, 1998). Google was based in a friend's (Susan Wojcicki) garage in Menlo Park, California. And Craig Silverstein, a fellow PhD student at Stanford, was hired as the first employee. [94]

Early in 1999, while still graduate students, Brin and Page decided that the search engine they had developed was taking up too much of their time from academic pursuits. They went to Excite CEO George Bell and offered to sell it to him for $1 million. He rejected the offer, and later criticized Vinod Khosla, one of Excite's venture capitalists, after he had negotiated Brin and Page down to $750,000. On June 7, 1999, a $25 million round of funding was announced, with major investors including the venture capital firms Kleiner Perkins Caufield & Byers (KPCB) and Sequoia Capital. [94]

After foundation in 1998 Larry Page and Sergey Brin channeled their energy into its free search product and left much of the business planning to a 22-year-old Stanford graduate named Salar Kamangar, Google's ninth employee. Larry Page and Brin managed the company up until it reached more than 200 employees in 2001, when they handed over the CEO position to Dr. Eric Schmidt [Woopidoo]. Eric Schmidt joined Google as chairman and chief executive officer – and revenues jumped tremendously for 2001-2002 (Figure I.159, Figure I.160).

"Kamangar joined Google after graduating from Stanford University in 1999, five years before the initial public offering, and his meteoric rise mirrored the company's own comet-like trajectory. In seven years, Kamangar has gone from newbie to key player in one of the most remarkable corporate success stories of the decade. Among his accomplishments were writing the first business plan, becoming a founding member of the Google product team, and leading the engineering team that launched *AdWords*, Google's proprietary method for tailoring Web ads to search terms." [DeBruicker 2006]. Google generated profit primarily from advertising through its *AdWords* program.

In 2000, against Page and Brin's initial opposition toward an advertising-funded search engine, Google began selling advertisements associated with search keywords. [94]

Google's rapid growth since its incorporation triggered further developments, acquisitions and also partnerships. In particular, there was another piece of software important for Google, "one they stumbled into when they bought Applied Semantics." [Altucher 2009] Sergey Brin has long been friends with Applied Semantics co-founder Gil Elbaz, Google pointed out. Interestingly, both Google and Applied Semantics had similar beginnings, as search engines with funky names launched in the late 1990s.

However, Applied Semantics moved more properly into the contextual advertisement, when it launched its AdSense program [Searchenginewatch 2003].

Google did not invent AdSense. Instead, they acquired the AdSense technology lock, stock and barrel – including the AdSense name – from Applied Semantics that Google purchased in April 2003 which was known as Oingo Inc. "Applied Semantics tooted their own Adsense horn well before Google picked them up." An Oingo press release mentioned AdSense already on December 4, 2000 and an application for a trademark on October 22, 1999 showed the name already [FirstMention].

AdSense [95] is a proprietary search algorithm that was based on word meanings and built upon an underlying lexicon called WordNet, which was developed by researchers at Princeton University. The AdSense program places paid listings into Web pages, by analyzing the content of those pages and then selecting ads that seem most appropriate [Altucher 2009].

"Applied Semantics is a proven innovator in semantic text processing and online advertising," said Sergey Brin, Google's co-founder and president of Technology. "This acquisition will enable Google to create new technologies that make online advertising more useful to users, publishers, and advertisers alike." [96]

When Google acquired Applied Semantics in April 2003 Susan Wojcicki, director of product management, explained more benefits. "Bringing on additional engineering support is also a key component." [Searchenginewatch 2003]. With the acquisition Google gained additionally employees already versed in the contextual ad space, an engineering team with its own unique ideas and methods of powering contextual ads plus a few existing partnerships [Sullivan 2003].

Furthermore, the acquisition of Applied Semantics gave Google new traffic for its paid listings, new strengths in the contextual advertising space, which Google entered and also potentially hurt then major Google-competitor Overture [SearchEngineWatch 2003]

Google's core business of selling search-based advertising, which allows companies to purchase ads tied to specific keyword searches, became one of the most lucrative and rapidly growing markets in the high-tech sector. A detailed description of Google's contextual ad business, Googlenomics, is described by Ley [2009].

Google's early years were during the Dot-Com Recession and the disruptive jump in revenues in 2001/2002 cannot be attributed straightforwardly. However, reference to the origins to Google's profits (search services versus advertisement contributions) provides the answer: It is the explosion based to a large extent on its AdSense approach (Figure I.159).

As a summary, given the original intention when founding the firm on focusing on Web search Google's later shift to contextual advertisement including the enforcement of the ad orientation and the related change of the business model, the role of Salar

Kamangar for developing AdWords and by taking over the firm Applied Semantics with its AdSense program, stumbling into that as Altucher [2009] described it, the explosive growth of Google (Figure I.159, Figure I.160) could not be expected considering founding intentions and orientations, but could not even expected looking at the development over the first two to three years.

What and How to Expect

For NTBFs there are two situations for which survival can be expected and the lowest level concerning growth:

- Non-growth or very low growth by intention, for instance, growth in line with or slightly above the inflation rate.

Basically, for *configurationally similar young companies* according to class properties outlined in Figure I.128 with similar growth aspirations (Figure I.122), one can assume that they face and solve similar developmental problems and will be exposed to similar *unbalanced* internal and/or external factors or events and will go through similar phases of decision-making and actions which will affect their growth states.

Therefore, we propose to generate expectations by comparisons (and call the related approach "*ex comparatio*ne"):

- Expectation of growth referring to the configurationally "nearest" case (growth parallel to the nearest similar, analogous or competitive firm).

The fundamental variable of the initial configuration of a startup for dealing with growth developments of a firm reduces basically to "measure" (ask for) the intention or aspiration of the entrepreneur whether he/she wants the startup to grow or is satisfied with "non-growth" (ch. 4.1).

However, it may well be that an original non-growth attitude may change over time, if the NTBF has the "potential" to grow (Figure I.97). The entrepreneur's intended mode of growth has an important consequence for expectations. If entrepreneurs do not rule out non-organic growth for the early development phases, there is, at best, only a qualitative statement of future growth possible: will (likely) survive and grow.

In a GST context, not knowing about the entrepreneur's growth intention or aspiration (Figure I.122) has an important consequence. Whenever there is a researcher's (or any third party's) assessment of a startup based on the initial configuration he/she is in the situation that his/her expectation actually induces a statement with conditionals (Figure I.2), a "reason for thinking that" ("… will grow significantly unless the founder(s) do not want to grow", "… unless a serious event/bracket prevents that.").

Following an *ex comparatione* approach the fundamental statements with regard to a firm's growth, for instance, in terms of revenues, would refer to "more, equal, less" or "higher growth, equal growth, less growth" compared with a "standard." This can be extended numerically to an "average growth," if a class of "nearest cases" is considered or even to semi-quantitative statements, if variables and parameters of the related configuration are considered and mapped. Correspondingly, concerning expectations referring to growth we may have "growth above average" or "growth below average." An average-related expectation of growth could refer to (statistical) averages of (non-VC controlled) NTBF types, such as TVT or HVT or both combined (Table I.1, Table I.70, Table I.71).

As the definition of an entrepreneurial configuration contains the firm's environment, particularly the market it operates in, statements about growth of a firm *ex comparatione* can become statement with conditions concerning the market or industry, respectively.

The related approach would assess the capability of a new firm to capture a share of a developing market and focus expectations on relations of the firm's growth to its markets using the following level of qualitative scores. Such an approach does not usually apply to disruptive innovations, as there may be no markets (exception: a Holy Grail; ch. 1.2.5.1). The assessment would have to focus on the firm's potential to create a new market (cf. also ch. 4.3.5.3 for assessments related to "growth factors").

- Growth less than the market/industry
- Growth with the market/industry
- Exceeding the growth of the market/industry
- Exceptional high growth ("super growth").

However, one must admit that expecting super-growth at or shortly after NTBF foundation is almost impossible. One situation one can think of will again refer to a Holy Grail of an industry (ch. 1.2.5.1).

The *ex comparatione* approach represents essentially a *measurement of order* against a fixed point or constellation, in particular, against a "*standard*" (as for personality disposition, Figure I.60). *Ex comparatione* relies on knowing the class properties of the "standard." And establishing the level of match between the new firm with the "standard" is a way of "*multidimensional case-based reasoning*" to generate expectations. For very close matches one can assume expectation to fit realities with acceptable reliability.

However, an external, scientific observer of a startup may miss a bracket (Figure I.136) for the standard or the target firm. Thus the observer may lack some information, fact or event to fully understand developments and outcomes. That means, the observer may not be aware that, for his/her understanding and explaining outcomes, rather than dealing with the "reason why", he/she is about to deal with "reason why thinking that" (Figure I.2).

A semi-quantitative *ex comparatione* approach referring to a "standard" is presented by The Venture Alliance (TVA) which targets *fundability of startups by investors* (Box I.17).

TVA addresses *fundability of firms and entrepreneurs* [TVA]. They intend to help entrepreneurs in their quest for funding by providing how an investor would view their offering (ch. 4.1) and what areas of their business need improvement.

TVA's methodology measures how well a company conforms to "fair and acceptable" funding criteria at a specific point in time and results in three products and associated services which represent a stepwise process to the Fundability Assessment.

1. TVA's Qualifier is a specialized test designed to function as a reverse business plan. It is occasionally offered for free but, has a variety of upgraded options. Its 65-question test covers the aspects of the business and will give applicants a clear look at what their business is providing to a potential investor and where it needs work.
   The Qualifier is the starting point for determining which companies make the Forbes list "America's Most Promising Companies" (AMPC; Box I.17).
2. The Vulnerability Feedback is a follow-up to the Qualifier. During this process, professional analysts at TVA work one-on-one with applicants to explain the results of the Qualifier, what the Qualifier results reveal about their business.
3. The Fundability Assessment which also serves for the Forbes AMPC list is a 1,000+ question questionnaire. It enables TVA to produce a 12-15 page report on the business that will enhance every aspect of the applicant's business plan as presented to investors.

TVA has a consistent methodology. It developed a scoring algorithm based on a vast range of variables that determine a company's potential – ultimately, its worth to investors defined by the fundability. Aggregation of measurements – conversion of a 1,200 point scoring system – leads to one "Fundability Score" and to twelve categories for macro interpretations to be visualized by a "TVA Radar Graph" (Figure I.162). The categories present criteria investors typically focus on as part of their due diligence.

The overall Fundability Assessment provides a view of the assessed business as it *currently* stands and includes, apart from a report,

- The numerical TVA Fundability Score (required to continue the funding process).
- Two of TVA's trademarked Radar Graphs showing 1) how the company compares to an "ideal" model related to type and stage of company and 2) how the company *compares* to other companies that are "similar" ("competing" and "comparable").

As the entrepreneurs' situations usually change fast TVA places an expiration date to every analysis they do of six months.

For those TVA clients whose score exceeds 800, help with funding sources may be available. Those whose score exceeds 1,000 will have a broader range of funding opportunities including, but not limited to, being published in one of more of TVA's TIGER (Top Investment Grade EntrepreneuR) lists, such as the Forbes list.

Hence, TVA's approach has many features of the *ex comparatione* approach described previously. Specifically, TVA takes input by the entrepreneur and determines which of its twenty-four models ("standards") is the most appropriate one to measure the new firm against. Each model examines and weighs twelve key areas to reach an assessment against the perceived "ideal" represented by the template model. The Radar Graph of Figure I.162 is a visualized example of this approach. In this regard it is similar to "measurement of personality disposition" (Figure I.60).

Apart from "Investment Value" all the other TVA categories, by sense not terms, will play also a role in the current approach to expectations based on firms' configurations (Table I.80; A.1.6).

Typical questions related to each of the twelve categories are given below with score points in parentheses (Total: 1070). A corresponding listing with largely explanatory character of the categories is given by Nelson [2009].

1. **Management Team:** Who are they? What's their track record? Etc. (170)
2. **Founder Commitment:** What have the founder's done to show they are serious? (20)
3. **Entrepreneurial Experience:** Does the team have a track record we can believe in? (60)
4. **Accomplishments:** What has the team done to show progress? (100)
5. **Market Opportunity:** Is there a real market for the products? (170)
6. **Marketing/Sales Strategy:** Do we believe in their value proposition? (110)
7. **Competition:** What is their "unfair" competitive advantage? (140)
8. **Intellectual Property:** Can they protect their advantage? (100)
9. **Directors & Advisors:** Who does this team listen to? (40)
10. **Financial Performance:** What have they done since founding the company? (100)
11. **Corporate Structure & Ownership:** Does ownership and structure make sense and help or hurt the company? (20)
12. **Investment Value:** Is this a good investment for what is being offered at the anticipated risk level? (40)

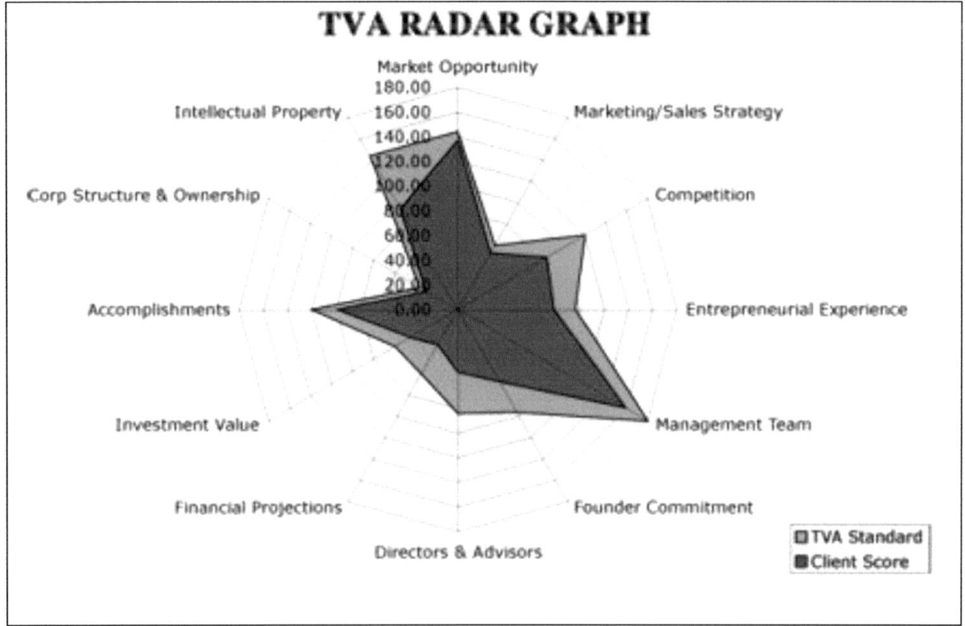

Figure I.162: The TVA Radar Graph to assess fundability of startups by investors – "The more (light) GREY you see (RED in the original), the greater the room for improvement." (Source: [TVA], courtesy of James W. Casparie).

A similar scorecard-like approach to assessing young firms is presented and illustrated for university "spin-out" companies (RBSUs) as compared with a group of companies arising from the community as a whole by De Coster [2004; 2005]. The assessment was based on key success criteria (Table I.79) and on information provided by the business plan of the new venture plus an interview with key personnel – plus secondary research. The scoring system focuses on criteria that have been identified in the research literature. The assessment methodology has been developed and operated over a period of four years by Clive Butler, HSBC Chair of Innovation at Brunel University, UK.

The methodology was primarily designed for the use of a major UK bank (HSBC) and now widely employed to assist in deciding whether *debt financing* is appropriate. Compared with VC investments, such funding tends to be early stage and the amounts relatively small. Such funding cannot bear the cost of in-depth due diligence procedures.

To improve the reproducibility of the assessment method the scoring method is seen to have the benefits that the assessments are more objective and there is less reliance on the individuals undertaking the assessment. A scorecard-type approach will usually also be used for business plan contests to judge all submitted business plans consistently.

To each key success criterion a verbal description is employed. The best fit determines the scoring number. The scales were developed by identifying the two endpoints presenting the extremes of expectations of successful development of an early stage NTBF (most negative and most positive expectation; for instance, scoring between 1 and 10). The mid point of the scale for the scale will represent a median state of an NTBF when seeking funding. The overall maximum achievable score represents the ideal business in their industry/stage of development [De Coster 2004; 2005].

Not all the criteria of this approach are considered to be of equal importance or independent of one another. Weightings (not shown) were assigned to each of the criteria to reflect the levels of importance.

Table I.79: Criteria for assessing RBSUs according to De Coster [2004; 2005].

No.	Criterion	Aim
1	Technological and Commercial Risk	To assess will it work
2	Level of Product Innovation	To assess the Unique Selling Proposition (USP)
3	Market Criteria – How it satisfies a market sector	To assess market demand
4	Market Criteria – Timeliness	To assess the market timeliness
5	Product Extensions – Longevity/ Repeat Orders	To assess whether it fits into a family of products to permit company establishment or development
6	Product Extensions – Family of Products	To assess the longevity of product or product line
7	Entrepreneurial Background	
8	Protecting Competitive Advantage – Sustainable	To assess the intellectual property rights

The proposed "*ex comparatio*ne" approach in this book to generate expectations about new firms' development referring to a configurationally "near" case will require a set of criteria to be matched similar to those discussed in this sub-chapter and simultaneously a match between comparable dynamically stable states of the two firms.

As indicated in Figure I.128 there will be several criteria to describe an entrepreneurial configuration by "major coordinates," which house a number of attributes ("specifications"). The last ones could provide values (usually numerals for scores) if a scoring procedure should be built. The gross criteria would be associated with weights. It must be admitted, however, that a scorecard approach is not in line with GST: It treats the significance of an individual factor as independent from all the other factors. The lacking systemuc effects would be partially mitigated by weighing the "major coordinates."

As an example of how to proceed with generating *ex comparatione* statements one can look at two *poster child companies* offering *nano-tools*, with *well growing markets ("scientific instruments"), proprietary technologies, real products and real customers.* Here, the earlier founded German firm WITec (Figure I.123) is the standard against which Cambridge Nanotech from the US will be matched.

Table I.80 provides the suggested configurational categories following largely the taxonomies discussed in Figure I.128. This makes the approach explicit and let emerge the close similarities of the two firms' configurations – simply by inspecting and comparing textual descriptions of related categories rather than performing a quantified approach based on scoring, such as the TVA Radar (Figure I.162).

Cambridge NanoTech founder and CEO Jill Becker turned her Harvard chemistry thesis research into a rapidly growing company whose revenue hit $17.6 million in 2010 (B.2). Due to strong roots in Atomic Deposition Layer (ALD) research, Cambridge NanoTech enjoys exceptional access to novel ALD applications and many great opportunities to nurture these applications to maturity.

The successful transition from serving academic customers to manufacturing customers has been in response to specific market needs and to developing products working closely with industry partners, research collaborators and key customers. Gross margins of the business are around 70 percent [Yang and Kiron 2010].

Atomic Deposition Layer originates in chemical nanotechnology and means a method of creating thin film materials by laying down a layer material a single atom's thickness at a time. ALD is an ideal *coating technology* because of its perfect, conformal, ultra-thin films that are scalable to large-area substrates. ALD simultaneously offers excellent thickness uniformity, film density, step coverage, interface quality, and low temperature processing, making ALD beneficial for both roll-to-roll flexible substrates and rigid substrates.

As both firms have very close configurations, apart from being sure about the firm's survival, Cambridge Nanotech would be assumed to further develop solidly and dynamically similar to WITec (Figure I.163). Furthermore, one can assume the development curve in terms of revenue to be described with the formula (Equation I.18) used for WITec's dynamically stable period (Figure I.156).

Though all this has turned out to be true from the start of the firm (2003) until 2008 and seemingly also until 2011 two *unexpected*, seriously negative brackets led Cambridge Nanotech to close doors by the end of 2012.

Table I.80: Matching entrepreneurial configurations of US Cambridge Nanotech against WITec GmbH to derive growth expectations for the former one.

Configurational Categories	**WITec GmbH (Germany; B.2)**	**Cambridge Nanotech, Inc. – CNT (US; B.2)**
Basic Firm Characteristics (Industry: Both in Nano-Tools, Scientific Instruments)		
Vision/Mission – Growth Strategy	"Focus Innovations," constantly introducing new technologies and a commitment to maintaining customer satisfaction through high-quality, flexible and innovative products; Organic growth (Box I.20; B.2)	Co-founder Jill Becker: "My fantasy was to marry science and business; sell a version of an ALD system, and evangelize this beautiful technology." [Yang and Kiron 2010] Organic growth
Firm Type	RBSU – university spin-out; direct commercialization of science (founded 1997)	RBSU – university spin-out; direct commercialization of science (founded 2003)
Legal Firm Form	Private, GmbH (LLC)	Private, Inc.
Special Externalities	Suffered early from Dot-Com Recession soon after start	
Initial Financing	Own resources and debts (bank loans)	Own resources, "bootstrapping" (plus loans?)
Further financing	Essentially cash flow	Essentially cash flow
Research (or R&D Intensity)	TVT (top value technology; Table I.1)	TVT (top value technology; Table I.1)
Networking	Ongoing contacts/coop with "home university"	Ongoing contacts/coop with "home university" and other universities
Founders' Personalities, Leadership and Corporate Culture		
Founders	Entrepreneurial Triple	Entrepreneurial Pair 1); 50:50 equity share in CNT

Motivation/Experience	Originally, team wanted to found an IT firm, grasped other opportunity; Father-in-law of one founder is self-employed (contributed also to initial funding)	Key founder Jill Becker always wanted to have own firm, marry science and business; father was serial entrepreneur [Yang and Kiron 2010] Entrepreneurial professor during doctoral thesis
Leadership Team: Managerial Roles and Execution	Personality and competency oriented distribution of management roles (Table I.41); Founders learning on demand; Strong business plan and financial planning [Koenen 2010]	Founder (Jill Becker) grasped first organizational skills managing university group, became Harvard Professor Roy Gordon's Chemistry lab *de facto* office manager; in charge of procurement, managing vendor relationships, tracking inventory and managing the lab's budget [Yang and Kiron 2010], gathered experienced team. By mid 2007 Becker hired Jay Ritter, a semiconductor industry veteran, as a manager
Technology, Innovation and Products/Services		
Technology	Enabling technology (nanotool); *nano-analytical* microscope systems (Raman, Atomic Force Microscope – AFM, Scanning Near-Field Optical Microscope – SNOM); with the first Confocal Raman Imaging system, WITec outperformed the existing Raman mapping techniques	Platform technology for *coatings*; Atomic Deposition Layer (ADL); creating *thin film* materials by laying down a layer material a single atom's thickness at a time.

Table I.80, continued.

Technology, Innovation and Products/Services		
Innovation Persistence	Steadily developing different types of instruments; first, combining SNOM and AFM in one single instrument; then, modular design allows the integration of Confocal Raman and Scanning Probe Microscopy (SPM) in one system; this innovation instigated a boom in combined Raman/SPM systems	Steadily developing different types of ADL systems and coatings devices
Technology Ownership (Own Development, Partnership, Licensed)	Own research and development at the University of Ulm, Germany	Developments based on Becker's dissertation; not known whether there is licensing with Harvard; presentation of Rogers and Mead [2011] induce some evidence for licensing
IP Protection	Own Patents	Patents
Regulatory Factors for Technology	None	None
Instrument Production	Utilizing university infrastructure to develop instruments including assembly of first instruments	Utilizing university infrastructure to develop instruments including assembly of first instruments; founder Jill Becker hand-assembled the first 13 units
Production	Own facilities	Contract manufacturers located in Massachusetts

Market and Opportunity		
First Customer Available (Startup Thrust Phase)	Sales of scientific (lab) instrument to (US) academics led to nano-tools startup: initial sales – research to research	Started with academic customer (sales to Stanford University); initial sales – research to research
Commercialization or Business Model	Research, development, production and sales / distribution of scientific instruments; user education and support; technical service; consulting concerning applications; analytical (nano)tools focusing on material and life sciences	Research, development, production and sales / distribution of scientific instruments and devices; user education and support; technical service; consulting concerning applications; providing coating services for a variety of materials the products are used in various applications, including optical, nanostructures, electronics, energy, biomedical, anti-corrosion, anti-stiction, chemical, etch resistance, internal tube liners, magnetic, roll to roll, semi and nanoelectronics, MEMS, and wear resistant.
Real Customers	From academic and industrial research, immediately ("real" products or services, e.g. no selling of licenses); international orientation	From academic and industrial research, immediately ("real" products or services, e.g. no selling of licenses); international orientation

Table I.80, continued.

Market Characteristics	Economic and policy-driven markets; global; almost recession proof; industry: scientific instruments (in a broad sense); analytics; Starting with customers primarily from academia; in 2005 the estimated SPM world market was €113 mio., Japan ca. 30%; in many cases, researchers order an SPM customized to their research purposes, and it is estimated that such a special SPM occupies 30% to 50% of the entire demand for SPMs	Economic and policy-driven markets; global; almost recession proof; industry: scientific instruments (in a broad sense) and surface coatings Almost none for the time of foundation! Starting with customers primarily from academia; currently: by 2012 observers expected the global ADL market to be $1 billion or 10% of the total market for deposition equipment. CNT estimated to have captured a 5-10% share of the current market for ADL equipment sold for R&D [Yang and Kiron 2010] – often customized products
Competition, Competitive Advantage	A few known "big guys" and a lot of "little guys;" however, small firms are in a better position to supply specialized instruments and accessories to the companies and research and educational institutions that work at the micro and nano levels; small firms look for lucrative niches (e.g. German JPK Instruments AG focuses on BioSMP, Figure I.141) Advantages: leadtime, Raman/SPM leader, innovation persistence	The top-five ALD equipment suppliers had an 81% share of the global ALD equipment market, middle tier of ALD equipment suppliers, most with less than $10 mil. CNT knows their most significant threats and competitors with the potential to become significant threats. [Yang and Kiron 2010]. Advantages: Innovation persistence

Marketing/Promotion and Customer Education	Customer workshops ("WITec Academy"); online webinars (home page); conferences, exhibitions and fairs	Online tutorials (home page); literature database (Knowledge Center); webinars, Customer Papers; conferences, exhibitions and fairs
Sales and Distribution	Subsidiaries: US, Singapore (US sales office 2002; Singapore sales office 2010); early on worldwide network of distributors (sales and customer support)	Subsidiary (UK 2009); early on worldwide network of distributors (sales and customer support)

1) One founder – Douwe Monsma – left CNT in 2008: company experienced serious friction between the co-founders, was resolved when in Oct. 2008, Monsma accepted a $5.4 mio. buy-out offer from Becker financed by a bank loan; Jill Becker further developed the firm – till its end.

Comparisons between CNT and WITec have to be made for equivalent dynamically stable states after the startup thrust phrase. However, immediately after the startup development of WITec was perturbed by the Dot-Com Recession. Therefore the <2004,2009> period may serve as a guide (Figure I.123, Figure I.156). The actual growth factor of Cambridge Nanotech for good fit is 0.51 (Figure I.163) for <2005,2008>. The introduction of the Phoenix instrument let a new bracket emerge in 2009 amd immediately after that two additional new instruments were introduced.

According to the bracket theory the data of 2010 comprise the occurrence of two new brackets. Indeed, the bracket is likely to be associated with launches of two new products which got a lift-off adding to revenues in 2010 (B.2) – despite the Great Recession. And also 2009 data should be affected by this recession. This situation of crowded brackets with negative effects being leveled off is similar to the situation of Microsoft depicted in Figure I.144.

Concerning technology there is plasma and thermal ALD. By mid of 2009 Cambridge Nanotech announced the launch of its first line of plasma ALD systems, the Fiji Series, but with the ability to conduct thermal ALD as well. According to Jill Becker "the Fiji is a breakthrough in ALD system design." "We built it from the ground up for the specific purpose of plasma ALD, but with the ability to conduct thermal ALD as well." Furthermore, in 2009 Cambridge Nanotech also launched its Tahiti system engineered for large-area manufacturing operations ensuring repeatable, exceptionally uniform, pinhole-free thin films on substrates.

Expectations (of the author) that CNT would follow a secure and successful development did not materialize due to frictions in the founder team in 2008. By the end of 2012 its auctioned assets and intellectual property were acquired by Ultratech, Inc.

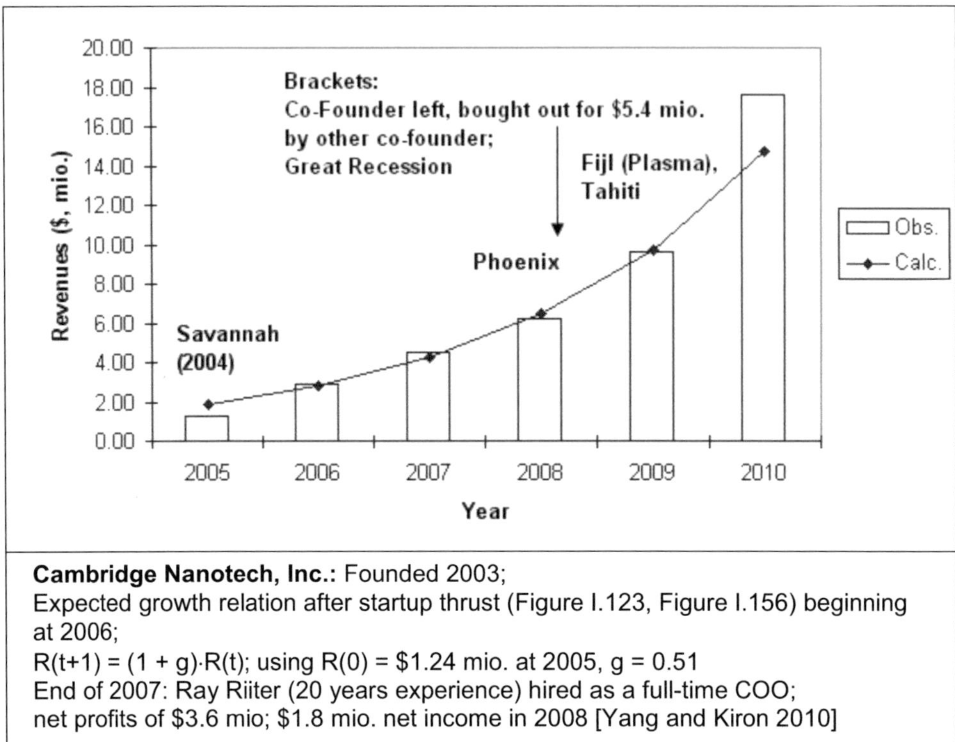

Figure I.163: Calculated and observed revenues of US Cambridge Nanotech and innovation persistence by launch of of instrument series.

Currently the broad analytical and life sciences instrumentation market enjoys considerable growth rates. It is reported, for instance, that the life sciences (academia and government) segment has a magnitude of $10 billion and annual growth rate of 8 percent, pharmaceutical and biotechnology is $9 billion with 5 percent growth rate, industrial (computers and semiconductors) has $5 billion and 4 percent growth rate and chemical & energy exhibits $2 billion and 5 percent growth [Thayer 2011].

Another technology startup with a configuration rather close to those discussed is the German firm Attocube Systems AG founded in 2001 (B.2). The then founders of the RBSU Attocube also had a sale of a university laboratory device to recognize the opportunity and after firm's foundation they grasped fast real customers.

Though founded during the Dot-Com Recession Attocube already made a profit in its first year of existence. It was financed essentially with own resources, used an enabling technology for its offerings to deliver international customers from academia and industry and proceeded with innovation persistence. The major differences to the above cases are that Attocube got early involvement of an angel investor who occupied a key role in the leadership team. Attocube is a private stock company.

Attocube's technology covers those of CNT ("moving atoms") and WITec (special microscopes). Attocube targets positioning of atoms which requires having them at rest with no fluctuations due to thermal energy – close to the absolute zero temperature 0 deg.K (which is -273.3 deg.C).

It provides the research market and industry with a reliable, compact, nano-precise and micro-precise positioning system that is capable of executing sample movement from the sub-nanometer to a centimeter range even in a large variety of extreme environments, such as ultra-high vacuum, extremely low temperatures (-273 deg.C) or at high magnetic fields. Simultaneously it provides tools to control and observe the positioning which are special microscopes allowing to operate under such extreme conditions.

In Figure I.164 the expectation to describe the growth of Attocube analytically by the same formula as used for WITec and Cambridge Nanotech with g between 0.35 and 0.51 is corroborated. Observed data for 2008 are incomplete, but, moreover, according to theory Attocube's dynamically stable state for the period <2003,2007> ended in 2008 due to the new front bracket of the "Wittenstein event."

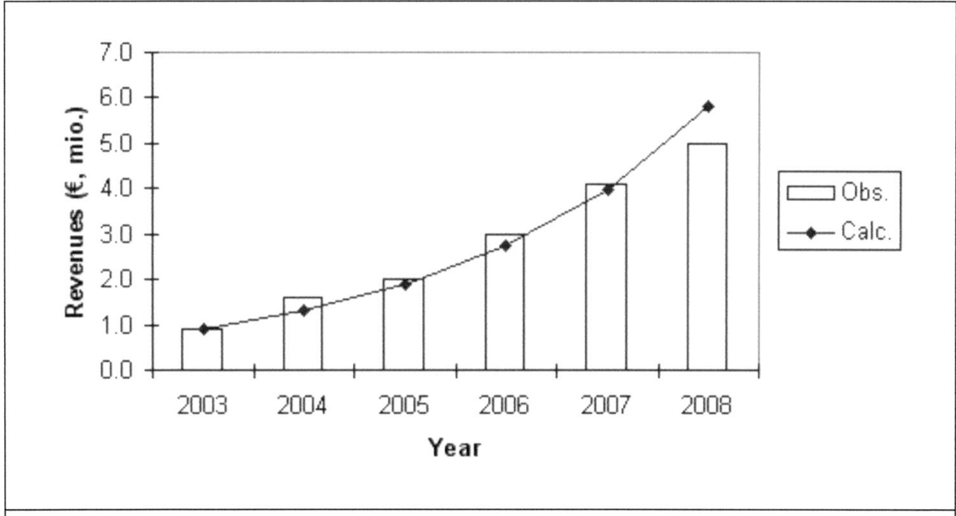

Attocube Systems AG

Founded 2001

Expected growth relation after startup thrust beginning at 2003;

R(t+1) = (1 + g)·R(t);

using R(0) = €0.9 mio. at 2003, g = 0.45; data for 2008 reported in September 2008

Productivity <2005,2007> almost constant, ca. €214,000 per employee

Figure I.164: Calculated and observed revenues of German Attocube Systems AG.

By September 2008 the German Hidden Champion Wittenstein AG acquired a 74 percent equity stake in Attocube keeping the two founders as executive managers and the angel investor as a member of the Supervisory Board.

Next we shall consider generating expectations *ex comparatione* for non-RBSUs, in particular, expectations for survival and growth related to *experience* of the founder(s) including expectations of *execution*. The emphasis will be on serial entrepreneurs and those with profound industry experience, relying on technical, commercial and managerial experience from previous jobs (Figure I.64).

Based on the fundamentals of US LED (Table I.81; B.2) one can expect the firm to successfully catch market share in its field of activity. Particularly strong arguments are the founder to be an experienced serial entrepreneur, initial financing being secured, detailed market knowledge and a favorable customer interface and distribution system and a focus on a strongly growing segment. Hence, there is little arguments why not expect US LED to grow at least with the market.

Table I.81: Key characteristics of a firm, here US LED Ltd., for generating expectations of firm growth.

Firm Type	Other NTBF (Table I.2, Figure I.128) (founded 2001)
Legal Firm Form	Private (Ltd., LLC); majority ownership/almost full control (Table I.74)
Initial Financing	Financed as a spin-off of US Signs
Research Intensity	HVT (high value technology; Table I.1)
Founder	Single entrepreneur (Ron Farmer)
Motivation/Experience	Serial entrepreneur; founded several companies but the most noteworthy are US Signs (founded in 1980) and US LED, both of which he still owns and participates in. Though relatively small with $22 million revenue, US Signs focused on neon lighting; ranks in the top 100 of the 30,000 sign companies in the US
Leadership//Management	As CEO for US LED , Ron Farmer helps manage the company and contributes to product development and sales

Technology and Application Focus	LEDs light emitting diodes; impact as a generic (and emerging) technology for the lighting area; US LED focuses primarily on "channel letter lighting" for illuminated signs for advertising and creating attention. Promoting LED relies on large cost savings as LED uses nearly 80 percent less electricity and lasts up to 16 years with no maintenance. Neon was and is still a strong illumination provider of channel letter lighting, but LEDs have already overtaken neon in 2008 in general-purpose channel letter signs in North America
Market and Opportunity	Economic market with strong components of ("green") attitudinal and policy-driven markets (Table I.15). In 2011 it was reported that the "high-brightness" LED market is forecast to grow to $19 billion in 2014 from last year's $5.3 billion, at an average annual rate of 29 percent. The fastest-growing segment is in displays and signs, which is predicted to grow 61 percent annually in the five years. It will also be the largest sector, accounting for 51 percent of the market, up from 36 percent this year. The general illumination segment is the next fastest-growing, at 45 percent a year, to $4.2 billion from $645 million during the period. Automotive applications are also projected to grow rapidly during the forecast period. LED lighting is one of the last analogue-to-digital transitions in the technical area. When so-called smart grids (computerized electricity distribution systems) are commercialized in 5 to 10 years' time, the energy-saving features will be magnified by demand-driven automatic dimming capabilities. In 2007 the worldwide market for "high-brightness" LEDs used in lighting applications reached $337 million, up from $205 million in 2006. *Haitz's Law*: the performance of an LED doubles every two years; this may explain the exponential growth of the LED lighting industry; every decade, the cost per lumen (unit of useful light emitted) falls by a factor of 10, the amount of light generated per LED package increases by a factor of 20, for a given wavelength (color) of light

Table I.81, continued.

Regulatory Factors for Technology	None
Attitudinal or Political Aspects Affecting Sales	Positive, energy efficiency (energy saving) aspects
Initial Customers, Real Customers	After its startup thrust phase US LED as a spin-off could rely on customers of US Signs. Generally, ca. 10-20 percent of customers are about retro-fitting neon with equivalent LED products and, thus, US LED could take advantage from existing customers of US Signs. LEDs are deployed in two areas of channel letter lighting; the first is *new signs* where the preferred lighting source is LEDs. The second area is *replacement for previously installed sign projects*, where for reasons of cost savings, the sign owners have opted to change out the neon with an LED system. US LED targets the home (US) market.
Innovation Persistence	LED as a platform technology; US LED started with lighting for signage, but expanded applications; it migrated into lighting for convenience store refrigeration, for parking lights, under canopy lights, industrial lighting, warehouse lighting, etc.; is working on the fluorescent tube lighting replacements for office lighting.
Competitive Advantage	US LED does not only provide LED for lighting, it also focuses on making it easy to install and easy to work with. US LED can rely on the widespread sales and distribution organization of US Signs, seems to be able to protect competitive advantage
Special Externalities	Ron Farmer: "…in 2009, we actually dropped back for the first time since we've been in business. We dropped back by about nine percent, but our piece count sales actually were up 50-percent."

According to the market data in Table I.81 one could expect US LED to grow with the market and have an average annual growth rate between 30 percent and 60 percent after its startup thrust phase, suggesting.to assume $g \approx 0.5$ in Equation I.18.

Actually it turned out US LED to grow in line with the maximum value estimated for the related markets (Figure I.165). It should be noted that according to the bracket theory calculated and observed values are restricted to the interval <2004,2008> stopping with the Great Recession. The nice fit for 2010 data is accidental; picking up previous growth occurs in a new firm's state!

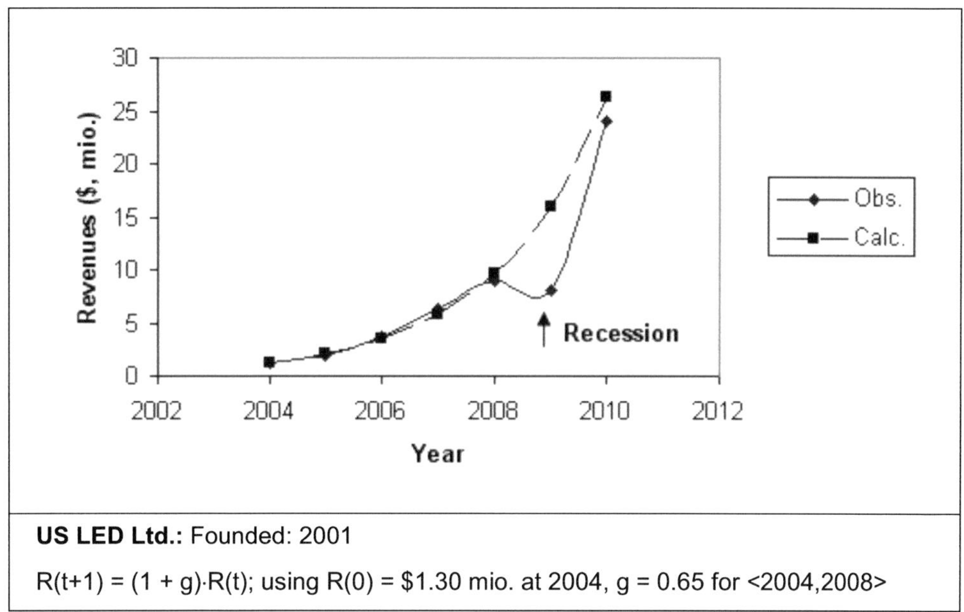

US LED Ltd.: Founded: 2001

R(t+1) = (1 + g)·R(t); using R(0) = $1.30 mio. at 2004, g = 0.65 for <2004,2008>

Figure I.165: Calculated and observed revenues of US LED.

US LED can serve as a model for expectable development of US Albeo Technologies Inc., founded 2004. Its firm type is "Academic Startup" (Table I.2) and is active in LED. The founder and members of the management team have 20+ years industrial experience, specifically also in the LED area ("veterans approach"). Founder and owner of the firm Jeff Bisberg spent more than 25 years developing and marketing innovative solid-state technologies (SST), with 20 years focused on light-emitting technologies, including developing novel organic LED, miniature LED print-heads, and an award winning laser printing system.

Albeo Technologies Inc. is a pure LED lighting company. It designs, manufactures and sells white LED lighting systems (intelligent lighting fixtures) for commercial and industrial indoor general lighting applications. Its products replace traditional fluorescent and high-intensity discharge lighting to decrease energy usage and maintenance.

Albeo targets the home (US) market. Albeo sells its lights for both new construction and retrofit projects, with the latter growing to 85 percent of sales. Albeo has designed a *flexible system* so that it is very easy to customize. Albeo delivers the "exact solution" to the clients ("solution provider"). It sees itself as a leader in the white-LED general-illumination fixture market.

Albeo's products and systems benefit from the "green" momentum as they reduce power consumption and maintenance for commercial and industrial facilities, and exhibits simultaneously environmental benefits (reduce carbon dioxide emissions). Its primary goal is to enable businesses to lower their total operating costs (total-cost-of ownership, TCO). The advantages over traditional lighting technologies, significant efficiency, lifetime and environmental advantages, mean also providing a short return on investment.

Expectation of Albeo's growth includes the applicability of Equation I.18, growth roughly comparable with that of US LED ($g \approx 0.7$) and an observable front bracket due to the Great Recession after a dynamically stable state for the <2005,2008> period. This can be seen in Figure I.166. The fit for 2010 must be viewed as accidental – not in line with theory.

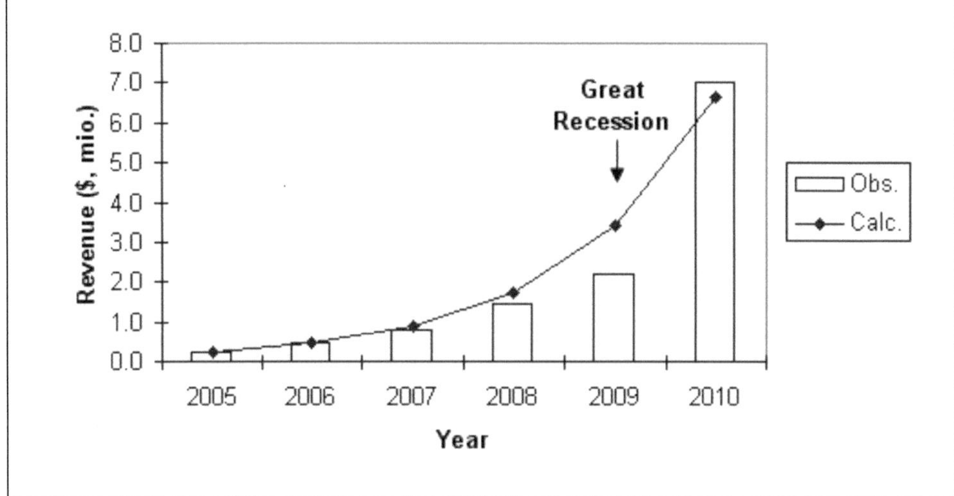

Albeo Technologies Inc.: Founded in 2004

$R(t+1) = (1 + g)\cdot R(t)$; using $R(0) = \$0.23$ mio. at 2005, $g = 0.96$ for <2005,2008>

Revenue data were obtained from values given for 2007 and 2010 in "The 2011 Inc. 5000 List of America's Fastest Growing Companies"; other data were derived from various statements of growth rates for the firm in the literature.

Figure I.166: Calculated and observed revenues of US Albeo Technologies.

Contrary to the situation of US LED the Great Recession shows up here only as a dip of revenue reflecting reduced continuous growth. The much larger absolute growth of Albeo Technologies ($g = 0.96$) is only partially surprising as for Albeo Technologies reference is made to the very early growth state of the firm which will exhibit generally much higher growth than later phases (Table I.71).

Actually, the previous discussions focused on the top 20 percent of NTBFs contributing most to job creation (Figure I.119) through related growth which makes sense from the point of view of economics of the country. We have dealt with "promising firms" or "promising NTBFs" (ch. 4.1, 4.3.1), respectively, which show good or high growth and the 33 – 46 percent of entrepreneurs who intended their firms to grow (Table I.63).

There is one fundamental issue of semi-quantitative expectations of growth for technical startups for the very early phase. Starting on a rather low level of revenues (200,000 – 300,000 dollars or euros per year) and catching just one big order valued double or triple the previous level of revenue in relation to the current revenue level would make any semi-quantitative expectation impossible. Moreover, the observed jump – similar to that in Figure I.137 – would mean a front bracket and a new firm state with different growth characteristics (Table I.77).

In line with US LED and Albeo Technologies Inc. privately-held MetroSpec Technology LLC (incorporated in Minnesota) also from the LED field represents a notable case to be discussed in the context of expectations utilizing the bracket model. "The 2011 Inc. 5000 List of America's Fastest Growing Companies" puts MetroSpec on #439 of the 2011 ranking with a "3-year growth" (2007, 2010 revenues) of 795 percent.

If one looks into the TECH{dot}MN company directory one will find [TECH{dot}MN]: "MetroSpec Technology manufactures FlexRad LED light sources exclusively *for light fixture manufacturers* for use in architectural lighting, streetlights, and signage. *The company has grown from a provider of engineering design and short-run production services to a high-volume manufacturer* of its patent-pending FlexRad LED technology." Three patent applications of MetroSpec from 2008/2009 have been converted to granted patents in 2011/2012.

Originally, MetroSpecc engineered products from concept through to production – involving and teaming with its customers at every step of the path to provide speed and quality in product development that can go hand in hand.

From Web-based job announcement ("Diversity Minnesota") of MetroSpec one learns that it "is the manufacturer of the "FlexRad™" brand of *high intensity* LED light systems. These systems are used by light fixture manufacturers to convert over their present incandescent and fluorescent products into highly reliable and lower operating cost products using LEDs."

Finally, on MetroSpec's Web home page one can read that its focus is on *high volume production* of *LED light circuits*. Utilizing its unique, patented FlexRad LED technology and working with "*customers* daily to fulfill all *their specific LED light circuit needs*." This allows providing superior LED light *circuits* and *services*. MetroSpec has an *out-*

standing market position by offering the only *high intensity flexible* LED light circuit which is *customizable to any shape and size*. It is not the typical low wattage flexible circuits which are limited in light output due to poor thermal performance. FlexRad is the exact opposite of these solutions; its flexible circuits are customizable to use 1W, 3W or even higher wattage LEDs. All this means the ability to perfect fit light fixtures.

Referring to the similarity with US LED as a manufacturer of LEDs and targeting also light fixture manufacturers for use in architectural lighting and signage we would expect for MetroSpec to exhibit growth in revenue in its LED segment in a comparable way as US LED (Figure I.165; g = 0.65) for the time period when it operated as a high volume LED circuit manufacturer.

However, considering the transition from an engineering firm to a manufacturer we would expect the appearance of a related bracket (Table I.76; "business model innovation") somewhere between 2001 (year of foundation) and 2010 showing up as either a jump or a steep increase in revenue extending over ca. two years.

This indeed is the case (Figure I.167). For Metrospec Technology one can see a jump in revenue from 2007 to 2008 which means a transition from the original "engineering state" into a new "manufacturer state." This coincides with the fact that MetroSpec reviewed its business plans in 2008 with SBDC consultants, "just as the company was migrating from a service engineering/design company to a manufacturing company." [SBDC 2012]

The growth of the "manufacturer state" for the period <2008,2011> with g = 0.45 is distinctly smaller than anticipated (g = 0.65), but can be accepted as a crude estimate on the basis of a correctly anticipated structure of the state equation. Based on comparisons with US LED and Albeo Technologies (Figure I.165, Figure I.166), however, it could not be expected MetroSpec to exhibit no or only a small recession dip at 2009.

The example of Metrospec Technology re-emphasizes the need of differentiating a *firm's states for growth periods* rather than focusing indiscriminately on "firm growth."

Metrospec Technology shares its path of migrating from a service engineering and design company to a manufacturing company with German PURPLAN GmbH (Box I.21) and British Quiet Revolution, Ltd. (Table I.60; B.2) which started as an engineering and design studio providing low carbon solutions for the urban environment.

As a conclusion the bracket model in the context of GST contributes to understanding and explanation of technology entrepreneurship. And even its (semi)quantitative use concerning expectations provides promising facets though, admittedly, it refers to only a relatively small number of case-based examples for corroboration.

As no surprise the bracket model is plagued by the fundamental issue of soft sciences. This is the occurrence of not expected or unexpectable events ("brackets") which impact a firm's development and the related observable indicator, respectively, significantly in a positive (Figure I.167) or a negative way (Figure I.163). In the last case the 2008/2009 bracket ("co-founder issue") as well as the coincident Great Recession do not even show up in the revenue curve.

As vision, mission, leaders and and culture (ch. 2.1.2.7; Figure I.120) provide the framework for attitudes, motivation, behavior and achievement of a new firm employee development (Figure I.121), organizational learning, resources and resource development, organization (particularly coordination) and systems of activities form the basis for sustainable development of the new venture (ch. 5.2).

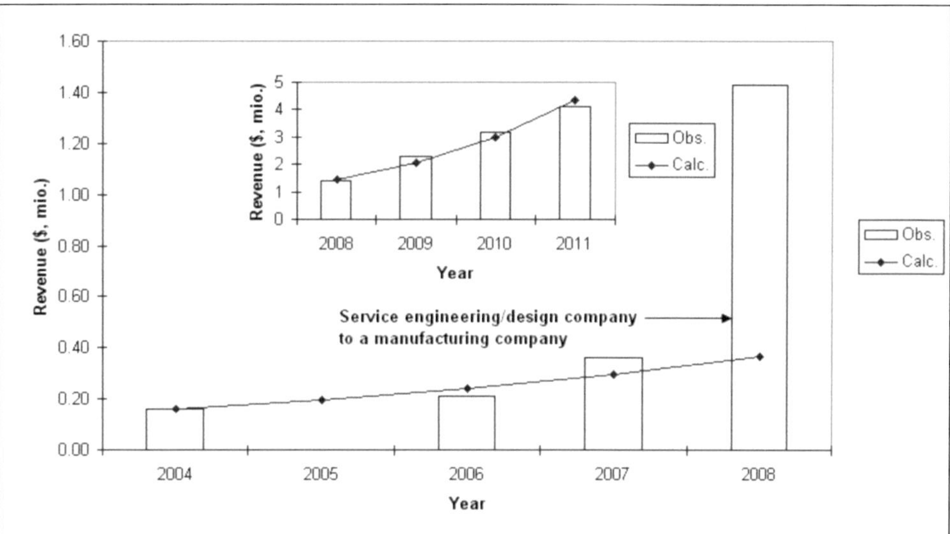

MetroSpec Technology LLC:

Founded in 2001

In 2008 the company was migrating from a service engineering and design company to a manufacturing company.

R(t+1) = (1 + g)·R(t); using R(0) = $0.16 mio. at 2004, g = 0.23 for <2004,2007>

R(t+1) = (1 + g)·R(t); using R(0) = $1.43 mio. at 2008, g = 0.45 for <2008,2011>

Revenue data were obtained from values given for 2007 and 2010 in "The 2011 Inc. 5000 List of America's Fastest Growing Companies," 2009 and 2006 data from Inc.'s 2010 list; 2006-2011 data were also obtained from PrivCo's database; 2004 data are from "Allbusiness" company profiles (no longer available on the Web).

Figure I.167: Calculated and observed revenues of US MetroSpec Technology.

5. PATHS OF TECHNOLOGY ENTREPRENEURSHIP

Paths are made by walking.
Franz Kafka

Wege entstehen dadurch, dass man sie geht.

By relying on General Systems Theory (GST) for technology entrepreneurship we are ultimately led to the question of the relation of founding and developing a technology venture and Systems Design or its partial revival in terms of Systems Thinking. In the context of GST the notion "design" is a process, an action, a verb not a noun. It is a protocol for *solving problems* and revealing new opportunities.

Systems Design can be seen as a *methodology of change* which proceeds essentially from the system *outward*, understanding the system and its relation to all other systems larger than it or interfacing with it (ch. 1.2.1).

In our context of Systems Design [Van Gigch 1974:2] there is currently in the US a new wave (and probably hype) with *Design Thinking* [Dziersk 2008; Wong 2009a, 2009b] which we shall critically consider for the framework of Systems Design.

In our context, apart from chance detection or serendipity, we have differentiated recognizing, identifying and discovering opportunities and ideas as different processes (ch. 3.2; Figure I.87).

As cited by Dziersk [2008] Herbert Simon, in the "Sciences of the Artificial" (MIT Press, 1969), has defined "*design*" as the "transformation of existing conditions into preferred ones" (p. 55). Design Thinking is, then, always linked to an "improved" future. Unlike critical thinking, which is a process of analysis and is associated with the "breaking down" of ideas, Design Thinking is a creative process based around the "building up" of ideas. Herbert Simon describes Systems Design by a seven steps process: Define, Research, Ideate, Prototype, Choose, Implement and Learn.

Designed systems necessarily always include *the goal of the designer* as the main driving instance. Corresponding artifacts are built as *purposeful systems* since the specification requires the dualism of *a priori* defining the components and their interactional relationship with the "environment" before the entire system starts to work. Systems Design for entrepreneurship focuses on *establishing a relation* of value creation between a new firm and its founders or owners, respectively, and its market(s) and customers as part of an all-embedding environment. Design will depend largely on *constraints*.

For the establishment of the relation there is a corollary:

> Go fast to market to learn about the market and as Kersting [2012] put it, "you don't know the market until the market knows you."

Systems Design is a creative process which for specific situations may be subjected to formalization. It means initiate and implement change in or through man-made things or entities which includes totally new things or entities. The focus is the *problem* at hand and the manner in which *problem-solving options* are considered, *ideas* are created and refined and *selections* are executed (Figure I.80, Figure I.87).

In our context we shall not consider design generally as a *prescribed* process. It is not consistently a rational process. It *may* begin with the identification and analysis of a problem or need and proceeds through a structured sequence in which information is researched and ideas explored and evaluated until the "optimum" or satisficing solution to the problem or need is devised.

In the context of social systems and entrepreneurship Systems Design is a *future-oriented,* partly disciplined inquiry. People engage in this inquiry in order to design a system that realizes their vision of the future, their own expectations, and the expectations of achieving them.

The *future environment of the system* has to be forecasted! If the design of the system has been set and is established, "*systems improvement* "refers to the process of ensuring that a system, or systems, perform according to expectations (ch. 1.2.1, ch. 3.2.1). Systems Design firms stand apart in their intention and willingness to cross the chasm to engage in and execute *continuously re-designing their business*. They do so with an eye to creating advances in both innovation and efficiency.

> The result of Systems Design has to pass the test of personal commitment – it requires conviction.

One question to be dealt with is how Systems Design and Systems Thinking affect foundation of new technology-based firms and interconnection of today's decisions and actions with past and future contextual factors. For the reasoning process underlying design one should be aware of whether following a line of "reasons why" versus "reasons for thinking that" (Figure I.2).

Systems Thinking will include past and future and that determines behavior in the present.

Past	**Present ("Today")**	**Future**
Analytic (Explain)	Context	Synthetic (Build))
Experiences, Perception		Imagination, Perceptiveness
Observations, Patterns		Opportunities, Possibilities
Extrapolations, Strategy		Trends, Stories
Achievements, Certainty		Expectations, Uncertainty

Systems Design questions assumptions on which old forms have been built or common or "standardized" recommendations on which new forms ("startups) have to be built. Correspondingly, the role of a system's *leader* is to *influence trends* rather than satisfying trends [Van Gigch 1974:9]. Intuition-oriented firms wax and wane with individual leaders.

In essence, Systems Design for value creation in terms of innovation or entrepreneurship requires bringing together two prevailing points of view on business today, analytically structured processes and intuitive originality.

Using GST means that instead of relying exclusively on analysis and deduction deeply ingrained into Western thinking we proceed also with synthesizing and being inductive (ch. APPROACH). Design Thinking means *intuitive thinking* – the art of subjectively knowing without reasoning and "*strategy logics*" (ch. 1.2.2, 2.1.2; Table I.33) being subjective logics. It relates to *perceptiveness*, a feeling of understanding.

According to mainstream approaches business organizations are dominated by *analytical thinking*. Strategy is based on rigorous, quantitative analysis. In this model analytical thinking harnesses the familiar Western forms of logic, deductive reasoning, to declare truths, facts and certainties about the (business) world. This model means mastery through formalized, continuously repeatable analytical processes. Judgment, bias and variation are the enemies. However, by sticking closely to the tried and true, organizations dominated by analytical thinking enjoy one very important advantage: they can *build size and scale*.

To summarize, neither rational reasoning nor intuition alone is enough. Using Systems Design and Design Thinking for entrepreneurship do not try to reconcile the two modes throughout the foundation of a new technology venture and its first dozen years of development. Both approaches will have different levels of significance for the various stages of firm development putting more emphasis on "designed" pre-start, foundation and early growth phases as illustrated in Figure I.1, but addressing particular sub-processes like financing in a rational and analytical manner.

As has been discussed (ch. 2.1.2.3) prevalence of one over the other mode of thought may be related to culture conditioned by higher education, in particular, for technology entrepreneurship by scientists versus engineers (Figure I.62) or scientists versus people for economics or business administration.

Engineers and application-oriented natural scientists tend to prefer rational, analytical thinking; they are "doers" (exception: software developers). They are used to plan, implement and execute "experiments" (or preparations) and measure outcomes. They are used to look for "recipes" and instructions and examples of how to do the experimental setups and what may go wrong. Hence, they appreciate "recipes" for founding and running a firm. And, therefore, writing a business plan matches often their education and culture.

The systems view provides some connections with "contingency theory" [124] which is a class of behavioral theory that claims that there is no best way to organize a corporation, to lead a company or to make decisions. Instead, the optimal course of action is contingent (dependent) upon the internal and external situation.

Key aspects close to GST are:

- Organizations are *open systems* that need careful management to satisfy and *balance internal needs and to adapt to environmental circumstances*. The design of an organization and its subsystems must "fit" with the environment.
- There is *no one best way of organizing*. The appropriate *form depends on the kind of task or environment* one is dealing with. An organizational/leadership style that is effective in some situations may not be successful in others.
- *Different types or species of organizations are needed in different types of environments*

In other words: The optimal organization/leadership style is contingent upon various internal and external constraints. And the needs of an organization are better satisfied when it is properly designed.

5.1 Firm's Foundation as Systems Design

Today perceptiveness is more important than analysis
Peter Drucker [Business Week 2005]

Who misses the first buttonhole
will not manage buttoning up.
Johann Wolfgang von Goethe

Wer das erste Knopfloch verfehlt,
kommt mit dem Zuknöpfen nicht zurande.

For technology entrepreneurship we have seen that for a significant amount of NTBF foundations started with customers and knowing or, at least, having a good estimate about the market(s) they may address. This means, the path to growth and success can rely to a large extent on planning (and having or writing business plans).

Apart from planned (goal-driven) approaches there are also *goal seeking approaches* based on a "seed of ideas," testing hypotheses and creating an options set of opportunities from open-ended business ideas, and also paths following *opportunistic adaptability* (ch. 1.2.1, 2.1.2.4).

Finally, we encountered situations where entrepreneurs created opportunities by creating new markets or even industries, in particular, in a "technology push" manner. This last aspect covers disruptive and sometimes discontinuous innovation and ultimately also intrapreneurship in existing large firms. All these situations can be assumed to take advantage from Systems Design and Design Thinking.

Opportunistic adaptability refers to a style of reasoning and behavior. It means to "adapt to unexpected circumstances in an opportunistic" fashion." It relates also to the fact that many ventures do not find success in their initial business idea. And, for instance, both the US firms YouTube and Yelp, Inc. learned a valuable lesson from

PayPal: The first idea is not always the best. YouTube started as a video dating play. After an aborted start as an email recommendation service Yelp, Inc. is a company that operates yelp.com, a social networking user review and local search Web site (A.1.7).

Furthermore, entrepreneurs with limited funds cannot afford to spend much time and efforts to prior research and planning. Sketchy planning and high uncertainty requires adapting to the many unanticipated problems and opportunities. They "cannot afford to sacrifice short-term cash for long-term profits" [Bhidé 2000:18] – and follow opportunistic adaptability.

Moreover, there is no point in engaging in elaborate strategic planning which is based solely on what the entrepreneur would wish to happen in a more or less ideal setting. It is necessary to adapt aspirations to what is achievable in terms of access to resources and to build in a capacity for opportunistic adaptability as external circumstances change. And "achievable" means the *goals and preferences* of the entrepreneur for financial (ch. Box I.20) and other resources are distinct variables before implementing and testing a strategy.

Preference for financial sources by entrepreneurs has been treated by the "Pecking Order Theory" (ch. 4.2.2). Preferences of an entrepreneur (Box I.20), for instance, rejecting venture capital, is part of the entrepreneur's disposition and thus part of the *decision environment* (Figure I.111).

Examples of the opportunistic adaptability are also observed for animals. There are birds which are always looking for new sources of food. They observe other birds eating and try it for themselves. The coyote's opportunistic adaptability emerges if one examines the coyote's feeding habits. Coyotes in the US, or foxes in Europe which are settling in large cities, will eat almost anything they can chew – demonstrating the enormous flexibility and opportunistic adaptability of biological systems.

For entrepreneurship and innovation aspects of Systems Design (ch. 1.2.1) for open (human-activity) systems key concepts or drivers, respectively, were distributed across the previous text in various contexts. These and some further concepts for firm's foundation shall be summarized and specified.

The System

The issue of systems and, hence, entrepreneurship, will be in how far to "compose" or "organize" the "entrepreneur system" into larger systems (Figure I.13, Figure I.16).

For composing a firm as a system one needs (ch. 1.2.1)

> to recognize or identify, respectively, "what is connected with what," "with what intensity/strength," and "what follows after what ("order," "function")?"

The system's (new firm's) environment fundamentally also includes the position of a new firm in a value system (Figure I.7), if it already exists.

For Systems Design the *The Environment-Modification Principle of General Systems Theory* states (ch. 1.2.1 – for instance, for technology push approaches): To survive, systems have to choose between two main strategies. One is to *adapt to the environment*, the other is to *change it*. And dealing with the future is in line with the old saying of Samuel Johnson (1709-1784), Peter Drucker (1909-2005) and Dennis Gabor (1900-1979) (the mottos of ch. 1.2).

Transferring these chapter mottos into a Systems Design paradigm for entrepreneurship and managing research and innovation they should read:

- Assess the businesses and current competencies;
- Find out how business and technology has changed recently and find the key factors for these changes;
- Get a sense and good feeling how things could change in the future;
- Develop a system that meets current needs and opportunities, identifies and responds to threats and positions the company for the future and supports future decision-making.

For composing systems and their stability note *The Variety-Adaptability Principle* (ch. 1.2.1): *Systemic variety enhances stability by increasing adaptability.* Variability refers also to paths to goal achievement for NTBFs and opportunistic adaptability versus goal persistence (Figure I.122).

In technology-based businesses dealing with the future means an emphasis on technology and commercial intelligence, which is "foreknowledge." A detailed overview on "intelligence" and intelligence systems is given by Runge [2006:520-531; 978-835; 934-970]. The systems approach is invariably bound to the foreknowledge part of intelligence (ch. 1.2.3; Box I.17), which has to position "technology forecasting," foresight, prognosis, technology trends, scenarios (Box I.19) etc. into appropriate contexts.

Accordingly, what distinguishes leaders from followers and laggards is the ability to have a unique imagination of what could be. Leadership is not to benchmark the competition and imitate its methods and offerings, but to develop an independent point of view about tomorrow's opportunities and devise a strategic architecture with which to implement and exploit them.

> In our world with an always increasing complexity and pace of change imagination and perceptiveness – a feeling of understanding and the "known unknowns" – become key for Systems Design.

Technology entrepreneurship occurs often in cross-industries environments with people from various scientific or engineering disciplines. In this book, in particular, the role of chemistry and material science and the chemical industry for co-evolutions with other industries were emphasized.

Systems Design suggests to put teams in situations where they are forced to synthesize meaningful opportunities out of incomplete and highly subjective information about the world (environment) and the future needs of end-users. Multi-disciplinary teams or persons interwoven in a multi-disciplinary environment are particularly appropriate to reveal opportunities across technologies or industries. Of special importance are *boundary spanners,* also called *gatekeepers* (ch. 1.2.3; Figure I.20).

The means, for firm founders having or hiring so-called *T-shaped individuals* may pay off. These tend to be professional in one area, but are skilled in many other areas. They are highly intuitive. The advantages of *"diverse experiences" for problem-solving* (ch. 3.2, Figure I.86) result from "mental restructurings," as the problem is only solved after someone asks a completely new kind of question.

According to GST decision-making (ch. 4.2.2) will occur in a *decision environment* (Figure I.111).

Business Ideas

As characterized previously (ch. 3.1) for entrepreneurship a "**business idea**" is actionable and associated with the following *processual* features

- A business idea acts as the basis for detailed considerations and targeted inquiries concerning a related commercial *opportunity* and a decision of firm's foundation and is associated with *expectation.*
- A business idea is associated with *implementation – execution.* For a startup the assessment of the significance or *value* of the idea is coupled to execution – with a disproportionate ratio which puts high weight on execution.
- Execution for the realm of technology entrepreneurship is coupled with 1) associated *contextual* insight in and options for *applications* in "real life" and 2) characteristics of the commercial *opportunity.*

Implementation and execution of a business idea means *testing a hypothesis.* It is in the same category as "strategy" as viewed generically in this book (ch. 1.2.1) and also the business model as a hypothesis to be tested (ch. 1.1.1.1). The "thrust approach" of exploratory research and technology explorations in the laboratory of large firms [Runge 2006:608, 726-727] are also testing hypotheses (thrust: research directions essentially given, more specifications needed).

Firms' foundation processes of technology ventures were specifically outlined for "bootstrapping startups" (ch. 4.3.3.1). Though accounting for only a small proportion of NTBFs (around 5 percent) VC-based NTBFs are at the center of MBA-style approaches in standard (text)books of technology entrepreneurship. Foundation processes and early phases of *VC-based NTBFs* can be seen as intermediate between bootstrapping and innovation and intrapreneurship in large firms (ch. 4.3.5, Figure I.134).

As described by Dorf and Byers [2007] foundation of a VC-based NTBF focuses on a business plan, its financial plan emphasizing financing by venture capitalists and the firm's IPO [Dorf and Byers 2007:379-399, 414-419, 428-436]. After "presenting the plan and negotiating the deal" details of executing the business plan are given [Dorf and Byers 2007:456-465].

If strictly applicable, firm's foundation and the early phase of a new firm based on principles of Systems Design can be boiled down to a six steps process (modifying the suggestion of Herbert Simon and Dziersk [2008]): Define the Problem; Create and Consider an Options Set, Refine and Test Selected Directions; Choose, Implement and Execute; Learn. And firm's foundation will have to ask: What are the critical success factors (CSFs) for a related "problem-solving project"?

Define the Problem

In our context an idea can be an expression of a hypothesis for solving a problem (Figure I.80, Figure I.87) and a problem is associated with an explicit or implicit need. Problems can be framed by questions, such as

- Whose problems shall be solved?
- Are there generics in the problem to be solved?
- Is the offering a complete solution for the customer or only part of the solution?
- What is the level of urgency to solve the problem?
- Can the addressees, the customers, state the problem explicitly without bias (specifications, latent needs)?

Talking to your end-users will bring fruitful ideas for later design as a response to a problem. Observation can discern what people really do as opposed to what you are told that they do. Getting into the field and involving oneself in the process and offering is fundamental. Cross-functional insight into each problem by varied perspectives as well as constant and relentless questioning – target the right problem to solve, and then to frame the problem – is a way that invites creative solutions.

Asking the right questions is fundamental to understand and solve a problem. Problem-solving by reference to existing information and knowledge has to

- review the history of the issue; revealing or remembering any existing obstacles and
- collecting examples of attempts to solve the same issue (Figure I.80).

Related issues of information overload for problem-solving and asking the right questions are described by Runge [2006:520].

Accessing information and the number of information resources has increased dramatically, and the information changes more rapidly than our ability to acquire or master it. And accordingly it is said that "The educated person used to be the one who could find information. Now, with a flood of data available, the educated mind is not the one that can master the facts, but the one able to ask the '*winnowing question*'." (Emphasis added).

Concerning the efforts that have to be put into *searching for the right questions* one can refer to Einstein: "Einstein once remarked that if he were to be killed and had only one hour to figure out how to save his life, he would devote the first 55 minutes of that hour to searching for the right question. Once he had that question ... finding the answer would take only five minutes."

Problem solving and generating related (business) ideas is also seen as an issue of *creativity* (Table I.9; ch. 2.1.2.2) which can be related essentially to recognize and interrelate that which is not obvious – whether it is chunks of data and information, patterns of entities, similarities, associations, metaphors etc., such as mavericity (ch. 2.1.2.2).

Cognition will be used to deal with revealing solutions of problems and opportunities as will rational processes. We tend to order our experience and perception in a manner that is regular, orderly, symmetric and simple within borders. Boundaries differentiate essentially thinking and perception in Western from Eastern cultures.

Concerning Western reductionism as a basis for recognition and reasoning Aristotle wrote about Pythagoreans that for them *emptiness* serves to divide things and define their boundaries. Crossing the borders of a (real or conceptual) system for problem-solving represents a fundamental barrier for cognition.

The nine dot puzzle of Figure I.81 introduced as stepping out of perceived boundaries of a system defined by the nine dots can be more generally interpreted comparing Western and Eastern thinking concerning *emptiness* (or "void") serving to divide entities and define their *boundaries*.

Referring to Innocentive (ch. 3.2; Figure I.86) and the posted problems people created a solution for it was found that often "problem solvers" were most effective at the margins of their own fields of expertise ("outsider thinking"), not inside their field of expertise and thus avoiding to run into the same stumbling blocks that held back their more expert peers by their "*cognitive frameworks*" (ch. 3.2). Furthermore, there is a trap of solving a problem the same way every time. Especially when successful results are produced and time is short.

Many times we are not aware of the filters we may be burdened with when we create answers to problems. In this stage opportunities may appear. The trick is to recognize them as opportunities. Multiple perspectives ("diverse experiences") and teamwork are crucial to overcome the barriers.

Create and Consider an Options Set

When we have needs to satisfy or a problem to solve decision-making is required (ch. 4.2.2). Systems or Design Thinking requires that no matter how obvious the solution may seem, many solutions be created for consideration. Generate as many ideas as possible to serve identified needs. Ideation may be used to create an options set for problem-solving or opportunities (Figure I.80; Box I.13) as well as researching ("exploration") in the laboratory or experimenting in a workshop. A model for the *search process* by which the mind generates alternatives is given in Figure I.113.

Seeking opportunities for Systems Design may rely on "technology intelligence" and "commercial intelligence" to generate *a "choice set"* of alternative market opportunities, an *"opportunity landscape."* Assessing the choice set to reveal the most promising ones may induce consideration of *opportunity cost* (ch. 1.2.5.2).

For the "choice set"

- Reserve judgment (ch. 4.2.2; Box I.17) and maintain neutrality.
- Seek feedback from a diverse group of people; include your *end-users* (customers).

Refine and Test Selected Directions

A handful of promising results need to be embraced and nurtured for hatching protected from the idea-killers (Table I.46) of previous experience and cognitive frameworks. On the other hand, decision-making is often constrained because the *time and effort* to gain information or identify alternatives are limited. The *time constraint* simply means that a decision must be made by a certain time ("*urgency of decision*") utilizing accessible resources (ch. 4.2.2).

Design Thinking requires allowing the potential of business ideas to be realized by creating an environment conducive to growth and experimentation, and the making of mistakes in order to achieve out of the ordinary results. This means not only refining and expanding ideas but also combining promising ideas.

In terms of Systems Thinking evaluation of outcomes means:

- The envisioned outcomes of a decision are evaluated for fit with the objectives, but also more possible positive and negative consequences (Figure I.111).
- The decisive actions are taken, and additional actions may be taken to prevent any adverse consequences from becoming problems and starting both problem analysis and decision-making all over again.

"Prototypes of solutions" representing hypotheses shall be tested to find the "best" path to problem-solving – similar to the "thrust" approach to exploratory research (ch. 2.2).

> Test hypotheses! – "If they are confirmed, you have learned something; if they have failed, you also have learned something." [Kersting 2012]

The importance of testing ideas on customers using rough-and-ready prototypes has been emphasized also based on the following argument: Customers "will be more willing to give honest opinions on something that is clearly an early-stage mock-up than on something that looks like the finished product." [The Economist 2011]

Furthermore, R. Hoffman, the co-founder of LinkedIn (B.2), pointed out that entrepreneurs should take "intelligent risks" comprising how he or she sees something others do not and emphasized the importance of holding and testing contrarian views.

Choose, Implement and Execute

After enough paths have been traveled to expect success (Figure I.122) it is the time to decide which one to follow, implement the design and finally commit resources to achieve the early objectives. The by-product of the process is often other unique ideas and strategies that are tangential to the initial objective. Key activities have to be balanced for a fledgling NTBF (Figure I.131).

- Plan processes, tasks and roles; make task descriptions; determine necessary resources (cf. overt strategy logics, Table I.33)
- Assign tasks/roles.

Feasibility and workability of designs is an important distinguishing feature of a business design. Designers repeatedly ask, "Does it work?" and "Does it work better than what we have now?" According to Figure I.87 feasibility interconnects the revealed opportunity options set and opportunity evaluation and the related offering option(s). Feasibility does not only refer to have or to access, respectively, resources, but should also include first contacts with (potential) customers testing the offering(s) as outlined for creating and considering the options set.

Feasibility relates the designed firm to the addressees of its offerings for problem-solving. Operational feasibility is a measure, an expectation, of how well a proposed system solves the problems, and takes advantage of the opportunities identified during scope definition and how it satisfies the requirements identified in the requirements analysis phase of system development. [122]

> As a final assessment let outsiders know why this will work and be able to support what you believe in!

"Implementation is the utilization or adoption of change," the actions of accomplishing some goal or executing some order where "utilizing" will refer to find a practical or effective use for something, especially to find a profitable or practical use for (ch. 1.2.1).

The success of implementation have been found to depend on the extent to which goals are "operational," that is (cf. Figure I.10), when a means of testing actions is perceived to relate a particular goal or criterion with possible courses of action. For implementation of change to occur it has to be timely ("Window of Opportunity"; Figure I.4, Figure I.92). Execution comprises

- Determine if the solution of the problem met its goals.
- Gather feedback from the customers.
- Discuss what could be improved.
- Measure goal achievement by collecting relevant data. (Kersting [2012]: "Know your numbers"; ch. 4.3.3.1).

Failures and Learning

As we have seen (false starts of 3M or NanoScape AG; ch. 4.3.2) failure to meet initial goals is a poor indicator of success to come. "Success and failure are not polar opposites: you often need to endure the second to enjoy the first. Failure can indeed be a better teacher than success. It can also be a sign of creativity." "One must do is distinguish between productive and unproductive failures." [The Economist 2011]

On the other hand, poor preparation of firm's foundation (including poor design, implementation and execution) leading to the startup's disaster is definitely an unproductive failure if insights of the founder(s) were lacking that it would be better not to start, that their proper self-assessment would reveal that they do not have what is needed to get the job done or that founders are not in the position to convert their business idea into a functional company design.

Additionally, a further challenge is notable, *getting a good handle on the competitive environment that the business will face*. Many startups make the major mistake of dismissing the competition.

When to stop or terminate a venture may also be important as knowing when to start. The particular aspect of **sunk cost** [123] may enter decision-making. Sunk cost is cost that has already been incurred and thus cannot be recovered. Sunk costs are independent of any event that may occur in the future and cannot be affected by any present or future decision. Sunk costs greatly affect actors' decisions, because many humans are loss-averse and thus normally act irrationally when making economic decisions.

When making business or investment decisions, organizations typically look at the future costs that they may incur, by following a certain strategy (logics) and plan. For instance, investment in a plant for manufacturing a particular material by a process which is entirely dependent on just one particular intermediate (irreversibly committing resources) may become sunk cost if, for instance, a legal regulation prohibits the use of that intermediate. It represents a total loss of the original expenditure. Similarly sunk cost may emerge if a firm exits a particular business (ch. 1.2.5.3).

Generally a more tolerant attitude to failure can help entrepreneurs and companies to avoid destruction. Here national cultural difference are important (ch. 2.1.2.3). Among the countries by many metrics the US actually has a bad education system. Why is it that the US has a superb entrepreneurship and innovation track record?

It is because, the author believes, culturally: The US has a lot of people who have no fear of failure and failure is not "punished" by society. The advantage of the US culture is that people will rebound from failure. And as from a nation's or a firm's point of view entrepreneurship and innovation is a numbers game (ch. 4.1, last paragraph) this bodes well for the US with overall very many firms' foundations (and very many failures).

Concerning failure there is the old saying "Fail early, fail often" [Kersting 2012], but there is no point in failing fast if you fail to learn from your mistakes [The Economist 2011] – "fail early, fail often – and learn." "Failing fast" is also a principle of innovation projects of large firms – to minimize innovation cost [Runge 2006:787].

> Though "failure is the opportunity to begin again, more intelligently" (Henry Ford) it is better to do it right from the start of the firm (ch. 4.3.2).

Learning: occurs on the individual or the group level (ch. 1.2.1; Table I.7). In the context of Systems Design *organizational learning*, which is a systems characteristic, must be emphasized. Conditions under which organizational learning occurs, its five disciplines and learning for the future, are relevant (ch. 1.2.3). As a venture evolves the entrepreneurs will transform resources, largely through organizational learning, into valuable and ideally unique organizational resources.

Learning business activities and processes during firm evolvement occurs on the individual and the organizational level. For instance, J. Koenen, a co-founder of German WITec GmbH (Table I.41) learned management and administration on the job and on demand (ch. 2.1.2.4) and N. Fertig and the team of Nanion Technologies GmbH learned professional project management for developing a new very complex product for growth after having launched already successfully products.

Hence, one can differentiate resources and *skills* which are related to learning. An entrepreneur who wants to be successful must learn to develop his or her skill set. Once learned, the skill set is something the entrepreneur can use routinely.

5.2 The Startups' Evolvements for Growth

The next question concerning the role of Systems Design after firm's foundation concerns to look into the process of firm's growth in more detail (Figure I.130).

So far some systemic processes associated with foundation and growth (or decline) were described. These include, for instance, the phased formation of the founder team by self-reinforcement (Figure I.68, Figure I.70) or decline and "death" of startups by reinforcing sub-processes leading to a non-viable financial state (Figure I.114) or setting up corporate culture by the founder or founder team, respectively, by the founder's leadership by self-replication and leading by example (Figure I.120). The leadership team and corporate culture represent two important resources for a firm's development.

For explaining or understanding, respectively, evolvement for growth of a micro NTBF (1-9 employees) to a small (10-49 employees) and then to a mid-sized company (50-249 employees; Table I.4) one can assume that the starting resource endowments are qualitatively not so different for "similar" ventures, but the *resource development* pathway (Figure I.130, Figure I.131) will be affected to a considerable extent by systemic principles embedded in the concept of the "initial configuration" of the NTBF (ch. 4.3.2; ch. 4.3.3). The initial configuration provides a way to differentiate evolvement in terms of relevant variables and parameters.

In the framework of the bracket model (ch. 4.3.5) venture development is described as a sequence of a firm's dynamically stable states, interrupted by transition states induced by internal or external events (Table I.76), such as setbacks, challenges or particular management tasks, which may lead to new growth states. The bracket model implicitly emphasizes *adaptability* of an NTBF/SME and related responsiveness to any effect that influences its competitive position as an important systemic resource.

Resources (and input) are fundamental for the *conversion processes* ("throughput") of the firm by which elements in the system change state ("systems dynamics", Figure I.5). This brings in the resource-based view (RBV) of venture growth (ch. 4.3.3) for the early stage of startups.

> The essence of the resource-based perspective is a *knowledge-based view* and in the context of GST this will include foreknowledge which overall means *intelligence* perspectives (ch. 1.2.3) which are important for adaptability.

While it is important to know the industry sector of a business to assess which type of market (Table I.15) and growth factors come into play even within sectors there will be differences in the type (Table I.12) and stage of technology and processes used. For instance, in biotechnology and the related biofuels sector (A.1.1), at the greatest level of industry and market sector detail, there remains a great deal of heterogeneity.

The timing for entrepreneurs to enter a particular life-cycle stage of the industry (ch. 1.2.4; Figure I.32) has revealed the significance of the markets' birth and emergence phases for SAP (Figure I.143), Microsoft (Figure I.144), Cisco (Figure I.145) and Google (Figure I.159).

The entrepreneur is the primary resource, and his or her expectations about the future of the venture are central to its strategic direction. Each entrepreneur begins with a personal resource endowment at the start of the resource building process. The first resources (for instance, education, experience, credibility and reputation, network contacts, knowledge of the industry etc.) exist in the entrepreneur rather than the new venture.

According to Bhidé [2000:17] the founders' capacity to differentiate their offerings through their personal efforts seem to be an important reason for profitability (cf. CEOs' knowledge of customers with Hidden Champions; ch. 4.1.1). The entrepreneur, rather than a product or technology, represents the source of the startup's profits – and the firm's productivity and performance (ch. 4.3.3).

However, there are a number of technology startups, RBSUs, whose foundation is induced because the to-be founders can start already with a product and customers (to name some: ChemCon GmbH, WITec GmbH, Nanion Technologies GmbH, Cambridge Nanotech, Inc., all in B.2; IoLiTec GmbH (A.1.5) and PURPLAN GmbH – Box I.21).

Founding a firm means a decision which is often made after consulting with other people and taking also *advice* from others into account. In particular, firm's foundation of new technology ventures is often associated with the establishment of an *Advisory Board*. The Advisory Board does not only provide advice and consulting to the leadership team of the new firm, but, depending on "names," may add reputation and credibility for the leadership team which may affect funding and hiring talented people.

The Advisory Board or other stakeholders of the new firm and their advices are part of the decision-environment of the firms' founders and in this way represent also a systemic aspect (Figure I.111; ch. 1.2.6.2, ch. 2.1.2.4).

Many technology entrepreneurs begin with a rather complex and often instrumental human and social capital that they have developed in another professional enterprise or work setting. They achieved industrial experience or management experience in large firms or gained experience of funding by public sources in public research institutions or acted as a serial entrepreneur (Figure I.64). They leverage these resources to acquire financial and physical resources, and to hire and develop qualified individual personnel.

Notably, serial entrepreneurs use often a totally different resource base when they start another firm (for instance, Klaas Kersing with Gameforge – B.2; Lars Hinrichs with Xing – B.2; Reid Hoffman with LinkedIn (B.2)) and can additionally utilize extended networked resources (cf. the PayPal Mafia – A.1.7).

> Ventures that are unable to transition from reliance on the individual resources of the founder(s) and extend those to organizational resources will be constrained in growth. To make the transition happen a strategic resource development plan may involve creating systems and routines, defining policies by which people work, and creating incentive systems for employees [Brush et al. 2001].

The process of building an initial resource base from scratch and transforming it for growth is a complex task – and even more for technology entrepreneurship. And its description is too – due to the broad variety of existing resources and usable resources by the startups – depending on their proper classification according to many dimensions (Table I.74, Figure I.128) and the potential to combine these dimensions in a meaningful way.

There are fundamental types of technology startups referring to taxonomies of financial structures of technical startups' initial architectures in terms of ownership and control by founders versus VC-based firms (Table I.74), RBSUs and Other (academic) NTBFs with/without large-scale production, the industry the startup is operating in (Figure I.128), networking (Figure I.51, Figure I.127), and organic versus non-organic growth (Figure I.127).

Describing entrepreneurial efforts striving for growth as a teleological relation ("startup $\rightarrow$ mid-sized firm"; Figure I.78) allows to describe and discuss a development path from a given NTBF configuration to a configuration of successful mid-sized firms, such as Hidden Champions (ch. 4.1.1) or by mapping an NTBF's growth of a given configuration to a known, successful NTBF with a compatible configuration and its growth path in terms of the *ex comparatione* approach (ch. 4.3.6; Table I.80, Figure I.164).

A more prescriptive approach would look for an NTBF to fit or implementing CSFs of mid-sized firms (ch. 4.3.6; Table I.78). This would be a quasi *top down* approach for a startup to SME path by a perspective from the end result.

Such an approach is oriented toward *growth factors* which may be interdependent as is done by Bordt et al. [2004] who looked into characteristics of technology ventures, mainly from biotechnology as well as electronics, information and communication, that grow from small to medium size. Growth factors may correspond to resources, such as IPR, but also processes, such as R&D and combined "innovation and investment persistence" and managing finances.

RBV sees *companies as different collections of tangible and intangible assets and capabilities*, which determine how effectively and how efficiently a company performs its functional activities. To apply RBV it is essential to *identify the firm's potential "key" resources*. However, the processes of *combining, organizing and leveraging resources let "new systemic resources" emerge.*

> According to GST resources are dependent on interactions and combinations with other resources and therefore no single resource or a set of individual resources can become the most important one for a firm's performance (ch. 4.3.3).

Moreover, s*ystemic effects* in small or large firms do not only emphasize interactions of resources, but also feedback and reinforcement mechanisms, *largely out of the control of leaders/managers, which will affect the firms' development/growth* (ch. 4.3.3).

For NTBFs one typically encounters multi-dimensionality of resources. Concerning systemic effects this can be lucidly seen, for instance, considering the roles of angel investors or corporate venturing companies (CVC) which can be treated primarily as financial resources. However, angel investors may act additionally as a member of the leadership team or an adviser of founders for various aspects of the business. In these role angels may act *reflexivety* affecting the financial situation of the firm, adding not only money to the company, but increasing the monetary return ("smart money") of the firm (ch. 1.2.7.2) and of themselves.

Corporate venturing (ch. 1.2.7.2) may result in leveraged startup resources in the sense of a "networked economy" (Figure I.125) inducing a three-way interaction between the NTBF, the investing large firm which is supporting financially or cooperatively and public research institutes (Figure I.51). With regard to financial resources the large firm may not only provide a particular investment sum, but additionally increase the NTBF's credibility and become a lead investor inducing more investments.

Furthermore, the NTBF may have access to other, non-monetary resources of the large firm, such as access to analytical or information services and advice, thus saving expenses compared with getting these otherwise. And, finally, there are often established joint research or development alliances (JRAs, JDAs), production alliances or contract production, sales and marketing agreements or the large firm acts as a customer for the NTBF.

Qualitatively, NTBF growth depends on financial resources (Table I.30). In the context of organic growth generating profit and cash may initiate a *self-reinforcing cycle* of innovation and investment (*innovation and investment persistence*; Figure I.127).

According to RBV entrepreneurs in emerging organizations must first assemble and acquire or access resources to meet a perceived opportunity, then combine them to build a resource platform that will yield *distinctive capabilities* before they are allocated to fit an offering and market strategy. But this does not cover the non-negligible set of NTBFs whose foundation is initiated by customers as described above providing cash to the to-be entrepreneurs.

Furthermore, RBV suggests that, related to NTBF growth, building an initial resource base in a new venture and then further develop the resource base may not only lead to growth but also achieve *sustainable competitive advantage* – a quasi *bottom up* view concerning the path "from startup to SME."

Strategies for attaining competitive advantages emphasize developing and configuring existing resource strengths into a valuable and *unique* resource base. Such a process requires for technology-based startups three fundamental differentiations:

- Tangible and intangible resources (Table I.8) and
- Internal and external resources in the sense of a "networked economy" (Figure I.51).

The third important differentiator for applying RBV to NTBF growth concerns ownership and control of the startup (Table I.74) where in VC-based startups venture capitalists may decide on types of resources to be used and the way to develop a resource base rather than the founder(s) who may have different ideas how to initiate and keep venture growth. Venture capitalists may also establish a leadership or management team whose strategies are not in line with those of the founders.

Following Brush et al. [2001] we differentiate resources by their application to the productive process as

- *Utilitarian* resources that are applied directly to the productive process or combined to develop other resources and
- *Instrumental* resources that are used specifically to provide access to other resources.

Financial resources are considered instrumental because they can be used to obtain other resources, such as people or equipment (or needed technology licenses).

Proprietary technology may be either utilitarian or instrumental depending on whether it resides in an individual (intangible; tacit technology - Table I.12) where it might be instrumental, or whether it is, for instance, a patented process applied directly to the production process of the firm – or sold on the basis of Intellectual Property Right (IPR) in terms of licenses.

Applying RBV to growth of a startup means also considering instrumental resources over a period of time. If, for instance, financial resources relate to public funding, such as research or development projects which are funded by government or research associations for some few (two to four) years, the time restriction may not only affect the financial situation, but at the end of the funding period may mean laying off important personnel. And there are more constraints (ch. 4.3.3).

Generally, each resource type will have different dimensions along a scale of complexity ranging from the simple and discrete to the complex and systemic. Furthermore, we regard resources that can be acquired by learning as simple, such as learning behavior as well as processes (like financial management and controlling) by outside education and training courses as well as inside employee development (Figure I.121) or learning by imitation focusing on "best practice" or learning by example, doing or mistakes (trial and error) (Table I.7).

The set of resource dimensions, simple to complex and utilitarian to instrumental, provides a basis for mapping possible combinations and applications of resources at the launch of a new venture in a quadrant and additionally specifying the resource type as tangible versus intangible and firm-internal versus external (Figure I.168).

One of the qualitative differences between small and medium-sized firms is the degree of functionalization, specialization, formalization of their organization and planning. The medium-sized firm will exhibit management specialization (development, marketing, human resources, administration, etc.) as well as more formal business planning (Figure I.72, Table I.69) with an emphasis on coordination and communication in the organization, and complementing leadership by various levels of management (Figure I.118) upon growth. Issues of coordination as a resource in growing ventures were emphasized by the "10 - 25 - 150" rule of thumb (ch. 4.3.1; Table I.72).

A challenge for a successful entrepreneur(s) will be to transfer the personal resource base into organizational resources to grow the enterprise. Correspondingly, resource development of startups entails building and transferring knowledge by creating a shared understanding among employees of the venture's direction and focus (vision, mission,goals). The entrepreneur may need to engage in intense and frequent communication to develop this shared understanding.

But a necessary condition to transfer the resource base into organizational resources is to obtain buy-in from employees, associates, management, suppliers and customers. Knowledge and reputation are not sufficient to gain organizational buy-in.

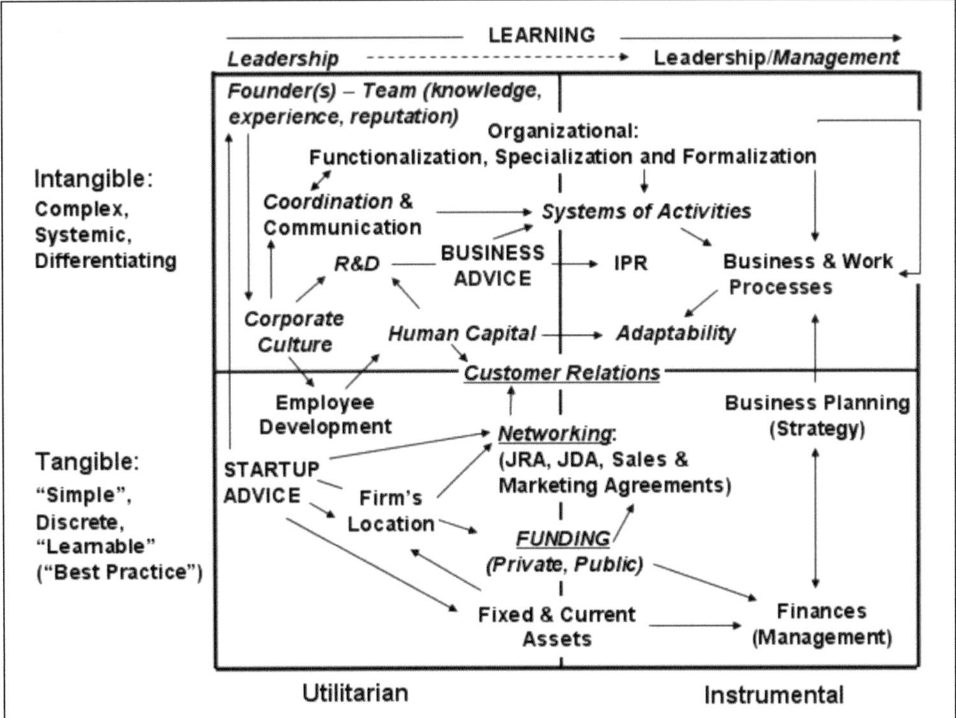

Figure I.168: Types of resources to be acquired or developed with selected interrelations to have an NTBF grow (underline – external; italics – systemic; capital – internal or external – except R&D and IPR).

Understanding the resource development pathway in terms of initial inputs (types of resources) and early uses (application of resources) is central to efficient, effective, and timely management of the resource building process as well as development of a competitive advantage.

In both exercises, one can sort resources into several different bins: *human* (individual traits and skills, knowledge and experience), *social* (team work, external relationships, networks, communication), *financial* (personal wealth, access to funds), *physical, technological, and organizational* (internal structures, processes, coordination and relationships).

It should be re-emphasized that we regard *coordination capability* as a key resource for NTBF development and view the competency for coordination as a "critical meta-asset of long-lived companies" (ch. 1.2.1).

In Figure I.168 resources assumed to be relevant for technology venture growth are displayed. But specification of the enterprise's needs will include not only estimates of types of resources needed, but also of quantity, quality, timing, and sequence of delivery. This enables the entrepreneur to stage resource acquisition and development.

While consulting and advice to found and grow a firm can be purchased from external services or received from friends, families or other persons of the social environment of the founder(s), the Advisory Board of a new NTBF is viewed as an entity with systemic features. Networking and customer and supplier relations viewed as "*organizational capital*" (Table I.8) also have systemic characteristics.

For working with a set of tangible and intangible resources (assets) two different directions seem to be advisable which have to be related to *execution*.

- "Tangible assets" are *managed efficiently according to "best practice"*!
- Working with "intangible assets," the founder's personality and the firm's employees (the human resource), firm culture, relations with customer or suppliers, networking etc., can create *a fundamental differentiator for sustainable competitive advantage and growth*. They can be transformed into *core competencies.*

Human capital (Table I.8) is likely to be particularly important in the context of technology entrepreneurship. A significant percentage of the value of technology-based new ventures is likely to be determined by the quality of the company's employees, especially the top management team (TMT) [Shrader and Siegel 2007].

When we ask whether one can *identify prototypical paths for developing a resource base of NTBFs* referring essentially to those resources given in Figure I.168 we shall look at a resource development pathway that allows the entrepreneur to *begin with starting (initial) endowments* and connect the specification or identification steps to acquisition of resources or to having access to specific ones.

The entrepreneur must consider how one type of resource can be leveraged to acquire another one (as exemplified in Figure I.125). Such an assessment allows for acquisition strategies specific to the situation at hand to be developed. As the period of building a resource base the emphasis will be on the NTBF's first five to six years of existence, which cover more than the startup thrust phase (ch. 4.3.2, Figure I.125).

In the context of GST, to account for interrelations of resources, we propose a process of resource acquisition and development to be represented as a layered rather than a simply staged process. The development process of the resource base will be visualized by *layered phases* and "cyclic arrows" indicating interdependencies and feedback between resources of the layers as is also used for displaying a firm and its supersystems by shells of relevant interacting systems (Figure I.13).

Figure I.169 provides a first example of privately held NTBFs of the TVT industry (all using nanotechnology) with an almost compatible taxonomy (Figure I.128) producing scientific instruments or devices for (first) academic and (then) industrial research customers – WITec GmbH, Cambridge Nanotech, Inc. (Table I.80, Figure I.163, B.2) and Nanion Technology GmbH (B.2) as well as Attocube AG (Figure I.164, B.2).

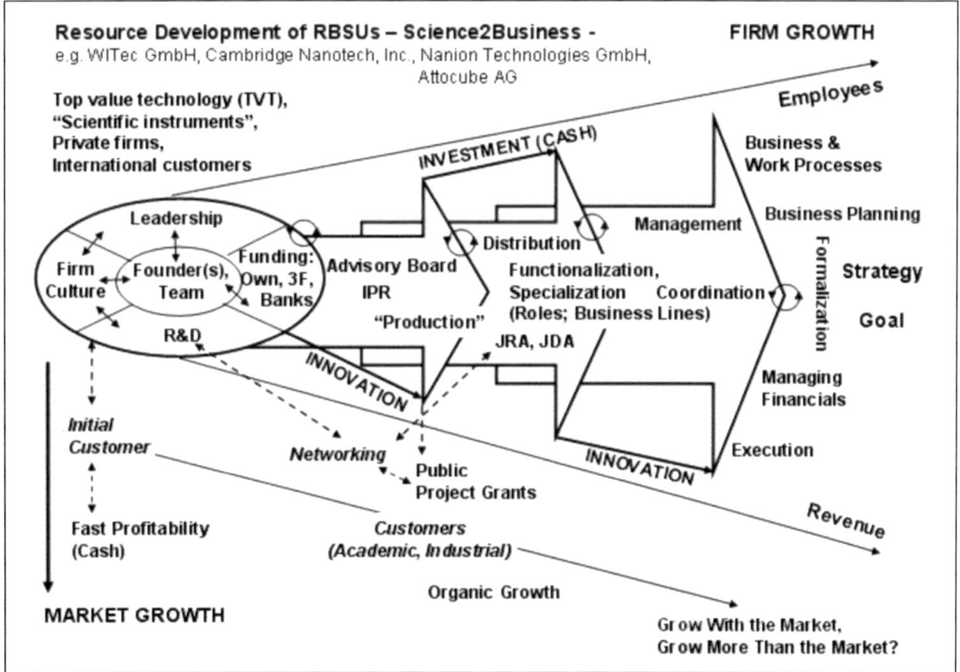

Figure I.169: Building a resource base of structurally compatible RBSUs in the field of nanotechnology-related scientific instruments or devices.

Nanion used capital from public investment firms, partially with a silent partnership, and Attocube an angel investor with no interferences in the firms' leadership in addition to 3F and bank loans as sources of financing. Firms' foundation occurred as university spin-outs (a science2business process) based on research of the founders.

Specialization in terms of business lines here means essentially types of instruments or devices and their related applications in the markets.

Firm development followed essentially a *bootstrapping* process (ch. 4.3.3) for an RBSU and all the above firms as well as IoLiTec (Figure I.170) generated a business plan. All these RBSUs were *located* close to their parent universities and *started with customers* (Figure I.124) and achieved profitability fast. This means, also cash was generated as a resource and innovation and investment persistence characterize the firms' *organic growth* process. For instance, WITec was always profitable (from year 1 on) and Nanion founded in 2002 became profitable in 2004. Having already customers (*market pull*) facilitated considerably access to bank loans for these RBSUs to complement own funds.

Further monetary resources, mainly for R&D, were pursued catching project grants of related ministries of government or public non-governmental organizations (NGOs) and capital-equivalent support by using university infrastructure free-of-charge. Sometimes grants were for cooperative projects.

The *initial endowments* are essentially associated with the founder team in terms of R&D and technology knowledge and financing contributions and its leadership capability to establish the firm culture. For Cambridge Nanotech there were serious frictions in the entrepreneurial pair with one of the founders eventually left the firm (Table I.80, [Yang and Kiron 2010]).

It is to be noted that *industry growth* has been found to have a positive effect on both profitability and sales growth of new technology ventures [Shrader and Siegel 2007].

For building a resource base by a different approach we shall consider an academic NTBF, the German IoLiTec (Ionic Liquids Technologies) GmbH founded in 2002/2003 (Figure I.170), which followed also a science2business path, but is not a typical university spin-out. Only one of the three founders came directly from a university.

IoLiTec entered and survived the entry into the new technology area of ionic liquids for which commercial interest emerged around 2000 though the technology has been known for decades (A.1.5). This case may also shed some light on the reasons why two other startups in ionic liquids, one from Germany and one from the UK, did not survive: One (Solvent Innovation GmbH) was purchased by a large firm and the other went bankrupt (Bioniqs Ltd.).

Ionic liquids represent a platform and enabling technology. The basic approach for entrepreneurship in the new class of materials is looking for their applications and related potential markets. Correspondingly, there was much need of educating (potential) customers or even generating new markets. For instance, IoLiTec promoted the new technology and its applications since 2005 by a free-of-charge newsletter "Ionic Liquids Today" (news and need-to-know facts) with currently globally ca. 6,500 subscribers.

Hence, for its development IoLiTec is essentially on a *technology push* path and furthermore, it has a major *barrier for market entry* of a new technology associated with the *high cost of ionic liquids*. It has a strong emphasis on research and development activities and its R&D intensity is 40-50 percent (A.1.5).

IoLiTec's technology is protected by *IPRs*. Between 2004 and 2007 IoLiTec submitted six patent applications, most of them citing two of the firm founders (A. Bösmann and T. Schubert) as inventors and some of them in cooperation with the University of Freiburg (Germany).

A key resource of IoLiTec is *technology intelligence*, emphasizing tracking scientific developments in the field, state-of-the-art and current awareness, by searching the scientific literature and tracking technology development and applications and protection of technology and applications by patent searches including competitor and market tracking.

This means organizing and managing the flood of data and information by appropriate data processing systems and, furthermore, utilizing the existing information and knowledge base for consulting and revealing optimum offerings to meet customer specifications on the basis of the existing data and information set using structure-property and structure-activity relations and corresponding software for data processing.

Furthermore, technology intelligence can also help identify not only commercial opportunities, but also spot opportunities for financing the startup's development by "public money" (Figure I.59).

In Figure I.170 the building process of a resource base for IoLiTec GmbH (A.1.5, B.2) is displayed. The formal foundation of IoLiTec (legal status: German GbR) occurred in November 2002 in Cologne/Aachen (Germany;). The search for a proper *location* anywhere in Germany (criteria: cost and networking) led to the BioTechPark in Freiburg with the near Freiburg University providing a potential for networking with academic ionic liquids experts.

Operation started in May 2003 (changed legal status to GmbH & Co. KG). In October 2003 the first employee was hired. For needed expansion in June 2005 the firm moved to the industrial area (in German "Gewerbegebiet") in Denzlingen near Freiburg as renting cost in the BioTechPark, particularly for laboratories, was very high.

IoLiTec started with an *entrepreneurial triple* (three chemists with experience in the technical field) and having a customer. One of the co-founders who brought in the customer left already after three months. *Having a customer* (a manufacturer of *sensors*) who placed an €180,000 order had a decisive influence to get a *bank loan* of €200,000. In 2005 and 2006 further funding occurred by loans from banks [Schubert

2008a]. IoLiTec was profitable from day 1 on. Hence, IoLiTec could rely also on *cash* for its operations.

In January 2004 it began offering custom synthesis of ionic liquids. During the startup phase IoLiTec used its Web site as a resource to promote its offerings. It could take advantage from the general interest in ionic liquids in academia and industry which ultimately led firms to directly contact IoLiTec via telephone or email. But due to its technology push approach it had often to create ("grow") its market.

In 2004 the "residual team" was complemented by a *business angel* (focused on financial, tax and law). Funding by an angel did not only affect the financial structure of the startup, but, as the angel took over administrative and financial responsibilities, IoLiTec could rely *very early on professional financial management* as a resource (B.2 – IoLiTec – Figure 1, Figure 3).

In 2006 IoLiTec's original leadership team lost again one of its founders. A. Bösmann, the CSO with broad technical experience in ionic liquids based on his dissertation at the Technical University of Aachen with Germany's key scientist in the field, left the firm to return to academic research. T. Schubert who drove the foundation and acted as a Managing Director (CEO) remained in his position (B.2 – Figure 1, Figure 3).

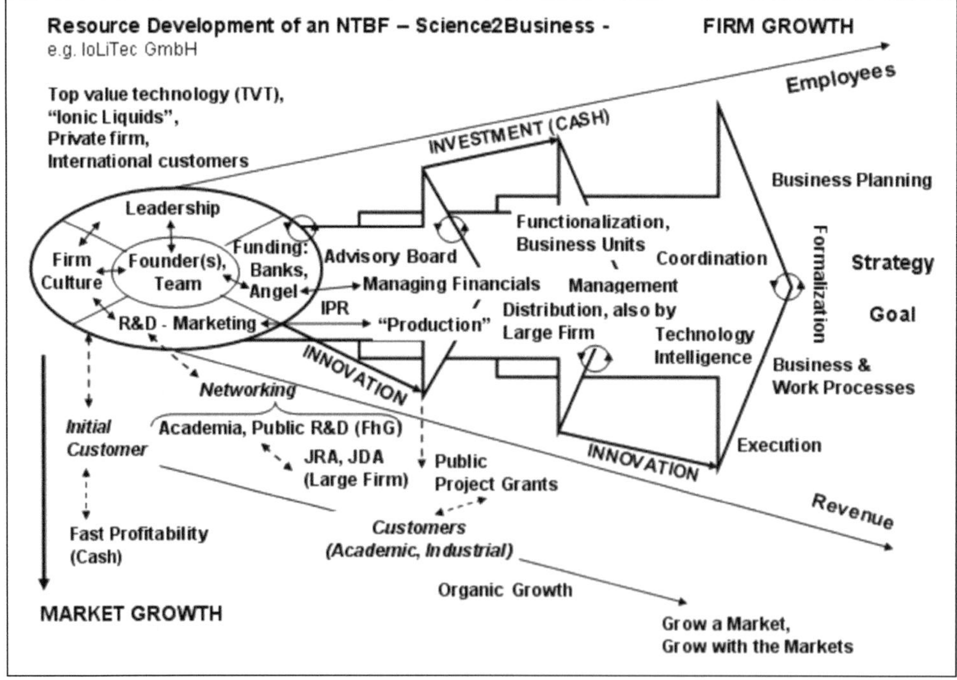

Figure I.170: Building a resource base of NTBF IoLiTec GmbH in the field of ionic liquids.

Schubert earned his doctoral degree in the group of Prof. Berkessel of the University of Cologne in bioorganic chemistry; the latter one became a member of the Scientific Board (B.2 – Figure 1, Figure 3).

Before founding IoLiTec T. Schubert gained experience (2001-2003) as a Sales Manager responsible for the technical synthesis and marketing of ionic liquids working for the startup Solvent Innovation GmbH in Cologne being also active in ionic liquids which actually meant a competitor for his startup.

Hence, Schubert could contribute a variety of resources to IoLiTec, technical, market, marketing and sales knowledge in the field and having networking contacts with large firms. Therefore, Schubert could *earlier complement leadership by management* (Figure I.118) and establish faster *formalized business processes* for running the firm's growth.

IoLiTec GmbH founder T. Schubert [2004, 2008a] lists twenty six specific applications clustered into six different general areas for applications of ionic liquids. This broad spectrum show what choices IoLiTec could make from existing options (A.1.5) and which ones it finally made.

Concerning organization in 2004 IoLiTec decided to focus on five areas of *specialization* [Schubert 2005]:

1. Contract R&D services
2. Special chemistry (ionic liquids)
3. Sensor technology
4. Energy
5. Nanotechnology.

Corresponding product lines according to these five areas are given by Schubert [2008a:slide 27].

The plan was to develop each focus into an independent division each one relying on the core platform technology "ionic liquids." Various fields of applications for ionic liquids let *business lines* emerge which showed up in 2006 (B.2 – Figure 2).

For the science2business path "out of the ivory tower" [Short 2006] into the market IoLiTec put much emphasis on its Advisory Board which functions as a "Scientific Board." Leveraging the Scientific Board occurs for consulting activities and analytical services for customers mediated through IoLiTec (B.3).

The very intense networking of IoLiTec with academia is reflected by defined links between IoLiTec's researchers with members of the Scientific Board or other professors from universities in related scientific fields (B.2 – Figure 3). Networking with public research institutes (B.2) as "partners" is essentially with relevant institutes of the Fraunhofer Society (FhG). FhG institutes became often partners in publicly funded cooperative projects of consortia which represented a significant factor of financing IoLiTec's R&D activities.

In June 2004 IoLiTec organized distribution of its products via Merck KGaA and JDA (cooperation contracts) with Degussa AG (now Evonik Industries) as strategic partners.

Apart from the above NTBFs which started from a unique technology or design as described above many NTBFs take advantage of market disequilibrium, "catch a wave," such as biofuels (A.1.1) or CleanTech, to anticipate profits. But they have to share the markets with many new entrants – small and large, attracted by potential profits. Related NTBFs belong largely to the class of VC-based startups and, specifically in biofuels, are addressing policy-driven markets (Table I.15).

In biofuels corresponding NTBFs often target large-scale manufacturing, but most of them are still struggling with the associated scale-up process. Financing these firms is often from the beginning or in an early stage of the scale-up process. Venture capitalists do not only infuse *money* into these firms but establish simultaneously an extended management team with *massive experience in target or related markets and in leading a firm* – and even having *ties to policy* concerning plant and production permits ("*veterans approach*"; A.1.1). The original founders of these firms, when being scientists or engineers, usually took the roles of a CSO or CTO in the management team.

Biofuel firms additionally take advantage from "public money" in terms of loan guaranties and tax exemption by public entities, cooperative R&D projects with national or federal research institutes and seek cooperative projects with large firms of the oil or chemical industry (Figure I.179, Figure I.183).

Such NTBFs *behave like a large sophisticated company* very early on or even from the start. For the innovation and firms' development processes they follow essentially the highly formalized processes of large firms (Figure I.134), in particular, the phase gate process or RD&D staged path for scale-up of production (Figure I.79, Figure I.180).

In particular for biofuels, venture capitalist V. Khosla refers to an innovation architecture which he calls an "innovation ecosystem at work," solving large problems by harnessing the power of ideas fueled by entrepreneurial energy of scientists, technologists, and entrepreneurs – very bright people working on solving a problem. Khosla's innovation architecture is often to be characterized as a *VC-based spin-out* (RBSUs) with *experienced managers* ("professional managers"; Figure I.118) from almost the point of firm's foundation (A.1.1.5).

Building the resource base in such an NTBF is driven essentially by connecting founder/management team *experience and strategy* – and execution. Fast growth in the number of employees often leads to *issues of coordination* as an important re source.

Shrader and Siegel [2007] assessed the role of human capital (Table I.8) in the growth and development of new technology-based ventures via an analysis of a large sample of publicly traded, technology-based new ventures using information available on initial public offerings (IPOs) in IPO prospectuses. They suggest that the fit between strategy and team experience is a key determinant of the long-term performance of these high-tech entrepreneurial ventures.

In particular Shrader and Siegel [2007] found that experience with previous startups helped ventures pursuing broad strategies (for instance, numerous customers, segments and products) achieving higher sales growth. As expected, marketing experience was shown to be highly significantly related to the pursuit of marketing-based differentiation.

Their findings provide also striking evidence of a clear and consistent fit between team backgrounds and competitive strategy among their sample. More specifically, they believe that *specialized experience in functional areas relates to the strategies* pursued by a new venture.

5.3 Some Concluding Remarks

When we embarked a journey following a purpose to outline the not-well and not adequately explored territory we have observed that Applied General Systems Theory (GST) and its principles and approaches provide a framework to tackle technology entrepreneurship which is characterized by high complexity in terms of numbers of variables and parameters that determine the developments of new technology ventures.

GST as an all embracing framework allowed a consistent, context-sensitive exploration and treatment of the subject to increase understanding and explanation. Here, we targeted rationalization to differentiate "reasons why" versus "reasons for thinking that." However, presented claims and propositions remain partly tentative. But many tentative propositions were driven to the point that will allow or induce, respectively, further inquiries.

For technology entrepreneurship rather than leveling off differences by methods that enforce compatibility of what does not fit human activities in various contexts the emphasis was on sources of variations which conventional theories and approaches dismiss as random. Variations were studied to reveal patterns of structure and dynamics within the variation ("Variations on a Theme"; A.1.6).

We think our APPROACH is a response to Schumpeter's [1939:44] criticism of research based on only aggregate data:

"It keeps analysis on the surface of things and prevents it from penetrating into the industrial processes below, which are what really matters. It invites a mechanistic and formalistic treatment of few isolated contour lines and attributes to aggregates a life of their own and a causal significance that they do not possess."

We did not neglect the entrepreneurs' personalities and the important role of individual entrepreneurs or founders in a "team system." And we did not neglect the time span the entrepreneurs are active in requiring detailed knowledge about related constellations of technologies, markets, industries and national economies. This is expressed by William Hewlett, one of the founders of HP (Hewlett – Packard, founded in 1939):

> "As I talk about the start of the company, it is important to remember that both Dave {Packard} and I {William Hewlett} were products of the Great Depression." [Hewlett]

First steps have been presented to handle development and growth of NTBFs. Focusing on growth states of firms and *equations of state* give up the idea that data can be forced into a one-size-fits-all model. Concepts and regularities in firms' evolutions were presented, some require refinement, but more regularity await discovery.

We did not just observe! We filtered and organized "observable" data, information and facts – whether from the literature or own observations. And epistemologically following Einstein that "it is the theory that decides what we can observe" we introduced "The Bracket Model of New Technology Venture Development" and emphasized expectations rather than predictions. The Bracket Model deals with processes of change paying equal attention to expected observations and events and to exceptional events or phenomena.

Finally, we do not claim to have developed a standard or a direction for further scholarly inquiry. We – with a strong background of natural sciences – made a suggestion how to approach the exciting real world issues of human activities leading to economic change based on changes of technology and utilization of new scientific results.

Concerning the comparative approach in this book, apart from basic inter-cultural and socio-economic differences, technology entrepreneurship in Germany and the US is generically closest for privately held or controlled new firms.

It is very different for VC-based startups which is largely related to the fundamentally different financial systems, in particular, based on availability of a gigantic amount of "loose money" in the US due to the dollar still being the world's reserve currency and the Federal Reserve being in a position that the world's economic, financial and policy systems accept the Fed to print as much dollars as the US is perceived to need. Hence, how VC-based startup emerge and develop in the US cannot be compared and be a general model for Germany's VC-based startups. In particular, the level of related technology speculation is negligible in Germany compared to that in the US.

The German national innovation and entrepreneurship system has emergeed with high systemic performance due to a rather organized, smooth and strong interplay of the Science & Technology, Economic, Financial and Policy Systems.

REFERENCES

A123 (2011): *Annual Report 2010.* http://files.shareholder.com/downloads/ABEA-3DGNI7/1400969767x0x492428/4CAEC6AA-6BDF-4417-A285-DA6D6C4472F6/AONE_2010_10-K.pdf (last access 9/14/2011); *Annual Report 2011.* http://files.shareholder.com/downloads/ABEA-3DGNI7/2084852290x0x559734/6496992D-F3B6-4634-ABB4-C86C8AE1F364/A123_2011_Annual_Report.pdf (last access 8/22/2012).

Abrahamson, E. (2000):*Change Without Pain: Dynamic Stability.* HBS Working Knowledge, 9/11/2000. http://hbswk.hbs.edu/archive/1683.html (last access 5/12/2011).

Abramson, H. N.; Encanação, J.; Reid, P. P.; Schoch, U. Eds. (1997): *Technology Transfer Systems in the United States and Germany – Lessons and Perspectives.* National Academy Press, Washington, D.C., 1997. http://books.nap.edu/openbook.php?record_id=5271&page=R1 (last access 12/10/2009).

Accel Partners (2005): *Accel Partners Invests in thefacebook.com.* Portfolio Company New, May 26, 2005. http://www.accel.com/news/news_one_up.php?news_id=1 (last access 10/5/2011).

Ackerman, R. (2009): *Valero Sees Dollar Signs In Ethanol.* Forbes Online, Mar. 18, 2009. http://www.forbes.com/2009/03/18/valero-energy-ethanol-markets-equity-verasun.html (last access 11/9/2009).

ACS – American Chemical Society (2011): *Innovation, Chemistry, and Jobs – Meeting the Callenges of Tomorrow.* ACS Presidential Task Force on Innovation in the Chemical Enterprise. http://web.2.c2.audiovideoweb.com/va92web25028/InnovationChemistryJobsReport-PDFs/InnovationChemistryandJobs.pdf (last access 9/25/2011).

ADAC (2011a): *125 Jahre Automobil – Motoren des Fortschritts (125 years automobile – Engines of progress).* ADAC Motorwelt, Jan. 2011, 21-29.

ADAC (2011b): *Platz da, für alle*! (Have space, for all!). ADAC Motorwelt, May 2011, 24-29.

Admin (2012): *2012 Merger mania gets underway: Green Biologics, Butylfuel merge.* Biofuels Digest, Jan. 20, 2012. http://www.biofuelsdigest.com/bdigest/2012/01/20/2012-merger-mania-gets-underway-green-biologics-butylfuel-merge/ (last access 4/27/2012).

Admin (2011a): *Coskata's $100 million IPO: The 10-Minute Version.* Biofuels Digest, Dec. 20, 2011. http://www.biofuelsdigest.com/bdigest/2011/12/20/coskatas-100-million-ipo-the-10-minute-version/ (last access 4/23/2012).

Admin (2011b): *Cobalt taps Mayer as new CEO.* Biofuels Digest, Nov. 29, 2011. http://www.biofuelsdigest.com/bdigest/2011/11/29/cobalt-taps-mayer-as-new-ceo/ (last access 5/3/2012).

Admin (2011c): *Mascoma's IPO: The 10-Minute version*. Biofuels Digest, Sep. 20, 2011. http://www.biofuelsdigest.com/bdigest/2011/09/20/mascomas-ipo-the-10-minute-version/ (last access 1/26/2012).

Admin (2011d): *KiOR files $100M IPO – the complete Digest analysis*. Biofuels Digest, Apr. 13, 2011. http://www.biofuelsdigest.com/bdigest/2011/04/13/kior-files-100m-ipo-the-complete-digest-analysis/ (last access 6/2/2011).

Admin (2011e): *KiOR raises $150M in IPO, 30 percent below latest forecast*. Biofuels Digest, Jun. 24, 2011. http://www.biofuelsdigest.com/bdigest/2011/06/24/kior-raises-150m-in-ipo-30-percent-below-latest-forecast/ (last access 6/30/2011).

Admin (2011f): *Genomatica's $100M IPO: The 10-Minute Version*. Biofuels Digest, Aug. 25, 2011. http://www.biofuelsdigest.com/bdigest/2011/08/25/genomaticas-100m-ipo-the-10-minute-version/ (last access 11/23/2011).

Admin (2011g): *The Magic Word is…Adipic. Looking at Verdezyne and its bull markets*. Biofuels Digest, Dec. 21, 2011. http://www.biofuelsdigest.com/bdigest/2011/12/21/the-magic-word-is%E2%80%A6adipic-looking-at-verdezyne-and-its-bull-markets/ (ast access 1/9/2012).

Administrator (2010): *Amyris – Farnesene and the pursuit of value, valuations, validation and vroom*. Biofuels Digest, Jun. 25, 2010. http://www.biofuelsdigest.com/biotech/2010/06/25/amyris-the-pursuit-of-value-valuations-and-validation/ (last access 8/13/2011).

Agarwal, R.; Echambadi, R.; Franco, A. M.; Sarkar, MB (2004): *Knowledge Transfer Through Inheritance: Spin-out Generation, Development and Survival*. Academy of Management Journal 47 (4), 501–522. http://www.rotman.utoronto.ca/facbios/file/Agarwal%20Echambadi%20Franco%20and%20Sarkar%20AMJ%202004.pdf (last access 7/6/2010).

Agrinova: http://an.agricola-tech.de/index.php?id=49 (last access 3/16/2010).

Ahlm, P. (2012): *Algenol*. Presented at DOE's 2012 Biomass Conference, July 10, 2012. http://www1.eere.energy.gov/biomass/pdfs/b12_ahlm_1-b.pdf (last access 5/7/2013).

Ahrens, R. (2005): *Flower-Power – Bodo Wolf und sein Sonnensprit aus Biomasse* (Flower-Power – Bodo Wolf and his Sunfuel from biomass). Radio WDR5, Leonardo, Broadcasted Nov. 22, 2005.
http://www.wdr5.de/sendungen/leonardo/sendungsdetailseite.html?tx_wdr5ppfe_pi1%5BshowUid%5D=175768&tx_wdr5ppfe_pi1%5BbeitragsUid%5D=2102&cHash=c8bde7b440&tx_rlmpflashdetection_pi1%5Bhtml%5D=1 (last access Nov. 30, 2005; no longer accessible);
Choren Industries: History.
http://www.choren.com/en/choren_industries/company/history/ (last access 12/17/2009).

Ainsworth, S. J. (2008): *Entrepreneurial Trailblazers – Women build businesses around their passion for science*. Chemical & Engineering News 86 (44), 50-54. http://pubs.acs.org/cen/coverstory/86/8644cover2.html (last access 1/29/2010).

Ainsworth, S. J. (2010): *Empowered Entrepreneurs – Downsized out of big companies, scientists are creating satisfying job opportunities in their own firms.* Chemical & Engineering News 88 (28), 57-60. http://pubs.acs.org/cen/employment/88/8828employment1.html (last access 7/14/2010.

Algae Industry Magazine (2011): *HR BioPetroleum to Acquire Shell's Share in Cellana Algae JV.* AlgaeIndustryMagazine.com, Jan. 27, 2011. http://www.algaeindustrymagazine.com/hr-biopetroleum-to-acquire-shells-share-in-cellana-algae-jv/ (last access 7/7/2011).

Altucher, J. (2009): *The Internet Is Dead (As An Investment).* Wall Street Journal, Jul. 17, 2009. http://online.wsj.com/article/SB124784696163158721.html (last access 7/18/2009).

Alvarez, S. H.; Busenitz, L. W. (2001): *The entrepreneurship of resource-based theory.* Journal of Management 27 (6), 755-775 (2001). http://www.udec.edu.mx/BibliotecaInvestigacion/Documentos/2009/Mayo/The%20entrepreneurship%20of%20resource-based%20theory.pdf (last access 1/19/2011); http://faculty-staff.ou.edu/B/Lowell.W.Busenitz-1/pdf_pro/2001%20jom%20the%20entrepreneurship%20of%20rb.pdf (last access 12/12/2010).

Amyris (2012): *Annual Report 2011.* http://files.shareholder.com/downloads/ABEA-4QL2IU/1899075180x0x559486/A7E56B05-7866-4E3C-9A7D-67A4E4001BC3/2011AnnualReport.pdf (last access 5/6/2012).

Anderson, C. W.; Moran, T. (2009): *Ethanol, biofuel industry in flux – Plants shut down in Modesto, Stockton.* Central Valley Business Journal 2009. http://www.cvbizjournal.com/local-news/ethanol-biofuel-industry-in-flux.html?option=com_content&catid=98&id=370&lang=en&view=article&option=com_content&catid=98&id=370&lang=en&view=article&fontstyle=f-smaller (last access 1/2/2010).

Andreessen, M. (2011): *Why Software Is Eating the World.* Wall Street Journal, Aug. 20, 2011. http://online.wsj.com/article/SB10001424053111903480904576512250915629460.html?mod=WSJ_hp_us_mostpop_read (last access 8/24/2011).

Amit, R.; Schoemaker, P.J.H. (1993): *Strategic assets and organizational rent.* Strategic Management Journal 14(1), 33-46.

Anonymus (2004): *Was bewegt ... Dietmar Hopp? (What drives ... Dietmar Hopp?).* Die Zeit, Oct. 21,.2004. http://www.zeit.de/2004/44/Portr_8at_Hopp (last access 8/3/2010).

Anonymus (2006a): *Put A Termite In Your Tank.* Business Week Online, Dec. 18, 2006. http://www.businessweek.com/magazine/content/06_51/b4014081.htm?chan=tc&chan=technology_technology+index+page_best+of+the+magazine (last access 12/23/2009).

Anonymus (2007a): *Nuthetaler Institut mit Partnern in den USA (Nuthetal institute with partners from the USA).* Potsdamer Neueste Nachrichten, Jan. 6, 2007. http://www.pnn.de/pm/56144/ (last access 11/10/2009).

Anonymus (2007b): *Kampf gegen den Klimawandel: Der Hoffnungsträger ist grün und glitschig (Fighting climate change – The object of hope is green and slippery).* Handelsblatt, Nov. 2, 2007. http://www.handelsblatt.com/technologie/forschung/der-hoffnungstraeger-ist-gruen-und-glitschig;1346018 (last access 1/25/2010).

Anonymus (2008): *DuPont and BP targetingmultiple bio-butanol molecules; biofuel allows for highwe blending ratios.* Biopact, Feb.14, 2008. http://news.mongabay.com/bioenergy/2008/02/dupont-and-bp-targeting-multiple-bio.html (last access 1/26/2010).

Ante, S. E. (2009): *'Super Angels' Shake Up Venture Capital.* Business Week, May 21, 2009. http://www.businessweek.com/magazine/content/09_22/b4133044585602.htm?chan=technology_technology+index+page_best+of+the+magazine (last access 5/ 28/2009).

Ardichvili, A.; Cardozob, R.; Ray, S. (2003): *A theory of entrepreneurial opportunity identification and development.* Journal of Business Venturing 18, 105–123. http://www2.sa.unibo.it/summer/testi/11_grimaldi_tagliaventi/ardichvili_2003.pdf (last access 4/5/2010).

Ardichvili, A., Harmon, B., Cardozo, R.N., Reynolds, P.D., Williams, M.L. (1998): *The New Venture Growth: Functional Differentiation and the Need for Human Resource Development Interventions.* Human Resource Development Quarterly 9 (1), 55-70.

Arenius, P; Ehrstedt, S. (2008):*Variation in the Level Of Activity Across the Stages of the Entrepreneurial Startup Process-Evidence From 35 Countries.* Estudios de Economía. Vol. 35 - N° 2, Diciembre 2008. Págs. 133-152. http://www.scielo.cl/scielo.php?pid=S0718-52862008000200002&script=sci_arttext (last access 8/10/2009).

Ashby, R. W. (1957): *An Introduction to Cybernetics. Second Impression.* Chapman & Hall, London, 1957, 1999. http://pespmc1.vub.ac.be/books/IntroCyb.pdf (last access 9/22/2009).

Associated Press (2009a*): BP and Martek form biodiesel partnership.* Business Week, Aug. 11, 2009. http://www.businessweek.com/ap/financialnews/D9A0ND9O1.htm (last access 8/11/2009); cf. also http://www.topix.net/content/prweb/2009/08/bp-and-martek-biosciences-enter-joint-development-agreement-to-deliver-advanced-biodiesels (last access 11/9/2009).

Associated Press (2009b*): Race to turn pond scum and other algae into affordable fuel heats up with new money, research.* , WGN Radio, Dec. 2, 2009. http://www.wgnradio.com/lifestyle/green/resourceguide/sns-green-affordable-fuel,0,7069152.story (last access 1/24/2010), cf. also http://bioenergyuiuc.blogspot.com/2009/12/race-to-turn-pond-scum-and-other-algae.html.

Aulet, B. (2006): *Generating Exponential Revenue Growth.* Presentation at HCT MIT Entrepreneurship Conference. Dubai, United Arab Emirates, September 20, 2006. http://www.hctmit.com/Conf2/Presentation/Day%202/AULET%20Generating%20Expo

nential%20Revenue%20Growth%20HCT.ppt#420,8,Technology Adoption Over History (Last access 7/24/2009). Slide 8 – no longer accessible; http://www.authorstream.com/Presentation/Irvette-15652-AULET-Generating-Exponential-Revenue-Growth-HCT-Desired-Outcomes-Presentation-Upfront-Caveat-Executive-Summary-Recommended-Reading-Material-Techn-trb2007-ppt-powerpoint/ (last access 10/30/2013).

Austin, A. (2009): *BlueFire, Arkenol process cofounder form agreement*. Ethanol Producer, posted Mar. 11, 2009. http://www.ethanolproducer.com/article.jsp?article_id=5492 (last access 1/7/2009).

Autio, E. (2007): *Global Report on High-Growth Entrepreneurship – 2007 Global Report on High-Growth Entrepreneurship.* http://www.gemconsortium.org/download/1285757314657/GEM%20High%20Growth%20Report%2007b.pdf (last access 9/28/2010).

Autio, E. (2005): *Global Entrepreneurship Monitor – High-Expectation Entrepreneurship 2005. Summary Report.* http://www.gemconsortium.org/download/1285693721657/18436%20GEM%20summary.pdf (last access 9/28/2010).

Autio, E., Garnsey, E. (1997): *Early Growth and External Relations in New Technology-Based Firms.* http://usasbe.org/knowledge/proceedings/proceedingsDocs/USASBE1997proceedings-P183Autio.PDF (last access 1/19/2011);
Autio, E., Garnsey, E. and Yli-Renko, H. (1997) "*Early growth and external relations in new technology-based firms.*" In Kunkel, S. and Meeks, M. (Eds.) Entrepreneurship: the engine of global economic development: Proceedings of the World Conference of the International Council for Small Business (42nd), 21st-24th June 1997, San Francisco, California, USA. St Louis, Mo.: ICSB;
Yli-Renko, H.; Autio, E. (1996): *The catalyzing impact of new, technology-based firms in industrial technological renewal: Developing a systemic evolution model.* Proceedings of the 41st ICSB World Conference, Stockholm.

AUTM – The Association of University Technology Managers (2008): *THE BETTER WORLD REPORT PART ONE: The Art of Collaboration: The Relationships That Bring Academic Innovations to the Marketplace.* http://www.enterprise.cam.ac.uk/uploads/File/Publications/AUTM_BWR1_08.pdf (last access 12/1/2009); THE BETTER WORLD REPORT PART TWO: Technology Transfer Works: 100 Innovations from Academic Research to Real-World Application.. http://www.enterprise.cam.ac.uk/uploads/File/Publications/AUTM_BWR2_08.pdf (last access 12/1/2009).

Ball, P. (2006): *Chancing upon chemical wonders.* Chemstry World, June 2006. http://www.rsc.org/chemistryworld/Issues/2006/June/ChancingUponChemicalWonders.asp (last access 10/20/2010).

Banathy, B.H. (1968): *Instructional Systems.* Fearon Publishers: Belmont, Palo Alto, CA.

Banathy, B.H. (1996): *A Taste of Systemics.* An activity of the Primer Group – The First International Electronic Seminar on Wholeness. A Special Integration Group (SIG) of the International Society for the Systems Sciences (ISSS) and IISII

INTERNATIONAL INSTITUTE for SYSTEMIC INQUIRY AND INTEGRATION, December 1, 1996; to December 31, 1997. http://www.newciv.org/ISSS_Primer/asem04bb.html (last access 9/21/2009).

Bandyk, M. (2009): *Why We Should Be Worried That Washington Is the New Job Creator.* U.S. News & World Reports, posted December 08, 2009 04:07 PM ET. http://www.usnews.com/mobile/blogs/capital-commerce/2009/12/8/why-we-should-be-worried-that-washington-is-the-new-job-creator.html (last access 1/12/2010).

Baron, R. A. (2004): *Opportunity Recognition – Insights from a Cognitive Perspective.* Chapter 3 in Butler, John E., Ed. (2004): Opportunity Identification and Entrepreneurial Behavior. A Volume in Researrch in Entrepreneurship and Management. Information Age Publishing, Inc. 2004.

Baron, R. A.; Ensley, M. D. (2006): *Opportunity Recognition as the Detection of Meaningful Patterns: Evidence from Comparisons of Novice and Experienced Entrepreneurs.* Management Science 52 (9), Sep. 2006, 1331-1344.

Barrett, A. (2009): *Venture Capitalists Head to Washington.* Business Week, Aug. 7, 2009. http://www.businessweek.com/magazine/content/09_68/s0908019273796.htm?chan=smallbiz_smallbiz+index+page_small+business+policy (last access 8/10/2009).

Barron, R. (2007): *Are Biofuels Pushing VCs Into a New Role?* GreenTech Media Newsletter, Sep. 14, 2007. http://www.greentechmedia.com/articles/read/are-biofuels-pushing-vcs-into-a-new-role-67/ (last access 11/30/2009).

BASF: *BASF History 1865 – 1901 – The birth of the chemical industry and the era of dyes.* http://www.basf.com/group/corporate/en/about-basf/history/1865-1901/index; http://www.basf.com/group/corporate/de/about-basf/history/1865-1901/index (last access 2/23/2010); http://www.basf.com/group/corporate/en/function/conversions:/publish/content/about-basf/history/1865-1901/images/1865-1901_en.pdf.

BASF (2013): *BASF plans production of butanediol from renewable feedstock using Genomatica technology.* Joint News Release of BASF and Genomatica, May 10, 2013. http://www.basf.com/group/pressrelease/P-13-268 (last access 5/14/2013).

BASF (2012a): *BASF participates in US American technology company Renmatix.* Press Release, Jan. 3, 2012. http://www.basf.com/group/pressrelease/P-12-101 (last access 1/30/2012).

BASF (2012b): BASF, *Cargill and Novozymes target commercial bio-based acrylic acid process.* Joint Press Release, Aug. 17, 2012. http://www.basf.com/group/pressrelease/P-12-363 (last access 8/23/2012).

BASF (2012c): *BASF and CSM establish 50-50 joint venture for biobased succinic acid.* Joint News Release of BASF and CSM, Oct 5, 2012. http://www.basf.com/group/pressrelease/P-12-444 (last access 10/16/2012).

Bates, T. (1995). *Self-employment entry across industry groups.* Journal of Business Venturing 10(2), 143-156.

Baumol, W.; Litan, R. E.; Schramm, C. (2007): *Sustaining Entrepreneurial Capitalism.* Capitalism and Society 2 (2), Article 1. Available at: http://www.bepress.com/cas/vol2/iss2/art1.

Baumol, W.; Litan, R. E.; Schramm, C. (2008): *Good capitalism and bad capitalism.* The Japan Time Online, Mar. 13, 2008. http://search.japantimes.co.jp/cgi-bin/eo20080313a2.html (last access 3/2/09).

BBC (2007): *Quick guide: Biofuels.* BBC News Jan. 24, 2007. http://news.bbc.co.uk/2/hi/science/nature/6294133.stm (last access 10/25/2009).

BBC: *Richard Trevithick and the First Railway Locomotive.* http://www.bbc.co.uk/wales/southeast/halloffame/historical_figures/richard_trevithick.shtml (last access 2/13/2013); http://news.bbc.co.uk/dna/place-lancashire/plain/A14113838 http://en.wikipedia.org/wiki/Richard_Trevithick; http://en.wikipedia.org/wiki/London_Steam_Carriage.

Bellis, M.: *The History of MP3 - Fraunhofer Gesellschaft and MP3.* ©2010 About.com, a part of The New York Times Company. http://inventors.about.com/od/mstartinventions/a/MPThree.htm (last access 6/18/2010).

Benda, C. (2003): *A Kernel of an Idea.* University of Minnesota Alumni Association., May 12, 2003. http://www.chem.umn.edu/alumni/patrick_gruber.pdf (last access 6/6/2012).

Berkum, S. (2008): *Thoughts on Google's 20% time.* Mar. 12, 2008. http://www.scottberkun.com/blog/2008/thoughts-on-googles-20-time/ (last access 9/1/2010).

Bessant, J.; Möslein, K.; von Stamm, B.: *In search of innovation.* http://businessonmain.msn.com/browseresources/articles/print.aspx?cp-documentid=25253702#fbid=61V6MW9dNrq; http://userhome.brooklyn.cuny.edu/irudowsky/EmergingTechnology/lectures/InSearchOfInnovation.ppt (last access 7/15/2011).

Bevill, K. (2012): *Target for Innovation.* Ethanol Producer Magazine, Mar. 2012, 40-45. http://www.greenbiologics.com/pdfs/gblinthenews1.pdf (last access 4/25/2012).

Bhidé, A. V. (2000): *The Origin and Evolution of New Businesses.* Oxford University Press, New York.

Bidaux, T. (2009): *Free to Play games in Europe – 2009.* Presentation Develop, Brighton, Jul. 14-16, 2009. http://www.slideshare.net/ICOPartners/free-to-play-games-in-europe-2009 (last access 7/31/2009).

Bigelow, B. V. (2008): *Sapphire Energy, Backed by Bill Gates, Tries to Tone Down the Hype as it Makes Gasoline From Algae.* Xconomy, Nov. 13, 2008. http://www.xconomy.com/san-diego/2008/11/13/sapphire-energy-backed-by-bill-gates-tries-to-tone-down-the-hype-as-it-makes-gasoline-from-algae/ (last access 3/14/2010).

Bigelow, B. V. (2009a): *Big Energy Collaborations Seen to Jump-Start Emerging Biofuels Technologies.* Xconomy, from Nov. 19, 2009. http://www.xconomy.com/san-diego/2009/11/19/big-energy-collaborations-seen-to-jump-start-emerging-biofuels-technologies/ (last access 11/21/2009).

Bigelow, B. V. (2009b): *Two Things I Learned During My Tour of Sapphire Energy.* Xconomy, Oct. 27, 2009. http://www.xconomy.com/san-diego/2009/10/27/two-things-i-learned-during-my-tour-of-sapphire-energy/ (last access 3/13/2010).

Bruce V. Bigelow (2009c): *After Re-Engineering Itself, Verdezyne Sets Course to Develop Biofuels and "Green" Industrial Chemicals.* Xconomy, Nov. 20, 2009. http://www.xconomy.com/san-diego/2009/11/20/after-re-engineering-itself-verdezyne-sets-course-to-develop-biofuels-and-green-industrial-chemicals/2/ (last access 3/23/2010).

Bingham, F.G.; Quigley, Jr. C.J. (1989). *Venture team performance to new product development.* The Journal of Business and Industrial Marketing, 4 (2), 49-59.

Biofuels Digest (2012): *USDA gives Zeachem conditional commitment of $232M loan guarantee.* Biofuels Digest, Jan. 27, 2012. http://www.biofuelsdigest.com/bdigest/2012/01/27/usda-gives-zeachem-conditional-commitment-on-232m-loan-guarantee/ (last access 3/9/2012).

Biofuels Digest (2008): *Germany to reduce biofuels target from 17 percent to 12 percent in 2020.* Biofuels Digest, Jul. 21, 2008. http://www.biofuelsdigest.com/blog2/2008/07/21/germany-to-reduce-biofuels-target-from-17-percent-to-12-percent-in-2020/ (last access 11/2/2009).

Bilton, N; Rusli, E. M. (2012): *From Founders to Decorators, Facebook Riches.* The New York Times, Feb. 1, 2012. http://www.nytimes.com/2012/02/02/technology/for-founders-to-decorators-facebook-riches.html?google_editors_picks=true (last access 1/2/2012).

BioFuelWatch (2009): *Shell's investments in agrofuel.* BioFuelWatch, Apr. 2009. http://www.biofuelwatch.org.uk/docs/Shell.pdf (last access 5/20/2009).

BioMCN (2011): *Dutch consortium takes initiative to build the world's largest biomass refinery in Delfzijl.* BioMCN New, Feb. 11, 2011. http://www.biomcn.eu/news/news/128-dutch-consortium-takes-initiative-to-build-the-worlds-largest-biomass-refinery-in-delfzijl.html (last access 12/12/2011).

BioMCN (2009): *BioMCN Annual Report 2008.* http://www.biomcn.eu/images/stories/BioMCN_2008_annual_report.pdf (last access 1/19/2010).

BioMCN (2008): *BioMCN Annual Report 2007.* http://www.biomcn.eu/images/stories/downloads/biomcn_annual_report_2007.pdf (last access 1/21/2010).

Bird, B. J. (1992): *The Operation of Intentions in Time: The Emergence of the New Venture.* Entrepreneurship: Theory and Practice 17 (1), 11-20. FindArticles.com. 08 Jul, 2009. http://findarticles.com/p/articles/mi_hb6648/is_n1_v17/ai_n28620138/.

Birds Eye Foods: *The Birds Eye Story.* http://www.birdseyefoods.com/ourcompany/story.aspx (last access 9/11/2009).

BIU – Der Bundesverband Interaktive Unterhaltungssoftware e. V. (2011): Association of Interactive Entertainment Software. *Market Data.* http://www.biu-online.de/.

Blanco, S. (2009): *China turns to methanol for flex fuel cars.* Autobloggreen, Nov. 10, 2009. http://green.autoblog.com/2009/11/10/china-turns-to-methanol-for-flex-fuel-cars (last access 1/19/2010).

Bläske, G.; Kiani-Kreß, R. (2010): *Wie OHB zum Raumfahrt-Konzernschreck aufstieg (How OHB emerged as a scare for space companies).* Wirtschaftswoche Jun. 30, 2010. http://www.wiwo.de/unternehmen-maerkte/wie-ohb-zum-raumfahrt-konzernschreck-aufstieg-433997/ (last access 7/4/2010); http://de.wikipedia.org/wiki/OHB_AG;
OHB AG: http://www.ohb.de/; http://www.ohb.de/historymilestones.html.

Blochwitz-Nimoth, J. (2008): *Novaled – von der Idee zum Produkt oder was macht eine erfolgreiche Ausgründung? (Novaled – from the idea to the product or what makes a successful spin-out?).* In Zukunft Ingenieurwissenschaften - Zukunft Deutschland. Beiträge einer 4ING-Fachkonferenz und der ersten Gemeinsamen Plenarversammlung der 4ING-Fakultätentage am 14. und 15.07.2008 an der RWTH Aachen, 51-60. Springer, Berlin-Heidelberg, 2008.

Blodget, H. (2009): *Mark Zuckerberg – The Man Behind The Company.* Business Insider, Oct. 8, 2009. http://www.businessinsider.com/henry-blodget-zuckerberg-personal-story-2009-10 (last access 9/10/2009), including linked videos to an interview with Zuckerberg.

BlueFire Ethanol Biofuels (2004): *Izumi Biorefinery – NEDO's Application of Arkenol's Concentrated Acid Hydrolysis Technology for the Conversion of Biomass to Ethanol.* http://bluefireethanol.com/images/IZUMI_Status_2004_for_BlueFire_051606.pdf (last access 12/23/2009).

BMBF – Bundesministerium für Bildung und Forschung (2007): *Entstehungspfade von Nachhaltigkeitsinnovationen (Development paths of sustainability innovations).* Expert Workshop of the research project nova-net, Feb. 13, 2007. http://wiki.iao.fraunhofer.de/images/novanet/Exp3.pdf (last access 7/4/2007).

BMFB (2008): *Schwerpunkt: Gründungen in außeruniversitären Forschungseinrichtungen (Center of gravity: Foundations in non-university research institutes)* – EXIST 2, 2008, 7-11. http://www.exist.de/imperia/md/content/exist-news/exist-news_2-2008.pdf (last access 7/12/2010).

BMBF (2011): *Federal Ministry of Education and Research – Innovation Alliance Carbon Nanotubes: Innovation for industry and society.* http://www.cnt-initiative.de/en/ (last access 2/22/2011).

BMWi – Federal Ministry of Economics and Technology (2011): *IGV Institut für Getreideverarbeitung GmbH – Profile.* Innovationskatalog,: Jan. 17, 2011 http://www.inno-watt.de/ikat/katalog/unternehmen.php?dataid=68&navanchor=2210003 (last access 6/5/2011).

BMWi - Federal Ministry of Economics and Technology – Germany: EXIST-Gründerstipendium: *Gründungsbeispiel - Cyano Biotech GmbH (Foundation Example – Cyano Biotech GmbH). Interview mit (with) Dan Kramer.* http://www.exist.de/exist-gruenderstipendium/gruendungsbeispiele/00156/index.php (last access 2/7/2010).

Böhm, U. (2005): *Kompetenz für deutsch-amerikanische Geschäftsbeziehungen (Competency for German-American business relations).* Lift Report, 07/01/2005. http://www.lift-report.de/index.php/news/57/328/Kompetenz-fur-deutsch-amerikanische-Geschaftsbeziehungen; English version http://www.lift-report.de/index.php/news/142/384/Competence-in-German-American-business-relationships (last access 9/21/2011).

Böttcher, C. (2006): *Füttern mit Treibhausgas (Feeding with green house gas).* Financial Times Deutschland, Apr. 27, 2006. http://www.ftd.de/wissen/:fuettern-mit-treibhausgas/68484.html (last access 3/15/2010).

Bomgardner, M. (2012a): *Biofuels Bloom In The South.* Chemical & Engineering News 90 (21), May 21, 2012. http://cen.acs.org/articles/90/i21/Biofuels-Bloom-South.html (last access 5/13/2012).

Bomgardner, M. (2012b): *LanzaTech Buys Range Fuels Facility.* Chemical & Engineering News 90, Jan. 11, 2011. http://cen.acs.org/articles/90/web/2012/01/LanzaTech-Buys-Range-Fuels-Facility.html (last access 1/14/2012).

Bomgardner, M. (2012c): *Money Flows For Biobased Chemicals.* Chemical & Engineering News 90, Feb. 28, 2012. http://cen.acs.org/articles/90/web/2012/02/Money-Flows-Biobased-Chemicals.html (last access 3/10/2012).

Bomgardner, M. (2012d): *Bioplastics Venture Is Off.* Chemical & Engineering News 90 (4), Jan. 23, 2012, 5. http://cen.acs.org/articles/90/i4/Bioplastics-Venture-Off.html (last access 1/28/2012).

Bomgardner, M. M. (2011a): *Genomatica wheels and deals with bio-based chemicals – Renewables: Firm inks butanediol partnership, makes bio-based butadiene.* Chemical & Engineering News, Aug. 22, 2011. http://pubs.acs.org/cen/news/89/i35/8935news.html (last access 9/6/2011).

Bomgardner, M. M. (2011b): *Advanced Biofuels: pipedream or solid investment?* Chemical & Engineering News, Dec. 16, 2011. http://cenblog.org/cleantech-chemistry/2011/12/advanced-biofuels-pipedream-or-solid-investment/ (last access 12/3/0/2011).

Bomgardner, M. (2011c): *US Cleantech firms in white-knuckle mode.* Chemcial & Engineering News, Dec. 7, 2011. http://cenblog.org/cleantech-chemistry/ (last access 12/202011).

Bomgardner, M. (2011d): *Algae Plans Bloom.* Chemical & Engineering News, Jul. 18, 2011. http://cen.acs.org/articles/89/i29/Algae-Plans-Bloom.html (last access Jul. 26, 2011).

Bomgardner. M. (2011e): *Making Rubber From Renewables.* Chemical & Engineering News 89 (50), Dec. 12, 2011, 18-19. http://cen.acs.org/articles/89/i50/Making-Rubber-Renewables.html (last access 12/15/2011).

Bomgardner, M. (2011f): *Genomatica To Go Public.* Chemical & Engineering News 89 (35), Aug. 29, 2011, 8. http://pubs.acs.org/cen/news/89/i35/8935notw7.html (last access 9/2/2011).

Bomgardner, M. (2011g): *BioAmber Files For $150 Million IPO*. Chemical & Engineering News 89, Nov. 15, 2011. http://cen.acs.org/articles/89/web/2011/11/BioAmber-Files-150-Million-IPO.html (last access 11/30/2011).

Bomgardner, M. M. (2011h): *Renewable Chemicals: OPX collaborates with Dow for biobased acrylic acid.* Chemical & Engineering News 89 (16), Apr. 18, 2011, 9. http://pubs.acs.org/cen/news/89/i16/8916notw5.html (last access 4/22/2012).

Bomgardner, M.(2011i): *Intermediates Start-ups Star in Green Chemistry*. Chemical & Engineering News, Jun. 20, 2011. http://cenblog.org/cleantech-chemistry/2011/06/intermediates-start-ups-star-in-green-chemistry/ (last access 8/28/2011).

Bonstetter, R.; Bonnstetter, B. J.; DeAnne, P. (2010): *Do you have what it takes to be a serial entrepreneur?* Target Training Systems, Ltd. 2010. http://www.perpetualdevelopment.com/files/serial-entrepreneurs.pdf (last access 11/5/2010.

Bordt, M.; Earl, L.; Lonmo, C.; Joseph, R. (2004): *Characteristics of Firms that Grow from Small to Medium Size – Growth Factors-Interviews and Measurability*. Statistics Canada, Science, Innovation and Electronic Information Division, Working Paper, Catalogue no. 88F0006XIE – No. 021, December 2004. http://publications.gc.ca/Collection/Statcan/88F0006X/88F0006XIE2004021.pdf (last access 7/7/2008).

Bosma, N.; Acs, Z. J.; Autio, E.; Coduras, A.; Levie, J. (2009): *Global Entrepreneurship Monitor – 2008 Executive Report.* http://www.gemconsortium.org/download/1250171096411/GEM_Global_08.pdf (last access 8/13/2009).
The Global Entrepreneurship Monitor (GEM): http://www.gemconsortium.org/.

Bosma, N.; Levie, J. (2010): *Global Entrepreneurship Monitor – 2009 Executive Report.* http://www.gemconsortium.org/download/1280664967554/GEM%202009%20Global%20Report%20Rev%20140410.pdf (last access 7/31/2010).

Bosma, N.; Harding, R. (2007): *Global Entrepreneurship Monitor – GEM 2006 Summary Results.* http://www.gemconsortium.org/download/1319374624384/GEM_2006_Global_Results_Summary_V2.pdf (last access 12/1/2007).

Böhme, H. (2008): *The New OLED Generation*. Deutsches Eigenkapitalforum 2008 Technology & Industrial Forum, file of 11/11/2008. http://www.eigenkapitalforum.com/082/presentations/f2/Company%20Presentation%20Novaled.pdf (last access 12/12/2010).

BP (2006): *BP Pledges $500 Million for Energy Biosciences Institute and Plans New Business to Exploit Research*. Release Jun. 14, 2006. http://www.bp.com/genericarticle.do?categoryId=2012968&contentId=7018719 (last access 11/10/2009).

BP (2007): *BP and D1 Oils Form Joint Venture to Develop Jatropha Biodiesel Feedstock*. Press Release, Jun. 29, 2007.

http://www.bp.com/genericarticle.do?categoryId=2012968&contentId=7034453 (last access 11/7/2009).

BP (2009): *BP and Verenium Form Leading Cellulosic Ethanol Venture to Deliver Advanced Biofuels.* http://www.bp.com/genericarticle.do?categoryId=2012968&contentId=7051362 (last access 12/7/2009).Birgeneau, R. (2007): *Energy Biosciences Institute.* Berkeleyan, Feb. 08, 2007. http://cio.chance.berkeley.edu/chancellor/Birgeneau/remarks/2-8-07-ebi.htm (last access 11/10/2009).

Brandon, G. – European Commission – Enterprise Directorate (2003): *General European Trend Chart on Innovation.* Thematic Report: Start-up of Technology-Based Firms. Covering period: October 2002 – September 2003. From INBIS Ltd, File TR_ntbfs_september_2003.pdf (last access 5/11/2010).

Brightman, J. (2011): *Online Games Market to Nearly Double to $29 Billion by 2016.* Industry Gamers, posted Jun. 23, 2011. http://www.industrygamers.com/news/online-games-market-to-nearly-double-to-29-billion-by-2016/ (last access 9/26/2011).

Breuer, R. (2009): *Carpe Diem – Interview mit Klaus Tschira.* Spektrum der Wissenschaft, Apr. 2009, 66-67. File sdw_2009_4_S66[1].pdf download from http://www.google.com/url?sa=t&source=web&cd=3&ved=0CBoQFjAC&url=http%3A%2F%2Fwww.wissenschaft-online.de%2Fartikel%2F986552&rct=j&q=%22carpe%20diem%22%20tschira%20wissenschaft%202009&ei=06ZbTOeiBYiBOOKO8dYP&usg=AFQjCNG0wOHTJJwPcZi1rq93xEmzkmNdlg&cad=rjt (last access 8/3/2010).

Brooks, D. (2011): *The Modesty Manifesto.* Aspen Ideas Festival in Aspen, Colo., Jun., 29, 2011. http://www.aifestival.org/session/modesty-manifesto (last access 7/1/2011); The New York Times, Mar. 10, 2011. http://www.nytimes.com/2011/03/11/opinion/11brooks.html (last access 3/11/2011).

Brooks, D. (2009a): *Yanks in Crisis.* The New York Times, Apr. 23, 2009. http://www.nytimes.com/2009/04/24/opinion/24brooks.html?_r=2&ref=todayspaper (last access 4/24/2009).

Brooks, D. (2009b): *The Nation of Futurity.* The New York Times, Nov. 16, 2009. http://www.nytimes.com/2009/11/17/opinion/17brooks.html?_r=1&hp (last access 11/17/2009).

Brosin, A. (2011): *Erschießen erlaubt, betäuben verboten - Aus Stasi-Akten: Forscher schlug dem MfS 1986 Betäubung von Mauer-Flüchtlingen vor, aber der Stasi war scharfe Munition lieber (Shooting allowed, to stun banned – Researcher suggested to the Stasi in 1986 stunning wall-refugees, but the Stasi preferred live ammunition).* Der Bundesbeauftragte für die Unterlagen des Staatssicherheitsdienstes der ehemaligen Deutschen Demokratischen Republik (BStU), http://www.bstu.bund.de/DE/BundesbeauftragteUndBehoerde/Aktuelles/2011_05_27_ardenne_betaeubung.html (last access 9/9/2012).

Brush, C.G.; Gieene, P. G.; Hait, M. M. (2001): *From initial idea to unique advantage: The entrepreneurial challenge of constructing a resource base.* IEEE Engineering Management Review 30 (1), :64-78 (2001)

Buchanan, L. (2011). *How Great Entrepreneurs Think*. Inc. Magazine, Feb. 1, 2011. http://www.inc.com/magazine/20110201/how-great-entrepreneurs-think.html (last access 4/5/2011).

Buckman, R. (2008): *Venture Capital's Coming Collapse*. Forbes, Dec. 18, 2008, Forbes Magazine, Jan 12, 2009. http://www.forbes.com/forbes/2009/0112/066.html (last acces 1/12/2009).

Buley, T. (2010): *Mark Zuckerberg – Richer Than Ever With An Estimated $4 Billion*. Forbes, Mar. 10, 2010. http://blogs.forbes.com/velocity/2010/03/10/mark-zuckerberg-richer-than-ever-with-an-estimated-4-billion/ (last access 4/2/2010).

Bullis, K. (2006): *The Methanol Economy – Forget about the hydrogen economy. Methanol is the key to weaning the world off oil. George Olah tells us how to do it.* Technology Review, Mar. 2, 2006. http://www.technologyreview.com/biztech/wtr_16466,296,p1.html?a=f (last access 1/22/2010); cf. also http://en.wikipedia.org/wiki/Methanol_economy.

Bullis, K. (2007): *Will Cellulosic Ethanol Take Off? Fuel from grass and wood chips could be big in the next 10 years – if the government helps.* Technology Review, Feb. 26, 2007. http://www.technologyreview.com/Energy/18227/?a=f (last access 11/27/2009).

Bullis, K. (2008a): *Fuel from Algae – A startup's new process could make fuel from algae as cheap as petroleum*. Technology Review, Feb. 22, 2008. http://www.technologyreview.com/Biztech/20319/ (last access 12/28/2009)

Bullis, K. (2008b): *Cheaper Butanol from Biomass. A startup has raised $25 million for inexpensively producing biofuel*. Technology Review, Oct. 27, 2008. http://www.technologyreview.com/business/21602/ (last access 12/30/2009).

Bullis, K. (2012a*): Sapphire Energy Raises $144 Million for an Algae Farm.* Technology Review, Apr. 6, 2012. http://www.technologyreview.com/energy/40073/ (last access 5/3/2012).

Bullis, K. (2012b): *Amyris Gives Up Making Biofuels: Update*. Technology Review, Feb. 10, 2012. http://www.technologyreview.com/blog/energy/27570/ (last access 2/19/2012).

Burmeister, K.; Schade, C. (2007)): *Are entrepreneurs' decisions more biased? An experimental investigation of the susceptibility to status quo bias*. Journal of Business Venturing 22, 340–362. http://opim.wharton.upenn.edu/risk/library/J2007JBV_KBur,CSchade_Entrepreneurs.pdf (last access 7/12/2010).

Burkeman, O. (2012): *The Power of Negative Thinking*. Wall Street Journal, Dec. 7, 2012. http://online.wsj.com/article/SB10001424127887324705104578147333270637790.html?mod=WSJ_MostPopular_Europe (last access 12/20/2012).

Burns, E. (2006): *Online Games Market to Hit $4.4 Billion by 2010*. ClickZ, Aug. 30, 2006. http://www.clickz.com/clickz/stats/1707790/online-games-market-hit-usd44-billion-2010 (last 9/26/2011).

Burtis, P. A. with Epstein, B. and Parker, N. (2006): *Creating Cleantech Clusters – 2006 Update*. E2 and cleantech VENTURE NETWORK, May 2006. http://www.e2.org/ext/doc/2006%20National%20Cleantech%20FORMATTED%20FINAL.pdf (last access 5/17/2010).

Business Week (2006): *Mistakes Made On The Road To Innovation.* Business Week, Nov. 27, 2006. http://www.businessweek.com/magazine/content/06_48/b4011421.htm (last access 7/31/2010).

Business Week (2005): *The Man Who Invented Management – Why Peter Drucker's ideas still matter*. Business Week, Front Cover, Nov. 28, 2005. http://www.businessweek.com/magazine/content/05_48/b3961001.htm (last access 5/30/2006).

Busse, J. (2011): *U.S. Firms, China Are Locked in Major War Over Technology* . Wall Street Journal, Feb. 2, 2011. http://online.wsj.com/article/SB10001424052748703439504576116152871912040.html (last access 2/2/2011).

Butler, J. E., Ed. (2004): *Opportunity Identification and Entrepreneurial Behavior.* A Volume in Research in Entrepreneurship and Management. Information Age Publishing, Inc. 2004.

Cain Miller, C. (2009): *Venture Capitalists Look for a Return to the A B C's.* The New York Times, Jul. 6, 2009. http://www.nytimes.com/2009/07/07/technology/start-ups/07venture.html?ref=todayspaper (last access 7/7/2009).

Campbell, J. (2007): *Science, Engineering, and Systems.* Lecture 2, Part 1, CSC340 - Winter 2007. http://www.cdf.toronto.edu/~csc340h/winter/lectures/w2/L2part1-4up.pdf (last access 8/11/2010.

Canada, A. (2009): *Company Growth Through Opportunity Maps*. Business Week, Feb. 9, 2009. http://www.businessweek.com/innovate/content/feb2009/id2009029_618352.htm?chan=innovation_innovation+%2B+design_innovation+strategy (last access 5/18/2010).

Carayannis, E. G.; Alexander, J.; Hausler, D.: *Government-University-Industry Partnerships*. Reference for Business. http://www.referenceforbusiness.com/management/Ex-Gov/Government-University-Industry-Partnerships.html (last access 6/4/2010).

Cardwell D. (2012): *Solazyme and Bunge Plan Factory to Make Oil From Algae*. The New York Times, Apr. 3, 2012. http://green.blogs.nytimes.com/2012/04/03/solazyme-and-bunge-plan-factory-to-make-oil-from-algae/ (last access 4/20/2012).

Carey, J. (2009): *The Biofuel Bubble*. Business Week, Apr. 16, 2009. http://www.businessweek.com/magazine/content/09_17/b4128038014860.htm?chan=top+news_top+news+index+-+temp_top+story (last access 4/18/2009).

Carlson, N. (2010): *How Mark Zuckerberg Hacked Into Rival ConnectU In 2004*. Business Insider, Mar. 5, 2010. http://www.businessinsider.com/how-mark-zuckerberg-hacked-connectu-2010-3 (last access 7/27/2010);
Carlson, N. (2010): *At Last – The Full Story Of How Facebook Was Founded.*

Business Insider, Mar. 5, 2010. http://www.businessinsider.com/how-facebook-was-founded-2010-3 (last access 7/27/2010).

Carroll, D. J. (2005): *Distributed R&D Case Study: InnoCentive. Managing Innovation: Emerging Trends.* Massachusetts Institute of Technology, Mar. 3, 2005. http://ocw.mit.edu/courses/sloan-school-of-management/15-352-managing-innovation-emerging-trends-spring-2005/lecture-notes/mit_mar07_2005.pdf (last access 9/24/2010).

Carter, R.; van Auken, H. (2006): *Small Firm Bankruptcy.* Journal of Small Business Management 44 (4), 493-512. http://www.allbusiness.com/management-companies-enterprises/3897245-1.html (last access 12/13/2010).

Casselman, B. (2009): *U.S. Gas Fields Go From Bust to Boom.* Wall Street Journal, Apr. 30, 2009. http://online.wsj.com/article/SB124104549891270585.html (last access 4/30/2009).

CeGaT (2012): *CeGaT– Center for Genomics and Transcriptomics.* Presentation, August 2012. http://www.cegat.de/files/CeGaT_Presentation.pdf; cf. also the German Founders' Award ("Deutscher Gründerpreis"), http://www.deutscher-gruenderpreis.de/owx_1_3619_1_5_0_00000000000000.html (last access 6/25/2011).

Chandler, G. N.; DeTienne, D.; Lyon, D. W. (2004): *Outcome Implications of Opportunity Creation / Discovery Processes.* Babson College, last updated May 2004. http://www.babson.edu/entrep/fer/BABSON2003/XVI/XVI-P1/XVI-P1.html (last access 10/8/2009).

Chandler, G. N.; Hanks, S. H. (1993): *Market attractiveness, resource-based capabilities, venture strategies, and venture performance.* Journal of Business Venturing 9, 331-349.

Charlton, B. (2005): *From inspiration to reality.* The Plain Dealer, Aug. 3, 2005. http://www.nanofilmtechnology.com/news/pdf/05AU01.pdf (last access 8/25/2009).

ChemSystems (2010): *Adipic Acid.* ChemSystems PERP Program, PERP 08/09, Feb. 2, 2010. http://www.chemsystems.com/reports/search/docs/abstracts/0809_2_abs.pdf (last access 6/2/2012).

Chen, M.-C.; Chang, Y.-C.; Chen, T.-M. (2004): *Entrepreneurial Leadership, Team Creativity, and New Venture Performance in Taiwan's SME Start-Ups.* Created: Nov. 23, 2004. http://www.yzu.edu.tw/admin/rd/files/rdso/G04/93/G04061(1).doc (last access 7/27/2010), and references therein.

Chernova, Y.; Ramsey, M. (2013): *How the Wheels Came Off for Fisker.* Wall Street Journal, updated April 24, 2013. http://online.wsj.com/article/SB10001424127887324874204578440640183079794.html?KEYWORDS=fisker+automotive (last access 4/29/2013).

Childs, D. (2007): *Diversa and Celunol become Verenium.* Cleantech News, Jun. 21, 2007. http://cleantech.com/news/1355/diversa-and-celunol-become-verenium (last access 12/1/2009).

Choia, K. S.; Deekb, F. P.; Imc, I. (2009): *Pair dynamics in team collaboration.* Computers in Human Behavior 25 (4), Jul. 2009, 844-852. Abstract and References:

http://portal.acm.org/citation.cfm?id=1551441&dl=GUIDE&coll=GUIDE&CFID=98428695&CFTOKEN=58683947.

Choia, K. S.; Deekb, F. P.; Imc, I. (2008): *Exploring the underlying aspects of pair programming: The impact of personality.* Information and Software Technology 50 (11), 1114-1126. Abstract and References: http://portal.acm.org/citation.cfm?id=1405313&dl=GUIDE&coll=GUIDE&CFID=98428695&CFTOKEN=58683947.

CHOREN (2010): *Launch of first BTL project in France using Carbo-V technology from CHOREN – Signature of EPC agreement for 45-MW synthesis gas plant.* Press Release, Jan. 29, 2010. http://www.choren.com/en/choren_industries/information_press/press_releases/?nid=216 (last access 5/6/2010).

Christensen, C. M. (1997): *The Innovator's Dilemma: When New Technologies Cause Great Firms to Fail.* Harvard Business School Press.

CI – Corporate Information (2009): *Bluefire Ethanol Fuels, Incorporation, Company Snapshot.* http://www.corporateinformation.com/Company-Snapshot.aspx?cusip=09622L102 (last access 1/8/2010).

Clanton, B. (2009): *Biofuels – CEO envisions a Gulf Coast algae boom.* Houston Chronicle, Aug. 8, 2009. http://www.chron.com/disp/story.mpl/business/energy/6565247.html (last access 2/2/2010).

CleanTech Group (2009): *Clean technology venture investment reaches record $8.4 billion in 2008 despite credit crisis and broadening recession.* Jan. 6, 2009 http://cleantech.com/about/pressreleases/010609.cfm (last access 5/18/2010); CleanTech Group (2009): *Clean technology venture investment totaled $5.6 billion in 2009 despite non-binding climate change accord in Copenhagen*, The Cleantech Group and Deloitte. Jan. 6, 2009 http://cleantech.com/about/pressreleases/20090106.cfm (last access 5/18/2010).

CNN (2010): *Algenol Partners With Lee County as Commission Votes to Approve Incentive Funding for Florida-Based Biofuels and Green Chemistry Facility.* CNNMoney.com, Feb. 3, 2010. http://money.cnn.com/news/newsfeeds/articles/prnewswire/DC48946.htm; http://www.news-press.com/article/20100202/GREEN/2020356/1075/Lee-County-approves-biofuel-facility (last access 2/2/2010).

Cocke, S. (2011): *HR BioPetroleum buying Shell's shares in Cellana.* Pacific Business News, Jan. 27, 2011. http://www.bizjournals.com/pacific/news/2011/01/27/hr-biopetroleum-buying-shells-shares.html (last access 7/7/2011).

Colao, J. J. (2013): *How Bad Is It At LivingSocial? PrivCo Claims 94% Of Value Wiped Out.* Forbes, Feb. 21, 2013. http://www.forbes.com/sites/jjcolao/2013/02/21/how-bad-is-it-at-livingsocial-privco-claims-94-of-value-wiped-out/ (last access 2/23/2013).

Collins, J.; Porras, J. (1996): *Building your company's vision.* Harvard Busimess Review, September 1996, 65-77.

http://www.intonz.dk/filoadmin/uoor_upload/Danoko/Downloado/built_to_laot_ _sammendrag_paa_engelsk.pdf (last access 8/23/2010).

Conner, K. R.; Prahalad, C. K. (1996): *A resource-based theory of the firm: Knowledge versus opportunism.* Organization Science 7(5), 477-501.

Contag, P. (2008a): *The Challenge for Biofuels.* Innovation, Vol. A (1), Special Issue 2008/2009. http://www.innovation-america.org/archive.php?articleID=495 (last access 1/29/2010).

Contag, P. (2008b): *Cobalt Biofuels.* In The 21st Industry Growth Forum. Transitioning Today's Energy to Tomorrow's Clean Energy. Denver, Colorado October 28-30, 2008. http://www.nrel.gov/technologytransfer/pdfs/igf21_cobalt_biofuels.pdf (last access 1/29/2010).

Cooney, T. M. (2005): *Editorial: what is an entrepreneurial team?* International Small Business Journal 23 (3), 226-235. http://goliath.ecnext.com/coms2/gi_0199-4370493/Editorial-what-is-an-entrepreneurial.html (last access 8/5/2010), and references herein.

Coval, J. D.; Jurek, J.; Stafford, E. (2008): *The Economics of Structured Finance.* Harvard Business School, Working Paper 09-060; 2008. http://www.hbs.edu/research/pdf/09-060.pdf (last access 9/12/2011).

Cowling, M.; Murray, G.; Fryges, H.; Licht, G. (2007): *The survival and growth of "adolescent" high-tech firms in Germany and the UK, 1997-2003.* Anglo-German Foundation for the Study of Industrial Society, January 2007. http://www.employment-studies.co.uk/summary/summary.php?id=agfsis07 (last access 7/10/2009).

Cramer, B. (2008): *More start-ups using credit cards as financing.* Published Jun. 27, 2008. http://canada.creditcards.com/credit-card-news/credit-cards-startup-funds-1269.php (last access 9/15/2009).

Creditreform – KfW – ZEW (2009): *Fahrt aufnehmen in stürmischen Gewässern - Chancen und Herausforderungen in der Entwicklung junger Unternehmen (Speeding in a stormy sea – Chances and challenges of developing young firms).* KfW – ZEW Gründungspanel 2, Oct. 2009. ftp://ftp.zew.de/pub/zew-docs/gruendungspanel/KfW_ZEW_Gruendungspanel_102009.pdf (last access 7/17/2010).

Crooks, E. (2009): *Shell reins back expectations for use of biofuels.* Financial Times, Nov. 30, 2009. http://www.ft.com/cms/s/0/d963b164-dd4e-11de-ad60-00144feabdc0.html (last access 12/18/2009).

Crowley, I. F.; Trudeau, J. R. (2011):*Wernher von Braun – An Ethical Analysis.* Document of 12/20/2011. http://www.wpi.edu/Pubs/E-project/Available/E-project-121811-161339/unrestricted/von_Braun_IQP_12_20_2011_bw_final.pdf (last access 1/9/2012).

CrunchBase (2009a): *Khosla Ventures.* Last edited Dec. 5, 2009. http://www.crunchbase.com/financial-organization/khosla-ventures (last access 11/30/2009).

CrunchBase (2009b): *Solazyme.* Last edited 12/8/2009. http://www.crunchbase.com/company/solazyme (last access 1/24/2010.

CrunchBase (2009c): *Sapphire Energy*. Last edited 12/7/2009. http://www.crunchbase.com/company/sapphire-energy (last access 3/13/2010).

Cyert, R.M.; March, J. G. (1963): *A Behavioral Theory of the Firm*. Prentice-Hall, Englewood Cliffs NJ.

Dankbar, C. (2011): *Familienfirma – Preußisch Blau für Fukushima (Family firm – Prussian Blue for Fukushima)*. Berliner Zeitung, Apr. 11, 2011. http://www.berlinonline.de/berliner-zeitung/berlin/339892/339893.php (last access 5/4/2011); also http://www.handelsblatt.com/technologie/forschung-medizin/medizin/ein-medikament-gegen-caesium/3980270.html (last access 3/23/2011).

Daly, J. W. (2010): *China spy scare: hypocrisy is spelled Echelon – Or, why allies spy on each other*. TechEye.net, Feb. 22, 2010. http://news.techeye.net/security/china-spy-scare-hypocrisy-is-spelled-echelon (last access 4/23/2011).

Daniel Meier Medienteam (2010): *Mikroalgen – klimafreundliche Alleskönner (Climate-friendly all-rounders)*. Landesinitiative Ernährungswirtschaft, Feb. 22, 2010. http://www.ernaehrungswirtschaft.de/index.php?con_cat=49&con_art=716&con_lang=1&sid=6r1iqoqav7ap75b03dld2965v3 (last access 3/16/2010).

DAPD (2011): *Linde kauft Biomasse-Sparte von insolventer Freiberger Firma Choren (Linde purchases the biomass business of the insolvent firm CHOREN in Freiberg)*. Freie Presse, Feb. 9, 2012. http://www.freiepresse.de/NACHRICHTEN/SACHSEN/Linde-kauft-Biomasse-Sparte-von-insolventer-Freiberger-Firma-Choren-artikel7900426.php (last access 2/15/2012).

Das, A. (2009): *Venture capitalists see clean-tech opportunity*. The New York Times, Jan. 9, 2009. http://www.nytimes.com/2008/01/09/business/worldbusiness/09iht-rtrinvest10.1.9096682.html (last access 11/20/2009).

Davis, A.; Russell. G. (2009): *U.S. Biofuel Boom Running on Empty*. Wall Street Journal, Aug. 27, 2009. http://online.wsj.com/article/SB125133578177462487.html (last access 8/27/2009).

De Guzman, D. (2012a): *LS9 prepares fatty alcohol sales* . ICIS Green Chemicals, Feb. 23, 2012. http://www.icis.com/blogs/green-chemicals/2012/02/ls9-prepares-fatty-alcohol-sal.html (last access 5/2/2012).

De Guzman, D. (2012b): *BASF invests in Renmatix*. ICIS Green Chemicals, Jan. 4, 2012. http://www.icis.com/blogs/green-chemicals/2012/01/basf-invests-in-renmatix.html (last access 1/ 30/2012).

De Guzman, D. (2012c): *First 2012 post: OPXBio update*. ICIS Green Chemicals, Jan. 1, 2012. http://www.icis.com/blogs/green-chemicals/2012/01/first-2012-post-opxbio-update.html (last access 2/5/2012).

De Guzman, D. (2012d): *Genomatica Withdraws IPO*. ICIS Green Chemicals, Aug. 2, 2012. http://www.icis.com/blogs/green-chemicals/2012/08/genomatica-withdraws-ipo.html (last access 17/10/2012).

De Guzman, D. (2011a): *Bio-lubes in the spotlight*. ICIS Green Chemicals, Apr. 13, 2011. http://www.icis.com/blogs/green-chemicals/2011/04/bio-lubes-in-the-spotlight.html (last access 8/13/2011).

De Guzman, D. (2011b): *BioAmber picks Ontario for new plant*. ICIS Green Chemicals, Sep. 1, 2011. http://www.icis.com/blogs/green-chemicals/2011/09/bioamber-picks-ontario-for-new.html (last access 10/23/2011).

De Guzman, D. (2011c): *Getting to know Renmatix*. ICIS Green Chemicals, Dec. 2, 2011. http://www.icis.com/blogs/green-chemicals/2011/12/getting-to-know-renmatix.html (last access 6/3/2012).

De Guzman, D. (2009): *Industrial Biotech Interview: Verdezyne*. ICIS Green Chemicals, Aug. 17, 2009. http://www.icis.com/blogs/green-chemicals/2009/08/industrial-biotech-interview-v.html (last access 9/3/2010).

Del Cardayre, S. (2009): *Introduction to LS9*. Conference Presentation, 2009. http://cleantech.com/documents/forums/2009/xxi_san_francisco/presentations/sector_leaders/LS9ConferencePresentation_gnp.pdf (last access 2/18/2010).

DeBruicker, J. (2006): *No. 13 Google: The Search Ended Here*. Business Week, Sep. 18, 2006. http://www.businessweek.com/magazine/content/06_38/b4001608.htm (last access 2/7/2008).

DeFrancesco, L. (2004): *Company founders: voices of experience – Interviews with nine leading scientists who founded startup companies*. Nature, published online Mar. 22, 2004. http://www.nature.com/bioent/2004/040301/full/bioent796.html (4/24/2008).

Della Vedova, B. (2009): *Bio-Fueling a Trade War. In Europe, protectionism passes as environmentalism.* Wall Street Journal, Aug. 3, 2009. http://online.wsj.com/article/SB10001424052970204313604574328533995657764 (last access 8/5/2009).

Den Hartigh, E.; Langerak, F.; Commandeur, H. R. (2002): *The Effects of Self-Reinforcing Mechanisms on Firm Performance*. April 2002. http://publishing.eur.nl/ir/repub/asset/200/erimrs20020517092944.pdf (last access 10/20/2010).

Dencker, J. C.; Grubery, M.; Shah, S. K. (2007): *Knowledge Acquisition Activities, Prior Knowledge and Experience & the Survival of New Firms*. In Frontiers of Entrepreneurship Research 27 (5), Chapter V – Entrepreneur Characteristics. Digital Knowledge at Babson - http://digitalknowledge.babson.edu/cgi/viewcontent.cgi?article=1011&context=fer (last access 9/21/2010).

Denk, J. E. (2012): *This Is Your Brain On The Goldberg Variations*. March 22, 2012. http://www.npr.org/blogs/deceptivecadence/2012/03/22/149148529/this-is-your-brain-on-the-goldberg-variations (last access 5/15/2013)/;
Variations in Music: http://en.wikipedia.org/wiki/Goldberg_Variations;
http://en.wikipedia.org/wiki/Piano_Sonata_No._11_(Mozart).

Dew, N. (2009): *Serendipity in Entrepreneurship*. Organization Studies 30 (7), 735-753.

Dinjus, E.; Dahmen, N.; Koerber, R. (2008): *Synthetic Fuel from biomass – The bioliq process*. IFAT China 2008, Shanghai. http://media.nmm.de/34/ic08_fuelfrombiomass-thebioliq-process_dr.rainerkoerber_21062234.pdf (last access 12/18/2009).
Dahmen, N (2009): *Combining industrial scale biosynfuel production with economic*

biomass logistics. ITC-CPV, Jan. 13, 2009. http://www.biofuelstp.eu/spm2/pdfs/Nicolaus_Dahmen_Armin_Gunther.pdf (last access 12/18/2009).

DiStefano, J. N. (2009): *PhillyDeals: Solazyme gets grant for biorefinery project in Pa.* Philadelphia Business Today, Dec. 8, 2009. http://www.philly.com/philly/business/78748952.html (last access 1/23/2010).

Djokovic, D.;Souitaris, V. (2008): *Spinouts from academic institutions – A literature review with suggestions for further research*. The Journal of Technology Transfer 33 (3), 225-247.

DOE – US Department of Energy (2006): *Breaking the Biological Barriers to Cellulosic Ethanol: A Joint Research Agenda*. A Research Roadmap Resulting from the Biomass to Biofuels Workshop, December 7–9, 2005, Rockville, Maryland. Publication Date: June 2006; DOE/SC-0095. http://genomicsgtl.energy.gov/biofuels/2005workshop/b2blowres63006.pdf (last access 12/9/2009).

DOE – US Department of Energy (2009): *DOE Issues – RFI on Draft National Algal Biofuels Technology Roadmap*. Green Car Congress, Jun. 4, 2009. http://www.greencarcongress.com/2009/06/algal-roadmap-20090604.html (last access 12/23/2009); Draft: National Algal Biofuel Roadmap: https://e-center.doe.gov/iips/faopor.nsf/UNID/79E3ABCACC9AC14A852575CA00799D99/$file/AlgalBiofuels_Roadmap_7.pdf (last access 12/23/2009).

Dolan, K. A. (2008): *Mr. Ethanol Fights Back*. Forbes Magazine, Nov. 24, 2008. http://www.forbes.com/forbes/2008/1124/052.html (last access 10/24/2009).

Donner, S. (2011): *Biosprit aus Kohlendioxide (Biofuels from carbon dioxide*). Technology Review, Feb. 16, 2011. http://www.heise.de/tr/artikel/Biosprit-aus-Kohlendioxid-1190439.html (last access 11/25/2011).

Dorf, R. C.; Byers, T. H. (2007): *Technology Ventures: From Idea to Enterprise*. McGraw Hill, New York (2nd Ed. with DVD; January 11, 2007) – emphasis on information & communication technology including software and electronics.

Downing, J. (2007): *Natural entrepreneur - Her dream is creating, selling biopesticides for organic farming.* The Sacramento Bee, Jun. 17, 2007. http://www.sacbee.com/103/story/225408.html (last access 7/15/2007); http://marronebioinnovations.com/pdf/TheSacBeeJune17-2007-NaturalEnt.pdf (last access 8/17/2010).

Dröscher, M. (2008): *Chemical Entrepreneurship – New Business Development in der Industrie – Beispiel Evonik Chemicals (Chemical Entrepreneurship – New Business Development in Industry – Example Evonik Chemicals)*. Presentation Technology Entrepreneurship, Jan. 9, 2008. http://ce.ioc.KIT.edu/download/Droescher_Karlsruhe20080109.pdf.

Drucker, P. F. (1964): *Managing for Results*. New York: Harper and Row Publishers, New York.

Drucker, P. F. (1995): *Innovation and Entrepreneurship: Practice and Principles.* Reprint from 1985, Harper & Row Publishers, New York.

Drucker, P. F. (2002). *Managing the Next Society*. The Four Entrepreneurial Pittfalls, 96-103. St. Martins Press, “Truman Talley Books,” 2002.
Rich Karlgaard (2004): *Peter Drucker On Leadership – Copied from Forbes with some format editing and notes added.* Forbes, Nov. 19, 2004. http://www.forbes.com/2004/11/19/cz_rk_1119drucker.html?partner=rss (last access 12/21/2010).

DTB – Deutsche Biogas AG (2011): *DTB – Deutsche Biogas AG baut Forschung und Entwicklung aus (Deutsche Biogas AG expands research and development*). http://www.presseportal.de/pm/82738/2004391/dtb-deutsche-biogas-ag-baut-forschung-und-entwicklung-aus (last access 5/23/2011).

Dziersk, M. (2008): *Design Thinking ... What is That*? Fast Company, Jul. 8, 2008. http://www.fastcompany.com/resources/design/dziersk/design-thinking-083107.html (last access 10/10/2009).

E.ON (2008): *Pilotanlage in Hamburg-Reitbrook*. E.ON Hanse, Press Release http://www.eon-hanse.com/pages/eha_de/Engagement/Mikroalgenprojekt_Hamburg/Pilotanlage_in_Hamburg-Reitbrook/index.htm (last access 1/25/2010).

Ebert, J. (2008): *Biobutanol: The Next Big Biofuel?* Biomass Magazine, May 2008. http://www.biomassmagazine.com/article.jsp?article_id=1605&q=&page=all (last access 10/30/2009).

Eckhaus, C. (2009): *Solazym*. Independent Search Consultant / Director Human Resources, Activities at Solazyme. http://www.linkedin.com/pub/christine-eckhaus/4/447/6a4 (last access 1/25/2010).

Eckmann, T. (2008): *Bootstrapping Your Start-Up*. File from 7/26/2008. http://bschool.washington.edu/cie/PDF/Bootstrapping_Your_Startup-Eckmann.ppt (last access 11/23/2009).

Ecofusion: Case Study: *Cobalt Biofuels – Introducing a New Fuel Solution.* http://www.ecofusion.com/case-studies/cobalt-biofuels.html (last access 1/28/2010).

EcoFys (2006): *Methanol plant takeover to bring bio-methanol production to Delfzijl*. Press release, Nov. 2, 2006. http://ecofys.com/com/news/pressrelease2Nov2006.htm (last access 1/19/2010).

Econo Editor WST (2007): *Bokela - Wer hat’s erfunden*? *Die Karlsruher* (Who did invent it – People from Karlsruhe). Econo, 1/2007 – Dec. 29, 2006, Technologieregion Karlsruhe, page 9. http://www.econo-online.de/dynamic/bildgalerie/08_trkx5.pdf (last access 11/13/2010).

Edible Apple (2011): *From Alan Kay’s Dynabook to the Apple iPad*. Edible Apple, Apr. 30, 2010. http://www.edibleapple.com/from-alan-kays-dynabook-to-the-apple-ipad/ (last access 2/17/2011).

EERE – U.S. Department of Energy – Energy Efficiency and Renewable Energy (2008): Oil *Companies, Universities Investigate Alternate Paths to Biofuels*. EERE News, Apr. 16, 2008. http://www1.eere.energy.gov/industry/distributedenergy/printable_versions/news_detail.html?news_id=11710 (last access 11/23/2009).

Effectuation.org: *About Effectuation*. http://www.effectuation.org/about-effectuation; http://en.wikipedia.org/wiki/Effectuation; http://de.wikipedia.org/wiki/Effectuation.

Efrati, E. (2013): *At 17, App Builder Rockets to Riches From Yahoo Deal*. Wall Street Journal, updated Mar. 26, 2013. http://online.wsj.com/article/SB10001424127887323605404578382243461824094.html?mod=WSJ_MostPopular_Europe (last access 3/26/2013).

Egeln, J.; Gottschalk, S.; Rammer, J.; Spielkamp, A. (2002): *Spinoff-Gründungen aus der öffentlichen Forschung in Deutschland (Spin-off foundations from public research in Germany)*. Zentrum für Europäische Wirtschaftsforschung GmbH (ZEW), Gutachten für das Bundesministerium für Bildung und Forschung (Report to BMBF), Document 03/02, Nov. 2002. ftp://ftp.zew.de/pub/zew-docs/docus/dokumentation0302.pdf (last access 6/3/2010).

Egeln, J.; Falk, U.; Heger, D.; Höwer, D.; Metzger, G. (2010): *Ursachen für das Scheitern junger Unternehmen in den ersten fünf Jahren ihres Bestehens (Reasons for failure of young enterprises within the first five years of their existence)*. ZEW – Zentrum für Europäische Wirtschaftsforschung, March 2010. ftp://ftp.zew.de/pub/zew-docs/gutachten/Scheitern_junger_Unternehmen_2010.pdf (last access 11/18/2010).

Eicher, C. C. (2012): *Bio hat Zunkunft (Bio has a future*). ADAC Motorwelt 6, June 2012, 45.

EIM (2009): *Study on the representativeness of business organizations for SMEs in the European Union*. EIM Business & Policy Research, Final report, May 2009. http://www.ifm-bonn.org/assets/documents/ENSR_Study_070709.pdf (last access 9/29/2010).

ElectroIQ (2011): *GE crashes CdTe party with record panels, US capacity, M&A*. Published Apr. 8, 2011. http://www.electroiq.com/index/display/photovoltaics-article-display/4741264984/articles/Photovoltaics-World/thin-film_solar_cells/cadmium-telluride/2011/4/ge-crashes_cdte_party.html (last access 5/8/2011).

EnBW (2008): *EnBW untersucht neuartige Algenzucht zur CO2-Bindung (EnBW investigates new algae culture for CO2 binding))*. Press Release, Jul. 16, 2008. http://www.enbw.com/content/de/presse/pressemitteilungen/2008/07/PM_20080716_FE_mw01/index.jsp;jsessionid=0232C3216B17114C0A0DC4B3B7C53DE2.nbw05 (last access 3/17/2010), also http://www.enbw.com/content/de/presse/pressemitteilungen/2009/07/pm_2009_07_17_enbw_subitec_cu_ys1/PM_Eutingen_EnBW_Subitec.pdf (last access 3/16/2010).

Energy Matters (2013): *Bosch To Cease Manufacturing Solar Panels*. Renewable Energy, News. http://www.energymatters.com.au/index.php?main_page=news_article&article_id=3655 (last access Mar. 28, 2013).

Engeler, G. (2008): *Finding Newer, Cleaner Ways to Meet Transportation Energy Needs*. Chevron – Global Downstream Technology, Hong Kong, Sep. 2008. http://www.hkbu.edu.hk/~hkesc/asianenergy/Session%202%20-paper%202.pdf (last access 1/23/2010).

Ensley, M. D.; Carland, J. W.; Carland, J. C. (1998): *The Effect of Entrepreneurial Team Skill Heterogeneity and Functional Diversity on New Venture Performance*.

Journal of Business and Entrepreneurship, Mar. 1998. http://www.sbaer.uca.edu/research/asbe/1998/pdf/09.pdf; http://findarticles.com/p/articles/mi_qa5424/is_199803/ai_n21417164/pg_2/ (last access 8/5/2010).

Erneuerbare Energien – Renewable Energy: *Besondere Windenergiesysteme – Vertikalachsen-Rotoren (Special wind energy systems – Vertical axis turbines).* http://www.energyprofi.com/jo/Besondere-Windenergiesysteme.html (last access 9/10/2010).

Ernst & Young (2011): *Mittelstandsstudie von Ernst & Young – Mit Innovationen kraftvoll an die Spitze (Study of medium-sized firms of Ernst & Young*). Aachen-euregio.businesson – Das regionale Wirtschaftsportal, Jan. 28, 2011. http://www.business-on.de/aachen-euregio/young-ernst-mitarbeiter-zielen-maerkte-angebot-_id1693.html (last access 2/11/2011).

Esposito, F. (2013): *Biopolymer maker BioAmber raises $80 million in IPO.* PlasticsNews, May 17, 2013. http://www.plasticsnews.com/article/20130517/NEWS/130519929/biopolymer-maker-bioamber-raises-80-million-in-ipo# (last access 5/17/2013).

EU (2006): *Biofuels in the European Union – A vision for 2030 and beyond.* Final draft report of the Biofuels Research Advisory Council, Mar. 3, 2006. http://ec.europa.eu/research/energy/pdf/draft_vision_report_en.pdf (last access 3/31/2010).

European Biodiesel Board (EBB) (2009): *Statistics 2009.* http://www.ebb-eu.org/stats.php (last access 3/11/2009).

European Commission: SME Definition: http://ec.europa.eu/enterprise/enterprise_policy/sme_definition/index_en.htm; http://ec.europa.eu/enterprise/enterprise_policy/sme_definition/sme_user_guide.pdf (last access 7/10/2009).

European Commission – Innovation Directorate (2003): "*Breeding business success.*" Euroabstracts 41(1): 6 (published February 2003). ftp://ftp.cordis.europa.eu/pub/euroabstracts/docs/archive1-03.pdf (last access 02/15/2008); reference to European Communities. 2002. European entrepreneurship training curriculums (ISBN 92 894 3967 X), Luxembourg.

Everts, S. (2008): *Max Planck Society Moves Stateside - Germany's prestigious research society sets up shop in Florida even as it faces criticism at home.* Chemical & Engineering News 86 (9), 12-17. http://pubs.acs.org/cen/coverstory/86/8609cover.html (last access 6/1/2010).

Fahrendorf, T. (2008): *Mikroalgen zur Steigerung der Wertschöpfung an Biogasanlagen durch Aquakultur (Microalgae to increase value capture of biogas plants by aquaculture*). Novagreen Presentation, Jun. 13, 2008.http://www.3-n.info/download.php?file=pdf_files/Vortraege/080613_5-2_fahrendorf_novagreen.pdf (last access 3/16/2010).

Faltin, G. (2007): *Erfolgreich gründen – Der Unternehmer als Künstler und Komponist (Founding Successfully – The entrepreneur as an Artist and Composer).* © DIHK – Deutscher Industrie und Handelskammertag, April 2007.

Farrell, I, (2013): *Recycling rare earth elements using ionic liquids*. Chemistry World, Mar. 15, 2013. http://www.rsc.org/chemistryworld/2013/03/recycling-rare-earth-neodymium-and-samarium-ionic-liquids (last access 3/27/2013).

Fehrenbacher, K. (2007): *10 Khosla Biofuel Bets*. Earth2tech, posted Oct. 12, 2007. http://earth2tech.com/2007/10/12/10-khosla-biofuel-bets/ (last access 11/24/2009).

Fehrenbacher, K. (2008): *Ethanol Startup Coskata Launches, Backed By General Motors and Khosla*. Earth2tech, posted Jan.13, 2008. http://earth2tech.com/2008/01/13/ethanol-startup-coskata-launches-backed-by-general-motors-and-khosla/ (last access 11/24/2009).

Fehrenbacher, K. (2009): *Cobalt Biofuels Opens Plant; Biobutanol the Next Thing?* Earth2tech, posted Jan. 12, 2010. (last access 1/28/2010).

Fehrenbacher, K. (2011): *Some red flags & numbers in Mascoma's IPO filing*. Gigaom – Earth2tech, Sep. 16, 2011. http://gigaom.com/cleantech/some-red-flags-numbers-in-mascomas-ipo-filing/ (last access 1/24/2012).

Fehrenbacher, K. (2012): *Biofuel startup Renmatix brings on BASF as backer*. Eartth2tech, Jan. 3, 2012. http://gigaom.com/cleantech/biofuel-startup-renmatix-brings-on-basf-as-backer/ (last access 1/ 30/2012).

Feldenkirchen, W.; Hilger, S. (2009): *Menschen und Marken – 125 Jahre Henkel (Men and brands – 125 years Henkel)*. File from 11/7/2009. http://www.henkel.de/de/content_data/177905_MenschenundMarken_125JahreHenkel.pdf (last access 3/18/2011).

Fergerson, W. (2007):*Using Credit Cards: Watch the Fine Print*. http://www.entrepreneurship.gov/ResourcesCenter/Topics/FinanceAndAccounting/13988.html (last access 9/15/2009).

Fertig, N. (2012): *Nanion Technologies – Von der Grundlagenforschung zum global erfolgreichen Unternehmen (Nanion Technology – From basic research to a globally successful enterprise*). Presentation Technology Entrepreneurship, Jan. 15, 2012. http://ce.ioc.KIT.edu/downloads/TH_Karlsruhe_Dr_Fertig1.pdf.

Fier, A.; Fryges, H. (1999): *Tendenz zur Internationalisierung junger Technologieunternehmen* (Tendency of Internationalization of Young Technical Enterprises). Zentrum für Europäische Wirtschaftsforschung, Mannheim. ZEW News, March 1999. ftp://ftp.zew.de/pub/zew-docs/zn/zn0399.pdf (last access 10/4/2010).

Fiet, J. O.; Clouse, V. G. H.; Norton, W. I. (2004): *Systematic Search by Repeat Entrepreneurs*. Chapter 1 in Butler, John E., Ed. (2004): Opportunity Identification and Entrepreneurial Behavior. A Volume in Researrch in Entrepreneurship and Management. Information Age Publishing, Inc. 2004.

Finfacts (2010): *US venture capitalists invested $18bn in 2009; Their European counterparts €4bn*. Finfacts, Sep. 15, 2010. http://www.finfacts.ie/irishfinancenews/article_1020572.shtml (last access 10/24/2010).

FirstMention: *Adsense 1999*. http://firstmention.com/Adsense.aspx (last access 1/29/2012).

Flaig, G.; Stadler, M. (1994): *Success breeds success. The dynamics of the innovation process*. Empirical Economics 19 (1), 55-68.

Forbes (2011): *Forbes 400 Richest Americans. Jim Breyer*. http://www.forbes.com/profile/jim-breyer/ (last access 10/30/2011).

Foss, N. J.; Klein, P. G.; Kor, Y. Y.; Mahoney, J. T. (2006): *Entrepreneurship, subjectivism, and the resource – based view: towards a new synthesis*. Working Paper. http://www.business.illinois.edu/Working_Papers/papers/06-0121.pdf (last access 9/28/2009).

Franco, A. M.; Filson, D. (2006): *Spin-outs: Knowledge Diffusion through Employee Mobility*. The Rand Journal of Economics; Winter 2006, 37 (4), 841-860. http://www.rotman.utoronto.ca/facBios/file/Rand.pdf (last access 7/20/2009).

Freundorfer, S. (2010): *Boombranche Browser-Game – Die kleinen Könige der Spielebranche (Booming browser games – The little kings of the games industry)*. Der Spiegel, Feb. 27, 2010. http://www.spiegel.de/netzwelt/games/0,1518,680100,00.html (last access 1/22/2011).

Friedman, T. (2008):*Time to Reboot America*. The New York Times, Dec. 23, 2008. http://www.nytimes.com/2008/12/24/opinion/24friedman.html?_r=1 (last access 12/23/2008).

Frishammar, J.; Florén, H.; Wincent, J. (2009)*: Patterns of Uncertainty and Equivocality during Predevelopment: Findings from Process-Based Firms.* Conference Paper, Conference: International Conference on Management of Technology, IAMOT 2009, No. 18, Orlando, United States, April 5, 2009 - April 9, 2009. https://dspace.hh.se/dspace/bitstream/2082/3388/1/Conference_paper_JF_HF_JW_IAMOT.pdf (last access 10/14/2009). (last access 10/14/2009).

Fröhlich, C. (2013): *Nick D'Aloisio verkauft App an Yahoo: Plötzlich Millionär (Nick D'Aloisio sells app to Yahoo: suddenly millionaire*). Der Stern, Mar. 26, 2013. http://www.stern.de/digital/online/nick-daloisio-verkauft-app-an-yahoo-ploetzlich-millionaer-1989636.html (last access 3/26/2013).

Fröndhoff, B. (2008): *Was Mittelständler erfolgreich* macht (What makes medium-sized firms successful). Handelsblatt, Jul. 15, 2008. http://www.handelsblatt.com/unternehmen/mittelstand/was-mittelstaendler-erfolgreich-macht/2989122.html?p2989122=all (last access 1/11/2011).

Fryges, H.; Gottschalk, S.; Licht, G.; Müller, K. (2007): *Hightech-Gründungen und Business Angels (Hightech Foundations and Business Angels).* ZEW – Zentrum fü Europäische Wirtschaftsforschung GmbH, Oct. 2007. ftp://ftp.zew.de/pub/zew-docs/gutachten/businessangel-endbericht.pdf (last access 10/12/2010).

Fuhrmans V. (2012): *Germany's New Export: Jobs Training*. Wall Street Journal, updated Jun. 14, 2012. http://online.wsj.com/article/SB10001424052702303665904577452521454725242.html?mod=WSJ_business_whatsNews (last access 6/15/2012).

Fyfe, D.; Townsend, R. (2005): *Chemical Science Spin-outs from UK Universities - Review of Critical Success Factors*. The Chemistry Leadership Council and the Royal Society of Chemistry (RSC), 2005.

http://www.rsc.org/images/ChemSciSpinOuts_tcm18-12643.pdf (last access 7/24/2010).

Galbraith, K. (2008): *Assessing the Value of Small Wind Turbines*. The New York Times, Sep. 3, 2008. http://www.nytimes.com/2008/09/04/business/04wind.html?_r=1&em (last access 9/9/2010).

Gage, D. (2012): *The Venture Capital Secret: 3 Out of 4 Start-Ups Fail*. Wall Street Journal, September 19, 2012. http://online.wsj.com/article/SB10000872396390443720204578004980476429190.html?mod=WSJUK_hpp_MIDDLEThirdNews (last access 9/21/2012).

Gangemi J. and Di Meglio, F. (2005), *Teaching the Startup Mentality*. Business Week, Oct. 25, 2005. http://www.businessweek.com/technology/content/oct2005/tc20051025_795457.htm?chan=technology_special+report+2005+-+young+entrepreneurs+of+tech (last access 05/07/2008).

Gangemi J. (2007): *A Weeklong Festival for Entrepreneurship*. Business Week, Feb. 22, 2007. http://www.businessweek.com/smallbiz/content/feb2007/sb20070222_645291.htm?chan=smallbiz_smallbiz+index+page_today%2527s+top+stories (last access 8/24/2009).

Garberson, J. (2011): *Precious Assets Hindered by Bureaucracy: Lab Scrutinized by National Academy*. The Independent, Apr. 30, 2011. http://www.independentnews.com/news/article_7a30ea08-72bd-11e0-bd4d-001cc4c002e0.html (last access 7/2/2011).

Garcia, M. L.; Bray, O. H. (1997):*Fundamentals of Technology Roadmapping*. Strategic Business Development Department, Sandia National Laboratories. Report SAND97-0665. http://prod.sandia.gov/techlib/access-control.cgi/1997/970665.pdf; file 10.1.1.31.1385.pdf accessed via Google search "Garcia Bray 'Fundamentals of Technology Roadmapping'" (last access 9/24/2010).

Garnsey, E.; Stam, E.; Heffernan, P. (2006): New *Firm Growth: Exploring Processes and Paths.* Industry and Innovation 13 (1), Mar. 2006, 1-20. http://econ.geo.uu.nl/stam/GarnseyStamHeffernan2006.pdf (last access 1/20/2011).

Gardner, T. (2009): *Virent biofuel sees financing round up to $40 million*. Reuters. http://www.reuters.com/article/idUSTRE55M6NS20090623 (last access 12/14/2009).

Garthwaite, J. (2009): *CHEAT SHEET: Heavy Hitters in Algae Fuel Deals*. Earth2tech, Jul. 14, 2009. http://earth2tech.com/2009/07/14/cheat-sheet-heavy-hitters-in-algae-fuel-deals/ (last access 1/2/2010).

Gartner, W. B. (2007): *Psychology, Entrepreneurship, and the "Critical Mess."* Chapter 14 in Baum, J. R.; Frese, M.; Baron, R. A. (eds.): The Psychology of Entrepreneurship. A Publication of the Society for Industrial and Organizational Psychology, Lawrence Erlbaum Associates, Publishers, 2007.

GBL – Green Biologics (2012): *Green Biologics Announces Merger with Butylfuel™*. Press Relase, Jan. 20, 2012. http://www.greenbiologics.com/pr-butylfuel.php (last access 4/23/2012).

Gellene, D. (2009): *A Sapphire Energy Co-Founder Sees Solutions in Algae for Drugs as Well as Biofuels.* Xconomy, Dec.2, 2009. http://www.xconomy.com/san-diego/2009/12/02/a-sapphire-energy-co-founder-sees-solutions-in-algae-for-drugs-as-well-as-biofuels/ (ast access 3/14/2010).

Gelsi, S. (2008): *Algenol takes fresh approach to ethanol production.* MarketWatch, last update, Sep. 24, 2008. https://www.sharebuilder.com/sharebuilder/Research/MarketUpdate/StockNews.aspx?GUID=%7BE36349E5-0713-4468-A8BA-FE5DBA1ADF0C%7D&Source=IndividualStock&Symbol=ADM (last access 2/2/2010).

Gelsi, S. (2009): *Exxon Mobil eyes bigger role for natural gas - Oil giant forecasts trillions for world energy projects.* MarketWatch, Dec. 8, 2009. http://www.marketwatch.com/story/exxon-sees-trillions-needed-for-energy-growth-2009-12-08?siteid=rss&rss=1 (last access 12/8/2009).

Gendron, G. (1996): *Flashes of Genius.* Inc. Magazine, May 15, 1996. http://www.inc.com/magazine/19960515/2083.html# (last access 3/23/2010).

Gergen, C.; Vanourek, G. (2009): *Fending off the Recession with "Adaptive Persistence."* HBR Blogs Christopher Gergen and Gregg Vanourek, Apr. 7, 2009. http://blogs.hbr.org/gergen-vanourek/2009/04/fending-off-the-recession-with.html (last access 10/31/2010); Perman, S. (2008): *Intrapreneurs and Adaptive Persistence.* Business Week, Apr. 2, 2008. http://www.businessweek.com/smallbiz/content/apr2008/sb2008042_307474.htm (last access 4/30/2008).

Gevo, Inc. (2009): *Brownfield: Ag News for America – Company plans to convert ethanol plants to biobutanol.* Gevo News, Oct. 9, 2009. http://www.gevo.com/news_Ag-News_100909.php (last access 12/10/2009).

Ghaemi, N. (2008): *The psychology of ambition. The price of ambition.* Psychology Today, August 21, 2008. Blogs – Mood Swings, a Psychiatrist Surveys the Mind and the Wider World 8/7/2009. http://www.psychologytoday.com/blog/mood-swings/200808/the-psychology-ambition (last access 8/7/2009).

Giebel, H. (2013): *TimberTowers for multi-megawatt wind energy plants.* Presentation Technology Entrepreneurship, Dec. 11, 2013. http://ce.ioc.kit.edu/downloads/TimberTower_130827_KIT.pdf.

Gibson, N. (2008): *German Persistence – Our friends in Deutschland have found a mass-market gaming goldmine.* develop, Sep. 22, 2008. http://www.developmag.com/interviews/276/German-Persistence last access 9/20/2010).

Gillies, A. T. (2008): *Business In The Beltway – Taking On Congress' Favorite Biofuel.* Fortune, Mar. 12, 2008. http://www.forbes.com/2008/03/11/virent-ethanol-gasoline-biz-cz_atg_0312beltway.html (last access 12/14/2009).

Giménez, E. H. (2007): *The biofuel myths*. The New York Times, Jul. 10, 2007. http://www.nytimes.com/2007/07/10/opinion/10iht-edholt.1.6588231.html (last access 7/11/2007).

GiSeung Kim (2008): *Entrepreneurship and Self-Employment: The State-of-the-Art and Directions for Future Research*. New England Journal of Entrepreneurship 11 (1), Apr. 1, 2008, 39-52. http://www.sacredheart.edu/download/1924_neje_spring08.pdf (last access7/18/2009).

Gladwell, M. (2000) *The Tipping Point: How Little Things Can Make a Big Difference*. Little, Brown (Boston).

glassdoor.com (2010): *Los Alamos National Laboratory Employee Review*. glassdoot™ - an inside look at jobs and companies, Mar. 22, 2010. http://www.glassdoor.com/Reviews/Employee-Review-Los-Alamos-National-Laboratory-RVW439215.htm (last access 6/19/2010).

Glocalist (2008): *Forschung: Biokraftstoff und Wasserstoff aus Bakterien (Research: Biofuels and hydrogen from bacteria)*. Glocalist Daily News, Aug. 12, 2008. http://www.glocalist.com/index.php?id=20&tx_ttnews%5Btt_news%5D=3925&tx_ttnews%5Bcat%5D=1&cHash=2e3d1ba4d5 (last access 12/16/2009).

Gold, R. (2009a): *Size Matters: Biobutanol is Latest (Small) Biofuel Aspirant*. Wall Street Journal Blog, Sep. 30, 2009. http://blogs.wsj.com/environmentalcapital/2009/09/30/size-matters-biobutanol-is-latest-small-biofuel-aspirant/ (last access 11/9/2009).

Gold, R. (2009b): *In Energy Innovation, Everything New Is Old Again – Most Alternative Fuel Technologies Have Roots in Long-Ago Discoveries; Scarcity of 'Killer Apps' Slows Progress*. Wall Street Journal, Dec. 11, 2009. http://online.wsj.com/article/SB126048948482786623.html (last access 12/11/2009).

Golon, J. (2006): *Algae to Biomass*. Presentation, Winter 2006/2007. http://www.ecology.uni-kiel.de/~bburkhard/seminar/material6/AESA_041207_Algae.pdf (last access 3/162010).

Gompers, P. A. (2002): *Corporations and the Financing of Innovation – The Corporate Venturing Experience*. Harvard Business School and National Bureau of Economic Research. Presentation, May 3, 2002. https://www.frbatlanta.org/filelegacydocs/policy_gompers.pdf (last access 10/29/2010).

Gompers, P.; Kovner, A.; Lerner, J.; Scharfstein, D. (2008): *Performance Persistence in Entrepreneurship*. Draft, July 2008. http://www.hbs.edu/research/pdf/09-028.pdf (last access 1/24/2011).

González, Á (2008): *Weyerhaeuser, Chevron form biofuels venture*. The Seattle Times, originally published Feb. 29, 2008. http://seattletimes.nwsource.com/html/businesstechnology/2004250829_webchevron01.html (last access 11/24/2009).

Gorlick, R. (2009): *Fiese Karrierefallen – So ticken amerikanische Chefs (Nasty career traps – how American bosses proceed)*. Der Spiegel, Dec. 12, 2009.

http://www.spiegel.de/unispiegel/jobundberuf/0,1518,663018,00.html (last access 04.12.2009).

Gorman, S. (2011): *U.S. Report Cites 'Persistent' Chinese, Russian Spying for Economic Gain*. Wall Street Journal, Nov. 3, 2011. http://online.wsj.com/article/SB10001424052970203716204577015540198801540.html?mod=WSJEUROPE_hpp_MIDDLEThirdNews (last access 11/3/2011).

Gottschalk, S.; Fryges, H.; Metzger, G.; Heger, D.; Licht, G. – ZEW (2007): *Start-Ups zwischen Forschung und Finanzierung: Hightech-Gründungen in Deutschland*. Mannheim, June 2007. (*Start-Ups between Research and Financing* : *Hightech Foundations in Germany. June 2007*). ftp://ftp.zew.de/pub/zew-docs/gutachten/hightechgruendungen07.pdf (last access 7/10/2009).

Gould, P.: *General History – The Steam Bus 1833-1923*. http://www.petergould.co.uk/local_transport_history/generalhistories/general/steambus.htm; http://en.wikipedia.org/wiki/Steam_bus (last access 1/6/2011).

Green Car Congress (2005): *Rentech Tracking to Startup Coal-to-Liquids Pilot Plant in Q4 2006*. Green Car Congress, Dec. 7, 2005. http://www.greencarcongress.com/2005/12/rentech_trackin.html (last access 2/20/2010).

Green Car Congress (2007a): *European Biodiesel Production Doubles in Two Years; EBB Warns on Need to Eliminate "Legislative and Standards Desert."* Published Jul. 18, 2007. http://www.greencarcongress.com/2007/07/european-biodie.html (last access).

Green Car Congress (2007b): *Range Fuels to Build First Wood Cellulosic Ethanol Plant in Georgia*. Green Car Congress, Feb. 7, 2007 (2/19/2010). http://www.greencarcongress.com/2007/02/range_fuels_to_.html.

Green Car Congress (2012): *BlueFire Renewables subsidiary and Caltex to build cellulose to sugar plant in Korea*. Green Car Congress, Feb. 15, 2012. http://www.greencarcongress.com/2012/02/clatex-20120215.html (last access 3/5/2-12).

Greene, S. (2009): *An Overview of the US Electric Bike Market*. Sperlingreene PR and Marketing Communications, Jul. 8, 2009. http://www.scribd.com/doc/17166410/Overview-of-the-Ebike-and-Electric-Bike-Market (last access 5/14/2011).

Greiner, L. E. (1998): *Evolution and Revolution as Organizations Grow*. Harvard Business Review, May-June 1998, 55-67;
Greiner, L. E. (1972): *Evolution and Revolution as Organizations Grow*. Harvard Business Review, July-August 1972, 37-46.
http://www.ils.unc.edu/daniel/131/cco4/Greiner.pdf (last access 1/31/2011).

Griesar, K. (2008): *Management von Forschung und Entwicklung (Management of Reseach and Development)*. Presentation Technology Entrepreneurship, Nov. 26, 2008. http://ce.ioc.KIT.edu/downloads/Praesentation_Griesar.pdf.

Grushkin, D. (2012): *The Rise And Fall Of The Company That Was Going To Have Us All Using Biofuels*. Fast Company, Aug. 8, 2012.

http://www.fastcompany.com/3000040/rise-and-fall-company-was-going-have-us-all-using-biofuels (last access 10/4/2012).

Gruber; M.; MacMillan, I. C.; Thompson, J. D. (2008): *Look Before You Leap - Market Opportunity Identification in Emerging Technology Firms*. Management Science 54 (9), September 2008, 1652-1665.

Guardian (2008): *Biofuels: The hottest clean technology companies in Europe*. The Guardian, Sep. 18, 2008. http://www.guardian.co.uk/environment/table/2008/sep/18/cleantech100biofuels.cleantechnology100 (last access 1/6/2010).

Hamilton, T. (2009): *Dow to Test Algae Ethanol – Startup Algenol partners with Dow Chemical on a demonstration ethanol plant*. Technology Review, Jul. 16, 2009. http://www.technologyreview.com/business/23009/?nlid=2182 (last access 2/8/2010).

Handelsblatt (2012): *Die größten Solarzellenhersteller der Welt (The largest solar cell manufacturers of the world)*. Handelsblatt Apr. 4, 2012. http://www.handelsblatt.com/unternehmen/industrie/solarbranche-die-groessten-solarzellenhersteller-der-welt/6127028.html (last access 4/7/2012).

Hanks S. H.; Watson, C. J.; Jansen, E.; Chandler, G. N. (1993): *Tightening the Life-Cycle Construct: A Taxonomic Study of Growth Stage Configurations in High-Technology Organizations*. Entrepreneurship: Theory and Practice, Dec. 22, 1993. http://findarticles.com/p/articles/mi_hb6648/is_n2_v18/ai_n28633506/pg_14/?tag=content;col1 (last access 1/21/2011).

Harris & Harris (2005): *Harris & Harris Group Invests in Solazyme, Inc*. Harris & Harris Group, News & Press Releases, Nov. 30, 2004. http://files.shareholder.com/downloads/TINY/828794935x0x42850/d9296dfa-a2fd-4c57-95cb-3998007c90be/TINY_News_2004_11_30_General.pdf (last access 1/25/2010).

Hauger, N.; Kaizl, M. (2010): *In virtuellen Welten* (*In virtual worlds*). Inside B Business Magazine 4, 2010, 38-41. http://www.insideb.de/upload/Files/038_039_IB_0410.pdf (last access 9/27/2011), http://www.insideb.de/upload/Files/040_041_IB_0410.pdf (last access 9/27/2011).

Hayes, T. J. (1997): *The Creative Entrepreneurs Organization: Developing Innovative Products and Businesses*. Thesis submitted to the Faculty of the Virginia Polytechnic Institute and State University. http://scholar.lib.vt.edu/theses/available/etd-111897-141039/unrestricted/TJ_s_Thesis.pdf (last access 7/8/2009).

Haynes, S. D. (2000): *The Systems Thinking Approach to Strategic Planning and Management*. CRC Press; Haynes, S. D. (2005): *The Science of Systems Thinking. The Natural Way the World Works*. http://www.stephenhaines.com/keynote/pdfs/MSST-01.pdf (last access 9/30/2009).

Headline News (2009): *Developing the Whole Biochain*. Headline News, Mar. 7, 2009. http://newsheadlinenews.blogspot.com/2009_02_09_archive.html (last access 1/22/2010).

Heathcote, M.; Fryer, C. (2008): *Alternatives to Petroleum Feedstocks : a Leading Role for Asia*. Tecnon OrbiChem Marketing Seminar at APIC 2008, The World

Chemical Industry Focuses on Asia Singapore, May 27, 2000. http://www.tecnon.co.uk/gen/uploads//rr4pcz3icabofgzuk54ocr5529052008123433.pdf (last access 1/22/2010).

Heffernan, M. (2007): *Men Dominate, Women Orchestrate.* Business Week, Small Biz, Mar. 13, 2007. http://msnbci.businessweek.com/smallbiz/content/mar2007/sb20070313_555883.htm?chan=smallbiz_smallbiz+index+page_today%27s+top+stories (last access 3/14/2007).

Heger, D.; Höwer, D.; Licht, G.; Metzger, G.; Sofka, W. (2009): High-Tech-*Gründungen in Deutschland – Optimismus trotz Krise (High-tech foundations in Germany – Optimism despite crisis*). Zentrum für Europäische Wirtschaftsforschung (ZEW) GmbH. Jul. 2009. ftp://ftp.zew.de/pub/zew-docs/gutachten/hightechgruendungen09.pdf (last access 7/17/2010).

Heiling, C. (2008):*Sriya Innovations Inc.* Wellford Energy Advisors, Lobbying Report, File of 4/14/2008. http://soprweb.senate.gov/index.cfm?event=getFilingDetails&filingID=E7B69370-1DE5-49A9-BA13-6DD88885FDDE (last access 6/4/2012).

Helmut Kaiser Consultancy (2010): *Ionic Liquid Technologies, Markets, Applications, Companies and Developments Worldwide 2008 to 2020 and 2030*. Market research report, http://www.hkc22.com/ionicliquids.html (last access 2/19/2012); http://www.prlog.org/10698069-ionic-liquids-markets-worldwide-will-reach-34-billion-us-by-2020-from-300-million-today.html (last access 6/6/2011).

Herndon, A.; Martin, C.; Goossens, E. (2011): *Total to Buy 60% of SunPower for $1.38 Billion in Solar Bet.* Bloomberg Business Week, Apr. 29, 2011. http://www.businessweek.com/news/2011-04-29/total-to-buy-60-of-sunpower-for-1-38-billion-in-solar-bet.html (last access 4/29/2011).

Herron, D. (2012): I*t's Wanxiang or bankruptcy for A123 Systems as the companies sign agreement*. Torque News, Aug. 16, 2012. http://www.torquenews.com/1075/its-wanxiang-or-bankruptcy-a123-systems-companies-sign-agreement (last access 9/3/2012).

Heskett, J. (2007): *Has Managerial Capitalism Peaked?* Harvard Business School Working Knowledge. http://hbswk.hbs.edu/item/5794.html (last access 6/20/2009).

Hess, G. (2007): *Chevron, Texas A&M Will Research Biofuels – Partnership seeks conversion of nonfood crops into renewable transportation fuel*. Chemical & Engineering News, News, May 31, 2007. http://pubs.acs.org/cen/news/85/i23/8523news5.html (last access 11/24/2009).

Hesselbein, F.; Goldsmith, M.; Beckhard, R. (Eds.), (1996): *The Leader of the Future*, foreword by Peter F. Drucker. Jossey-Bass Publishers, San Francisco.

Hewlett, W.: *The Human Side Of Management*. SMECC Vintage Electrics Vol #3 Issue #1. http://www.smecc.org/the_human_side_of_management_-_bill_hewlett.htm (last access 7/27/2010), cf. Ed Sharpe, E.: *Hewlett-Packard, The Early Years.* http://www.smecc.org/hewlett-packard,_the_early_years.htm (last access 7/27/2010).

Hiersemenzel, P. A. (2003): *Rumsfeld und Clausewitz*. Die Zeit 14/2003 – Mar. 27, 2003. http://pdf.zeit.de/2003/14/hiersemenzel2803.pdf (last access 10/9/2009).

Higgins, T.: *Casual Games as a Business*. UNITY. http://unity3d.com/support/documentation/articles/CasualGamesBusiness (last access 10/2/2011); http://en.wikipedia.org/wiki/Casual_game

Hills, G. E.; Shrader, R. C. (1998): *Successful Entrepreneurs' Insights into Opportunity Recognition.* Frontiers of Entrepreneurship Research. Wellesley, MA: Babson College. last updated January. 1999. http://www.babson.edu/entrep/fer/papers98/I/I_A/I_A_text.htm (last access 9/22/2010).

Hipple, J.; Hardy, D.; Wilson, S. A. Michalski, J. (2001): *Can corporate innovation champions survive?* Chemical Innovation 31 (11), 14-22. http://pubs.acs.org/subscribe/journals/ci/31/i11/html/11hipple.html (last access 8/12/2010).

Hitt, M., Ireland, R., and Hoskisson, R. (2005): *Strategic Management Competitiveness and Globalization Concepts.* Thompson, South-Western, 6th edition, 2005.

Hof, R. D. (2008): *How Google Fuels Its Idea Factory*. Business Week, Apr. 29, 2008. http://www.businessweek.com/magazine/content/08_19/b4083054277984.htm?chan=innovation_innovation+%2B+design_innovation+strategy (last access 9/1/2010).

Hoffman, R. (2012): *Entrepreneurs Take Intelligent Risks*. Stanford Technology Ventures Program, Entrepreneurial Thought Leaders Lecture Series, Feb. 22, 2012. http://ecorner.stanford.edu/authorMaterialInfo.html?mid=2907 (last access 4/22/2012).

Hoffman, R. (2010): *LinkedIn's Reid Hoffman: Five Lessons for Entrepreneurs*. WST Blogs, Nov. 18, 2010. http://blogs.wsj.com/tech-europe/2010/11/18/linkedins-reid-hoffman-five-lessons-for-entrepreneurs/ (last access 11/19/2010); http://en.wikipedia.org/wiki/PayPal.

Hoffman, R. (2007a): *Core Ideas for a Social Networking Company*. Stanford Technology Ventures Program, Entrepreneurial Thought Leaders Lecture, Feb. 7, 2007. http://ecorner.stanford.edu/authorMaterialInfo.html?mid=1689 (last access 9/19/2011).

Hoffman, R. (2007b): *Facing Competition through Innovation*. Stanford Technology Ventures Program, Entrepreneurial Thought Leaders Lecture, Feb. 7, 2007. http://ecorner.stanford.edu/authorMaterialInfo.html?mid=1688 (last access 8/2/2008).

Hoffmann, D. (2009): *Manfred von Ardenne – Der rote Baron (Manfred von Ardenne – The Red Baron*). FAZ, Nov. 10, 2009. http://www.faz.net/s/Rub163D8A6908014952B0FB3DB178F372D4/Doc~E842A62CD9ECB4D0BA32383FBA0DFA8B9~ATpl~Ecommon~Scontent.html (last access 11/11/2009).

Hoffritz, J. (2013): *Farben für die ganze Welt (Dyes for the whole world*). Handelsblatt, Jan. 12, 2013. http://www.handelsblatt.com/unternehmen/industrie/basf-bayer-und-co-farben-fuer-die-ganze-welt-seite-all/7611332-all.html (last access 1/20/2013); cf. also http://de.wikipedia.org/wiki/Carl_Friedrich_Wilhelm_Meister.

Hofmann, S. (2010): *Die Früchte des loyalen Ungehorsams (Fruits of loyal disobedience). Handelsblatt, Jul. 16, 2010.* http://www.handelsblatt.com/technologie/forschung/pharmakonzern-merck-die-fruechte-des-loyalen-ungehorsams;2618923 (last access 7/17/2010).

Holt, K, (2011): INFOGRAPHIC: *1 In 5 Facebook Users Play 'FarmVille,' Social Gamers Are 55% Female*. scribbal, Jun. 21, 2011. http://www.scribbal.com/2011/06/infographic-1-in-5-facebook-users-play-farmville-social-gamers-are-55-female/ (last access 10/24/2011).

Howe, J. B. (2006): *Celunol Corporation – An Introduction*. Presentation, Massachusetts Cellulosic Ethanol Workshop, Nov. 8, 2006. http://www.mass.gov/Eoeea/docs/doer/alternative_fuels/cew061108-celunol-howe.pdf (last access 12/30/2009).

Hoyos, C. (2010): *Shell to create $12bn biofuels JV in Brazil.* Financial Times, Feb. 1, 2010. http://www.ft.com/cms/s/0/7133e6a6-0f47-11df-8a19-00144feabdc0.html (last access 2/1/2010).

Huber, G. P. (1991): *Organizational learning: The contributing processes and the literatures*. Organization Science, 2: 88–115.

Humboldt Innovation (2009): *Cyano Biofuels*. SpinoffZone 2009. http://www.humboldt-innovation.de/HumboldtInnovation_spinoffdetail_de.html?mCtrl=SpinOff&mOp=View&vwflwid=2ux8cz5bvu&m_SpinOff%5BspinOffID%5D=22 (last access 12/16/2009).

ICB Americas (2004): *Technology Watch: Ionic Liquids Offer New Solutions*. ICIS, Jan. 19, 2004. http://www.icis.com/Articles/2004/01/16/549791/technology-watch-ionic-liquids-offer-new-solutions.html (last access 5/7/2005).

Iijima, S. *The discovery of carbon nanotubes – Guided by serendipity*. NEC Corp. http://www.nec.co.jp/rd/en/innovative/E1/01.html (last access 9/10/2009).

Imbler, J. *(2009): Mission: Not Impossible Successful Strategies for Cellulosic Ethanol Deployment.* Pacific Rim Summit on Industrial Biotechnology & Bioenergy, Nov. 10, 2009. http://www.zeachem.com/press/pdf/PacRim111009.pdf (last access 1/7/2010).

Ingersoll, E.; Gallagher, D. C.; Vysatova, R. A. (1998): *Industry Development Strategy for the PV Sector*. REPP - Special Report. http://www.repp.org/repp_pubs/articles/pv/2/2.html (last access 8/25/2009).

INITIAL LIGHT BULB (2009): *Small biz tech-startup: OPX Biotechnologies Inc*. WiesnerMedia, Coloradobiz, Aug. 17, 2009. http://www.cobizmag.com/articles/small-biz-tech-startup-opx-biotechnologies-inc (last access 11/11/2010).

Itim International: *An international consulting organization utilizing Prof. Hofstede's concepts, as well as those of other leading sociologists*. http://www.geert-hofstede.com/geert_hofstede_resources.shtml (last access 8/10/2009); Country Data: http://www.geert-hofstede.com/hofstede_dimensions.php (last access 8/10/2009).

ISEE – International Sustainable Energy Exchange (http://www.hart-isee.com/index.php). *Global Fuel Solutions – Fuel Technology*. http://www.hart-isee.com/index.php?page=synthetic-fuel-btl-technologies (last access 10/30/2009).

Issig, P. (2006): *Hochschulen als Unternehmerschmiede – Förderprogramm ermöglichte fast 100 Firmengründungen (Universities as forges for enterprises – Public funding facilitates almost 200 foundations)*. Die Welt, Mar. 12, 2006. http://www.welt.de/print-wams/article139605/Hochschulen_als_Unternehmerschmiede.html (last access 10/1/2008).

Jacquot, J.E. (2008): *Biofuel Comparison Chart: The "Good," the Bad and the (Really) Ugly*. Treehugger – Science & Technology, May 10, 2008. http://www.treehugger.com/files/2008/05/biofuel-comparison-chart.php (last access 11/28/2009).

Jahn, D. (2007): *Firmengründungen in der Chemie (Firm's foundation in chemistry)*. Presentation Technology Entrepreneurship, Nov. 21, 2007. http://ce.ioc.KIT.edu/downloads/Praesentation_Jahn.pdf (last access 6/21/2010).

Jahn, D. (2008): *The Chemical Industry – Engine of innovation with the other industries.* Presentation VCW, Jun. 12, 2008. http://www.gdch.de/strukturen/fg/wirtschaft/vcw_va/jahn.pdf (last access 6/23/2010).

Jan, T. (2009): *Harvard workers stunned by layoffs*. The Boston Globe, Jun. 24, 2009. http://www.boston.com/news/education/higher/articles/2009/06/24/harvard_workers_stunned_by_layoffs_of_275/ (last access 6/24/2009).

John Brown: *Gustave Whitehead – Pioneer Aviator.* http://www.gustave-whitehead.com/ (last access 3/31/2013); cf. also http://www.npr.org/2013/03/19/174634237/historian-propels-connecticut-to-claim-first-in-flight; U.S. Air Force Fact Sheet – GUSTAVE WHITEHEAD: http://www.hill.af.mil/library/factsheets/factsheet_print.asp?fsID=5948&page=1; Gustave Whitehead: http://en.wikipedia.org/wiki/Gustave_Whitehead.

Johnson, J. (2011): *U.S.-Backed Solar Panel Maker Folds – Bankruptcy: Solyndra got loan guarantee but can't make a profit.* Chemical & Engineeering News 89 (36), 12. http://pubs.acs.org/cen/news/89/i36/8936notw2.html (last access 9/5/2011).

Johnson, J. (2010): *Potholes On The Road To Renewable Fuels – Corn-kernel-based ethanol hits the fast lane, but cellulosic ethanol is still stuck in first gear.* Chemical & Engineering News 37, Sep. 13, 2010. http://pubs.acs.org/isubscribe/journals/cen/88/i37/html/8837gov1.html?featured=1 (last access 9/13/2010).

Johnson, J. A. (2001): *Reconsidering the Causal Efficacy of Situations*. Invited Talk for the Biannual Kurt Lewin Institute Congress. Zeist, The Netherlands, May 11, 2001. http://www.personal.psu.edu/faculty/j/5/j5j/papers/KLI.html (last access 4/16/2010).

Johnson, K. (2009): *Brewing Biofuels: Shell, Codexis Team Up to Drive Cellulosic Ethanol Research*. WSJ Blogs – Environmental Capital, Mar. 10, 2009. http://blogs.wsj.com/environmentalcapital/2009/03/10/brewing-biofuels-shell-codexis-team-up-to-drive-cellulosic-ethanol-research/ (last access 11/24/2009).

Johnson, L. (2011): *The dismal science is bereft of good ideas*. Financial Times, Jun. 7, 2011. http://www.ft.com/intl/cms/s/0/121b216c-914a-11e0-b1ea-00144feab49a.html (last access 6/8/2011).

Johnson, W. C. *(2005): Changing Interfaces Between the Research University and , Industry*. Presentation, 2005 Engineering Research Council Workshop and Forum, Washington DC, 2/27/05. http://www.asee.org/conferences/erc/2005/upload/wayne-c-johnson.ppt (last access 6/21/2010).

Jonietz, E. (2007): *Oil from Wood – Startup Kior has developed a process for creating "biocrude" directly from biomass*. Technology Review, Nov. 9, 2007. http://www.technologyreview.com/read_article.aspx?ch=specialsections&sc=biofuels&id=19694 (last access 12/30/2009).

Kaiser, N. (2004): *Interview with Reid Hoffman, CEO of LinkedIn*. NPost, Mar. 17, 2004. http://www.npost.com/?s=hoffman (last access 11/20/2011).

Kanellos, M. (2008): *Algae: Another way to grow edible oils*. The Pulse of Technology, Jan. 25, 2008. http://www.pulseoftechnology.com/?p=4 (last access 1/12/2010).

Kanellos, M. (2009a): *Chevron Invests in LS9; Microbe Diesel by 2011*. Green Tech Media, Sep. 5, 2009. http://www.greentechmedia.com/articles/print/chevron-invests-in-ls9-microbe-diesel-possible-by-2011/ (last acess 12/11/2009).

Kanellos, M. (2009b): *GreenFuel Technologies Closing Down*. Green Tech Media, May 13, 2009. http://www.greentechmedia.com/articles/read/greenfuel-technologies-closing-down-4670/ (last access 1/2/2010).

Kaplan, J. (2003a): *Five Biggest Mistakes That Entrepreneurs Make.* Stanford Technology Ventures Program, Entrepreneurial Thought Leader Speaker Series, Oct. 1, 2003. http://ecorner.stanford.edu/authorMaterialInfo.html?mid=364 (last access 12/29/2010).

Kaplan, J. (2003b): *Five Critical Skills That Entrepreneurs Need*. Stanford Technology Ventures Program, Entrepreneurial Thought Leader Speaker Series, Oct. 1, 2003. http://ecorner.stanford.edu/authorMaterialInfo.html?mid=366 (last access 9/23/2010).

Kaplan, K.; Litoff, L.; Wright, D (2010*): Inside Zynga: Now the Creators of 'CityVille'*. ABC News, Nov. 17, 2010. http://abcnews.go.com/Nightline/inside-zynga-creators-farmville/story?id=12169767 (last access 10/5/2011).

Karlsruhe (2001): *Karlsruher Bokela erschließt sich neue Märkte (Bokela of Karlsruhe enters new markets*). Karlsruhe, Wirtschaftsförderung, Oct. 17, 2001. http://www1.karlsruhe.de/Wirtschaft/php_news/zo1_news_detail.php?auswahl=58&archiv=1 (last access 11/13/2010).

Kemna, V. (2007): *Blühende Landschaften (Flourishing landscapes).* Deutschlandfunk, Jan. 30, 2007. http://www.dradio.de/dlf/sendungen/campus/588154/ (last accesss 8/18/2010).

Kempkens, W. (2009): *Biosprit – Deutschland verpasst den Anschluss (Biofuels – Germany misses catch-up)*. Wirtschaftswoche, Apr. 28, 2009. http://www.wiwo.de/technik/deutschland-verpasst-den-anschluss-393963/ (last access (11/24/2009).

Kersting, K. (2012): *Building a startup (in ten easy steps*). Presentation Technology Entrepreneurship, Feb. 15, 2012. http:// http://ce.ioc.KIT.edu/downloads/Kersting_Vorlesung.pdf.

Keßler, A.; Frank, H.; Korunka, C.; Manfred Lueger, M. (2010): *Predicting Founding Success and New Venture Survival: A Longitudinal Study.* http://www.kmu.unisg.ch/rencontres/Renc2010/Topics_2010/C/Rencontres_2010_Topic_C_Frank_Kessler_Korunka_Lueger_f.pdf (last access 10/31/2010).

Kho, J. (2009): *The Algae Fuel Backlash: Here Come the Skeptics*. Earth2tech, Sep. 25, 2009. http://earth2tech.com/2009/09/25/the-algae-fuel-backlash-here-come-the-skeptics/ (last access 1/18/2010).

Khosla, V. (2006): *Biofuels: Think outside the Barrel*. KPCB – Experienced Team Member – Vinod Khosla., April 2006. http://www.khoslaventures.com/presentations/Biofuels.Apr2006.ppt (last access 12/7/2009).

Khosla, V. (2008a): *Emerging Energy Solutions Must Compete on Price*. Stanford Technology Venture Program, Entrepreneurial Thought Leader Lecture, Oct. 22, 2008. http://ecorner.stanford.edu/authorMaterialInfo.html?mid=2080 (last access 12/4/2009).

Khosla, V. (2008b): *The Innovation Ecosystem*. In THE BETTER WORLD REPORT PART ONE: The Art of Collaboration: The Relationships That Bring Academic Innovations to the Marketplace, 13-15. http://www.enterprise.cam.ac.uk/uploads/File/Publications/AUTM_BWR1_08.pdf (last access 12/1/2009).

Khosla, V. (2009a): *Renewable Energy – Maintech, not Cleantech*. Presentation, April 2009. http://www.khoslaventures.com/presentations/Google-4-22-09.ppt (last access 12/12/2009).

Khosla, V. (2009b): *Our portfolio – Khosla Ventures.* http://www.khoslaventures.com/presentations/Flower_Chart.ppt (last access 11/24/2009).

Kim, W. C.; Mauborgne, R. (1997): *Value Innovation – The Strategic Logic of High Growth*. Harvard Business Review, January-February 1997, 103-112; republished in 2004. HBR (July-August), 172-180. http://www.adaptive-edge.com/vi_hbr.pdf (last access 7/30/2010); http://csom.umn.edu/assets/2435.pdf (last access 1/17/2013); http://media.library.ku.edu.tr/reserve/resspring10/mktg451_AKok/kim.pdf (last access 5/17/2013).

KiOR (2012): *Annual Report 2011*. http://files.shareholder.com/downloads/ABEA-6B1XEW/1896825919x0x571270/2FFBA31D-5AB7-435B-969C-2578ACE1B273/KIOR_2011_AR.pdf (last access 5/3/2012).

Kiplinger Washington Editors (2007): *Biobutanol: On the Road to Overtaking Ethanol*. Kiplinger's Biofuels Market Alert 1 (1), 1-2, Jul. 2, 2007. http://www.biontech.com/news/print-air/KiplingerBiofuels.070702.pdf (last access 12/23/2009).

Kirchhoff, B. A.; Spencer, A. (2008):*New High Tech Firm Contributions to Economic Growth*. File of 7/22/2008. http://www.icsb.org/documents/New_High_Tech_Firms.pdf (last access 5/23/2009).

KIT (2008): *Professor and Parallel Work at Industry – Dr. Gisela Lanza is Holder of the First Shared Professorship at the KIT*. Press Release 11/2008. http://www.fzk.de/fzk/idcplg?IdcService=KIT&node=4211&document=ID_062922&lang=en (last access 7/372010).

KIT (2009): *Thin Films with a great Impact - Three Companies Are Sharing the Professorship of Thin Film Technology*. Press Release 085/2009. http://www.kit.edu/english/55_356.php (last access 7/372010).

Klann, A. R. (2007): *BlueFire Ethanol*. American Coalition for Ethanol Conference, Aug. 21, 2007. http://www.ethanol.org/pdf/contentmgmt/Arnie_Klann_BlueFire.pdf (last access 1/10/2010).

Klepper, S. (2001): *Employee start-ups in high tech industries*. Industrial and Corporate Change, 10 (3), 639–674.

Klausmann, P. (2011): *Vitracom AG – Von der Gründung zur Marktführerschaft (From Foundation to Market Leader)*. Presentation Technology Entrepreneurship, Jan. 19, 2011. http://ce.ioc.KIT.edu/downloads/Praesentation_Klausmann.pdf (last access 2/1/2011).

Koenen, J. (2010): *Unternehmer? Warum nicht! (Entrepreneur? Why not?)*. Presentation Technology Entrepreneurship, Jan. 27, 2010. http://ce.ioc.KIT.edu/downloads/Praesentation_Koenen.pdf (last access 7/20/2010).

Koenig, C. (2005): *Bugs waste no time*. The Oxford Times, March/April 2005. http://www.greenbiologics.com/documents/Oxford_Times_article.pdf (last access 1/28/2010).

Köhler, K. (2005): *Katja Heppe produziert mit Chitosan Wirkstoffe für Medizin und Kosmetik (With chitosan Katja Heppe produces ingredients for medicine and cosmetics)*. Brikada, Oct. 25, 2005. http://www.brikada.de/cgi-bin/con.cgi?action=pub&id=557&channel=Frauen%20heute (last access 8/20/2010).

Kraft, T. E. (2005): *REVIEW The Wright Way: Seven Problem-Solving Principles from the Wright Brothers That Can Make Your Business Soar By Mark Eppler*. Journal of Industrial Teacher Education 42 (1), 67- 71. http://faculty.ksu.edu.sa/Alhassan/MAster_Students_Research_Readings/The%20Wright%20Way,%20Seven%20Problem-Solving%20Principles%20From%20the%20Wright%20Brothers%20That%20Can%20Make%20Your%20Business%20Soar.pdf (last access 4/11/2010).

Kramer, D. (2007): *Project: "Systems Biology of Cyanobacterial Biofuel Production."* http://www.forsys.net/joomla/images/partner_en/dan_kramer_eng_03-04.pdf (last access 12/16/2009);
FORSYS - *Research Units for Systems Biology in Germany*. Federal Ministry of of Education and Research (BMBF). http://www.forsys.net.

Krauter, B. (2007): *Biopesticide pioneer covers globe – Pam Marrone leads staff in finding natural solutions to big problems*. Capital Press, Oct. 5, 2007. http://www.marroneorganicinnovations.com/news/news10172007.html (last access 11/25/2007).

Krauss, C. (2009): *Ethanol, Just Recently a Savior, Is Struggling*. The New York Times, Feb. 12, 2009. http://www.nytimes.com/2009/02/12/business/12ethanol.html?ref=todayspaper (last access 11/9/2009).

Kunkel, S. W. (2001): *Toward a Typology of Entrepreneurial Activities*. Academy of Entrepreneurship Journal, January 2001. http://findarticles.com/p/articles/mi_hb6005/is_1_7/ai_n28892970/?tag=content;col1 (Last access 11/28/2010), http://home.sandiego.edu/~kunkel/articles/typent.pdf.

Kunz, M. (2008): *TV- und Atombomben-Pionier ohne Abi (TV and atomic bomb pioneer without a high-school degree)*. Focus, May 26, 2008. http://www.focus.de/wissen/bildung/biografien/manfred-von-ardenne-tv-und-atombomben-pionier-ohne-abi_aid_303718.html (last access 11/11/2009); cf. also http://de.wikipedia.org/wiki/Manfred_von_Ardenne.

Kuo, I. (2011): *On the GreenBeat: Solyndra's star fades, admits to mistakes; Khosla says new biofuel IPO coming*. GreenBeat, Mar. 4, 2011.http://venturebeat.com/2011/03/04/on-the-greenbeat-solyndras-star-fades-admits-to-mistakes-khosla-says-new-biofuel-ipo-coming/ (last access 3/30/2011).

Lachmann, H. (2007): *Wir haben fünf Jahre Vorsprung (We have a Leadt Time of Five Years)*. VDI Nachrichten, Oct. 19, 2007. http://www.vdi-nachrichten.com/vdi-nachrichten/aktuelle_ausgabe/akt_ausg_detail.asp?cat=3&id=35504&doPrint=1 (last access 12/17/2009).

Lachmann, L. M.: Cited in Garrison, R. W. : *From Lachmann to Lucas – On Institutions, Expectations, and Equilibrating Tendencies*. http://www.auburn.edu/~garriro/j6lachlucas.htm#N_7_ (last access 7/30/2010).

Lacy, S. (2009): *VC Markets: No Easy Answers*. Business Week, May 20, 2009. http://www.businessweek.com/technology/content/may2009/tc20090520_516821.htm?chan=technology_technology+index+page_tech+investing (last access 5/22/2009).

Lakhani, K. R.; Jeppesen, L. B.; Lohse, P. A.; Panetta, J. A. (2006): *The Value of Openness in Scientific Problem* Solving. Harvard Business School, Working Paper 07-050, October 2006. http://www.hbs.edu/research/pdf/07-050.pdf (last access 4/6/2008).

Lane, J. (2007): *Shell, HR Biopetroleum form JV to produce algae oil in Hawaii*. Biofuels Digest, Dec. 11, 2007. http://www.biofuelsdigest.com/blog2/2007/12/11/shell-hr-biopetroleum-form-jv-to-produce-algae-oil-in-hawaii/ (last access 11/24/2009); cf. also http://www.bloomberg.com/apps/news?pid=20601102&sid=ahSsmDmNPDXk&refer=uk.

Lane, J. (2008): *Cellana*. Biofuels Digest, Oct. 8, 2008. http://www.biofuelsdigest.com/blog2/2008/10/08/cellana/ (last access 12/27/2009).

Lane, J. (2009a): *BP to establish joint venture with Verenium for cellulosic ethanol; invests additional $22.5 million*. Feb. 20, 2007. http://www.biofuelsdigest.com/blog2/2009/02/20/bp-to-establish-joint-venture-with-verenium-for-cellulosic-ethanol-invests-additional-225-million/ (last access 11/12/2009).

Lane, J. (2009b): *DuPont Danisco: 50 Hottest Companies in Bioenergy candidate profile.* BioFuels Digest, Nov. 04, 2009. http://www.biofuelsdigest.com/blog2/2009/11/04/dupont-danisco-50-hottest-companies-in-bioenergy-candidate-profile/ (last access 11/20/2009).

Lane, J. (2009c): *BP to commence cellulosic ethanol production in Brazil by 2013.* BioFuels Digest, Oct. 28, 2009. http://www.biofuelsdigest.com/blog2/2009/10/28/bp-to-commence-cellulosic-ethanol-production-in-brazil-by-2013/ (last access 11/23/2009).

Lane, J. (2009d): *Gevo.* Biofuels Digest, Oct. 7, 2009. http://www.biofuelsdigest.com/blog2/2009/10/07/gevo-50-hottest-companies-in-bioenergy-candidate-profile/ (last access 12/1/2009).

Lane, J. (2009e): *Mascoma's CEO steps down as company's focus on commercialization steps up.* Biofuels Digest, Aug. 21, 2009. http://www.biofuelsdigest.com/blog2/2009/08/21/mascomas-ceo-steps-down-as-companys-focus-on-commercialization-steps-up/ (last access 12/9/2009).

Lane, J. (2009f): *Amyris raises $24.7 million in first tranche of Series C funding; aims for $62 million in total for commercialization drive.* Biofuels Digest, Aug. 24, 2009. http://www.biofuelsdigest.com/blog2/2009/08/24/amyris-raises-247-million-in-first-tranche-of-series-c-funding-aims-for-62-million-in-total-for-commercialization-drive/ (last access 12/13/2009).

Lane, J. (2009g): *50 Hottest Companies in Bioenergy: #9, Verenium.* Biofuels Digest, Dec. 28, 2009. http://www.biofuelsdigest.com/blog2/2009/12/28/50-hottest-companies-in-bioenergy-9-verenium/ (last access 12/30/2009).

Lane, J. (2009h): *BP, DuPont butanol JV, Butamax, heads for commercialization; BP = "butanol play"? How will Gevo, Cobalt counter?* Biofuels Digest, Jul. 14, 2009. http://www.biofuelsdigest.com/blog2/2009/07/14/bp-dupont-butanol-jv-is-named-butamax-heads-for-commercialization-bp-stands-for-butanol-play-how-will-gevo-cobalt-counter/ (last access 1/27/2010).

Lane, J. (2009i): *BP outlines 3-pronged biofuels strategy, says biobutanol plant will open at commercial scale in 2013.* Biofuels Digest, Nov. 6, 2009. http://www.biofuelsdigest.com/blog2/2009/11/06/bp-outlines-3-pronged-biofuels-strategy-says-biobutanol-plant-will-open-at-commercial-scale-in-2013/ (last access 1/26/2010).

Lane, J. (2009j): *50 Hottest Companies in Bioenergy for 2009/10: #3, Amyris Biotechnologies – Amyris Biotechnologies.* Biofuels Digest, Dec. 16, 2009. http://www.biofuelsdigest.com/blog2/2009/12/16/50-hottest-companies-in-bioenergy-for-200910-3-amyris-biotechnologies/ (last access 12/30/2009).

Lane, J. (2009k): *50 Hottest Companies in Bioenergy: #5, Sapphire Energy.* Biofuels Digest, Dec. 18, 2009. , http://www.biofuelsdigest.com/blog2/2009/12/18/50-hottest-companies-in-bioenergy-5-sapphire-energy/ (last access 2/10/2010).

Lane, J. (2012l): *ZeaChem and the Renewable Fuel Standard.* Renewable Energy World, Jan. 17. 2012. http://www.renewableenergyworld.com/rea/news/article/2012/01/zeachem-and-the-renewable-fuel-standard (last access 3/25/2012).

Lane, J. (2010m): *50 Hottest Companies in Bioenergy: #30, Cobalt Biofuels – Cobalt Biofuels*. Biofuels Digest, Jan. 26, 2010. http://www.biofuelsdigest.com/blog2/2010/01/26/50-hottest-companies-in-bioenergy-30-cobalt-biofuels/ (last access 2/1/2010).

Lane, J. (2010n): 50 *Hottest Companies in Bioenergy for 2009-10: #24, Algenol*. Biofuels Digest, Jan. 19, 2010. http://www.biofuelsdigest.com/blog2/2010/01/19/50-hottest-companies-in-bioenergy-for-2009-10-24-algenol/ (last access 2/10/2010).

Lane, J. (2009o): *Coskata: 50 Hottest Companies in Bioenergy candidate profile*. Biofuels Digest, Sep. 23, 2009. http://www.biofuelsdigest.com/blog2/2009/09/23/coskata-50-hottest-companies-in-bioenergy-candidate-profile/ (last access 11/30/2009).

Lane, J. (2012p): *Chain, Chain, Chain – Chain of Fuels: Mascoma, Sud-Chemie on short path to parity with gasoline*. Biofuels Digest, Mar. 2, 2012. http://www.biofuelsdigest.com/bdigest/2012/03/02/chain-chain-chain-chain-of-fuels-mascoma-sud-chemie-on-short-path-to-parity-with-gasoline/ (last access 3/21/2012).

Lane, J. (2012q): *Genomatica delays commercial-scale production to 2013*. Biofuels Digest, Feb, 14, 2012. http://www.biofuelsdigest.com/bdigest/2012/02/14/genomatica-delays-commercial-scale-production-to-2013/ (last access 2/27/2012).

Langreth, R. (2008): *Beaker Fuel*. Forbes Magazine, Nov. 24, 2008. http://www.forbes.com/forbes/2008/1124/058.html (last access 10/24/2009).

Lane, J. (2012r): *Virdia ups the ante in the Race for the New Sugars*. Biofuels Digest, Mar. 6, 2012. http://www.biofuelsdigest.com/bdigest/2012/03/06/virdia-ups-the-ante-in-the-race-for-the-new-sugars/ (last access Jun.24, 2012).

Lardon, L.; Hélias, A.; Sialve, B.; Steyer, J.-P.; Bernard, O. (2009): *Life-Cycle Assessment of Biodiesel Production from Microalgae*. Environmental Science & Technology 43 (17), 6475-6481. http://pubs.acs.org/doi/pdfplus/10.1021/es900705j (last access 1/3/2009).

Lechler, T. (2001). *Social Interaction: A Determinant of Entrepreneurial Team Venture Success*. Small Business Economics 16, June, 263-278.

Lécuyer, C. (2001): *Technology and Entrepreneurship in Silicon Valley*. Article from Dec, 3, 2001. http://nobelprize.org/nobel_prizes/physics/articles/lecuyer/index.html (last access 10/12/2009).

Lemos-Stein, M. (2009): *The Summer Of Algae*. Wall Street Journal Blog, Jul. 14, 2009. http://blogs.wsj.com/venturecapital/2009/07/14/the-summer-of-algae/ (last access 12/16/2009).

Lenk Krueger, D. (1983): *Pragmatics of dyadic decision making: A sequential analysis of communication patterns*. Western Journal of Speech Communication 47 (2), 99-117. http://www.tandfonline.com/doi/abs/10.1080/10570318309374110 (last access 9/13/2010).

LeVine, S.; Aston, A. (2009): *Betting Big on a Boom in Natural Gas*. Business Week, Oct. 8, 2009. http://www.businessweek.com/magazine/content/09_42/b4151046053399.htm (last access 10/10/2009).

Levenson, M. (2008): *Facebook, ConnectU settle dispute – Case an intellectual property kerfuffle*. The Boston Globe, Jun. 27, 2008. http://www.boston.com/business/technology/articles/2008/06/27/facebook_connectu_settle_dispute/ (last access 7/19/2009).

Levy, S. (2009): *Secret of Googlenomics: Data-Fueled Recipe Brews Profitability*. Wired Magazine, Jun. 17, 2009. http://www.wired.com/culture/culturereviews/magazine/17-06/nep_googlenomics?currentPage=all (last access 2/12/2011).

Liedtke, M. (2008): *Google is world's most powerful 10-year-old*. AP, Sep. 5, 2008. http://www.msnbc.msn.com/id/26560994/ (last access 10/6/2008).

Lieske, P. (2004): *Marktvorteil für Klötzer Alge (Market advantage for algae from Klötze)*. Altmarkzeitung, Oct. 2, 2004. http://www.algomed.de/media/presse/11-45-14-25-01-2008-Kloetzer%20Rundschau%2002.10.04.pdf (last access 2/7/2008).

Locke, E. A.; Baum, J. R. (2006): *Entrepreneurial Motivation*. Chapter 5, 93-112. In Baum, J. R.; Frese, M.; Baron, R. A. (Eds.). The Psychology of Entrepreneurship. August 2006. Taylor & Francis, Inc.

Lopez, M. R. (2009): *Bio-methanol producer BioMCN receives €36 million in private equity from Waterland*. Ecoseed, Mar. 10, 2009. http://www.ecoseed.org/en/general-green-news/features/green-living/941 (last access 1/19/2010).

Loten, A. (2011): *Firms Face Hurdles Overseas – Small Companies Seek Growth Abroad, but Lack Resources of Bigger Rivals*. Wall Street Journal, Aug. 25, 2011. http://online.wsj.com/article/SB10001424053111904787404576528451448313010.html?mod=WSJEurope_hps_MIDDLE_Video_second (last access 8/25/2011).

Loui, M. C. (2009): *Ethics and Social Responsibility for Scientists and Engineers*. Friday Forum, University YMCA, Champaign, Illinois, Feb. 6, 2009. https://netfiles.uiuc.edu/loui/www/fridayforum09.pdf (last access 9/1/2012).

Lovell, N. (2011): *Gamescom Hosts Fight for Games Market*. Wall Street Journal, Aug. 17, 2011. http://blogs.wsj.com/tech-europe/2011/08/17/gamescon-hosts-fight-for-the-games-market/ (last access 8/18/2011).

Lowry, A. (2011): *Culture as a Sustainable Advantage*. Stanford Technology Ventures Program, Entrepreneurial Thought Leaders Lecture Series, Nov. 9, 2011. http://ecorner.stanford.edu/authorMaterialInfo.html?mid=2839 (last access 2/1/2012).

Luggen, M.; Tschirky, H. (2003): *A conceptual framework for technology and innovation management in new technology-based firms (NTBF)*. Management of Engineering and Technology, PICMET '03. Technology Management for Reshaping the World. Portland International Conference on; 20-24 July 2003, Page(s): 342-347.

Luggen, M.; Tschirky, H. (2004): *An Entrepreneurial Approach to Technology and Innovation Management in New Technology-Based Firms (NTBF)*. International Conference on Management of Technology, IAMOT Conference, Washington D.C. 2004. http://www.iamot.org/conference/index.php/ocs/4/paper/viewFile/958/386 (last access 7/11/2009).

Lux Research (2009): *Green Biologics Limited*. Lux Research, updated on Oct. 27, 2009. https://portal.luxresearchinc.com/research/briefing_excerpt/5577 (last access 1/6/2010).

Lux Research (2007): *Synthetic Genomics*. The Cleantech Report. http://www.luxresearchinc.com/pdf/CompanyProfile_SyntheticGenomics.pdf (last access 5/17/2010).

Luxshan (2009): *Lecture 12 –Innovation Timing & Innovation Communities*. http://luxshan.com/swi/12.pdf (last access 5/28/2010).

Mäki, K.; Hytti, U. (2008): *"Pitching Out the Lost Cases" – A Longitudinal Study of the Precedents of Non-Growth of Recent Incubator Graduates*. Paper submitted in the ICSB World Conference 2008. http://www.smu.ca/events/icsb/proceedings/entk6f.html (last access 7/10/2009).

Ma, W. (2013): *China's LDK Solar Defaults on Bond Payment*. Wall Street Journal, updated Apr. 16, 2013. http://online.wsj.com/article/SB10001424127887324030704578426241915931194.html?KEYWORDS=solar+chinese (last access 4/16/2013).

Machatschke, M. (2008): *Enercon – Vom Winde verwöhnt (Blessed by the winds)*. Manager Magazin, Dec. 8, 2008. http://www.manager-magazin.de/magazin/artikel/0,2828,586223,00.html (last access 1/15/2011).

Madrigal, A. (2009): *How Algal Biofuels Lost a Decade in the Race to Replace Oil*. Wired Science, Dec. 29, 2009. http://www.wired.com/wiredscience/2009/12/the-lost-decade-of-algal-biofuel/?utm_source=feedburner&utm_medium=feed&utm_campaign=Feed%3A+wired%2Findex+%28Wired%3A+Index+3+%28Top+Stories+2%29%29 (last access 1/2/2010).

Madrigal, A. (2007): *Supercrude Scum: The Algae-To-Biodiesel Battle*. Earth2tech, Oct. 3, 2007. http://gigaom.com/cleantech/supercrude-scum-the-algae-to-biodiesel-battle/ (last access 11/6/2008).

Mah, J. (2011): *Are You Building Products Customer Want?* Stanford Technology Venture Network, Entrepreneurial Thought Leaders Lecture Series, Nov. 30, 2011. http://ecorner.stanford.edu/authorMaterialInfo.html?mid=2853 (last access 6/23/2012).

Malone, T. W. (1988): *What is Coordination Theory?* Paper presented at the National Science Foundation Coordination Theory Workshop, Massachusetts Institute of Technology, Feb. 19, 1988. SSM WP # 2051-88. http://citeseerx.ist.psu.edu/viewdoc/download?doi=10.1.1.86.8170&rep=rep1&type=pdf (last access 12/5/2011).

Mandaro, L. (2007): *Monsanto, BASF in $1.5 bln genetic seeds venture*. MarketWatch, Mar. 21, 2007. http://www.marketwatch.com/story/monsanto-basf-in-15-bln-union-to-develop-seeds-for-biofuels (last access 10/30/2009).

Marine Biological Laboratory (2009). *Biofuel Displacing Food Crops May Have Bigger Carbon Impact Than Thought*. ScienceDaily, 2009. http://www.sciencedaily.com-/releases/2009/10/091022141117.htm (last access 10/25/2009).

Markman, G. D.,Gartner, W. B. (2002). *Is extraordinary growth profitable? A study of Inc. 500 high-growth companies*. Entrepreneurship: Theory and Practice – Fall, 2002. http://www.entrepreneur.com/tradejournals/article/110811955.html (last access 3/11/2011).

Marvel, M. R.; Murphy, P. J. (2007): *Human Capital and High Tech Entrepreneurial Discovery: Beyond the Search Serendipity Continuum*. Frontiers of Entrepreneurship Research 27 (5), Article 14. Available at: http://digitalknowledge.babson.edu/fer/vol27/iss5/14 (last access 9/10/2009), http://digitalknowledge.babson.edu/cgi/viewcontent.cgi?article=1097&context=fer (last access 9/10/2009).

Mascoma Corp. (2009): *Mascoma Announces Feedstock Processing and Lignin Supply Agreement with Chevron Technology Ventures*. News, Sep. 14, 2009. http://www.mascoma.com/download/Mascoma%20Chevron%20Agreement%20Release%20FINAL%20091409.pdf (last access 11/24/2009).

Marsh, P. (2008): *Prudent partnerships*. The Financial Times, Dec. 23, 2008. https://www.ft.com/cms/s/0/1d7312a6-d10b-11dd-8cc3-000077b07658,dwp_uuid=02e16f4a-46f9-11da-b8e5-00000e2511c8.html (last access 12/23/2008); http://www.cbnl.com/news/pdf/ftdotcom_231208.pdf (last access 10/30/2010).

Marsh, P. (2003): *Inside Track: Thinking big in nanoscience*. Financial Times; Jan 27, 2003. http://www.attocube.com/company/download/FT_270103.pdf (last access 1/13/2006).

Marshall, M. (2006): *Amyris gets $20M from high-profile investors to find alternative fuel*. Venture Beat, Oct. 12, 2006. http://venturebeat.com/2006/10/12/amyris-gets-20m-from-high-profile-investors-to-find-alternative-fuel/ (last access 5/6/2009).

Marshall, R. (2012): *LinkedIn considers launching intranet service for businesses*. V3.co.uk, Feb. 15, 2012. http://www.v3.co.uk/v3-uk/news/2152671/linkedin-considers-launching-intranet-service-businesses (last access 3/23/2012).

Martin, J. (2010): *Bioethanol: Cyano Biofuels setzt auf die Herstellung aus Algen (Bioethanol – Cyano Biofuels Focuses on Production by Algae)*. CleanThinking, Jan. 3, 2010. http://www.cleanthinking.de/biothenaol-cyano-biofuels-algen/3789/ (last access 1/4/2010).

Martin, D.; Bloch, P. (2004): *Patent Donations and the Problem of Orphan Technologies*. A Policy Forum on The Intangible Economy. Held at the Woodrow Wilson International Center for Scholars, Washington, DC, April 14, 2004. http://www.athenaalliance.org/issuedialogue/orphan_tech.html (last access 6/11/2009).

Mauer, R.; Brettel, M. (2008): *Learning From Technology Entrepreneurs: Effectuation as a New Toolbox In Contexts of High Uncertainty*. In Iandoli, L. (2008) Technology-driven entrepreneurship – how technology shapes the way entrepreneurs start and run their businesses. 5th Inter-RENT Online Publication, European Council for Small Business and Entrepreneurship, Mar. 2008. File interrent_2008_final.pdf, created 1/8/2009 (last access 10/20/2010); search Google: "mauer brettel Effectuation toolbox pdf."

Mayer, J. D. (1993-1994): *A System-Topics Framework for the study of personality.* Imagination, Cognition, and Personality, 13 (2), 99-123. http://www.thepersonalitysystem.org/PFAAssets/sysframework/systopics.pdf (last access 7/23/2009);
Mayer, J. D.: *Personality Psychology: A Systems Approach.* http://www.thepersonalitysystem.org/index.htm (last access 7/23/2009); John D. Mayer::*The Systems Framework for Personality in Focus.* More literature references herein. http://www.thepersonalitysystem.org/Systems%20Framework%20in%20Focus/pfasysframeandsystheory.htm (last access 7/23/2009).

MBI: Marrone, P. (Marrone Organic Innovations): *Hire People Who Share Your Values.* Business Leaders – 50 Lessons. http://www.50lessons.com/peoplefinder.asp?l=M&s=118 (last access 1/14/2010).

MBI (2007): *Marrone Organic Innovations Closes $3.75 Million Series A Financing* http://marronebioinnovations.com/pdf/MOISeriesAFinal.pdf (last access 1/14/2010).

McCann, J. E. (1991): *Patterns of growth, competitive technology, and financial strategies in young ventures.* Journal of Business Venturing 6 (3), May 1991, 189-208; available online 11 April 2002.

McCline, R. L.; Bhat, S.; Baj, P. (2000): *Opportunity Recognition: An Exploratory Investigation of a Component of the Entrepreneurial Process in the Context of the Health Care Industry.* Entrepreneurship: Theory and Practice, Winter, 2000. http://www.entrepreneur.com/tradejournals/article/74524634.html (last access 5/26/2010).

McCoy, M. (2006): *Biomass Gambit – Shell, Codexis Sign Biofuels Pact.* Chemical & Engineering News 84 (47), 21. http://pubs.acs.org/cen/news/84/i47/8447notw9.html (last access 11/24/2009).

McDermott, M.(2012): *Shell Builds Drop-In Biofuels Plant In Texas, While Revving Up Arctic Drilling Plans.* Treehugger, Mar. 1, 2012. http://www.treehugger.com/renewable-energy/shell-builds-drop-in-biofuel-plant-while-revving-oil-exploration-arctic.html (last access 5/27/2012).

McDermott, M. (2009): *BP Backs Out of Jatropha Joint Venture with D1 Oils.* Treehugger, Jul. 20, 2009. http://www.treehugger.com/files/2009/07/bp-backs-out-jatropha-joint-venture-d1-oils.php (last access 11/7/2009).

McDonald, T. (2011): E•*BOOM CAPITAL: Gevo and Amyris Take Similar Routes to Advanced Biofuel.* Energyboom, Feb. 22, 2011. http://www.energyboom.com/biofuels/e%E2%80%A2boom-capital-gevo-and-amyris-take-similar-routes-advanced-biofuel (last access 6/23/2011).

McLeod R.; Winsor, B. (2003): *3M Case Study: Entrepreneurship.* Scottish Institute for Enterprise. http://www.sie.ac.uk/File/12R0.aspx (last access 8/29//2009), http://www.venturenavigator.co.uk/content/3m_case_study?s=true (last access 8/292009).

MacMillan, D. (2013): *LivingSocial Falls to a Quarter of 2011 Value in Latest Funding.* Bloomberg News, Feb 22, 2013. http://www.bloomberg.com/news/2013-02-

22/livingsocial-falls-to-a-quarter-of-2011-value-in-latest-funding.html (last access 2/23/2013).

Meeker, M. (2001): *Technology and Internet: The Technology & Internet IPO Yearbook: 7th Edition – 21 Years of Tech Investing* Morgan Stanley Deam Witter, April 2001. http://www.morganstanley.com/institutional/techresearch/pdfs/2001ipoyb.pdf (last access 5/2/2010); Meeker, M. (2003): *Technology: Cross-Industry Insights: The Technology IPO Yearbook – 9th Edition – 23 Years of Tech Investing.* Morgan Stanley Equity Investment, Oct. 3, 2003. http://www.morganstanley.com/institutional/techresearch/pdfs/2003_tech_ipo_yearbook.pdf (last access 5/2/2010).

Melendez, L. (2009): *Green jobs may help CA out of recession.* Peninsula News, Sep. 18, 2009. http://abclocal.go.com/kgo/story?section=news/local/peninsula&id=7021996 (last access 1/25/2010).

Merton, R. K. (1968): *The Matthew effect in Science.* Science (New Series) 159 (3810), 56-63. http://www.unc.edu/~fbaum/teaching/PLSC541_Fall06/Merton_Science_1968.pdf (last access 11/3/2010).

Metzger, G. (2007): *On the Role of Entrepreneurial Experience for Start-up Financing – An Empirical Investigation for Germany.* ZEW, Discussion Paper No. 07-047, ftp://ftp.zew.de/pub/zew-docs/dp/dp07047.pdf (last access 7/11/2009).

Metzger, G.; Niefert, M.; Licht, G. (2008): *Hightech-Gründungen in Deutschland: Trends, Strukturen, Potenziale.* Mai 2008 (*Hightech Foundations in Germany: Trends, Structures, Potentials. May 2008*). ZEW. ftp://ftp.zew.de/pub/zew-docs/gutachten/hightechgruendungen08.pdf (last access 7/10/2009).

Metzger, G.; Heger, S.; Höwer, D.; Licht, G. (2010a): *High-Tech-Gründungen in Deutschland – Hemmnisse junger Unternehmen.* ZEW – Zentrum für Europäische Wirtschaftsforschung, September 2010. ftp://ftp.zew.de/pub/zew-docs/gutachten/hightechgruendungen2_10.pdf (last access 11/182010).

Metzger, G.; Heger, D.; Höwer, D.; Licht, G. (2010b): *High-Tech-Gründungen in Deutschland - Zum Mythos des jungen High-Tech-Gründers (High-tech foundations in Germany – On the myth of the young high-tech founder).* ZEW – Zentrum für Europäische Wirtschaftsforschung, February 2010. ftp://ftp.zew.de/pub/zew-docs/gutachten/hightechgruendungen10.pdf (last access 11/18/2010).

Meuser, F. (2011): *50 Jahre Institut für Getreideverarbeitung im Spiegel der Zeit (50 years of the Institute of Cereal Processing through the ages*). In [IGV 2011], Annual Report 2010, 38-45. http://www.igv-gmbh.de/images/stories/docs/IGV_Jahresbericht_2010_6.pdf (last access 5/8/2012).

Michel, R. (2009): *"Green gas" innovator Virent sees value in manufacturing expertise.* Manufacturing Business Technology, Jul. 29, 2009. http://www.mbtmag.com/blog/Operation_Green/20614-_Green_gas_innovator_Virent_sees_value_in_manufacturing_expertise.php (last access 12/17/2009).

Milgrom, L. (2003): *Spin doctors. Chemistry* in Britain, Jul. 2003. http://www.chemsoc.org/chembytes/ezine/2003/milgrom_jul03.htm (last access 7/31/2003).

Miller, K. (2007): *Silly Enough to Try Again?* Business Week, Jun. 1, 2007. http://www.businessweek.com/smallbiz/content/jun2007/sb20070601_336379.htm (last access 8/18/2010); *Great Advice from Serial Entrepreneurs. Pam Marrone, 50.* Business Week. http://images.businessweek.com/ss/07/06/0602_serial_entp/index_01.htm?popupWidth=770&popupHeight=660.

Mills, P. (2008): *Uncle Sam Wants U.V.* SpecialChem, Sep. 24, 2008. http://www.specialchem4coatings.com/resources/articles/article.aspx?id=10040&lr=mci08166&li=20024489 (last access 10/12/2009).

Minshall, T.; Wicksteed, B. (2005): *University spin-out companies: Starting to fill the evidence gap.* A report on a pilot research project commissioned by the Gatsby Charitable Foundation, Jan. 2005. http://www.insme.org/documents/University%20spin-out%20companies.pdf (last access 7/12/2010.

Mitchell P. C.; McKeown, A. E. (2004): *Importance of peer support and tutor involvement in entrepreneurship education for overseas Bioscience students.* BEE-j 3, May 2004. http://www.bioscience.heacademy.ac.uk/journal/vol3/beej-3-2.pdf (last access 05/07/2008).

Mitra, S.(2009): *U.S. And India's Tech Dilemma.* Forbes, Feb. 19, 2009. http://www.forbes.com/2009/02/19/india-engineer-computer-technology-enterprise-tech_india.html (last access 2/19/2009).

Molitor, J.-P. (2012): W*hite Biotech Strategies of Large Corporations: Hopes & Challenges*. SpecialChem, Jan. 30, 2012. http://www.specialchem4coatings.com/resources/articles/article.aspx?id=17704&lr=mci12337&li=20024489 (last access 2/9/2012).

Moore, G. (1999): *Solid-State Physicist William Shockley.* Time Magazine, Mar. 29, 1999. http://www.time.com/time/magazine/article/0,9171,990623-1,00.html (last access 1/7/2011).

Morais, R. (2004): *OTB-Group Makes The World Smaller.* Forbes, Apr. 23, 2004. http://www.forbes.com/2004/01/23/cz_rm_0123davosotb.html (last access 10/12/2009).

Moresco, J. (2008): *Khosla Ventures Partners With Hawaiian Utility.* Redherring, Dec. 17, 2008. http://www.redherring.com/Home/25665 (last access 12/29/2009).

Morris, M. H., Kuratko, D. F. and Covin J. G. (2008): *Corporate Entrepreneurship and Innovation: Entrepreneurial Development within Organizations* (2nd ed.). Thomson-South-Western.

Morrison, C. (2009): *Sapphire Energy gets "open checkbook" from investors for algae-based gasoline.* GreenBeat, May 29, 2008. http://green.venturebeat.com/2008/05/29/sapphire-energy-gets-open-checkbook-from-investors-for-algae-based-gasoline/ (last access 3/14/2010).

Morrissey, S. R. (2009): *High Risk Energy Projects Get Funds. Grants: Big winners include Foro Energy, DuPont, and MIT*. Chemical & Engineering News 87 (44), 9. http://pubs.acs.org/cen/news/87/i44/8744notw1.html (last access 11/4/2009).

Moustras, M. – The Royal Society of Chemistry (2003), *Spin out companies from UK chemistry departments*. http://www.rsc.org/pdf/indusdiv/spinout.pdf (last access 02/18/2008).

MPG (2009): *Max Planck Gesellschaft – Annual Report 2008*. May 2009. http://www.mpg.de/pdf/jahresbericht2008/jahresbericht2008.pdf (last access 6/12/2010);
MPG (2010): Max Planck Gesellschaft – *Annual Report 2009*. May 2010. http://www.mpg.de/180981/Jahresbericht_2009.pdf (last access 6/12/2011).

MSFC – US Marshall Space Flight Center: *Dr. Wernher von Braun – First Center Director, Jul. 1, 1960 - Jan. 27, 1970*. http://history.msfc.nasa.gov/vonbraun/bio.html.

Mueller, S.L.; Thomas, A.S. (2001): *Culture and entrepreneurial potential – A nine country study of locus of control and innovativeness*. Journal of Business Venturing, 16 (1), 51-75.

Müller-Jung, J. (2009): *Bioenergie – Der Frühling der grünen Sonnenkollektoren (Bioenergy – The spring of green sun collectors)*. FAZNET, Apr. 7, 2009. http://www.faz.net/s/Rub80665A3C1FA14FB9967DBF46652868E9/Doc~E4E0FED18D1D14969A12DBEB846362998~ATpl~Ecommon~Scontent.html (last access 11/10/2009).

Müller-Jung, J. (2010): *Wer zaubert Biosprit aus dem grünen Topf? (Who performs magic out of the green pot?)*. FAZNET, Feb. 17, 2010. http://www.faz.net/s/Rub80665A3C1FA14FB9967DBF46652868E9/Doc~E076358A5606E4CA2B741BFA3A61DE9EC~ATpl~Ecommon~Scontent.html (last access 2/17/2010).

Mugler, J. (1998): *Betriebswirtschaftslehre der Klein- und Mittelbetriebe*. Band 1, Dritte, überarbeitete Auflage, Springer Verlag Wien, New York. Kapitel: *Der Konfigurationsansatz (Business Administration for Small and Medium-Sized Enterprises* Vol. 1, Chapter: *The Configuration Approach).*

Mukhopadhyay, R. (2011a): *Biofuels: An Ethical Framework*. Chemical & Engineering News 89 (33) Web Exclusive, Aug.15,20011. http://pubs.acs.org/cen/coverstory/89/8933cover2.html (last access 8/18/2011).

Mukhopadhyay, R. (2011b): *Technology: Energy Department Offers Deep Discounts On Its Patents*. Chemical & Engineering News, May 6, 2011. http://pubs.acs.org/cen/news/89/i19/8919notw5.html (last access 5/6/2011).

Mukhopadhyay, R. (2011c): *DOE, USDA Biofuel Projects Get A Boost*. Chemical & Engineering News 89, May 10, 2011. http://pubs.acs.org/cen/news/89/i20/8920news4.html (last access 5/12/2011).

Mukhopadhyay, R. (2011d): *Clean Energy For The Military*. Chemical & Engineering News 89 (18), May 2, 2011, 8. http://pubs.acs.org/cen/news/89/i18/8918notw7.html (last access 5/4/2011).

Mukhopadhyay, R. (2011e): *Examining Biofuels Policy*. Chemical & Engineering News 89 (33), Aug. 15, 2011, 10-15. http://pubs.acs.org/cen/coverstory/89/8933cover.html?featured=1 (last access 8/17/2012).

Mullaney, T. (2011): *Why Mascoma's Shaky IPO Is Lehman's Fault*. The Motley Fool, Sep. 26, 2011. http://www.fool.com/investing/general/2011/09/26/why-mascomas-shaky-ipo-is-lehmans-fault.aspx (last access 1/23/2012).

Mullins, J. W. (2006): *What to Run BeforeYour Business Plan Course*. London Business School, 2006. http://www.stanford.edu/group/ree/archives/archive06/europe/presentations/REE_Europe_2006_JMullins.pdf (last access 5/26/2010); cf. also http://www.ieec.co.uk/2006/docs/Pre-BusinessPlanninCurricula.ppt.

Murphy, M. (2011): *Zynga Explains Method For Recording Real Revenue From Virtual Goods*. Wall Street Journal, Jul. 1, 2011. http://online.wsj.com/article/BT-CO-20110701-712257.html (last access 7/2/2011).

Murmann, J. P. (2003): *Knowledge and Competitive Advantage. The Coevolution of Firms, Technology, and National Institutions*. Cambridge University Press; The Coevolution of Industries and National Institutions: Theory and Evidence. Discussion Paper FS IV 02-14, Wissenschaftszentrum Berlin, 2002. http://skylla.wz-berlin.de/pdf/2002/iv02-14.pdf (last access 6/21/2010).

Muth, W. (2010): *Novaled AG – OLEDs for General Lighting*. Presentation, Nov. 2, 2010). http://www.rusnanoforum.ru/Document.aspx/Download/29189 (last access 2/12/2011).

Myers, S.; Majluf, N. S. (1984): *Corporate Financing and Investment Decisions when Firms have Information that Investors Do Not Have*. Journal of Financial Economics 13 (2) (June 1984), 187-221; http://en.wikipedia.org/wiki/Pecking_Order_Theory.

Nagappan, P. (2009): *Solazyme Lands U.S Navy Contracts for Algae Jet Fuel*. EnergyBoom – Biofuel Energy, Nov. 10, 2009. http://www.energyboom.com/biofuels/solazyme-lands-us-navy-contracts-algae-jet-fuel (last access 1/25/2010).

Naimark, N. M. (1995): *The Russians in Germany: A history of the Soviet Zone of occupation, 1945-1949*. The Soviet Use of German Science, p. 205-250. Harvard University Press, 1995.

Narvaes Wilmsen, E. (2006): *Colorado State Engines Lab Teams with Solix Biofuels Inc. to Mass Produce Oil from Algae as Diesel Fuel Alternative*. Colorado State University, News & Information, Dec. 7, 2006. http://www.news.colostate.edu/Release/4260 (last access 4/18/2007).

NanoScape (2002): *Nanozwiebeln bringen Styrol-Synthese auf Trab. (Nano-onions accelerate styrene synthesis*). Innovations-Report, Oct. 17, 2002. http://www.innovations-report.de/html/berichte/verfahrenstechnologie/bericht-13744.html (last access 5/5/2006).

NanoScape (2005): *Firm Profile – Design, Discovery, Development – NanoMaterialien für die Zukunft. (Firm Profile – Design, Discovery, Development – Nano-materials for the future*). File from 4/7/2005. http://www.nanoscape.de/pdf/History_NanoScape.pdf (last access 3/5/2006).

NASA (2003): *Remarks by NASA Deputy Administrator Gregory Centennial of Flight Commemoration.* Agency News, Dec. 17, 2003. http://www.nasa.gov/audience/formedia/speeches/fg_kitty_hawk_12.17.03_prt.htm (last access 4/11/2010).

NASDAQ (2011): *Gevo - Prospectus* - Filed Pursuant to Rule 424(b)(4), Registration No. 333-168792, Feb. 8, 2011 http://www.nasdaq.com/markets/ipos/filing.ashx?filingid=7365503 (last access 12/12/2011).

Nelson, B. (2009): *Explanation of the TVA Radar Graph.* Forbes Magazine Sep. 16, 2009. http://www.forbes.com/2009/09/08/radar-graph-explanation-entrepreneurs-finance-promising.html (last access 9/23/2010).

Nelson, J. W.(2002): *Color Me With Good Fortune*. Molecular Interventions 2 (3), 186-187 (June 2002). http://molinterv.aspetjournals.org/content/2/3/186.full (last access 2/18/2010).

NEnA: *Katja Heppe – Ich lebe für meine Idee* (*Katja Heppe* – I live for my idea). Nano-Entrepreneurship-Academy. http://www.nano-4-women.de/content/view/50/104/lang,en/ (last access 8/16/2010).

Neubauer, U. (2006): *Forscher in geheimer Mission (Researchers with secret mission).* Nachrichten aus der Chemie 54, 759-761. http://www.gdch.de/taetigkeiten/nch/inhalt/jg2006/forscher.pdf (last access 8/25/2010).

Neutraceuticals World (2009): *B&D Nutritional Ingredients Partners with Solazyme.* Published Dec. 1, 2009. http://www.nutraceuticalsworld.com/contents/view/15110 (last access 1/24/2010).

New York Times (1997): *Manfred von Ardenne, 90, Dies; Was German Nuclear Physicist.* The New York Times, May 29, 1997. http://www.nytimes.com/1997/05/29/world/manfred-von-ardenne-90-dies-was-german-nuclear-physicist.html (last access 3/24/2011).

NewNet – New Energy World Network: *NewNet Innovator Profile: Rick Wilson, Cobalt Biofuels.* NewNet, Oct. 12, 2009. http://www.newenergyworldnetwork.com/clean-energy-investor-profiles/newnet-innovator-profile-rick-wilson-cobalt-biofuels.html (last access 1/26/2010).

Newton D. (2001): *Using Credit Cards to Finance Your Startup. Think of it as a temporary solution.* Jun. 18, 2001. http://www.entrepreneur.com/money/financing/financingcolumnistdavidnewton/article41520.html (last access 9/15/2009).

Nicolaou, N.; Shane, S.; Cherkas, L.; Hunkin, J.; Spector, T. D. (2008): *Is the Tendency to Engage in Entrepreneurship Genetic?* Management Science 54 (1), January 2008, 167-179; cf. also
Nicolaou, N.; Shane, S.; Cherkas, L.; Hunkin, J.; Spector, T. D. (2007): *Is the Tendency to Engage in Entrepreneurship Genetic?* http://wsomfaculty.cwru.edu/shane/ge/GE1.pdf (last access 6/18/2009).

Niefert, M., Metzger, G.; Heger, D.; Licht, G. (2006): *Hightech-Gründungen in Deutschland.* Endbericht, Mannheim, Juni 2006. (*Hightech Foundations in Germany.* Final Report, Mannheim, June 2006). ZEW. ftp://ftp.zew.de/pub/zew-docs/gutachten/hightechgruendungen.pdf (last access 7/10/2009).

NNFCC (2009): *Review of Technologies for Gasification of Biomass and Wastes.* Final Report, NNFCC Project 09/008, Jun. 2009. NNFCC final report E4tech 090609[1].pdf; access via Google search "Review of Technologies for Gasification of Biomass and Wastes" (last access 2/202010).

Nonnast, T. (2006): SAP – *Langsam, aber gewaltig (SAP – slow but colossal).* Handelsblatt, Apr. 7, 2006. http://www.handelsblatt.com/sap-langsam-aber-gewaltig;1061581 (last access 7/29/2010.

NVCA (2009): *Venture Impact – The Economic Importance of Venture Capital-Backed Companies to the U.S. Economy.* The National Venture Capital Association (NVCA). http://www.nvca.org/index.php?option=com_content&view=article&id=255&Itemid=103; NVCA_venture_impact_5th_ed.pdf (file last access 10/10/2010).

NVCA: *Venture Capital 101 – What is Venture Capital.* Venture_Capital_101.pdf from www.NVCA.org; (file last accessed 10/11/2010).

O'Brien, J. M. (2007): *Meet the PayPal mafia.* Forbes Magazine, Nov. 26, 2007. http://money.cnn.com/2007/11/13/magazines/fortune/paypal_mafia.fortune/index.htm (last access 7/1/2011); http://en.wikipedia.org/wiki/PayPal_Mafia.

Obituary (1956): *Thomas J. Watson Sr. Is Dead; I.B.M. Board Chairman Was 82.* The New York Times, On this Day, Jun. 20, 1956. http://www.nytimes.com/learning/general/onthisday/bday/0217.html (last access 1/25/2011).

Oilgae Blog (2008): *Vinod Khosla Feels Algae Are not Yet Ready for Primetime?* October 31, 2008. http://www.oilgae.com/blog/2008/10/vinod-khosla-feels-algae-are-not-yet.html (last access 2/18/2010).

Ohff, H. (1981): *Karl Friedrich Schinkel – Preußische Köpfe (Prussian Heads).* Stapp Verlag, Berlin 1981; cf. also http://en.wikipedia.org/wiki/Karl_Friedrich_Schinkel; http://de.wikipedia.org/wiki/Karl_Friedrich_Schinkel; http://de.wikipedia.org/wiki/Christian_Peter_Wilhelm_Beuth.

Oilgae Blog (2009): *Biofuels Digest released summary of US venture capital investment in biofuels.* Posted on January 25, 2009 by GURU @ 8:23 PM. http://www.oilgae.com/blog/2009/01/biofuels-digest-released-summary-of-us.html (last access 10/24/2009).

Olson, P. D. (1986): *Entrepreneurship: opportunistic decision-makers.* Journal of Small Business Management, 11 (2), 25-31. Google cache - http://www.allbusiness.com/human-resources/employee-development-problem-solving/102133-1.html (last access 7/8/2009).

Oneal, M. (2006): *Entrepreneur putting his chips on ethanol.* Chicago Tribune, Oct. 13, 2006. http://www.chicagotribune.com/news/nationworld/chi-0610130129oct13,0,2550074.story (last access 2/20/2010).

Onuoha G. (2007): "*Entrepreneurship*" AIST International Journal 10.20 02 cited in Wikipedia (http://en.wikipedia.org/wiki/Entrepreneurship, last access 7/8/2009).

Openshaw, E.; Pentland, A. (2012): *Making Visible the Invisible: Meaning, Not Content, Matters in Social Data.* All Things, Sep. 28, 2012. http://allthingsd.com/20120928/making-visible-the-invisible-meaning-not-content-matters-in-social-data/ (last access 10/5/2012).

OPXBio (2011): *Dow and OPXBIO Collaborating on Renewable Route to Acrylic Acid.* News, Apr. 11, 2011. http://www.opxbiotechnologies.com/press/Dow_and_OPXBIO_Collaborating_on_Renewable_Route_to_Acrylic_Acid.php (last access 4/29/2011).

ORNL – Oak Ridge National Laboratory (2010): *DNP Green Technology, Inc. Exclusively Licenses Patents Invented at Argonne National Laboratory and Oak Ridge National Laboratory.* ORNL, Mar, 9, 2010. http://www.ornl.gov/adm/partnerships/success_stories/dnp.shtml (last access 3/4/2012).

Osterath, B.; Guggolz, E. (2013): *Wir waren immer kurz vom Abschuss (We were always just before the shooting down.) Interview with key researcher Ludwig Pohl.* Nachrichten aus der Chemie 61, 758-761.

Ozgen, E.; Baron, R.A. (2007): *Social sources of information in opportunity recognition: Effects of mentors, industry networks, and professional forums. Journal* of Business Venturing 22, 174-192.

Page, L.; Schmidt, E. (2002): *Video – Innovate in Technology and Business: The Founding of Google.* Innovate in Technology and Business: The Founding of Google Entrepreneurial Thought Leader Speaker Series, May 1, 2002. http://ecorner.stanford.edu/authorMaterialInfo.html?mid=1074 (last access 7/2011).

Pasztor, A. (2012): *Private Space Run Makes History.* Wall Street Journal, May 25, 2012. http://online.wsj.com/article/SB10001424052702304840904577426042171703270.html?mod=WSJEurope_hpp_LEFTTopStories (last access 5/26/2012); cf. also Andresen, T. (2012): *Elon Musk - Der Elektroauto-Pionier, der auf den Mars will (Elon Musk – The electrocar pioneer who want to reach the Mars).* Handelsblatt. May 25, 2012. http://www.handelsblatt.com/unternehmen/management/koepfe/elon-musk-der-elektroauto-pionier-der-auf-den-mars-will/6671546.html (last access 5/25/2012).

Patrashkova-Vozdolska, R. (2005): *Modeling teams: A general systems theory approach.* PhD Thesis, University of Massachusetts Amherst, 2005, 228 pages; AAT 3193932). http://gradworks.umi.com/31/93/3193932.html (last access 7/27/2010); Presentation with some details. http://home.kelley.iupui.edu/tatikond/webpage/AoM2005StudentPresentations/PatrashkovaPresentation.ppt.

PBS – Public Broadcasting Service (2009): *Timeline: History of the Electric Car.* Now on PBS, Week of Oct. 30, 2009. http://www.pbs.org/now/shows/223/electric-car-timeline.html (last access 5/28/2010).

Peach S. (2011): *Current Technology Could Reduce Global Energy Demand by 85%.* Chemical & Engineering News, Jan. 19, 2011. http://pubs.acs.org/cen/news/89/i04/8904scene1.html (last access 1/20/2011).

Peer, M. (2013): *Netzwirtschaft – Deutsche Gründer erobern Asien (Net economy – German founders conquer Asia)*. Handelsblatt, published on Apr. 2, 2013. http://www.karriere.de/karriere/deutsche-gruender-erobern-asien-165593/ (last access 3/6/2013); cf. also http://www.crunchbase.com/company/lazada.

Perlin, J. (2003): *From Space to Earth. The Story of Solar Energy.*, http://www.californiasolarcenter.org/pdfs/rahus/Panels_1-10_poster.pdf (last access 8/25/2009); also http://www.californiasolarcenter.org/fste/fste.html (last access 8/25/2009).

Perman, S. (2006): *Is There a Gene for Business?* Business Week, Oct. 30, 2006. http://www.businessweek.com/smallbiz/content/oct2006/sb20061030_540336.htm (last access 7/19/2009).

Phuong Le (2008): *Algae fuel science chalks up federal, private funding.* Lubbock Online, Nov. 5, 2008. http://www.lubbockonline.com/stories/110508/bus_352191682.shtml (last access 12/12/2009).

Pienkos, P. T. (2008): *Historical Overview of Algal Biofuel – Technoeconomic Analyses.* DOE Algal Biofuels Workshop, Dec. 9, 2008. http://www.nrel.gov/docs/fy09osti/45622.pdf (last access 12/23/2009).

Pienkos P. T.; Darzins, A. (2009): *The promise and challenges of microalgal-derived biofuels.* Biofuels, Bioprod. Bioref. 3, 431–440 (2009); DOI: 10.1002/bbb). http://www.afdc.energy.gov/afdc/pdfs/microalgal_biofuels_darzins.pdf (last access 12/23/2009).

pipeline (2008): *pipeline™ Exclusive – Fermentation Technology Developer Eyes £3.5 Million.* Clean Technologies & Renewable Energy II (27), 4-5 (25 November – 01 December 2008). http://www.mountroyalventures.com/assets/pdfs/VB%20Research%20Ingeligence%20Report.pdf (last access 1/28/2010).

Pisano, G. (2006): *Don't Wait Around for the Next Amgen.* Business Week, Dec. 20, 2006. http://www.businessweek.com/technology/content/dec2006/tc20061220_865132.htm?chan=technology_technology+index+page_science (last access 8/23/2010).

Pleschak, F.; Werner, H. (1998): *Technologieorientierte Unternehmensgründungen in den neuen Bundeländern – Wissenschaftliche Analyse und Begleitung des BMBF-Modellversuchs (Technology-oriented firms' foundation in the new federal states – Scientific analysis and accompanimentof the BMBF pilot).* Physica-Verlag, Heidelberg.

Poliakoff, M.; Howdle, S.; Lathbury, D. (2009): *Ten things chemical engineers should know about chemists.* Chemistry World, January 2009. http://www.rsc.org/chemistryworld/Issues/2009/January/Comment.asp (last access 7/1/2011).

Porretto, J. (2009): *Exxon makes first big investment in biofuels.* Forbes, Jul. 14, 2009. http://www.forbes.com/feeds/ap/2009/07/14/ap6650792.html (last access); also http://www.google.com/hostednews/ap/article/ALeqM5jORVy4DeZKdWVikICUewO'YPE nmhQD99E5HS03.

Prahalad, C.K.; Bettis, R.A. (1986): *The dominant logic: A new linkage between diversity and performance.* Strategic Management Journal, 7: 485-501.

PresseBox (2012): *PURPLAN GmbH wird Preisträger 2012 – Einmaliges Know-how (PURPLAN GmbH awarded in 2012 – A unique know-how*). Pressemeldung BoxID 535224, Sep. 8, 2012. http://www.pressebox.de/pressemeldungen/oskar-patzelt-stiftung/boxid/535224 (last access 10/23/2012).

PrivCo (2013): *Fisker's Road to Bankruptcy/Ruin.* Research Note, Apr. 4, 2013; Breaking News, Apr. 22, 2013; Breaking News, Apr. 23, 2013; Breaking News, Apr. 26, 2013; Breaking News, Apr. 29, 2013.

Procter, C. (2008): *Solix Biofuels hopes to generate green from algae.* Denver Business Journal, Jan. 25, 2008. http://www.bizjournals.com/denver/stories/2008/01/28/story4.html?page=all (last access 11/6/2008).

Pruitt, A. (2009a): *Gevo, Inc. Develops Revolutionary Butanol Biofuel.* EnergyBoom, Biofuel Energy, Nov. 25, 2009). http://www.energyboom.com/biofuels/gevo-inc-develops-revolutionary-butanol-biofuel (last access 1/11/2010).

Pruitt, A.; Kalish, A. (2009b): *U.S. Census data showing new and young firms as the primary source of new jobs come at a critical time for policymakers.* Kauffman Foundation, Nov. 5, 2009. http://www.kauffman.org/newsroom/kauffman-foundation-analysis-emphasizes-importance-of-young-businesses-to-job-creation-in-the-united-states.aspx (last access 3/23/2010).

Purac (2011): *BASF and CSM explore a bio-based Succinic Acid joint venture.* Purac, Press Release, Aug. 1, 2011. http://www.purac.com/EN/Biobased_chemicals/News/Press-release-BASF-and-CSM-explore-a-joint-venture.aspx (last accss 9/23/2011).

Pure Energy Systems (PES) Network: *Wind Turbine Systems for Home Power Generation.* http://peswiki.com/index.php/Directory:Home_Generation:Wind_Turbine (last access 9/9/2010).

Quon, K. (2012): *Solazyme: An Undervalued Investment With A Hidden Identity.* Seeking Alpha, May 24, 2012. http://seekingalpha.com/article/614271-solazyme-an-undervalued-investment-with-a-hidden-identity (last access 5/25/2012).

Quatar Airways (2009): *World's First Commercial Passenger Flight Powered By Fuel Made From Natural Gas Lands In Qatar.* Media Room, Oct. 12, 2009. http://www.qatarairways.com/global/en/newsroom/archive/press-release-12Oct09-2.html (last access 12/11/2009); cf. also 12/11/2009 http://www.qatarairways.com/global/en/newsroom/archive/press-release-12Oct09-2.html.

Radiomuseum: *History of the radio manufacturer Leybold und von Ardenne Oszillografengesellschaft.* http://www.radiomuseum.org/dsp_hersteller_detail.cfm?company_id=7507 (last access 9/12/2012).

Raice, S.; Ovide, S. (2012): *Groupon Investors Give Up*. Wall Street Journal, Aug. 20, 2012.
http://online.wsj.com/article/SB10000872396390443989204577599273177326912.html (last access 8/120/2012).

Raice, S.; Smith, R. (2011): *Groupon IPO Cheers Valley*. Wall Street Journal, Nov, 5, 2011.
http://online.wsj.com/article/SB10001424052970203716204577017773545604142.html?KEYWORDS=groupon+ipo (last access 11/6/2011).

Ramey, D. E. (2007):*Butanol: The Other Alternative Fuel*. In NABC 19: Agricultural Biofuels: Technology, Sustainability and Profitability. South Dakota State University, Conference May 22 – 24, 2007, 137-147.
http://nabc.cals.cornell.edu/pubs/nabc_19/NABC19_5Plenary2_Ramey.pdf (last access 12/23/2009).

Ramsey, M. (2012): *China's New Target: Batteries*. Wall Street Journal, Aug. 8, 2012.
http://online.wsj.com/article/SB10000872396390443991704577576881949308486.html (last access 8/8/2012).

Rapier, R. (2011): *What Happened at Choren?* The energy collective, posted Jul. 9, 2011. http://theenergycollective.com/robertrapier/60963/what-happened-choren (lass access 11/14/2011); Consumer Energy Report, Jul. 8, 2011.
http://www.consumerenergyreport.com/2011/07/08/what-happened-at-choren/ (last access 8/3/2011).

Rapier, R. (2010): *Broken Promises from Range Fuels*. Consumer Energy Report, Feb. 23, 2010, update 8/18/2010.
http://www.consumerenergyreport.com/2010/02/23/broken-promises-from-range-fuels/ (last access 11/25/2010); Rapier, R. (2010): *Diminishing expectations from Range Fuels*. Forbes, Feb. 25, 2010.
http://www.forbes.com/sites/energysource/2010/02/25/diminishing-expectations-from-range-fuels/ (last access 3/1/2010).

Rapier, R. (2009a): *Vinod Khosla at Milken Institute: Part II*. R-Squared Energy Blog Pure Energy, May 1, 2009. http://robertrapier.wordpress.com/category/cilion/ (last access 12/9/2009).

Rapier, R. (2009b): *GreenFuel Bites the Dust*. R-Squared Energy Blog, May 15, 2009. http://i-r-squared.blogspot.com/2009/05/greenfuel-bites-dust.html (last access 1/2/2010).

Rapier, R. (2009c): *Book Review: Green Algae Strategy: End Oil Imports And Engineer Sustainable Food And Fuel* by Mark Edwards. The Oil Drum, posted on Jun. 8, 2009. http://www.theoildrum.com/node/5465 (last access 1/12/2010).

Rapier, R. (2006): *Cellulosic Ethanol vs. Biomass Gasification*. R-Squared Energy Blog, Oct. 22, 2006. http://robertrapier.wordpress.com/2006/10/22/cellulosic-ethanol-vs-biomass-gasification/ (last access 2/19/2010).

Ravenstijn, J. (2011): *Switched on to biopolymers*. In Biowerkstoff Report – Report on Bio-based Plastics and Composites. Edition 8, March 2011. File 11-03_Biowerkstoff-Report_8.pdf. http://www.nova-institut.de/download/bwr8_2011 (last access 6/5/2012).

RedHerring (2010). *Cleantech Venture Break as a Smoggy Day*. RedHerring, Podcast on 30 September 2010. http://www.redherring.com/Home/26444 (last access 10/7/2010).

Reinhold, R.: *Understanding the MBTI ® and Personality Type – Type Dynamics: Interpreting the Myers Briggs MBTI ®Personality Type Code*. PersonalityPathways; Reinhold Development – Marketing & Internet Development for Small Business Daniel Laufenberg – Laufenberg Economic Quarterly. http://www.personalitypathways.com/article/type-dynamics.html (last access 8/12/2010).

Reisch, M. S. (2006): *Start-up Firms Pursue Biofuels – Entrepreneurs invigorate search for economical biomass-based fuels.* Chemical & Engineering News 84 (47), Nov. 20, 2006. http://pubs.acs.org/cen/coverstory/84/8447cover3a.html (last acces 2/20/2010). Reisch, M. S. (2006): *Fuels Of The Future – Chemistry and agriculture join to make a new generation of renewable fuels.* Chemical & Engineering News 84 (47), 30-32. http://pubs.acs.org/cen/coverstory/84/8447cover3.html (last access 11/11/2009).

Reisch, M. S. (2007a): *Suits and Lab Coats – Industry draws on academic know-how to help develop specialty chemicals and other new materials.* Chemical & Engineering News 85 (47), 15-20. http://pubs.acs.org/cen/coverstory/85/8547cover.html (last access 12/20/2007).

Reisch, M. (2007b): *BASF Commits $20 Million To Harvard Research – Five-year program will draw on Harvard's School of Engineering and other departments.* Chemical & Engineering News 85 (45), Oct. 29, 2007 http://pubs.acs.org/cen/news/85/i45/8545news1.html (last access 6/21/2010).

Reisch, M. S. (2010): *French chemical firm ramps up effort to commercialize acrylic acid based on glycerin.* Chemical & Engineering News, 88 (21), May 24, 2010, 20-21.

Reisch, M. S. (2011a): *Instrumental Efforts – Entrepreneurs take big risks to bring the latest scientific tools to market.* Chemical & Engineering News 89 (32), 14-19. http://pubs.acs.org/cen/coverstory/89/8932cover.html?featured=1 (last access 8/9/2011).

Reisch, M. (2011b): *Investors Bet On Cleantech*. Chemical & Engineering News 89 (43), Oct. 24, 2011, 9. http://pubs.acs.org/cen/news/89/i43/8943notw9.html (last access 10/27/2011).

Reisch, M. S.(2012): *A123 Systems Goes Bankrupt*. Chemical & Engineering News 90 (43), 8. http://cen.acs.org/articles/90/i43/A123-Systems-Goes-Bankrupt.html (last access 10/29/2012).

RenewableEnergyWorld.com Editors (2010): *Solyndra Feeling the Pressure.* RenewableEnergyWorld.com, Nov. 5, 2010. http://www.renewableenergyworld.com/rea/news/article/2010/11/solyndra-feeling-the-pressure (last access 3/30/2011).

Rensselaer Polytechnic Institute (2007): *Beyond Batteries: Storing Power in a Sheet of Paper.* Rensselaer Research Review Fall 2007. http://www.rpi.edu/research/magazine/fall2007/batteries-1.html (last access

7/21/2009). Rensselaer Polytechnic Institute. *Serendipity and the Paper Battery.* http://www.rpi.edu/research/biotechnology/report/battery.html (last access 7/21/2009).

Reportlinker (2011): *Virtual Goods in Social Networking and Online Gaming.* PRNewswire, Jun. 6, 2011. http://www.kfvs12.com/story/14845327/virtual-goods-in-social-networking-and-online-gaming?clienttype=printable (last access 7/4/2011).

Reuter, B. (2011): *Unternehmenspleite - Aus für den Holzdiesel (Company insolvency – Stop for the wood diesel*). Wirtschaftswoche, Jul. 15, 2011. http://www.wiwo.de/unternehmen/unternehmenspleite-aus-fuer-den-holzdiesel/5307676.html (last access 11/17/2011).

Reuters (2010): *Venture capital money drops to 1997 levels – report.* Reuters, Jan. 22, 2010. http://www.reuters.com/article/idUSN2213491820100122 (last access 1/22/2010).

RFA – Renewable Fuels Association: Statistics, *Historic U.S. fuel, Ethanol Production.* http://www.ethanolrfa.org/pages/statistics (last access 5/5/2012).

Ritch, E. (2010): *Solix shipping algae oil biofuel from demo plant.* Cleantech, Jan. 5, 2010. http://cleantech.com/news/5460/solix-shipping-algae-oil-biofuel-de (last access 10/2/2010).

Riddell, L. (2012): *Solazyme's IPO filing revealed.* San Francisco Business Times, Mar. 15, 2011. http://www.bizjournals.com/sanfrancisco/blog/2011/03/solazymes-ipo-turns-up-interesting.html (last access 4/1/2012).

Riddell, L. (2011): *Amyris, Total create joint biofuels venture.* San Francisco Business Times, Dec. 1, 2011. (last access 1/10/2012).

Riegler, A. (2002): *When Is a Cognitive System Embodied?* Cognitive Systems Research 3 (3), 339-348. http://pespmc1.vub.ac.be/riegler/papers/riegler01embodiment(old).pdf (last access 4/16/2010).

Riley, C. (2013): *Major Chinese solar company goes bankrupt.* CNNMoney Mar. 21, 2013. http://money.cnn.com/2013/03/21/news/suntech-solar-bankruptcy/index.htm (last access 4/16/2013).

Ripplinger, P (2009): *Industrial production of microalgae biomass with flat-panel airlift photobioreactors.* Presentation, ALGAE 2009, Hamburg. http://global.subitec.com/pdf/Subitec_ALGAE_2009.pdf (last access 3/17/2010).

Ritchey, T. (2011): *General Morphological Analysis – A general method for non-quantified modeling*. File from 4/1/2011. Adapted from the paper "Fritz Zwicky, Morphology and Policy Analysis," presented at the 16th EURO Conference on Operational Analysis, Brussels, 1998. http://www.swemorph.com/pdf/gma.pdf; http://www.swemorph.com/ma.html (last access 8/19/2011).

Rittel, H.; Webber, M. (1973): *Dilemmas in a General Theory of Planning.* Policy Sciences (4), 155-169. Elsevier Scientific Publishing Company, Inc., Amsterdam. http://www.uctc.net/mwebber/Rittel+Webber+Dilemmas+General_Theory_of_Planning.pdf (last access 11/16/2009).

Ritter, S. K. (2011): *Small-Business Award: BioAmber, Minneapolis*. Chemical & Engineering News 89, Jun. 27, 2011. http://pubs.acs.org/cen/news/89/i26/8926notw1c.html (last access 7/2/2011).

Ritter, S. K. (2004): *Biomass or Bust – Technology to use plant-derived sugars to produce chemical feedstocks is ready ... and waiting*. Chemical & Engineering News 82 (22), 31-34. http://pubs.acs.org/email/cen/html/060804150713.html (last access 11/3/2009).

Roberts, E. B. (1989): *Initial capital for the new technological enterprise.* MIT Working Paper, No 2103-1-89, September 1989. http://dspace.mit.edu/bitstream/handle/1721.1/2235/SWP-2603-20912627.pdf?sequence=1 (last access 10/14/2010).

Robinson, S. (2009): *Second-generation biofuels need work – Fuel trek: the next generation*. ICIS, Feb. 24, 2009. http://www.icis.com/Articles/2009/02/24/9195433/second-generation-biofuels-need-work.html (last access 11/9/2009).

Roe, W. (2009): *Coskata – The Leader in Feedstock Flexible Ethanol.* US DOE - Biomass 2009. http://www1.eere.energy.gov/biomass/pdfs/Biomass_2009_Adv_Biofuels_I_Roe.pdf (last access 2/19/2010).

Rogers, W. S.; Mead, J. L. (2011): *The next small step for mankind – nanotechnology's journey to commercialization*. File from 4/29/2011. http://www.daypitney.com/news/docs/dp_3562.pdf (last access 5/18/2011).

Rooney, B. (2011): *Bigpoint In $350 million Deal.* Wall Street Journal, Apr. 26, 2011. http://blogs.wsj.com/tech-europe/2011/04/26/bigpoint-in-350-million-deal/ (last access 4/26/2011).

Rosenfeld, E. (2007*): Credit Cards: A Small Business Financing Tool.* http://www.entrepreneurship.gov/ResourcesCenter/Entrepreneur/13500.html (last access 9/15/2009).

Ross Range, P. (2012): *Vorbild Deutschland – Ein Amerikaner entdeckt das deutsche Industriewunder (Model Germany – An American discovers the German industrial miracle*). Handelsblatt, Aug. 7, 2012. http://www.handelsblatt.com/politik/konjunktur/nachrichten/vorbild-deutschland-ein-amerikaner-entdeckt-das-deutsche-industriewunder/6966492.html (last access 8/8/2012).

Ruch, E. (1975): *The Diagram Lattice as Structural Principle*. Theoretica Chimica Acta 38, 167-183.

Ruhm, K. H.(2008): *Dynamic System in the Steady State Condition – State Space*. Internet Portal "Measurement Science and Technology"; Document d0000881; rev00. ETH Zurich, Switzerland, Institute for Dynamic Systems and Control, First Edition 12/11/2008.http://www.mmm.ethz.ch/dok01/d0000881.pdf; http://www.mmm.ethz.ch/dok01/d0000321.pdf (last access 5/6/2011).

Rumelt, R. (1996): *The Many Faces of Honda.* California Management Review 38 (4), 103-111– citation on page 110. cf. also http://www1.ximb.ac.in/users/fac/dpdash/dpdash.nsf/pages/BP_Honda (last access 6/3/2009).

Runge, W. (2006): *Innovation, Research and Technology Intelligence in the Chemical Industry – Integrated Business, Technical and Systems Approaches*. Fraunhofer IRB Verlag, Stuttgart, 2006.

Runge, W. (2006a): *Patentbewertung und Patentanalysen – Zweck, Adressaten, Kontext, Methoden and Technik (Patent Assessment and Patent Analyses – Purpose, Addressees, Context, Methods, and Technology)*. In: Conference Proceedings PATINFO 2006, Schramm R. und Milde, S. (Eds) 2006, 269 – 287.

Runge, W.; Bräse, S. (2008): *Education in Chemical Entrepreneurship: Towards Technology Entrepreneurship for and in Chemistry-Related Enterprises*. Karlsruhe University (TH). http://ce.ioc.KIT.edu/download/Chem_Entrepreneurship_Paper.pdf.

Runge, W. (2010): *Mit einer Idee zum Technologieführer (From an Idea to a Leader of Technology).* An Overview on Nanopool GmbH Based on a Guest Lecture and Literature for Technology Entrepreneurship. PresentationTechnology Entrepreneurship, Dec. 9, 2009.
http://ce.ioc.KIT.edu/downloads/Praesentation_Schwindt.pdf (last access 7/19/2010).

Russo, A. (2001): *Five Myths About Entrepreneurs. Understanding How Businesses Start and Grow*. National Commission on Entrepreneurship (of the US), March 2001. http://www.publicforuminstitute.org/nde/sources/reports/2001-five-myths.pdf) (last access 11/23/2007); http://www.renderx.com/~renderx/myths/en/art_letter.pdf (last access 10/23/2012).

RWE (2009a): *Das RWE Algenprojekt (The RWE Alga Project)*. RWE Power AG in Bergbau 2, 64 – 66, 2009 http://www.bergbau-zeitschrift.de/zeitung09/09-02-64-66.pdf (last access 3/15/2010), cf. also http://www.rwe.com/web/cms/de/213188/rwe-power-ag/klimaschutz/innovationszentrum-kohle/algenprojekt/ (last access 3/14/2010), http://www.rwe.com/web/cms/mediablob/de/2374/data/213188/1096/rwe-power-ag/klimaschutz/innovationszentrum-kohle/algenprojekt/brosch-algenprojekt.pdf.

RWE (2009b): *Klimafreundliche CO2-Nutzung: Algen binden Kohlendioxid problemlos in Biomasse ein (Climate-firiendly CO2 Utilization – Algae capture carbon dioxide in biomass without problems).* RWE Power, Sep. 11, 2009.
http://www.rwe.com/web/cms/de/2320/rwe-power-ag/pressemitteilungen/?pmid=4003947 (last access 3/16/2010).

RWE (2010): *RWE Power, BRAIN join forces in white biotech: co-operation on CO2 as raw material for new products*. RWE Power, Jan. 4, 2010.
http://www.rwe.com/web/cms/de/2320/rwe-power-ag/pressemitteilungen/?pmid=4004370 (last access 3/16/2010).

Sanderson, K. (2009): *The Great Beyond. Biofuel woes.* Posted by Katharine Sanderson on October 23, 2009.
http://blogs.nature.com/news/thegreatbeyond/2009/10/biofuel_woes.html (last access 10/25/2009).

SAP-1: *Geschichte der SAP: Mehr als 37 Jahre im Business mit dem E-Business History of SAP – More than 37 years in business with e-business).*
http://www.sap.com/germany/about/company/geschichte/index.epx (last access

7/30/2010); cf. also Dietmar Hopp:
http://www.whoswho.de/templ/te_bio.php?PID=1187&RID=1; Hasso Plattner:
http://www.whoswho.de/templ/te_bio.php?PID=1181&RID=1; Klaus Tschira:
http://www.whoswho.de/templ/te_bio.php?PID=2563&RID=1.

SAP-2: SAP Tutorial and Guide – SAP Definition and Brief History. http://saptutorial.net/sap-definition-and-brief-history/ last access 7/29/2010); cf. also http://en.wikipedia.org/wiki/SAP_AG; http://de.wikipedia.org/wiki/SAP.

Sasiadek, S. (2004): *Entrepreneurship Versus Intrapreneurship*. July 4, 2004. http://drsusansasiadek.com/yahoo_site_admin/assets/docs/Entrepreneurship_Versus_Intrapreneurship.86181859.htm (last access 7/13/2009).

Say, J.-B. (1803): Traité d'économie politique, ou simple exposition de la manière dont se forment, se distribuent, et se consomment les richesses – *A Treatise on Political Economy; or the Production, Distribution, and Consumption of Wealth*. Translated from the Fourth Edition of the French, By C. R. Prinsep, M. A. with Notes by the Translator. New American Edition. Containing a Translation of the Introduction, and Additional Notes by Clement C. Biddle, LL. D. Member of the American Philosophical Society. Philadelphia: Claxton, Remsen & Haffelfinger 624, 625, & 628 Market Street 1880; Batoche Books, Kitchener 2001. http://socserv.mcmaster.ca/econ/ugcm/3ll3/say/treatise.pdf (last access 6/19/2009).

SBDC – Minnesota Small Business Development Centers (2012): *MetroSpec Technology*. Annual Report 2010, 10. File of 1/9/2012. http://www.positivelyminnesota.com/About_Us/Goals_Results/Our_Progress/Program_Results/Small_Business_Development_Centers_2010_Annual_Report.pdf (last access 3/1/2012).

Schedl, H. L.; Vola, M. (2007): *Hidden Champions of the 21st Century – Presentation for the Conference "Top management: Challenges and Limits of Competition."* Minsk, Nov. 30 2007. http://www.krconsult.org/download_files/confa3/shedl_hidden_champions_presentation_minsk.pdf (last access 10/8/2010).

Scheffler, S. (2009): *Biosprit – Warum Bill Gates auf Algensprit setzt (Biofuels – Why Bill Gates Bets on Algae)*. Handelsblatt, Apr. 4, 2009. http://www.handelsblatt.com/technologie/energie_technik/warum-bill-gates-auf-algensprit-setzt;2232435 (last access 9/3/2009).

Schibilsky, M. (2008): *Treibstoff aus Algen – Energiewunder der Erdgeschichte (Fuel from Algae – The Energy Wonder of the Earth's History)*. RBB Broadcast – Ozon aus Wissenschaft und Umwelt, Mar. 17, 2008. http://www.orb.de/ozon/beitrag/2008/treibstoff_aus_algen.html (last access 1/4/2010).

Schild, R. (2009): *Business & Change in VON ARDENNE*. MBA Alumni Event, Sep. 15, 2009. http:// mbaurv.oasi.org/bd/business_and_change_in_von_ardenne1.pdf (last access 11/23/2009).

Schneider, J. J. (2010): *Akademia-Industrie-Austausch – ein Schuss Neugier und Offenheit (Academia-industry-exchange – a little bit of curiosity and openness)*. Nachrichten aus der Chemie 58 (7/8), 838-839.

Schonfeld, E. (2009a): *Interview: Vinod Khosla Is On The Hunt For Great Technologies*, TechCrunch, Sep. 12, 2009. (last access 12/8/2009).

Schonfeld, E. (2009b): *Khosla Ventures Raises $1.1 Billion. It's For More Than Just Clean Tech*. TechCrunch. http://www.techcrunch.com/2009/09/01/khosla-ventures-raises-11-billion-its-for-more-than-just-clean-tech/ (last access 11/24/2009).

Schrage, M. (2010): *The Delicate Art of Unauthorized Innovation*. Harvard Business Review, Mar. 24, 2010. http://blogs.hbr.org/schrage/2010/03/the-delicate-art-of-unauthoriz.html (last access 8/25/2010).

Schramm, C. (2006a): *The Entrepreneurial Imperative.* Philanthropy Magazine, Sep. 25, 2006 http://prt.timberlakepublishing.com/article.asp?article=1416&paper=0&cat=149 (last access 3/2/09).

Schramm, C. (2006b): *Entrepreneurial Capitalism and the End of Bureaucracy – Reforming the Mutual Dialog of Risk Aversion.* http://www.aeaweb.org/annual_mtg_papers/2006/0107_1015_0304.pdf (last access 6/20/2009 as html.

Schürmann, H. (2007): *Den Klimawandel bekämpfen: Mikroalgen retten die Umwelt (Fighting climate change – Microalgae to save the world)*. Handelsblatt, Nov. 12, 2007. http://www.handelsblatt.com/technologie/umwelt-news/mikroalgen-retten-die-umwelt;1350492 (last access 2/14/2010).

Schubert, T. (2008a): *Unternehmensgründung in der Chemiebranche – Ein Erfahrungsbericht (Firm's Foundation in the Chemical Industry – A Report of Experiences)*. Presentation Technology Entrepreneurship, Jan. 23, 2008. http://ce.ioc.KIT.edu/download/Schubert_Karlsruhe20080123.pdf (particularly slide 31).

Schubert, T. (2008b): *IOLITEC – A Material Supplier for the PV Industry*. 2nd International Conference on Solar PV Investments 19th & 20th February, Frankfurt (Germany). http://investments2008.epia.org/Files/PDF/WS_080220_3A-06_TSchubert.pdf (last access 3/10/2011).

Schubert, T. (2006): *IoLiTec – Ihr Partner für ionische Flüssigkeiten (Your Partner for Ionic Liquids)*. 2nd Partnering-Event Chemie Start-ups, Frankfurt am Main, 9/22/2006. http://events.dechema.de/index.php?id=1052&suffix=pdf&nonactive=1&lang=de&site=events_media (last access 5/3/2007).

Schubert, T. (2005): *A brief history of IoLiTec*. Company Newsletter, Ionic Liquds Today, 1/05, Mar. 3, 2005.

Schubert, T. (2004): *IoLiTec*. File 4-IoLiTec.ppt as pdf-from 7/6/2004, particularly slide 5 (last access on the Web 4/30/2010).

Schuetzle, D.; Tamblyn, G.; Tornatore, F.; MacDonald, T. (2007): *Range Fuels has BioConversion Technology as a partner for the Georgia plant*. In *Alcohol Fuels from Biomass – Assessment of Production Technologies*. Report Draft, Mar. 27, 2007, page 67. ASSESSMENT-BIOALCOHOL-tech[1].pdf, Google search "Alcohol Fuels from Biomass – Assessment of Production Technologies" (last access 2/20/2010).

Schuler, T. (1999): *Verrat unter Freunden (Betrayal among friends).* Die Zeit, 1999. http://www.zeit.de/1999/40/199940.nsa_2_.xml (last access 8/30/2010); poor English translation: Schröm, O. (2009): *Betrayal among friends.* http://groups.google.com/group/sci.geo.meteorology/msg/1617cdd82eedfcd3 (last access 3/16/2011).

Schumpeter, J. A. (1942): "*Creative Destruction.*" http://transcriptions.english.ucsb.edu/archive/courses/liu/english25/materials/schumpeter.html (last access 6/2/2011). From:
Joseph A. Schumpeter Capitalism, Socialism, and Democracy. New York: Harper, 1942, pp. 82-85.

Schumpeter, J. A. (1911): *Theorie der wirtschaftlichen Entwicklung.* Duncker and Humblot, Leipzig. (Translated by R. Opie. Cambridge: Harvard University Press, 1934. The Theory of Economic Development: An inquiry into profits, capital, credit, interest and the business cycle.

Schumpeter, J. (1939): *Business Cycles.* MacGraw-Hill, New York.

Schwartz, D. (2010): *The A.I.M Interview - Sapphire Energy's Tim Zenk.* Algae Industry Magazine, Feb. 9, 2010. http://www.algaeindustrymagazine.com/tim-zenk/ (last access 3/13/2010).

Schwedt, G. (2009): *Goethe – Der Manager (Goethe – The Manager).* Wiley-VCH, Germany.

Schweinsberg, M. (2007): *SusTech GmbH & Co. KG – Erfolg durch Public-Private-Partnership (SusTech GmbH & Co. KG – Success through Public-Private-Partnership).* In VDI: Identifizierung von Innovationshemmnissen in der Werkstoffforschung und –entwicklung. Ergebnisse des Workshops am 20.09.2007 in Bonn (Identification of innovation barriers for materials research and developments. Results of a workshop on 9/20/2007 in Bonn). http://www.techportal.de/uploads/publications/572/Workshopbericht_Innovationshemmnisse.pdf (last access 7/3/2010).

Scully, V. (2007): *Effects of the Biofuel Boom.* Business Week, Aug. 27, 2007. http://www.businessweek.com/investor/content/aug2007/pi20070824_143619.htm?chan=technology_technology+index+page_top+stories (last access 11/2/2009).

SearchEngineWatch (2003): *Google Buys Applied Semantics.* SearchEngineWatch.com, Apr. 24, 2003. http://searchenginewatch.com/article/2047715/Google-Buys-Applied-Semantics (last access 2/12/2011).

SEC (2009): *SEC-Filing Bluefire.* 10-K 1 f10k2008_bluefire.htm 2008 ANNUAL YEAR END REPORT. http://www.sec.gov/Archives/edgar/data/1370489/000121390009000636/f10k2008_bluefire.htm.

SEC (2010): *Amyris Biotechnologies, Inc.* United States – Securities and Exchange Commission, FORM S-1, REGISTRATION STATEMENT. http://www.sec.gov/Archives/edgar/data/1365916/000119312510085595/ds1.htm (last access 4/25/2010).

Seelye, K. Q. (2009): *What Happens to the American Dream in a Recession*? The New York Times, May 8, 2009. http://www.nytimes.com/2009/05/08/us/08dreampoll.html?_r=1&hp (last access 5/8/2009).

Seidler, C. (2009): *Alles auf Alge* (*Bet on Algae*). Der Spiegel, Dec. 16, 2009. http://www.spiegel.de/wissenschaft/natur/0,1518,665958,00.html (last access 16/12/2009.

Senge, P. M (1990): *The Fifth Dimension: The Art and Practice of the Learning Organization*. Doubleday, New York, NY.

Senge, P. M. (2006): *The Fifth Discipline – The Arts and Practice of the Learning Organization*. Second revised ed., Random House Business Books, 2006.

Seraphim, P. G. (2009): *Brazil's Ethanol-Enhanced History*. Ethanol Producer Magazine, September 2009. http://www.ethanolproducer.com/article.jsp?article_id=5906&q=&page=all (last access 10/30/2009).

Seymour, R. (2000): *Ms. versus Mr. Who is the Better Boss*? The Magazine for Canadian Entrepreneurs. V. 19, 1-2. http://www.profitguide.com/w100/2000/related.asp?ID=857 (last access 12/5/2002).

Shah, D. (2007): The *Dark Side of Startups: 5 Corrosive Co-Founder Conflicts*. OnStartups, posted Aug. 21, 2007. http://onstartups.com/tabid/3339/bid/2148/The-Dark-Side-of-Startups-5-Corrosive-Co-Founder-Conflicts.aspx (last access 12/17/2011).

Shane, S. (2000): *Prior Knowledge and the Discovery of Entrepreneurial Opportunities*. Organization Science (4), July-August 2000, 448-469.

Shell (2007): *Shell Hydrogen and Virent Energy Systems to Manufacture Hydrogen Using Biomass*. Press Release, May 24, 2007. http://www.shell.de/home/content/usa/aboutshell/media_center/news_and_press_releases/archive/2007/virent_biomass.html (last access 1/26/2020).

Shelley, C. E.; Zhou, J. (2008): *Organizational Creativity Research – A Historical Overview*. In Handbook of Organizational Creativity, Jing Zhou, Christina Shalley (Eds.), Chapter 1. Lawrence Erlbaum Associates New York 2008 (subsidiary of the Taylor & Francis Group, LLC).

Sheridan, C. (2009): *Making green*. Nature Biotechnology 27, 1074–1076. http://www.nature.com/nbt/journal/v27/n12/full/nbt1209-1074.html (last access 3/21/2010).

Sherman, E. (2011): *How LinkedIn Turns a Profit when "Users" Don't Use the Site*. CBS Money Watch, Aug. 5, 2011. http://www.cbsnews.com/8301-505124_162-43452177/how-linkedin-turns-a-profit-when-users-dont-use-the-site/?tag=bnetdomain (last access 8/12/2011).

Short, P. L. (2009a): *Tight Focus Helped German Clusters Compete*. Chemical & Engineering News 87 (3), Jan. 19, 2009. http://pubs.acs.org/cen/business/87/8703bus2a.html (last access 1/26/2009).

Short, P. L. (2009b): *Clustering In Heidelberg – German region wins competition to anchor one of five national R&D clusters*. Chemical & Engineering News 87 (3), Jan. 19, 2009. http://pubs.acs.org/cen/business/87/8703bus2.html (last access 1/26/2009).

Short, P. L. (2006): *Out of the Ivory Tower.* Chemical & Engineering News 84 (17), 15-21. http://pubs.acs.org/cen/coverstory/84/8417ionicliquids.html (last access 6/23/2006).

Shrader, R.; Siegel, D. S. (2007): *Assessing the relationship between human capital and firm performance: evidence from technology-based new ventures.* Entrepreneurship: Theory and Practice 31 (6), 893-908. http://www.entrepreneur.com/tradejournals/article/170729964.html (last access 10/14/2008).

Shubnikov, A. V.; Koptsik, V. A. (1974): *Symmetry in Science and Art.* Plenum Press, New York, 1974.

Sierra Angels (2008): *Astia Honors Pam Marrone with Clean Tech Innovator Award.* San Francisco, Nov. 17, 2008. http://www.sierraangels.com/news-111708-marrone-organic/ (last access 1/14/2010).

Simon, H. (1955): *A Behavioral Model of Rational Choice.* Quarterly Journal of Economics 69 (1) (February 1955), 99-118.

Simon, H. (1956): *Rational Choice and the Structure of the Environment.* The Psychological Review 63 (March 1956), 129-138.

Simon, H. (2008): *Hidden Champions des 21. Jahrhunderts: Die Erfolgsstrategien unbekannter Weltmarktführer (Hidden Champions of the 21st century – Success Strategies of Unknown Global Market Leaders*). Presentation, RKW-Forum 2008, Nov. 20, 2008. http://www.rkw-bw.de/pdf/RKW-Organisation/RKW-Forum2008_VortragProfSimon.pdf (last access 10/8/2010); http://www.businessconnect.com.au/Documents/Content/Hermann_Simon%20_HiddenChampions.pdf (last access 10/8/2010).

Simon, H. (2007): *Hidden Champions des 21. Jahrhunderts : Die Erfolgsstrategien unbekannter Weltmarktführer (Hidden Champions of the 21st century – Success Strategies of Unknown World Market Leaders*). Campus, Frankfurt a. M., 2007; Simon, H. (2009): *Hidden Hidden Champions of the 21st – Success Strategies of Unknown World Market Leaders*. Springer, New York, 2009.

Simon, H. (1996): *"Hidden Champions – Lessons from 500 of the World's Best Unknown Companies"* Harvard Business School Press, Massachusetts, 1996.

Skyttner, L. (2001): *General Systems Theory – Ideas and Applications.* World Scientific Publishing Company, Hackensack, NJ, USA; *General Systems Theory – Problems Perspectives Practice.* World Scientific Publishing Company, Singapore. 2nd ed 2005.

Smith, M. R. (1985): *Military Enterprise and Technological Change – Perspectives on the American Experience.* MIT Press 1985.

Smith, R.; Raice, S. (2011): *Groupon Shares Backpedal.* Wall Street Journal, Nov. 23, 2011.

http://online.wsj.com/article/SB10001424052970203710704577054613912424388.html (last access 11/23/2011).

Smith, R.; Grocer, S. (2011): *Wall Street Cools to Groupon.* Wall Street Journal, Dec. 15, 2011. http://online.wsj.com/article/SB10001424052970204026804577098804223481254.html (last access 12/15/2011).

Smith-Hunter, A.; Kapp, J.; Yonkers, V. (2003): *A psychological model of entrepreneurial behavior.* Journal of the Academy of Business and Economics , April, 2003. FindArticles.com. 08 Jun, 2009. http://findarticles.com/p/articles/mi_m0OGT/is_2_2/ai_113563670/.

Snyder, M. (2010): *3 Charts That Prove We Are Living In The Biggest Debt Bubble In The History Of Our World.* The Economic Collapse, Nov. 3, 2010. http://www.businessinsider.com/charts-of-the-debt-bubble-2010-11 (last access 11/3/2010).

Sonnete, D. (2011): *Lecture 5: The Tipping Point and Epidemics – of innovations and entrepreneurial successes.* D-MTEC Chair of Entrepreneurial Risks. Swiss Federal Institute of Technology Zurich, file of 5/20/2011. http://www.er.ethz.ch/teaching/Lect5_Tipping_point_epidemics_11-2.pdf (last access 10/28/2011).

SpecialChem (2012): *Enerkem Begins Production of Cellulosic Ethanol at its Québec Facility.* SpecialChem, Jun. 12, 2012. http://www.specialchem4adhesives.com/resources/latest/displaynews.aspx?id=5976&lr=mas12195&li=10030139 (last access 6/20/2012).

SpecialChem4Coating (2011): *The most important factor that enables to develop new products & solve technical problems.* SpecialChem, Jan. 1, 2011. http://www.specialchem4coatings.com/community-pulse/community-insight.aspx?id=15117&lr=wpc11606&li=20024489 (last access 2/8/2011).

Specialchem4polymers (2011): *LANXESS Plans to Commercially Produce EPDM from Bio-based Ethylene by End of Year.* SpecialChem, Sep. 30, 2011. http://www.specialchem4polymers.com/resources/latest/displaynews.aspx?id=6505&lr=wpa11107la15&li=70056708 (last access 10/15/2011).

Spiewak, M.; Thuswaldner, G. (2009): *US Hochschulen – In Harvard wird gespart (US Universities – Harvard has to save money).* DIE ZEIT, Jan. 8, 2009. (last access 1/9/2009).

St. John, J. (2009): *Mascoma Starts Up Cellulosic Ethanol Demo Plant.* Green Tech Media, Feb. 25, 2009. http://www.greentechmedia.com/articles/read/mascoma-starts-up-cellulosic-ethanol-demo-plant-5790/ (last access 12/28/2009).

St. John, J. (2008): *LS9 Seeks Up to $100M Under New CEO.* Green Tech Media, Oct. 8, 2008. http://www.greentechmedia.com/articles/read/ls9-seeks-up-to-100m-under-new-ceo-1550/ (last access 12/11/2009).

St. John, J. (2011): *Oil Giant BP Backs Biofuel Startup Verdezyne.* Earth2tech, May 9, 2011. http://gigaom.com/cleantech/oil-giant-bp-backs-biofuel-startup-verdezyne/ (last access 6/23/2011).

Stack, J.; Balbach,, J.; Epstein, B.; Hanggi, T. (2007): *Cleantech Venture Capital: How Public Policy Has Stimulated Private Investment*. Environmental Entrepreneurs and Cleantech Venture Network LLC, May 15, 2007. http://www.e2.org/ext/doc/CleantechReport2007.pdf (last access 10/31/2009).

Stam, E.; Garnsey, E. (2007): *Entrepreneurship in the Knowledge Economy*. File from 5/8/2007. http://www.dur.ac.uk/resources/dbs//faculty/centre_entrepreneurship/publications/ResearchPaper018.pdf (last access 1/15/2011);
Stam, E.; Garnsey, E. (2006) *New Firms Evolving in the Knowledge Economy; Problems and solutions around turning points*. In: Dolfsma, W. & Soete, L. (Eds.), Understanding the Dynamics of a Knowledge Economy, 102-128. Cheltenham, Edward Elgar.

Stangler, D. (2010): *High-Growth Firms and the Future of the American Economy*. Kauffman Foundation Research Series: Firm Formation and Economic Growth, March 2010 .http://www.kauffman.org/uploadedfiles/high-growth-firms-study.pdf (last access 3/11/2011).

Startup Life (2007): *Ubiquitous Startups and the VC – Biofuel Investment Jumps 4000%*. Posted on Wednesday, March 21, 2007. http://sookman.blogspot.com/2007/03/biofuel-investment-jumps-4000.html (last access 10/20/2009).

Stewart, C. (2006): *The CODA Genomics executives*. The Orange County Register, Jun. 26, 2006. http://www.ocregister.com/articles/favorite-40767-professor-school.html (last access 2/4/2011).

Steyer, C.-D. (2007): *Von Tempo-Erbsen bis Kosmetik-Algen (From speed-up peas to cosmetics algae*). Der Tagesspiegel, Oct. 26, 2010. http://www.tagesspiegel.de/berlin/Brandenburg;art128,2407052 (last access 2/14/2010).

Stone, B. (2008): *Judge Ends Facebook's Feud With ConnectU*. The New York Times, Jun. 26, 2008. http://bits.blogs.nytimes.com/2008/06/26/judge-ends-facebooks-feud-with-connectu/ (last access 7/19/2009).

Stone, B. (2009): *ConnectU's 'Secret' $65 Million Settlement With Facebook*. The New York Times, Feb.10, 2009. http://bits.blogs.nytimes.com/2009/02/10/connectus-secret-65-million-settlement-with-facebook/ (last access 7/19/2009).

Strategy Analytics (2007): *Games Industry Transformed As One-Third of Games Software Revenues will be Generated Online by 2011*. Strategy Analytics, Sep. 10, 2007. http://www.strategyanalytics.com/default.aspx?mod=PressReleaseViewer&a0=3569 (last access 9/26/2011).

Streb, J. (2002): *Internationale Zusammenarbeit und nationale Alleingänge: Die Entwicklung der Synthesekautschukindustrie in Deutschland und den USA vor und während des Zweiten Weltkriegs. (International cooperation and national independence initiative – The development of the synthetic rubber industry in Germany and the USA before and during the Second World War)*. http://www.uni-tuebingen.de/uni/wwl/streb-ht.doc (last access 2/24/2010).

Streb, J. (1999): *How to win Schumpeterian Competition. Technological Transfers in the German Plastics Industry from the 1930s to the 1970s*. Yale University - Economic Growth Center, Center Discussion Paper No. 811, Sep. 1999. http://www.econ.yale.edu/growth_pdf/cdp811.pdf (last access 2/23/2010).

Stromeyer, E. B. (2009): *Pathway to Clean Energy – LS9 Inc. Innovates with UltraClean Technology*. Suite 101, Jul. 23, 2009. http://oil-energy-companies.suite101.com/article.cfm/pathway_to_clean_energy (last access 12/30/2009).

Stroud, S. (2010): *Cobalt teams up with Fluor*. Sustainable Industries, Aug. 26, 2010. http://sustainableindustries.com/articles/2010/08/cobalt-teams-fluor (last access 5/11/2011).

Süd-Chemie (2010): *Süd-Chemie to build Germany´s largest plant for the biotechnological production of biofuels from straw*. Press Release, Jul. 21, 2010. http://www.sud-chemie.com/scmcms/web/page_en_7566.htm (last acces 8/22/2010).

Süd-Chemie (2009): *Süd-Chemie and Linde launch production of climate-friendly biofuels*. Press Release. May 7, 2009. http://www.sud-chemie.com/scmcms/web/content.jsp?nodeIdPath=2991,6599,7362,7440&lang=en (last access 8/22/2010).

succidia AG: *Prof. Dr. Dr. hc. Otto Pulz*. Labor&more. http://www.laborundmore.de/research/6538/Prof.-Dr.-Dr.-h.-Otto-Pulz.html (last access 5/2/2013).

Sugarcaneblog (2009):*Brazil's CTC and BASF to develop Biotech Sugarcane*. Posted on Aug. 6, 2009. http://sugarcaneblog.com/2009/08/06/brazils-ctc-and-basf-to-develop-biotech-sugarcane/ (last access 10/30/2009).

Sull, D. N.; Wang Y. (2005): *The Three Windows of Opportunity*. Harvard Business School – Working Knowledge, Jun. 6, 2005. http://hbswk.hbs.edu/archive/4835.html (last access 4/21/2010).

Sullivan, D. (2004): *Google IPO To Happen, Files For Public Offering*. SearchEngineWatch.com, Apr. 29, 2004. http://searchenginewatch.com/3347471 (last access 2/22/2011).

Sullivan, D. (2003): *What's Behind Google's Purchase of Applied Semantics*? ClickZ, Apr. 30, 2003. http://www.clickz.com/clickz/column/1699604/whats-behind-googles-purchase-applied-semantics (last access 2/12/2011).

Sutton, R. I. (2007): *Extraordinary Contempt and Defiance Beyond the Normal Call of Engineering Duty*. Bob Sutton Work Matters http://bobsutton.typepad.com/my_weblog/2007/04/extraordinary_c.html (last access 8/25/2010)

Sutton, R. I. (2004): *Two Weird Ideas That Work*. Entrepreneurial Thought Leader Speaker Series, Stanford University, STVP, Oct. 27, 2004. http://edcorner.stanford.edu/authorMaterialInfo.html?mid=1190 (last access 8/25/2010).

Svoboda, E. (2008): *Fueling The Future.* Fast Company Magazine 122, Feb. 1, 2008. http://www.fastcompany.com/magazine/122/fueling-the-future.html (last access 10/30/2009).

Tajeddini, K.; Mueller, S. L. (2009): *Entrepreneurial characteristics in Switzerland and the UK: A comparative study of techno-entrepreneurs.* Journal of International Entrepreneurship 7 (1), 1-25.

Tam, P.-W. (2010): *Venture-Capital Firms Caught in a Shakeout.* The Wall Street Journal, Mar. 9, 2010. http://online.wsj.com/article/SB10001424052748703915204575104221092909884.html?mod=WSJEUROPE_hpp_sections_tech (last access 3/10/2010).

Tam, P.-W.; Bensinger, G.; Sherr, I. (2012): *Startups Adjust to Web's Down Cycle.* Wall Street Journal, Dec. 23, 2012. http://online.wsj.com/article/SB10001424127887324296604578177541652032424.html (last access 12/16/2012); cf. also http://en.wikipedia.org/wiki/LivingSocial.

Tam, P.-W.; Ante, S. E. (2010): *Super Angels' Alight – No Longer Flying Solo, Big Investors Attract Others to Juice Start-Ups.* The Wall Street Journal, Aug. 16, 2010. http://online.wsj.com/article/SB10001424052748703321004575427840232755162.html (last access 10/27/2010).

Tamásy, C.; Otten, C. (2000): *Unternehmensgründungen aus Hochschulen – welche Faktoren begünstigen die Gründungswahrscheinlichkeit und den Gründungserfolg? (University-based start-ups – which factors favour the probability and success of new enterprises?).* BMBF – exist, Jan. 2000; 7- 11. http://www.bmbf.de/pub/EXIST-D.PDF, English version http://www.bmbf.de/pub/EXIST-EN.PDF (last access 4/1/2010).

Taylor, P. (2010): *Biosuccinic acid ready for take off?* Chemistry World, Jan. 21, 2010. http://www.rsc.org/chemistryworld/News/2010/January/21011003.asp (last access 3/9/2010).

Taylor, W. C. (2009): *MBAs vs. Entrepreneurs.* Harvard Business Online May 5, 2009. http://www.businessweek.com/managing/content/may2009/ca2009055_933788.htm (last access 5/3/2010).

Tech Crunchies (2010): *Google Profit And Revenues Timeline.* Go rumors, posted: Mar. 3, 2010. http://gorumors.com/crunchies/google-ad-revenue-profit-timeline/ (last access 8/6/2010).

TECH{dot}MN: *MetroSpec Technology – FlexRad.* http://tech.mn/directory/companies/metrospec-technology/ (last access 3/12/2012).

Technology Park Adlershof (2008): *Lebenswissenschaften im Adlershof Focus – Wirkstoffe aus blauen Bakterien und molekulare Angeln (Life science in focus of Adlershof – Bio-active compounds from blue algae and molecular hooks).* Adlershof News, Aug. 27, 2008. http://www.adlershof.de/newsview/?no_cache=1&tx_ttnews%5Btt_news%5D=3660 (last access 2/16/2010).

Technology Park Adlershof (2009): *Over the Pond.* Adlershof News, Nov 2. 2009. http://www.adlershof.de/newsview/?no_cache=1&L=2&tx_ttnews%5Btt_news%5D=7717&klTyp=0 (Lats access 1/4/20010).

Teresko, J. (2005): *From Confusion To Action*. IndustryWeek, Sep. 1, 2005. http://www.industryweek.com/ReadArticle.aspx?ArticleID=10650 (last access 7/26/2009).

Thayer A.. M. (2011): *Top Instrument Firms*. Chemical & Engineering News 89 (17), 13-19. http://pubs.acs.org/cen/coverstory/89/8917cover.html?featured=1 (last access 5/16/2011).

Thayer, A. M. (2008): *Building Businesses – Turning university research into products takes time, money, and initiative as four nanotechnology companies' experiences show.* Chemical & Engineering News 86 (13), 10-15. http://pubs.acs.org/cen/coverstory/86/8613cover.html (last access 7/7/2010).

The Energy Blog (2006): *Solix Biofuels/Colorado State Deleloping Novel Algae Production System*. The Energy Blog, Dec. 28, 2006. http://thefraserdomain.typepad.com/energy/2006/12/solix_biofuelsc.html (last access 4/18/2007).

The Economist (2011): *Fail often, fail well.* Economist.com/blogs/schumpeter, The Economist, Apr. 14, 2011. http://www.economist.com/node/18557776 (last access 12/9/2011).

The Kirton Adaption-Innovation Inventory: http://www.kaicentre.com/ (last access 8/12/2010); *Publications and Current Work Using – Adaption-Innovation Theory*. http://www.kaicentre.com/PubList.doc (last access 8/13/2010).

The Myers & Brigg Foundation: *MBTI® Basics*. http://www.myersbriggs.org/my-mbti-personality-type/mbti-basics/ (last access 8/12/2010); *The 16 MBTI® Types*. http://www.myersbriggs.org/my-mbti-personality-type/mbti-basics/the-16-mbti-types.asp (last access 8/12/2010); *Understanding MBTI® Type Dynamics*. http://www.myersbriggs.org/my-mbti-personality-type/understanding-mbti-type-dynamics/ (last access 8/12/2010); cf. also http://en.wikipedia.org/wiki/Myers-Briggs_Type_Indicator.

The New York Times (2009): *Do Women Make Better Bosses?* Editors of The New York Times, The New York Times, Aug. 2, 2009. http://roomfordebate.blogs.nytimes.com/2009/08/02/do-women-make-better-bosses/ (last access 8/3/2009).

Theil, S. (2005): *The Next Petroleum*. Newsweek, Aug. 8, 2005. http://www.newsweek.com/id/56334 (last access 10/30/2009).

Thurm, S.; Raice, S.; Emos, T. (2012): *Social-Media Frenzy Fizzles*. Wall Street Journal, updated Jul. 27, 2012. http://online.wsj.com/article/SB10000872396390443477104577553431459135876.html (last access 7/28/2012).

Thurm, S.; Tam, P.-W. (2012): *Prominent Investors Miss Web IPO Payoff*. Wall Street Journal, Jun. 18, 2012. http://online.wsj.com/article/SB10001424052702303836404577474422342954922.html?mod=WSJEUROPE_hpp_MIDDLESecondNews (last access 6/19/2012).

Tidd, J.; Bessant, J.; Pavitt, K. (2001): *Managing Innovation – Integrating Technological, Market and Organizational Change* (2nd Ed.). John Wiley, Chichester [3rd Edition 2005].

Thomas, A. (2010): *Peter Thiel: From Frankfurt to the World of Global Markets.* aTrader, Jun. 17, 2010. http://www.atrader.com/manager-profile/11-peter-thiel-frankfurt-world-global-markets (last access 7/1/2011).

Tozzi, J. (2009): *Think Twice About Being First to Market.* Business Week, May 19, 2009. http://www.businessweek.com/smallbiz/content/may2009/sb20090519_306313.htm?chan=smallbiz_smallbiz+index+page_getting+started (last access 5/23/2010).

Tracey, J. (2009): In Kohler, C. (2009): *The 15 Most Influential Games of the Decade.* Wired, Dec. 24, 2009. http://www.wired.com/gamelife/2009/12/the-15-most-influential-games-of-the-decade/ (last access 9/24/2010).

Tuckman, B. W. (2001): *Developmental sequence in small groups.* Psychological Bulletin 63 (6), 384-99 (1965). Copyright International Association of Facilitators Spring 2001. Provided by ProQuest Information and Learning Company. All rights Reserved FindArticles.com. 07 Jul, 2011. http://findarticles.com/p/articles/mi_qa3954/is_200104/ai_n8943663/?tag=mantle_skin;content (last access 7/7/2011).

Tullett, A. D. (1997): *Cognitive Style - Not Culture's Consequence.* European Psychologist 2 (3), 258-267.

Tullo, A. H. (2011): *Breaking In The New.* Chemical & Engineering News 89 (38), Sep. 19, 2011, 10-14. http://pubs.acs.org/cen/coverstory/89/8938cover.html?featured=1 (last access 9/23/2011).

Tullo, A. H. (2010): *Amyris Files For IPO – Industrial Biotechnology: Firm is devising biomass-based routes to chemicals, fuels.* Chemical & Engineering News, Apr. 21, 2010. http://pubs.acs.org/cen/news/88/i17/8817news2.html (last access 4/24/2010).

Tullo, A. H. (2009): *Two Companies Eye Renewable Polymers – Green Chemistry: Braskem and Novomer have big plans for polymers based on renewable resources.* Chemical & Engineering News, Dec. 15, 2009. http://pubs.acs.org/cen/news/87/i51/8751news6.html (last access 12/20/2009).

TVA - The Venture Alliance: *Fundability Assessment.* http://www.tvausa.com/fundability_assessment.html (last access 9/23/2010).

UC Davis (2006): *Chevron to Fund Major Biofuel Research Projects at UC Davis.* UC Davis News & Information, Sep. 19, 2006. http://www.news.ucdavis.edu/search/news_detail.lasso?id=7873 (last access 11/24/2009).

Ucbasaran, D.; Wright, M.; Westhead, P.; Busenitz, L. W. (2003): *The Impact of Entrepreneurial Experience on Opportunity Identification and Exploitation – Habitual and Novice Entrepreneurs.* Advances in Entrepreneurship, Firm Emergence and Growth 6, 231-263.

Ucbasaran, D.; Westhead, P.; Wright, M. (2004):*The focus of entrepreneurial research – contextual and process issues*. File modified Aug. 10, 2004. http://ibis.nott.ac.uk/uniei/papers/paper1.doc (last access 9/15/2010).

UFOP - Union zur Förderung von Oel- und Proteinpflanzen e.V. (2007): *BASF und DBE planen gemeinsame Biodieselanlage (BASF and DBE plan a joint biodiesel plant).* UFOP-Marktinformation Ölsaaten und Biokraftstoffe, January 2007. http://www.ufop.de/downloads/RZ_MI_0107.pdf (last access 10/30/2009).

Ulrich, K. T.; Eppinger, S. D. (2000): *Product design and development*. McGraw-Hill, 2000; Exhibit 14-12 shown by
Thonemann, U. W. (2005): *Lecture – Product Development*. University of Cologne, File from 11/4/2005. http://s3.amazonaws.com/zanran_storage/www.wiso.uni-koeln.de/ContentPages/42419587.pdf (last access 7/12/2010).

US Bureau of Labor Statistics (2011): *International Comparisons of Annual Labor Force Statistics, Adjusted to U.S. Concepts, 10 Countries, 1970-2009 - Employment by sector, 1970 and 2009.* http://www.bls.gov/fls/intl_labor_force_charts.pdf; http://www.bls.gov/fls/flscomparelf/employment.htm#table2-4.f.na (last access 3/11/2011).

Utsch, A.; Rauch, A.; Rothfuss, R.; Frese, M. (1999). *Who becomes a small scale entrepreneur in a post-socialist environment: On the differences between entrepreneurs and managers in East Germany*. Journal of Small Business Management, 37 (3), 31-42. http://www.allbusiness.com/management-companies-enterprises/291539-1.html (last access 4/25/2009).

Van Gigch, J. P. (1974): *Applied General Systems Theory*. Harper & Row, Publishers, New York.

Van Noorden, R. (2009): *Editorial*: *Sustainable connection*. Chemistry World, January 2009. http://www.rsc.org/chemistryworld/Issues/2009/January/Editorial.asp (last access 3/25/2009).

Van Noorden, R. (2008): *Biobutanol enters battle of the alcohols*. Chemistry World, Feb. 2008.
http://www.rsc.org/chemistryworld/Issues/2008/February/BiobutanolEntersBattleAlcohols.asp (last access 3/8/2009).

Van Zanten, M. (2007): *Persoonlijke aansprakelijkheid wordt in grote bedrijven geschuwd (Personal liability is avoided in large companies)*. In: Topmanagers op eigen benen (Top managers on their own feet). Management Scope, Oct. 26, 2007. http://managementscope.onzetestomgeving.nl/magazine/artikel/271-topmanagers-eigen-benen (last access 1/19/2010).

Vanden Plas, J. (2008): *Alternative energy Visions: Virent looks for a successor to Apfelbach.* Wisconson Technology Network – WTN News, Oct. 6, 2008. http://wistechnology.com/articles/5094/ (last access 12/15/2009).

VentureBeat: *Green Biologics Overview*. Venture Beat Profiles. http://venturebeatprofiles.com/company/profile/green-biologics (last access 1/5/2010).

Vercipia (2010): *Consolidated Financial Statement – Vercipia Biofuels – Highlands Ethanol LLC.*

http://www.myflorida.com/audgen/pages/nonprofit_forprofit%20rpts/2009%20highlands%20ethanol%20dba%20vercipia%20biofuels.pdf (last access 4/2/2012).

Verganti, R. (2010): *Having Ideas versus Having a Vision*. Harvard Business Review Blog – The Conversation, Mar. 1, 2010. http://blogs.hbr.org/cs/2010/03/having_ideas_versus_having_a_vision.html (last access 3/30/2010).

Verser, D. (2008): *Will we really have $1.50 Non Food Ethanol? An Interview with Zeachem's EVP R&D, Dan W. Verser*. Alternative Energy Blog, posted on Apr. 12, 2008. http://www.alternativeenergy.com/profiles/blog/show?id=1066929%3ABlogPost%3A26030 (last access 1/5/2010).

Verser, D. (2009): *Dan Verser – Co-Founder – ZeaChem*. Entrepreneur Trek – Stanford, Mar. 21, 2009. http://www.slideshare.net/EntrepreneurTrek/dan-verser-cofounder-zeachem (last access 1/5/2010).

Viering, K. (2009): *Die Treibstoff-Züchter (The Fuel Breeders*). Berliner Zeitung, Nov. 20, 2009. http://www.berlinonline.de/berliner-zeitung/archiv/.bin/dump.fcgi/2009/1120/mutmacherjetztundhier/0443/index.html (last access 12/16/2009).

Villanueva, J.; Forbes, D.; Zellmer-Bruhn, M.; Sapienza , H. (2005): *The Entrepreneurial Intentions of Academic Scientist-Inventors*. Babson Frontiers of Entrepreneurship Research 2005; to be published in Journal of Business Venturing. http://www.babson.edu/entrep/fer/2005FER/chapter_xxii/paper_xxii3.html (last access 7/8/2009).

Vitols, S. (2001): *The Origins of Bank-Based and Market-Based Financial Systems: Germany, Japan, and the United States*. Discussion Paper FS I 01 – 302, Wissenschaftszentrum Berlin für Sozialforschung, Jan. 2001. http://bibliothek.wz-berlin.de/pdf/2001/i01-302.pdf (last access 8/15/2010).

Vlasic, B. (2013): *Breaking Down on the Road to Electric Cars*. The New York Times, Apr. 23, 2013. http://www.nytimes.com/2013/04/24/business/fisker-broke-down-on-the-road-to-electric-cars.html?pagewanted=all&_r=0 (last access 4/25/2013).

Voigt, K.-P. (2008): *Mikroalgen aus der Altmark mindern CO2-Aufkommen (Micro-algae from the Altmark reduce CO2 rise*). Mitteldeutsche Zeitung, Aug. 15, 2008. http://www.mz-web.de/servlet/ContentServer?pagename=ksta/page&atype=ksArtikel&aid=1218774013904 (last access 5/9/2011).

Voith, M. (2009a): *Dow Plans Algae Biofuels Pilot – Project will test a process to turn CO2 into ethanol.* Chemical & Enfineering News 87 (27), 10. http://pubs.acs.org/cen/news/87/i27/8727notw5.html.

Voith, M. (2009b): *Up From The Slime - Start-up companies see algae as the new renewable fuel source*. Chemical & Engineering News 87 (4), 22-23. http://pubs.acs.org/cen/email/html/cen_email_cen_87_i04_8704bus1.html (last access 1/2/3/2010).

Voith, M. (2010a): *A Slowdown For Start-Ups*. Chemical & Engineering News 88, Oct. 18, 2010. http://pubs.acs.org/cen/news/88/i43/8843news2.html (last access 10/20/2010).

Voith, M. (2010b):*Cleantech Funding Falls*. Chemical & Engineering News 88 (41), 10. http://pubs.acs.org/cen/news/88/i41/8841notw6.html (last access 10/21/2010).

VON ARDENNE:*The life's work of Manfred von Ardenne*. http://www.vonardenne.biz/VON_ARDENNE/Company/Company/Manfred-von-Ardenne/Introduction.html; in German: http://www.vonardenne.biz/VON_ARDENNE/UNTERNEHMEN/VON-ARDENNE/Manfred-von-Ardenne/Einleitung.html; cf. also http://en.wikipedia.org/wiki/Manfred_von_Ardenne.

Von Bertalanffy, L. (1968). *General System Theory: Foundations, Development, Applications*. George Braziller; New York.

Von Hoerner, H. (2008): vH&S *– Ein mittelständiges Unternehmen erfolgreich auf dem Gebiet der Raumfahrt seit 1971 (vH&S – A medium-sized enterprise successful in arerospace since 1971)*. Presentation Technology Entrepreneurship, Jan. 16, 2008. http://ce.ioc.KIT.edu/downloads/Praesentation_vonHoerner.pdf, http://ce.ioc.KIT.edu/59.php (last access 6/8/2011).

Vosmek, S.(2009): *Marrone Bio Introduces Innovative iPhone Application*. Astia, Nov. 16, 2009. http://astianotes.blogspot.com/2009/11/marrone-bio-introduces-innovative.html (last access 1/14/2010).

Wadhwa, V. (2006a): *Venture Capital: The Good, Bad, and Ugly*. Business Week, Jul. 18, 2006. http://www.businessweek.com/print/smallbiz/content/jul2006/sb20060717_402629.htm (last access 10/21/2010).

Wadhwa, V. (2006b): *15 Things You Need to Score VC Funding*. Business Week, Aug. 1, 2006. http://www.businessweek.com/smallbiz/content/aug2006/sb20060801_119338.htm (last access 11/5/2010).

Wadhwa, V. (2009): *Why Less Is More for Startups*. Business Week, Jun. 23, 2009. http://www.businessweek.com/smallbiz/content/jun2009/sb20090623_890999.htm (last access 11/10/2009).

Wadhwa, V.; Aggarwal, R.; Holly, K. "Z"; Salkever, A. (2009): *The Anatomy of an Entrepreneur – Family Background and Motivation*. Kauffman Foundation Report, July 2009. http://www.kauffman.org/uploadedFiles/ResearchAndPolicy/TheStudyOfEntrepreneurship/anatomy-of-entrepreneur-family-background-and-motivation.pdf (last access 7/27/2009).

Wald, M. L. (2009a): *E.P.A. Says It Expects to Raise Amount of Ethanol Allowed in Fuel Blends to 15%*. New York Times, Dec. 2, 2009. http://www.nytimes.com/2009/12/02/business/energy-environment/02ethanol.html (last access 12/2/2009).

Wald, M. L. (2009b): *Industry Built From Scratch*. The New York Times, Oct. 15, 2009. http://www.nytimes.com/2009/10/15/business/energy-environment/15biofuel.html?_r=1 (last access 11/30/2009).

Walker, J. (2011): *For Tech's Elite, Mobile Gaming Is a Big Play*. Wallstreet Journal, Nov. 15, 2011. http://online.wsj.com/article/SB10001424052970204323904577038462736751448.html?mod=WSJEUROPE_hpp_MIDDLEThirdNews (last access 11/15/2011).

Wallace, A. (2012): *Boulder's OPX Biotechnologies reaches 'scale-up' milestone*. The Denver Post, Mar. 27, 2012. http://www.denverpost.com/business/ci_20267474/boulder-opx-biotechnologies-reaches-milestone-bioacrylic (last access 3/31/2012).

Waltz , E. (2009a): *Algae Energy Orgy*. Mother Jones, September/October 2009 Issue. http://motherjones.com/politics/2009/09/algae-energy-orgy (last access 1/26/2010).

Waltz , E. (2009b): *Biotech's green gold*? Nature Biotechnology 27, 15-18. http://www.nature.com/nbt/journal/v27/n1/full/nbt0109-15.html (last access 2/2/2010); Waltz, E. (2009b): *Box 1 The economics of closed bioreactors: a back-of-the-envelope calculation*. Nature Biotechnology 27, 15-18. http://www.nature.com/nbt/journal/v27/n1/box/nbt0109-15_BX1.html (last access 2/2/2010).

Wang, U. (2011): *Report: Range Fuels to Shut Down Plant*. Earth2tech, Jan. 14, 2011. http://gigaom.com/cleantech/report-range-fuels-to-shut-down-plant/ (last access 3/17/2011).

Wang, U. (2008): *Cobalt to Build Pilot Biobutanol Plant in 2009*. Green Tech Media, Oct. 21, 2008. http://www.greentechmedia.com/articles/read/cobalt-to-build-pilot-biobutanol-plant-in-2009-5037/ (last access 11/9/2009).

Wasch, M. (2010): *Bokela setzt auf China-Geschäft (Bokela focuses on business in China)*. Badische Neueste Nachrichten (BNN), Nov. 14, 2010.

Wasik, J. F. (2007): *Forget the Ethanol Myth – Avoid Biofuel Bubble*. Bloomberg New, Jul. 23, 2007. http://www.bloomberg.com/apps/news?pid=20601039&refer=columnist_wasik&sid=aOS8e5kvDESE (last access 10/7/2007).

Watzlawick, P.; Beavin, J.; D. Jackson, D. D. (1967): *Pragmatics of Human Communication. A Study of Interactional Patterns, Pathologies, and Paradoxes*. W. W. Norton and Co, New York;
Watzlawick, P.; Beavin, J. H.; Jackson, D. D. Jackson (1969): *Menschliche Kommunikation. Formen, Störungen, Paradoxien*. Verlag Hans Huber, Bern, Stuttgart, Wien.

Waters, R. (2011): *Technology: the Internet bubble*. Financial Times, Jun. 20 2011. http://www.ft.com/intl/cms/s/0/6fd71904-9b69-11e0-bbc6-00144feabdc0.html (last access 6/21/2011).

Watson, W. E.; Ponthieu, L. D.; Critelli, J. W. (1995): *Team interpersonal process effectiveness in venture partnerships and its connection to perceived success.* Journal of Business Venturing 10, (5), 393-411.

Weber, R. (2002): *Managing growth to achieve efficient coordination in large groups.* Carnegie Mellon University, Aug. 8, 2002. http://www.casos.cs.cmu.edu/publications/papers/GrowNew.pdf (last access 9/6/2010).

Weernink, J. (2009): *Product development efforts.* SpecialChem, Jul. 31, 2009. http://www.specialchem4coatings.com/community-pulse/community-insight.aspx?id=11754 (last access 7/31/2009).

Weidemann, T. (2009): *Free to Play Games – Essentials.* Consultant Online Entertainment, file last access 11/20/2009. http://kapitalrejsen.dk/sites/kapitalrejsen.dk/files/dokumenter/FREE%20Teut%20Weidemann%20f2p%20Market%20Foundations.pdf (last access 10/22/2011).

Weishaupt, G. (2012): *Solarbranche – Nur die Giganten dürfen sich sonnen (Solar industry – Only the giants may sunbathe).* Handelsblatt, Apr. 4, 2012. http://www.handelsblatt.com/unternehmen/industrie/solarbranche-nur-die-giganten-duerfen-sich-sonnen/6475028.html (last access 4/4/2012).

Wellsphere (2009): *BlueFire Wants to Be Algae's Sugar Daddy.* Posted May 27, 2009. http://stanford.wellsphere.com/green-living-article/bluefire-wants-to-be-algae-s-sugar-daddy/687352 (last access 12/23/2009).

Wernher von Braun – http://en.wikipedia.org/wiki/Wernher_von_Braun; http://de.wikipedia.org/wiki/Wernher_von_Braun.

Wesoff, E. (2009): *The Elephant in the Room.* Green Tech Media, Jun, 30, 2009. http://www.greentechmedia.com/green-light/post/algenol-the-elephant-in-the-room/ (last access 2/2/2010).

Wheeler, T. B. (2009): *Algae Power – Researchers In A Baltimore Lab Work To Tap The Energy Potential Of A Bacterium.* The Baltimore Sun, May 8, 2009. http://articles.baltimoresun.com/2009-05-08/news/0905070142_1_blue-green-algae-ethanol-carbon-dioxide (last access 2/5/2010).

Wheelwright, S. C.; Clark K. B. (1992): *Revolutionizing product development: quantum leaps in speed, efficiency, and quality. Free Press*, 1992. Figures shown on page 13 by Thonemann, U. W. (2005): Lecture – Product Development. University of Cologne, file from 11/4/2005. http://s3.amazonaws.com/zanran_storage/www.wiso.uni-koeln.de/ContentPages/42419587.pdf (last access 7/12/2010).

Wiele, J. (2011): *Facebook und die Folgen – Gesichtsverlust (Facebook and the consequences – Loss of face).* FAZ, Sep. 29, 2011. http://www.faz.net/aktuell/feuilleton/medien/facebook-und-die-folgen-gesichtsverlust-11373614.html (last access 9/30/2011).

Wikinvest (2009): *Bluefire Ethanol (BFRE).* http://www.wikinvest.com/stock/Bluefire_Ethanol_(BFRE) (last access 12/27/2009); Wikinvest (2013):

http://www.wikinvest.com/stock/Bluefire_Ethanol_(BFRE)/Data/Income_Statement (last access 12/27/2009).

Wilder, B. (2004): *Pushing for alternative fuel – Blacklick company's butanol production process seeks investors to grow*. Columbus Business Journal, Sep. 3, 2004. http://columbus.bizjournals.com/columbus/stories/2004/09/06/smallb1.html (last access 8/25/2008).

Wilson, L. (2008): *Kelp wanted: Biofuel firms turn to algae, Solazyme doubles space in South S.F.* San Francisco Business Times, Mar. 26, 2007. http://sanfrancisco.bizjournals.com/sanfrancisco/stories/2007/03/26/story11.html?i=76479&b=1174881600%5E1436302 (last access 1/12/2010).

Wintzenburg, J. B. (2009): *Erfolgreiche Frustgründung (Foundation resulting from frustration).* Der Stern, . Jun. 7, 2009. http://www.stern.de/wirtschaft/news/unternehmen/gruender-trotzen-wirtschaftskrise-die-unternehmenslustigen-705563.html (last access 2/7/2011); Deutscher Gründerpreis - *Purplan GmbH aus Wallenhorst gehört zu in Deutschland (German National Founders' Award – Purplan GmbH from is one of the Top-Climbers in Germany)*. Weser-Ems Business-on, Jun. 3, 2009. (last access 2/7/2011); http://www.deutscher-gruenderpreis.de/owx_medien/media17/1735.pdf (last access 2/7/2011).

Wirth (2009): *Prozessfiltration: schnelle Umsetzung akademischen Wissens in den Filterapparatebau (Process filtration – fast transfer of academic knowledge into construction of filter apparatuses*). ChemManager, Feb. 12, 2009. http://www.chemanager-online.com/news-opinions/interviews/prozessfiltration-schnelle-umsetzung-akademischen-wissens-den-filterapparat (last access 6/10/2010).

Witt, P. (2006): *Unternehmensgründer als ehrbare Kaufleute (Entrepreneurs as honest businesspersons).* http://www2.wiwi.hu-berlin.de/institute/im/publikdl/Witt.pdf (file created on 10/11/2006).

Witzmann, R.- P. (2006): *Kooperation zwischen FH Lausitz und IGV auf dem Gebiet der phototrophen Biotechnologie (Cooperation between FH Lausitz and IGV in the field of phototrophic biotechnology*). Innovations-Report, Nov. 13, 2006. http://www.innovations-report.de/html/berichte/biowissenschaften_chemie/bericht-73868.html (last access 3/28/2010).

WOC - World of Chemicals (2012): *BASF to invest $30 million in Renmatix*. World of Chemicals News, Jan. 3, 2012. http://www.worldofchemicals.com/media/basf-to-invest-30-million-in-renmatix/2076.html (last access 2/4/2012).

Wolcott, R. C.; Lippitz, M. J. (2007): *The Four Models of Corporate Entrepreneurship.* MIT Sloan Management Review 49 (1), 74- 82. http://www.business.ualberta.ca/mlounsbury/techcom/readings/wolcott.pdf (last access 7/21/2009).

WolframMathWorld: *Fibonacci Number.* http://mathworld.wolfram.com/FibonacciNumber.html; cf. also http://en.wikipedia.org/wiki/Fibonacci_number; *Fibonacci Numbers in Nature*: http://britton.disted.camosun.bc.ca/fibslide/jbfibslide.htm; Welcome to the Fibonacci Project: http://www.fibonacci-project.eu/.

Wong, V. (2009a): *How to Nurture Future Leaders*. Business Week, Sep. 30, 2009. http://www.businessweek.com/innovate/content/sep2009/id20090930_806435.htm?chan=innovation_innovation+%2B+design_top+stories (last access 10/10/2009).

Wong, V. (2009b): *How Business Is Adopting Design Thinking*. Business Week, Sep. 30, 2009. http://www.businessweek.com/innovate/content/sep2009/id20090930_853305.htm?chan=innovation_design+index+page_special+report+--+design+thinking (last access 10/10/2009).

Woods, P. (2009a): *Paul Woods on Algenol Biofuels*. Clean Skies News, Aug. 3, 2009. http://www.cleanskies.com/videos/paul-woods-algenol-biofuels (last access 2/2/2010).

Woods, P. (2009b): *Algenol Biofuels – Integrated Biorefinery for Producing Ethanol from Hybrid Algae*. Presentation, Dec. 4, 2009. File of 9/29/2009. http://lee.ifas.ufl.edu/AgNatRes/Sustainability/Algenol_Lee_county_climate_change_2009.pdf (last access 2/2/2010).

Woods, P. (2009c): *Algenol Biofuels – Integrated Biorefinery for Producing Ethanol from Hybrid Algae*. Presentation, The University of Florida, IFAS, Dec. 4, 2009. File of 11/17/2009. http://lee.ifas.ufl.edu/AgNatRes/Sustainability/Algenol_Lee_county_climate_change_2009.pdf (last access 2/11/2010).

Woo, C.Y.; Daellenbach, U.;Nicholls-Nixon. C. (1994): *Theory Building in the Presence of Randomness – The Case of Venture Creation and Performance*. Journal of Management Studies, 31 (4), 507-524.

Woopidoo: *Larry Page Biography (Lawrence E. Page): Google Cofounder*. http://www.woopidoo.com/biography/larry-page/index.htm (last access 1/21/2011).

Worthen, B. (2012): *Newest Tech IPO Doubles on Day One*. Wall Street Journal, Apr. 19, 2012. http://online.wsj.com/article/SB10001424052702303513404577353891795293210.html?mod=WSJEUROPE_hpp_MIDDLEFourthNews (last access 4/20/2012).

Wuttke, K. (2008): *Sächsische Firma entwickelt ethisch-ökologisch unbedenklichen Biosprit (Company from Saxony develops ethicics- and ecology-compliant biofuel)*. Deutschlandfunk, May 2, 2008. http://www.dradio.de/dlf/sendungen/firmen/778008/ (last access 12/20/2009).

Wüst, C. (2007): *Thanksgiving in the gas tank – New technologies from the former East Germany*. Der Spiegel, May 1, 2007. http://www.spiegel.de/international/world/0,1518,479922-5,00.html (last access 12/21/2009).

Yang, W.; Kiron, D. (2010): *Cambridge NanoTech*. Harvard Business Review Case 9-610-083, May 14, 2010. http://www1.hbr.org/product/cambridge-nanotech/an/610083-PDF-ENG?Ntt=cambridge%2520nanotech.

Yin, R. (1994). *Case study research: Design and methods* (2nd ed.). Beverly Hills, CA: Sage Publishing.

Zakaria, F. (2009): *Is America Losing Its Mojo?* Newsweek, Nov. 14, 2009. From the magazine issue dated Nov. 23, 2009. http://www.newsweek.com/id/222836 (last access 11/21/2009).

Zakaria, F. (2008): *There Is a Silver Lining*. Newsweek, Oct. 20, 2008. http://www.newsweek.com/id/163449/output/print (last access 10/21/2008).

Zbikowski, F.; Brandkamp, M. (2013): *Start-ups sind flexibler, schneller und risiobereiter (Start-ups are more flexible, faster and more ready to take risk)*. Nachrichten aus der Chemie 61, 982-985; http://www.high-tech-gruenderfonds.de/, http://www.en.high-tech-gruenderfonds.de/.

Zeller-Silva, G. (2007): *Zu schnell zu groß (Too fast too large)*. Die Zeit Online, Feb. 27, 2007. http://www.zeit.de/online/2007/08/Solarenergie-Aktien (last access 2/27/2007).

Zentner, B. (2004): *Merseburger sieht in Klötze grün (Person from Merseburg sees green in Klötze)*. Mitteldeutsche Zeitung, Oct. 2, 2004. http://www.algomed.de/media/presse/11-46-52-25-01-2008-MZ%20Artikel%20Merseburg.pdf (last access 5/12/2012).

Zezima, K. (2009): *Harvard Slows Its Growth in a Boston Neighborhood*. The New York Times, Feb. 18, 2009. http://www.nytimes.com/2009/02/19/us/19harvard.html?ref=us (last access 2/18/2009).

Zhao, Y. (2009):*Young Tableaux and the Representations of the Symmetric Group*. Student Article, MIT, File from 4/4/2009. http://web.mit.edu/yufeiz/www/papers/youngtab-hcmr.pdf (last access 3/21/2012).

Zolin, R.; Kautonen, T.; Kautonen, A. (2008):): *The Devil You Know? The Potential Downside of Strong And Weak Ties for Entrepreneurial Team Formation*. Swinburne University of Technology, Regional Frontierts 2008 – Conference Paper, AGSE 2008, 325-337. http://www.swinburne.edu.au/lib/ir/onlineconferences/agse2008/000059.pdf (last access 8/5/2010).

ZoomInfo (2008): *Solazyme , Inc. – Company Profile*. http://www.zoominfo.com/Search/CompanyDetail.aspx?CompanyID=104890649&cs=QGGQIEZIE (last access (1/12/2010); http://www.yatedo.com/p/Christine+Eckhaus/normal/dc5f4cc12f98ec1f1bed1272034d417c (last access 2/11/2011).

Zuber-Knost, E. (2009): *Hydrogen from Microalgae – Scientists Work on Highly Efficient Processes – KIT Coordinates Joint Project*. KIT Press Release 141/2009. http://www.kit.edu/english/55_604.php?PHPSESSID=37c3fa1b957025a2de2f590a0ecf7876 (last access 2/12/2010).

Zynga (2011): *Zynga, Inc.* S-1 Filing S-1 Filing, PROSPECTUS (Subject to Completion), Issued Jul. 1, 2011. http://www.sec.gov/Archives/edgar/data/1439404/000119312511180285/ds1.htm (last access 9/22/2011).

NOTES

1. Technology Entrepreneurship: http://ce.ioc.KIT.edu.
2. Webster's Online Dictionary: Entrepreneurship.http://www.websters-online-dictionary.org/definition/entrepreneurship (last access 7/12/2009).
3. Spin-out: http://en.wikipedia.org/wiki/Spin_out.
4. http://en.wikipedia.org/wiki/Leadership (last access 7/14/2009).
5. http://en.wikipedia.org/wiki/Political_economy (last accesss 9/11/2009).
6. Trade-off: http://en.wikipedia.org/wiki/Trade-off (last access 6/6/2011).
7. Supply chain: http://en.wikipedia.org/wiki/Supply_chain (last access 12/3/2010).
8. BOS (Balance-of-System) Cost:
   A term for the parts of a solar electric system besides the actual solar panels. It includes batteries, cables, inverter, safety and monitoring equipment, and the mounting racks that hold the panels;
   it may also represent costs of all components other than the PV modules including design, land, site preparation, system installation, support structures, power conditioning, operation and maintenance, batteries, indirect storage, and other related costs.
9. Click'n'vote: http://www.omnexus4adhesives.com/community-pulse/pastclicknvote.aspx?id=72 (last access 6/7/2011).
10. MSN Encarta: Utilize is more common in technical contexts. The device utilizes a special plug-in connection. It can also refer to using things in unusual or unintended ways, as a more formal equivalent of "make use of." http://encarta.msn.com/dictionary_/utilize.html; The Free Dictionary by Farlex: http://www.thefreedictionary.com/utilize.
11. "Kein Operationsplan reicht mit einiger Sicherheit über das erste Zusammentreffen mit der feindlichen Hauptmacht hinaus.
    Nur der Laie glaubt in dem Verlauf eines Feldzuges die konsequente Durchführung eines im Voraus gefassten, in allen Einzelheiten überlegten und bis ans Ende festgehaltenen, ursprünglichen Gedankens zu erblicken.......Es kommt darauf an, in lauter Spezialfällen die in den Nebel der Ungewissheit gehüllte Sachlage zu durchschauen, das Gegebene richtig zu würdigen, das Unbekannte zu erraten, einen Entschluss zu fassen und dann kräftig und unbeirrt durchzuführen."
12. Jack Welsh: *Clausewitz in Business*. http://www.clausewitz.com/business/index.htm; on the Clauswitz Homepage http://www.clausewitz.com/index.htm (last access 10/1/2009).
13. Magnetfeldtechnik Resonanz GmbH: http://www.deutscher-gruenderpreis.de/resonanz (last access 10/13/2009).
14. Experience: http://en.wikipedia.org/wiki/Experience (last access 6/10/2011).

15. Society for Organizational Learning (SOL): http://www.solonline.org/, http://www.solonline.org/organizational_overview/ (last access 10/21/2011).

16. http://en.wikipedia.org/wiki/Experience_curve_effects; http://en.wikipedia.org/wiki/Learning_curve (last access 11/4/2010).

17. Capitalism: http://en.wikipedia.org/wiki/Capitalism; http://en.wikipedia.org/wiki/Financial_capitalism.

18. Capitalism: http://www.websters-online-dictionary.org/definition/capitalism.

19. OECD: Glossary of Statistical Terms – Marketing Innovation. http://stats.oecd.org/glossary/detail.asp?ID=6871 (last access 4/10/2010).

20. Carl Benz (1910): "Aufgrund der beschränkten Verfügbarkeit von Chauffeuren wird es niemals mehr als 5000 Automobile geben." http://www-pu.informatik.uni-tuebingen.de/users/klaeren/sprueche/805.htm (last access 7/24/2009); Gottlieb Daimler ((1834-1900): "Die weltweite Nachfrage nach Kraftfahrzeugen wird eine Million nicht überschreiten – allein schon aus Mangel an verfügbaren Chauffeuren." http://manager-lounge.com/testimonials/mercedes/index.php (last access 7/24/2009).

21. Exchange value: http://en.wikipedia.org/wiki/Exchange_value (last access 6/14/2011).

22. Intangible assets: http://en.wikipedia.org/wiki/Intangible_asset (last access 6/14/2011).

23. Herd behavior: http://en.wikipedia.org/wiki/Herd_behavior (last access 6/14/2011).

24: Entry barrier: http://en.wikipedia.org/wiki/Barriers_to_entry; Exit barrier: http://en.wikipedia.org/wiki/Barriers_to_exit (last access 6/14/2011).

25. Business cycle: http://en.wikipedia.org/wiki/Business_cycle, http://en.wikipedia.org/wiki/Economic_downturn, http://en.wikipedia.org/wiki/Economic_bubble. (last access 6/11/2011).

26. An-Institute: http://de.wikipedia.org/wiki/An-Institut (last access 6/18/2011).

27: BMWi (2010): Jahresbericht 2010 | 2011 – Die Initiative Kompetenznetze Deutschland im Überblick (Annual Report 2010 | 2011 – Overview of the Initiative Competence Networks). http://www.kompetenznetze.de/service/bestellservice/medien/jahresbericht2010.pdf (last access 6/18/2010), Kompetenznetze Deutschland: http://www.kompetenznetze.de/.

28. Innovationsallianz: http://www.bmbf.de/pub/flyer_innovationsallianz.pdf (last access 6/19/2011).

29: CRADA: http://www.usbr.gov/research/tech-transfer/crada/whatcrada.html (last access 6/19/2011).

30. Nanosys data: http://www.hoovers.com/company/Nanosys_Inc/rfsffri-1.html (last access 6/19/2011).

31. In Germany (Austria and Switzerland) a "habilitation" (leading to the degree Dr. habil.) is usually one prerequisite to become a professor in a science discipline at a university.

32: *Inc.* is a monthly US magazine written for the people who run growing companies. The magazine publishes an annual list of the 500 (and also 5000) fastest-growing private companies in the US, the "*Inc.* 500" list.

33: German High-Tech Gründerfonds: conditions – requires founder's own contributions (up to 20 percent); up to €500,000 in a first round of funding, purchases 15% shares with a nominal value and provides a subordinated shareholder loan. In addition, the Fund keeps a further €500,000 for follow-on financing. The loan has a term of 7 years.
http://www.high-tech-gruenderfonds.de/finanzierung/finanzierungskonditionen/.

34: SBA: http://en.wikipedia.org/wiki/Small_Business_Administration (last access 6/23/2011).

35: Venture Capital: http://en.wikipedia.org/wiki/Venture_capital; http://de.wikipedia.org/wiki/Risikokapital (last access 6/23/2011).

36: Limited Partnership (LP): http://en.wikipedia.org/wiki/Limited_partnership; General Partnership (GP): http://en.wikipedia.org/wiki/General_partnership (last access 6/23/2011).
GP is similar to the German GbR (Gesellschaft bürgerlichen Recht). GbR is the basic form of partnership based on an agreement (or contract) of at least two persons intending to achieve a common goal. For any liabilities of the partnership the partners are liable personally and with all their wealth and assets. Liability cannot be limited in principle.

37: Penny stocks: http://en.wikipedia.org/wiki/Penny_stock
http://en.wikipedia.org/wiki/Pink_Sheets.
Pink Quote, informally known as the Pink Sheets, is an electronic quotation system operated by Pink OTC Markets that displays quotes from broker-dealers for many over-the-counter (OTC) securities. These securities tend to be inactively traded stocks, including penny stocks and those with a narrow geographic interest.

38. Locus of control: http://en.wikipedia.org/wiki/Locus_of_control (last access 6/27/2011).

39. August Horch: http://de.wikipedia.org/wiki/Audi, http://de.wikipedia.org/wiki/August_Horch; http://en.wikipedia.org/wiki/Audi; http://en.wikipedia.org/wiki/August_Horch.

40. Volition: the capability of conscious choice and decision and intention; an act of will.

41. Samuel Moore "Sam" Walton: http://en.wikipedia.org/wiki/Sam_Walton; http://www.fastcompany.com/magazine/77/walmart.html.

42. The Kirton Adaption-Innovation Inventory: http://www.kaicentre.com/ (last access 6/29/2011); The Myers & Brigg Foundation: http://www.myersbriggs.org/ (last access 6/29/2011)

43. Gedanken experiment: http://en.wikipedia.org/wiki/Gedanken_experiment. (physics) A hypothetical ("thought") experiment which is possible in principle and is analyzed (but not performed) to test some hypothesis. Also known as thought experiment.

44. "dm" and Götz Werner": Blocked careers or business ideas as the origin of entrepreneurship is all over. For instance, in Germany Götz Werner was employed with a large drugstore organization. In Germany a drugstore is a retail store featuring basically no drugs and medicines, but miscellaneous items for the home, such as household and baby care, personal care and cosmetics, hygiene and cleaning products, food additives and nutraceutials etc. After a reorganization of the sales unit he suggested to introduce the concept of a discount drugstore chain combined with competent advising services. This idea was rejected and, after leaving his employer, he founded in 1973 his own discount drugstore called "dm-drogerie markt GmbH + Co. KG" (dm is Drogerie Markt in German) as the kernel of the currently leading German discount drugstore chain. For 2009/2010 revenues were €5,6 billion with ca. 30 percent outside Germany and ca. 36,000 employees. ttp://de.wikipedia.org/wiki/G%C3%B6tz_Werner.

45. Group dynamics: http://en.wikipedia.org/wiki/Group_dynamics (last access 7/4/2011).

46. Ernst Werner von Siemens: http://www.ask.com/wiki/Werner_von_Siemens?qsrc=3044; Johann Georg Halske: http://www.ask.com/wiki/Johann_Georg_Halske?qsrc=3044.
Johann Georg Halske (1814 – 1890) was a German master mechanic who started his own workshop in Berlin in 1844, which he ran together with his partner F. M. Böttcher.
In 1847 Halske co-founded the Siemens & Halske Telegraph Construction Company together with Werner von Siemens (1816– 1892). Siemens took over the role of the developer, Halske focused on finishing and production of the telegraphs. Halske was particularly involved in the construction and design of electrical equipment such as the press which enabled wires to be insulated with a seamless coat of gutta-percha, the pointer telegraph and the morse telegraph and measuring instruments. In 1867 he withdrew from the company.
Siemens left school without finishing his education, but joined the Prussian army to undertake training in engineering, in mathematics, physics, chemistry and ballistics. He is known world-wide for his advances in various technologies, and chose to work on perfecting technologies that had already been established.

47. Enercon GmbH: http://en.wikipedia.org/wiki/Enercon; http://www.windsofchange.dk/WOC-usaturb.php (last access 7/12/2011).

48. Theoria – expectation: Wikipedia: http://en.wikipedia.org/wiki/Theoria; http://en.wikipedia.org/wiki/Praxis_(process); http://www.merriam-webster.com/dictionary/theory; en.wikipedia.org/wiki/Expectation_(epistemic).

49. Best practice: http://en.wikipedia.org/wiki/Best_practice.

50. Bootlegging: http://en.wikipedia.org/wiki/Bootleg.

51. Idea: http://www.merriam-webster.com/netdict/idea.

52. ConnectU.com lawsuit: http://en.wikipedia.org/wiki/Criticism_of_Facebook.

53. Electric vehicles: http://en.wikipedia.org/wiki/Electric_vehicle; http://en.wikipedia.org/wiki/Electric_car; http://de.wikipedia.org/wiki/Elektroauto; http://en.wikipedia.org/wiki/La_Jamais_Contente (last access 7/19/2011).

54. Gestalt: http://en.wikipedia.org/wiki/Gestalt_psychology.

55. Clarence Birdseye: http://en.wikipedia.org/wiki/Clarence_Birdseye, http://web.mit.edu/invent/iow/birdseye.html, http://www.usatoday.com/money/industries/technology/maney/2004-07-14-frozen_x.htm.

56. Drucker Quotes: http://www.1-famous-quotes.com/quote/927861.

57. First-Mover Advantage: Encyclopedia of Management, http://www.enotes.com/management-encyclopedia/first-mover-advantage (last access 7/24/2011); http://en.wikipedia.org/wiki/First-mover_advantage.

58. Kondratiev waves: http://en.wikipedia.org/wiki/Kondratiev_wave.

59. Environmental Entrepreneurs: http://www.e2.org/.

60. Lux Research: THE CLEANTECH REPORT™. http://www.luxresearchinc.com/cleantech.php; CleanTech Group: Cleantech definition. http://cleantech.com/about/cleantechdefinition.cfm (last access 5/18/2010).

61. Orphan drug: http://en.wikipedia.org/wiki/Orphan_drug.

62. Switching cost: http://en.wikipedia.org/wiki/Switching_barriers.

63. Critical success factor (CSF): http://en.wikipedia.org/wiki/Critical_success_factor.

64. SWOT analysis: http://en.wikipedia.org/wiki/SWOT_analysis.

65. Screwdriver: http://en.wikipedia.org/wiki/Screwdriver (last access 9/6/2011).

66. Technology trajectory: http://en.wikipedia.org/wiki/Technology_trajectory.

67. Paradigm: http://en.wikipedia.org/wiki/Paradigm.

68. Hybrid power plant: https://www.enertrag.com/projektentwicklung/hybridkraftwerk.html, https://www.enertrag.com/download/prospekt/hybridkraftwerk_kurzinfo_090417.pdf.

69. Massively Multiplayer Online Games (MMOGs): http://en.wikipedia.org/wiki/Massively_multiplayer_online_game; http://de.wikipedia.org/wiki/Massively_Multiplayer_Online_Role-Playing_Game; http://en.wikipedia.org/wiki/MUD.

70. Widget application: http://en.wikipedia.org/wiki/Software_widget.

71. AAA game: The gaming press tends to use AAA to mean a really high quality game. Marketing folks will use it to refer solely to the advertising budget (i.e. the actual quality of the game is irrelevant). Producers usually will use it to mean both (i.e. is high quality and has a big marketing budget).

72. Generally Accepted Accounting Principles (GAAP): http://en.wikipedia.org/wiki/Generally_Accepted_Accounting_Principles; http://en.wikipedia.org/wiki/Generally_Accepted_Accounting_Principles_(United_States).

73. Ambiguity: http://en.wikipedia.org/wiki/Ambiguity.

74. Scenario: http://en.wikipedia.org/wiki/Scenario.

75. Currency Volatility: http://sdw.ecb.europa.eu/quickview.do?SERIES_KEY=120.EXR.M.USD.EUR.SP00.A (last access 12/2/2011).

76. William Shockley: http://en.wikipedia.org/wiki/William_Shockley.

77. Intel: http://en.wikipedia.org/wiki/Intel.

78. Fairchild Semiconductor: http://www.fairchildsemi.com/company/history/.

79. Hanks et al. [1993] derived empirically a taxonomy of growth stage architectures in a 1988 sample of 126 high-technology organizations (R&D intensity on average 3.1%) reflecting a situation in 1987/1986. The firms represent 14 industry groups, have mean sales of $5,530,783, and employ a mean of 125 employees. Industries represented in the sample include computer software, electronic and communications equipment, chemicals, pharmaceuticals, aerospace equipment, lasers and optics, and analytical and measuring devices. The study allowed to characterize states of the young firms in terms of age and organizational features. Other variables were defined as follows.

    Structural form, or basis of organization, was self reported by respondents based on brief descriptions and coded as follows: simple structure, 1; by function, 2; by divisions, 3; and other, 4.

    Centralization was measured by giving respondents a list of five decision issues. They were then asked to indicate the level of management that must approve the decision before legitimate actions may be taken.

    Formalization was operationalized using a scale of eleven items. The first ten items used a 7-point Likert-type scale, ranging from strongly agree to strongly disagree. The eleventh item measured the formalization of the decision-making process in the organization. Typical formalization include accounting and financing.

80. Targeted Selection: http://www.pacificu.edu/offices/hr/training/interview/pdfs/TargetedSelection.pdf (last access 2/19/2011).

81. NanoGate AG: Runge, W.: *Technology Entrepreneurship Curriculum – Course Material.* Handout Lectures 0-4, page 37. http://ce.ioc.KIT.edu/downloads/Chem_Entrepreneur_0_4.pdf; Handout Lectures 5-9. page 34. http://ce.ioc.KIT.edu/downloads/Chem_Entrepreneur_5_9.pdf; Handout Lectures 10-13, pp. 9, 14, 17. http://ce.ioc.KIT.edu/downloads/Chem_Entrepreneur_10_13.pdf.

82. Resource-Based View (RBV): http://en.wikipedia.org/wiki/Resource-based_view.

83. Financial Bootstrapping: http://en.wikipedia.org/wiki/Entrepreneurship.
84. Ashby Memory: "Thus, suppose I am in a friend's house and, as a car goes past outside, his dog rushes to a corner of the room and cringes. To me the behaviour is causeless and inexplicable. Then my friend says, "He was run over by a car six months ago." The behaviour is now accounted for by reference to an event of six months ago. If we say that the dog shows "memory" we refer to much the same fact – that his behaviour can be explained, not by reference to his state now but to what his state was six months ago. If one is not careful one says that the dog "has" memory, and then thinks of the dog as having something, as he might have a patch of black hair. One may then be tempted to start looking for the thing; and one may discover that this "thing" has some very curious properties. Clearly, "memory" is not an objective something that a system either does or does not possess; it is a concept that the observer invokes to fill in the gap caused when part of the system is unobservable." [Ashby 1957:117]
85. Molecular Excited States: This is similar to issues of physical chemists or chemical physicists who study electronically excited states of molecules by ultraviolet (UV) absorption spectroscopy of circular dichroism (CD) spectroscopy in solution (wavelength: ca. 300 nm – 200 nm) or the gas phase (wavelength ca. 190 nm – 120 nm). The spectra are analyzed in terms of spectral bands and the band peaks corresponds to energies of particular molecular excited states. Overlapping bands sometimes show only up as shoulders of very strong bands or not at all; theoretical curve resolution (assuming bell-shaped Gaussian bands) may reveal more bands. Whereas UV spectroscopy exhibits only positive bands, for certain molecular configurations CD spectroscopy (using linearly polarized light) exhibits positive and negative bands and thus supports band resolutions or even may reveal very weak bands which cannot be observed with UV spectroscopy.
(cf. the author's list of publication:
http://www.riscnet.de/Files/Publications_WR.htm).
86. Heyl GmbH & Co. KG: Capture radioactive cesium-through Radiogardase-Cs® capsules which are produced by the firm Heyl Chemisch-Pharmazeutische Fabrik GmbH & Co. KG (Berlin). The firm was founded in 1734 as a trading company and became a paint factory and chemical wholesale firm in 1765. Particularly, Heyl focused on Prussian (Berlin) Blue which was developed in Berlin in 1704 (http://www.heyl-berlin.de). Currently one business of Heyl operates in a niche that focuses on drugs and medicaments which remove radioactive elements and compounds from the body. It has approval in Germany, Japan and the US (FDA) [Runge 2006:398-401].
87. Nanogate AG: [Runge 2006:550]; Runge, W.: Technology Entrepreneurship – Course Material. Slide Numbers 3.7 (http://ce.ioc.KIT.edu/55.php), 7.8 (http://ce.ioc.KIT.edu/downloads/Chem_Entrepreneur_5_9.pdf), 10.18, 10.28, 10.33 (http://ce.ioc.KIT.edu/downloads/Chem_Entrepreneur_10_13.pdf).
88. Microsoft – Revenues and Numbers of Employees (last updated on October 4, 2011): http://www.thocp.net/companies/microsoft/microsoft_company.htm (last access 8/3/2010).
89. Cisco Systems – Revenues and Number of Employees:
http://www.icmrindia.org/casestudies/catalogue/Business%20Strategy2/Cisco-

Acquisition-Strategy.htm (last access 2/8/2011); http://www.referenceforbusiness.com/history2/83/Cisco-Systems-Inc.html (last access 1/3/0/2011); http://www.icmrindia.org/casestudies/catalogue/Business%20Strategy2/Cisco-Acquisition-Strategy.htm (last access 2/8/2011); http://newsroom.cisco.com/dlls/corpinfo/corporate_overview.html.

90. The German Renewable Energy Act: http://en.wikipedia.org/wiki/Renewable_energy_in_Germany; http://en.wikipedia.org/wiki/Feed-in_tariffs_in_Germany.

91. Chernobyl disaster: http://en.wikipedia.org/wiki/Chernobyl_disaster.

92. Silicon Cycle: BizEsp – Solar Power – The Market 2nd May 2007: http://www.bizesp.com/uploads/SolarPower2.05.07.pdf (last access 3/14/2011).

93. Solarbuzz Reports World Solar Photovoltaic Market Grew to 18.2 Gigawatts in 2010, Up 139% Y/Y. Mar. 15, 2011. http://www.solarbuzz.com/our-research/recent-findings/solarbuzz-reports-world-solar-photovoltaic-market-grew-182-gigawatts-20 (last access 4/5/2011).

94. Google: http://en.wikipedia.org/wiki/Google.

95. AdSense: http://en.wikipedia.org/wiki/AdSense.

96. Acquisition of Applied Semantics: Press release, Apr. 23, 2003. http://www.google.com/press/pressrel/applied.html (last access 1/29/2012).

97. Biorefinery funding: http://www1.eere.energy.gov/biomass/financial_opps_news.html?sol_id=254; https://www.fedconnect.net/FedConnect/?doc=DE-FOA-0000096&agency=DOE; Federal Biomass Policy - Federal Legislation - The Energy Independence and Security Act of 2007, http://www1.eere.energy.gov/biomass/federal_biomass.html (last access 3/25/2010).

98. The 1970s oil crises: http://en.wikipedia.org/wiki/1973_oil_crisis; http://en.wikipedia.org/wiki/1979_energy_crisis; http://en.wikipedia.org/wiki/Price_of_petroleum.

99. Methanol: http://en.wikipedia.org/wiki/Methanol_fuel; http://de.wikipedia.org/wiki/Holzvergasung, http://de.wikipedia.org/wiki/Methanol.

100. Crude oil formation: http://en.wikipedia.org/wiki/Crude_oil#Formation.

101. NELHA: Natural Energy Laboratory of Hawai'i Authority Annual Report 2008. http://www.nelha.org/pdf/ARnelhaFY2008.pdf (last access 1/2/2010).

102. HR BioPetroleum: Shell and HR BioPetroleum build facility to grow algae for biofuel. News, Nov. 12, 2007 http://www.hrbp.com/News/121107.html; HR BioPetroleum: HR BioPetroleum, Alexander & Baldwin, Hawaiian Electric and Maui Electric to Develop Algae Facility for Biodiesel on Maui. News, Jul. 15, 2008. http://www.hrbp.com/News/071508.html; HR BioPetroleum: Pilot Facility. http://www.hrbp.com/Facilities/Pilot.html; http://investing.businessweek.com/research/stocks/private/person.asp?personId=4172513 Edward T. Shonsey.

103. Solazyme, Inc.: http://en.wikipedia.org/wiki/Solazyme.

104. Hydrothermal carbonization (HTC): http://www.dw-world.de/dw/article/0,,2071791,00.html (last access 3/18/2010); http://en.wikipedia.org/wiki/Hydrothermal_carbonization; http://www.rsc.org/Publishing/Journals/dt/News/b804644cpersp.asp.

105. Prof. Steinberg Produktions- und Vertriebs GmbH & Co KG and Algomed in Klötze/Sachsen-Anhalt: http://www.algomed.de/; http://www.algomed.de/index.php?op=algenfarm; http://www.algomed.de/index.php?op=algenfarm_anlage http://www.algomed.de/?op=presse&id=9 etc. (last access 5/10/2012).

106. Prof. Dr. Karl-Hermann Steinberg: http://www.algomed.de/index.php?op=algenfarm_koepfe (last access 2/8/2008).

107. Range Fuels: http://www.rangefuels.com/range-fuels-technology.html (last access 1/12/2010).

108. An Altenews Interview with Mark Emalfarb, Founder and CEO of Dyadic (Symbol: DIL), and Dr. Glenn Nedwin, PhD, Dyadic's Chief Science Officer. http://www.altenews.com/Dyadic%20Interview.pdf (last access 2/2/2010).

109. ButylFuel, LLC: Home Page http://www.butanol.com/, http://butanolc.startlogic.com/page7.html and http://butanol.com/docs/Weizman-Terre_Haute.doc, http://butanol.com/docs/SciAm_7-27.doc.

110. Samuel Morse: http://en.wikipedia.org/wiki/Samuel_Morse.

111. Biofuel: http://en.wikipedia.org/wiki/Biofuel; http://en.wikipedia.org/wiki/Cellulosic_ethanol; http://en.wikipedia.org/wiki/Cellulosic_ethanol_commercialization.

112. Compounding: The goal of compounding is to modify the properties of a basic raw materials for particular applications. Compounding means preparing mixtures or blends of pure-grade raw materials, where additionally fillers, reinforcing agents (e.g. fibers) or other functional additives (e.g. stabilizers, plasticizers, flame retardants etc.) are added in small amounts. A solution of the individual components will not take place. Hence, by compounding at least two materials are joined together (in a molten state) to form a homogeneous mixture. The special challenge is to avoid a possible separation of the compound over time.

113. Nylons-Srikanth, P. (Ed.) (2011): *Handbook of Bioplastics and Biocomposites Engineering Applications*, p. 411. John Wiley and Scrivener Publishing.

114. William Henry Perkin, FRS (1838 – 1907) http://en.wikipedia.org/wiki/William_Henry_Perkin; http://www.answers.com/topic/william-perkin.

115. Faraday Effect: The rotation of the plane of polarization of either a plane-polarized light beam passed through a transparent isotropic medium or a plane-polarized microwave passing through a magnetic field along the lines of that field. Also called Faraday rotation.

116. Catalysis Research Lab (CaRLa): http://www.carla-hd.de/.

117. InnovationLab GmbH (iL): http://www.innovationlab.de/en/homepage/.

118. HP – Innovation Research Program: http://www.hpl.hp.com/open_innovation/irp/index.html.

119. Young diagrams: http://en.wikipedia.org/wiki/Young's_lattice.

120. PayPal Mafia: http://en.wikipedia.org/wiki/PayPal_Mafia.

121. Yelp: http://en.wikipedia.org/wiki/Yelp,_Inc.

122. Feasibility: http://en.wikipedia.org/wiki/Feasibility_study.

123. Sunk cost: http://en.wikipedia.org/wiki/Sunk_costs.

124. Contingency theory: http://en.wikipedia.org/wiki/Contingency_theory.

125. http://en.wikipedia.org/wiki/Price_of_petroleum.

126. http://en.wikipedia.org/wiki/Methanol_fuel; http://de.wikipedia.org/wiki/Holzvergasung, http://de.wikipedia.org/wiki/Methanol.

127. Elon Musk: http://en.wikipedia.org/wiki/Elon_Musk.

APPENDIX A

A.1 Entrepreneur, Company and Market Cases

A.1.1 The Biofuels Bubble and the Related Outburst of Entrepreneurship and Intrapreneurship

In the new century industrialized countries ran into a biofuels boom by demanding ambitious renewable-fuel targets set by policy and legislation ("mandates"). For instance, biofuels were to provide 5.75 percent of Europe's transport power by 2010 and 10 percent by 2020. US Federal fuel standards increased the volume of renewable fuel required to be blended into gasoline from 9 billion gallons in 2008 to 36 billion gallons by 2022. (Box I.1).

In essence, the *biofuels market* has been created by policy and, together with other effects, became the origin of a *boom*. It is a *policy-driven market* (Table I.15). The following biofuel bubble effect is particularly striking for the US, though structurally similar events can be observed around the world including Germany.

Public policy has not only stimulated increased public funding of research and development but stimulated also huge private investment. Governmental grants, subsidies and loan guarantees as well as private investment were swamping universities and public research institutions. Government spurred also with grants the construction of new biofuel plants, and also with big per-gallon subsidies. Furthermore, dynamics was generated by hordes of startups and their quest for venture capital financing (Figure I.34). It was estimated that in the US blending mandates alone would provide over $33 billion in tax credits to the biofuels industry from 2009 through 2013 [Davis and Russell 2009].

Contrary to an investment in a software or Internet company, for a "*research-to-manufacturing*" startup one has to finance heavy in investments upfront in large scale plants and product introduction will take years. Thus for VCs the favorite business model is to get an industrial partner (corporate venturing) on board prior to the capital intensive investment stage and who eventually will take over the company years later.

The current description and discussion of entrepreneurship and intrapreneurship in biofuels will focus on the period 2000 – 2008/2009 and looks at significant developments after the Great Recession in 2007–2009. Concerning new technology ventures there are two fundamental orientations related to technology and type of NTBF:

1. The thermochemical and catalytic routes are preferentially followed by *engineering-type firms*, often with well experienced founders ("veterans") and often further developing a century old technology;

2. The biotechnology and bioengineering route is preferentially followed by *research-based startups* (RBSUs).

A.1.1.1 The Origins and the Drivers

The starting situation around 2000 for an industry was focused on *first generation biofuels*. It was characterized by a core of some well established food firms, giant to small, engaged in sugar, corn or soybeans and "*corn ethanol*" and vegetable oils-based biodiesel taking advantage from the big *incentive-driven opportunities*. That let farmers shift to crops to meet growing demand for vegetable-based fuels (biodiesel) and sugar-based fuels (bioethanol – *corn ethanol*). And farmers who bought shares in nearby ethanol facilities became wealthy, thanks to corn prices increasing by 65 percent in just two years, and the skyrocketing value of existing ethanol plants.

The US Department of Energy (DOE) expected in 2004 that ethanol could eventually supply 30 percent or more of US transportation fuel needs [Ritter 2004]. For instance, in the US it took ten years from 1980 to 1990 to increase the volume of corn ethanol for fuel five-fold (1980: 175 – 1990: 900 mio. gallons) [RFA]. But from 2000 (1,630 mio.) to 2008, with billions of dollars flooding into new facilities, the absolute volume shot up to 9,000 mio. gallons based on an ever increasing number of new plants or capacity extensions of existing plants.

Also food and agricultural giants like Archer Daniels Midland (AMD) and Cargill in the US (both large suppliers of bioethanol and biodiesel) or Südzucker in Germany [Runge 2006:188] which is Europe's largest producer of sugar and sugar products jumped onto the band-wagon now called bioethanol. ADM, the largest US (and global) grain processor, then got 25 percent of its operating profit from biofuels, including both ethanol and biodiesel, and its shares were being increasingly seen as an energy, as well as an agriculture, play [Scully 2007].

In 2007 ethanol production accounted for about 20 percent of the US corn crop [Scully 2007]. Similarly, in Europe in 2002, 2003 and 2004 biodiesel production rose already by a 30-35 percent rate. But there was a 65 percent record growth in 2005 over 2004. And production of biodiesel in Europe jumped 54 percent in 2006 to 4.89 million tons (about 1.5 billion gallons), up from 3.184 million tons (about 961 million gallons) in 2005 [Green Car Congress 2007a]. Total EU27 biodiesel production for 2008 was over 7.7 million metric tons, an increase, but significantly reduced growth rate in relation to previous ones, of 35.7 percent from the 2007 figures [European Biodiesel Board 2009].

In that line chemical giants like the German BASF and DuPont or Monsanto in the US which for decades have produced agricultural chemicals for plant and crop protection (pesticides, insecticides etc.) or genetically modified seeds to achieve resistance against pests entered the scene. For instance, BASF formed joint ventures with Monsanto [Mandary 2007] and partnered with the Brazilian Sugarcane Technology

Center (CTC) [Sugarcaneblog 2009] to develop genetically modified corn, soybean or sugarcane to increase crop yields. And additionally in 2007 BASF started together with DBE Deutsche Bioenergie AG to build a biodiesel plant [UFOP 2007].

As *crude-oil prices continually rose*, the arguments for alternative fuel sources grew stronger. With global oil prices shooting up from $40-50 per barrel by September 2004 to achieve $77 in July 2006 and, finally to peak at $147 by July 2008, public and private investments was flooding into the development of biofuel technology and facilities to produce it – just to counteract the exploding energy cost and the dawning end of the petro-oil age. [125]

The *sky-high energy prices* induced some far-reaching effects centering on the deep belief, short- to medium-term, biofuels to become an economically competitive power alternative to petro-fuel [Scully 2007]. And, furthermore, there was corroboration by a related success story for bioethanol. In the mid-1970s the world had been hit by two oil price explosions, caused by production restraints in OPEC countries, and oil prices soared from a few cents per gallon to a couple of dollars per gallon. Through *initiation by a governmental program Brazil* strove to have so-called flex-fuel cars which run on ethanol.

In 2003 *flexible fuel vehicles* (FFVs) appeared on the market which have engine systems that are able to run with a mix of gasoline and ethanol. Currently all major automotive firms of the world manufacture these FFVs for use in Brazil. And a growing fleet of new-generation (flex-fuel) cars can run on straight ethanol. Ethanol accounts for more than 50 percent of the whole consumption of light car fuels in the country [Seraphim 2009].

In the late 1990s, Brazil dropped its alcohol subsidies and now made biofuel so competitive that (in 2005) it could trump gasoline at $25 a barrel [Theil 2005]. Brazil, which produces 7 billion gallons of ethanol per year, has 15 million ethanol-based or flex-fuel cars [Lane 2009c].

It is interesting to note that in the US a fuel blend called E85, which is 85 percent ethanol and 15 percent gasoline, is being made available in many states. For instance, in California, there were more than 300,000 flex-fuel vehicles in 2006 that were designed to use E85, but because the E85 distribution system has not developed as fast as the vehicle fleet, most are operating on gasoline [UC Davis 2006]. Globally, General Motors produced more than 5 million flex-fuel vehicles by 2009. In the US alone, there were more than 3.5 million GM flex-fuel cars and trucks on the road: For the 2010 model year, there were seventeen E85-capable flex-fuel vehicles from the Chevrolet, Cadillac, Buick and GMC brands [Coskata 2009.].

Since around 2004 *too much loose cash has found its way into biofuels* as a special area of the big *policy-driven* market of renewable energy (Figure I.34). Expectations grew dramatically. Many people in the venture capital industry were betting huge amounts of money on the sector. And they expected to make a 10-to-1 return [Das

2009]. "There are so many people that this almost feels like the oil land rush of the mid-1800s" in the US" [Langreth 2008].

Venture capitalists backing clean technology entrepreneurs sensed that opportunity, in particular, with regard to *second generation biofuels*. And entrepreneurship was additionally associated with a number of *"me too" firms' foundation*. In 2006 in the US already 48 new ethanol plants and eight expansions of existing plants were under construction [Reisch 2006]. And there may have been the dream that someone is going to be the next Exxon of biofuels.

Basically, however, there is not enough suitable land for corn growing to make a significant dent in America's voracious energy needs. Even if every bushel of US corn, wheat, rice and soybean were used to produce ethanol, it was estimated that it would only cover about 4 percent of US energy needs on a net basis [Wasik 2007]. Yet that did not stop ethanol investors or a wave of irrational exuberance from Wall Street to Brazil. Venture capital investment in biofuels increased from less than $1 million in 2004 to $20.5 million in 2005 [Startup Life 2007] and in 2007 venture capitalists poured $637 million into biofuels.

As part of the 2005 Energy Act, the US Department of Energy granted six cellulosic facilities special financing of up to $385 million to help build their first production facilities that, in aggregate, should reach 130 million gallons per year [Stack et al. 2007]. For 2008, VCs poured $680.2 million into US biofuels, including $437 million for cellulosic ethanol, $175.9 million in microalgae, $42 million in butanol and 25.3 million into systems and infrastructure providers [Oilgae Blog 2009].

The rapid capitalization and concentration of power within the biofuels industry was extreme. Behind the scenes giant oil, chemical, agricultural, auto corporations and large enzymes and genetic engineering companies were forming partnerships and joint ventures, and they were consolidating the research, production, processing and distribution chains of food and fuel systems under one industrial roof.

The names of giants then showing up in the game included Shell, BP, Chevron, and at last ExxonMobil, ADM, Cargill, BASF, Bayer, DuPont, Dow Chemical, Monsanto, Syngenta, VW, Daimler, Ford and General Motors and large enzyme companies like Danish Novozymes and Danisco (with its US subsidiary Genecor; Danisco A/S was recently acquired by DuPont) and the US firm Diversa [Runge 2006:872-873]. Biofuels was estimated to "be a $150 billion industry" [Langreth 2008]. But, "*the industry is still pretty much a government creation.*" [Carey 2009]

When farm products were increasingly being converted to biofuels, the offered biofuels do act like energy products. As corn and other crops become increasingly important raw materials for biofuels, the companies that make and process them were starting to act more like energy companies.

But, the most powerful players to enter the scene was "*Big Oil*," not just because of the issues of oil supply and prices, but the perceived danger to loose control of the multi-trillion-dollar transportation fuel industry (cars, trucks, trains, ships and airplanes) not just to food and agricultural industries, but also to electricity providers through the push to electrocars – also initiated by governments. Royal Dutch Shell, BP and chemical giant DuPont were leading the industry's heavy counterattack (Table I.83) – though experts said biofuels will not replace all petroleum-derived gasoline or diesel. Instead, biofuels will only extend fossil fuel supplies.

For instance, Royal Dutch Shell, Europe's second biggest energy group, formed a $12bn joint venture with Cosan, the big Brazilian sugarcane processor, that will bring together the operations of sugar, ethanol and the distribution and marketing of fuels in Brazil. The joint venture will be the world's largest bioenergy operation. Both partners will contribute retail stations and fuel retail stations and establish a network of ca. 4,500 retail sites. Shell's deal followed that of BP, its closest European rival, which saw BP providing half the $1bn investment in two ethanol plants being prepared by Tropical BioEnergia, a venture it entered with Grupo Maeda, a Brazilian agribusiness group, and Santelisa Vale, a Brazilian sugar and ethanol producer [Hoyes 2010].

There is *a bewildering array of technologies for biofuels* (Figure I.184), pushed by startups and NTBFs which are often spin-outs from universities and public research institutes and funded by governmental grants and subsidies as well as venture capital. After second generation biofuels (bioalcohols) the next biobased input for biofuels turned to algae ("*third generation biofuel,*" Figure I.34).

Following these developments the *"heavyweights" from oil* were hedging their bets taking fundamental ethical, technical and commercial disadvantages of first generation biofuels into account (Box I.1) and established differentiated strategies relying often on genetically modified objects (GMOs) and genetic engineering or chemical and process engineering to transform non food-related input into biofuels. As their innovation strategies *corporate investors* have been drawn to invest in or acquire biofuels companies that fit neatly into their value chains or long-term strategies. But also giant automotive firms, such as VW, Daimler and General Motors, entered the scene in a corresponding move (Table I.83).

Already in 2005 [Theil] the question was: "*will biofuels be able to take hold without tax credits and subsidies, especially if oil prices head downward*? Then there is the politics of global trade." By mid of 2008 the biofuel bubble began to burst, at least with respect to agricultural first generation biofuels. Those who bet exclusively on bioethanol often suffered the same fate as those investors who took the plunge on the Internet dot.com companies around 2000. Awareness spread that the goals lawmakers set for the biodiesel and ethanol industry are in serious jeopardy.

However, misaligned political incentives are not sufficient to explain the bubble (and its burst). The *Great Recession* with its global credit crisis, a glut of capacity, lower oil prices by crushed fuel demand and delayed government rules changes on fuel mixes (blending mandates postponed) were threatening the viability of biodiesel and second-generation fuels derived from feedstocks other than food. Low oil prices had a numbing effect on consumers and their interest in this area.

“Ethanol, the largest biofuel sector, was also in financial trouble, although longstanding government support will likely protect it to a certain degree.” Plans were lagging for a new generation of factories that were supposed to produce ethanol from substances like wood chips and crop waste, overcoming the drawbacks of corn ethanol. But that nascent branch of the industry conceded it has virtually no chance of meeting political production mandates that kick in soon [Theil 2005].

Many biodiesel companies started operating in the red. Even ethanol producers, which have enjoyed government subsidies and growing federal requirements to blend it into gasoline, were operating at a loss over 2008. Numerous established producers in the US filed for Chapter 11 bankruptcy-court protection. By mid of 2009 two-thirds of US biodiesel production capacity were sitting unused.

GreenHunter Energy Inc., operator of the largest US biodiesel refinery, stopped production and in June 2009 said it may have to sell its Houston plant, only a year after politicians presided over its opening. GreenHunter's business model hinged on selling to a government-guaranteed buyer [Carey 2009].

Until the mandate kicked in, GreenHunter and other biodiesel makers counted on exporting their output to Europe, a much bigger user of diesel. GreenHunter opened in June 2008 as oil prices skyrocketed. By then, soybean oil prices were soaring, too. Dozens of other new biodiesel plants, which make a diesel substitute from vegetable oils and animal fats, stopped operating because biodiesel production was no longer economical. The European Union dealt the final blow when it slapped a tariff on US biodiesel, killing what had been the industry's main sales outlet [Davis and Russell 2009].

Furthermore, the shift in power to Big Oil was already showing effects in the traditional corn ethanol business in 2008, where low prices led to the idling of more than 20 percent of capacity. VeraSun Energy, one of the largest US ethanol companies, filed for Chapter 11 (bankruptcy) in October 2008 [Carey 2009].

The situation in 2009: “The ethanol industry is on its back despite the billions of dollars they have gotten in taxpayer assistance, and a guaranteed market.” There were overcapacities due to reduction in blending and/or production mandates. It was estimated that of the 150 ethanol companies and 180 plants in the US, 10 or more companies have shut down 24 plants during the last quarter of 2008. That idled about 2 billion gallons out of 12.5 billion gallons of annual production capacity. Furthermore, it was estimated that a dozen more companies were in distress [Krauss 2009].

And there was consolidation. As ethanol producers teeter on the brink of bankruptcy – slammed by high corn prices and low gasoline demand – larger refiners were looking for opportunities to buy their assets on the cheap. For instance, Valero Energy bought seven of VeraSun Industries' ethanol plants. VeraSun owned 16 biorefineries with the total capacity to produce 1.4 billion gallons of ethanol annually, or about 13.0 percent of the total US capacity [Ackerman 2009].

As a *typical response* to such difficulties of a policy-driven market interference by the corn ethanol industry association occurred and requested a change of the blending mandate in favor of corn ethanol, in the US from E10 to E15 [Wald 2009a].

In Germany, the Federal Environment Ministry announced plans to reduce the biofuels target by 2020, because of "changes in circumstance." The German blending target was 6.25 percent in 2009, and the government said that lifting the E10 mandate (blending 10 percent of biofuel into conventional gasoline) in the market meant that the 2009 quote would have to be lowered to 5 percent.

Already by mid of 2008 27 percent of German oilseed mills had shut down production entirely and 36 percent were running on less than 50 percent of capacity. In Germany, the biodiesel industry was additionally facing a new extinction threat as federal government carried forward with its plan to increase biodiesel taxes by 40 percent. The tax hike, from 15 Euro cents to 21 cents per liter, was part of a based increase of green fuel taxes until they are the same as conventional fuel taxes. A previous tax hike removed the price advantage of biodiesel over conventional diesel and resulted in a massive decline in biodiesel output. All this became finally effective in 2009. Increasing the blending proportion from 5 to 6.25 percent was postponed to 2011 [Biofuels Digest 2008]. Correspondingly, in the US in 2009 biodiesel plants started to be idled due to the expiration of the biodiesel tax credit.

One often hears "explanations" for the tough times for biofuel startups during the Great Recession, for instance:

> "We are closing doors. We are a victim of the economy," said a venture capitalist at Polaris Venture Partners, which invested in the algae firm GreenFuel [Kanellos 2009b], but the point is:
> we are going to be hearing "victim of the economy" every time one of these "hypesters" runs out of money. It is the convenient excuse [Rapier 2009b].

In the end biofuel carcasses were everywhere and it was an open question in how far governmental stimulus programs in CleanTech to ease the Great Recession will help survival of biofuel firms which came under scrutiny with regard to new studies on greenhouse gas emissions by second generation biofuels (Box I.1) [Carey 2009].

But irrespective of the mass of "biofuel firm deaths" there was an emerging algae biofuel boom with a lot of startups – as third generation biofuels based on algae obviously could be a response to lifting greenhouse gas emissions by second generation biofuels (A.1.1.4). Over the 2004-2010 period a large horde of biofuel startups exposed very many smart ideas. But "the winners likely will be Shell, BP, DuPont, and other majors" [Carey 2009] and ExxonMobile.

Finally and, in particular, with regard to "energy independence" of a country, one has to consider *biofuels in the context of the overall "energy mix,"* the various contributions of the different energy sources of a country. And for the US there is an indication that this will change, at least for the medium term markedly. There was an orientation in the US from oil-based fuels and coal towards natural gas [Gelsi 2009; LeVine and Ashton 2009].

In the US, but also Australia, massive natural-gas discoveries heralded a big shift in the energy landscape of the US – for power supply, heating, the petrochemical industry and also transportation fuel. After an era of declining production, the US is now swimming in natural gas. And 98 percent of the natural gas consumed in the US is produced in North America [Casselman 2009]. This will finally have effects on people's attitudes and behavior and thus on the biofuels scene. Furthermore, the distribution (via ships, railroads or trucks) of natural gas as "*liquefied natural gas*" (LNG) will induce a changed economy [Runge 2006:124].

And there is another key factor which may have an influence on biofuels policy. After, in early 2010 the US government provided several drilling permissions for oil in the Gulf of Mexico, the "Deepwater Horizon catastrophe" occurred which put drilling on hold and thus affected "energy independence" efforts of the US. Offshore drilling has been put forward by the Obama administration as one prong of a multi-prong approach to ending the foreign oil dependence. The Transocean Deepwater Horizon oil rig related to oil giant BP as the operator exploded, sank and masses of oil were leaking from the well. It triggered the worst oil spill in US history and seriously damaged the economy and the environment of the Gulf States. According to Obama "BP is responsible for this leak. BP will be paying the bill."

The Deepwater Horizon environmental catastrophe in the US was assumed to may exert an influence on societal attitudes and new political initiatives with regard to boosting renewable energy and, in particular, biofuels.

As a summary, the jumping-off situation for the biofuel bubble and the policy-driven market is described in Box I.1 and its structural layout is illustrated in Figure I.34. Its discussion requires a systems approach and, for entrepreneurship and intrapreneurship, an emphasis on the related value system and approaches to problem-solving of technical and commercial as well as cost issues. In the following we shall not discuss the issue whether sufficient land, for instance, in the US or Germany, is available to meet the quantitative requirements or mandates set by policy.

Companies, NTBFs and startups had to adapt themselves also to different legislation and the rest of realities of the sector, such as second and third generation biofuels, hybrids, prices of raw materials and petro-oil prices, margins and the increase in fuel demand in China and India.

Last but not least, companies will have to deal with governmental initiatives, incentives, subsidies and pressure which should guarantee energy security throughout diversification and local production. Finally, according to venture capitalist Vinod Khosla [2008a], "if a technology doesn't meet the '*Chindia test*' – meaning that it is cheaper than the current status quo in China and India – then it is not a viable, scalable, and cost-effective long-term alternative. Anything that will uproot the global reliance on oil or coal must be less expensive; else it will never gain traction in the global marketplace."

> Only the companies able to face those challenges will survive. The key factors to determine failure or success for these companies are – apart from concurrence and balance with food crops for biofuels, their ability to guarantee their raw materials, their collection, storage and distribution management as well as their cost efficient conversion to energy/power sources and the energy storage and their fit into existing or currently built energy distribution systems, and the approval of legislation in favor of biofuels to reduce CO2 emissions – in short, the value system of the transportation fuel segment and related technical and commercial hurdles (Figure I.171).

Interpreting the Chindia test as measure of price of products based on the same or a generic technology, also the policy-driven boom/bust developments of photovoltaic and solar cells markets, the Chindia effect and the Great Recession encountered extreme problems for once "shining" NTBFs in Germany and the US (ch. 4.3.5.2).

A.1.1.2 The Technologies and Products' Situation

Ultimate goals of the biofuels key players are either commercial large-scale production of biofuels and distribution into the transportation fuel market segments defined roughly by type of vehicle, usage for short- or long-distance transportation and the power source of the vehicles (Figure I.171) or to provide a platform for "*biorefineries*" [Runge 2006:849-873] targeting production of biobased fuel and additional biobased materials ("co-products", A.1.1.6). The essential classes of biofuel products comprise

- Biodiesel, where its production process may also provide glycerin (also named glycerol) and special biosolvents
- Bioethanol, first- and second-generation ethanol (also called corn and cellulosic ethanol), which can be used also for solvents or other basic chemicals
- Biobutanol, which can be used as a full replacement of (at least) petro-fuel for cars and simultaneously can be used for an important solvent or intermediate for the chemical industry and is currently produced by petrochemical processes

- Biogasoline, which is a mixture of hydrocarbon molecules like those produced at a petroleum refinery and can be used as gasoline, diesel fuel, and jet fuel in blending with or as a replacement of petro-fuel.

Some comparisons and differentiators of bioethanol, biobutanol and biogasoline are given in Table I.82. Here, in particular, for the butanol case it must be considered that (branched chain) isobutanol has even better energy efficiency as a transportation fuel than normal (straight chain) n-butanol.

Even though the commercial opportunity appears vast, the constraints within which biofuel producers must operate are extremely tight (Figure I.171). The biofuel industry is essentially characterized by *competition* related to

- *End-products* as well as *application* areas (where biobutanol additionally is overlaid by an n-butanol versus isobutanol or mixture of isomeric butanols competition),
- Corresponding *production technologies* and *raw materials input*, as expressed in the corn ethanol versus cellulosic ethanol positions and the types of biomass (various plant and waste-related biomass versus algae),
- Access to *financial and human resources*,
- The market segments, financial and political power of players and
- Corresponding political interferences in the policy-driven market through legislative, subsidizing and national protection measures which may affect the various types of products and the financial resources allocated to them differently.

For instance, in the US there is not only a quantitative setting for the overall biofuels market, but even a "supplier constraint" in terms of type of biofuels. The Energy Independence and Security Act of 2007 mandates the use of 36 billion gallons of biofuels by 2022. Some 15 billion of those gallons must come from corn-based ethanol. The rest falls under the category of "advanced biofuels." Within the latter category, 16 billion gallons of the mandate were reserved for fuels derived from cellulosic biomass, such as wood chips, straw and wheat. One billion gallons must be biomass-based diesel [Gillies 2008].

The emphasis on corn-ethanol and corn as an input reflects a *country-specific feature for directions of innovation and entrepreneurship by the country's "natural resources"* [Runge 2006:287]. Discussing entrepreneurship in biofuels in a global context one has to consider the relevance of national resources. This is for the US farmland, corn, soybean and cereals resources and the related economy of the US "Corn Belt" – and political implications – and forests' wood [Runge 2006:287]. Sugar as a natural resources has been discussed above for Brazil.

This means, national advantages in natural resources and traditional industries can be fused with related competencies in broad technological fields and thus provide the basis for technological advantages in new product fields and often new and strong and innovative companies. As a corollary, this situation often induces strong political effects of the particular industry through lobbying on the federal and state level.

With respect to *end-use of transportation biofuels* innovative and entrepreneurial activities are also to a large extent *country-specific*. For instance, considering cars in the US there is a preference for gasoline combustion engines and little use of Diesel engines and gas-powered engines. With regard to blending with bioethanol there is no broad use of FFVs, partly due to lacking infrastructure, whereas in Brazil FFVs dominate. Also for gas-powered engines infrastructure is lacking in the US. But in Europe and particularly Germany Diesel engines are broadly used and even gas-powered cars can rely on a relatively well distributed infrastructure.

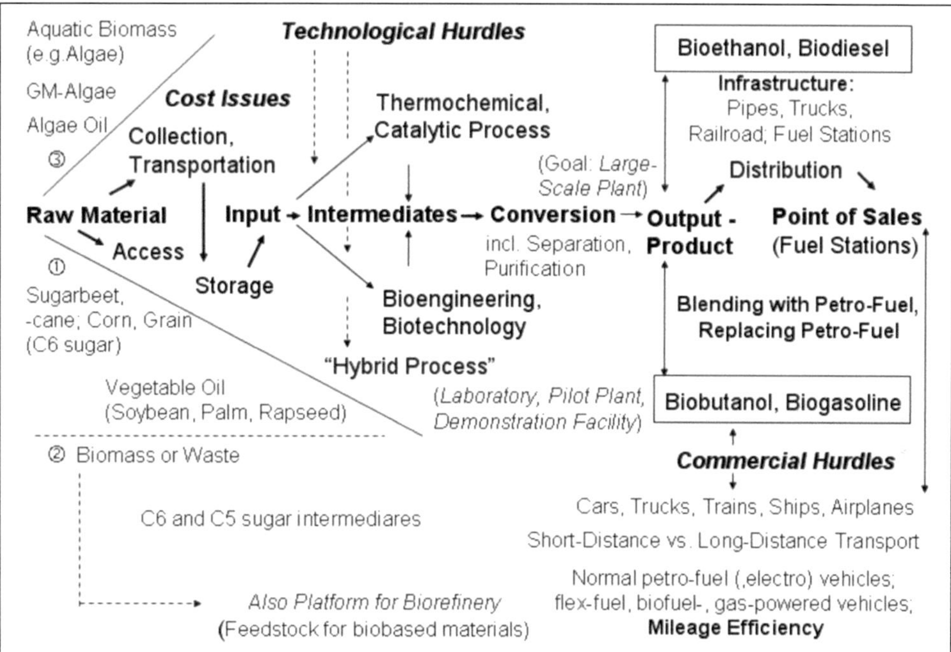

Figure I.171: The biofuels industry: The segments and technical and commercial hurdles for the race across the value system.

Concerning input (raw material and intermediates) in terms of the above discussions one might assume that the least expensive and most energy dense feedstock would be used to increase return on investment? No, it is not only the challenges to think about the logistics of the feedstock, but to consider the cost components and their mutual balances in the overall system. In tackling the competition with petro-fuel one must *carefully weigh all the costs before settling on a business model*.

Think of all your inputs including transportation, costs for equipment and operations, process design (for instance, recycling energy from other parts of the process), needs for license-in technologies, ultimate products and co-products etc. And, finally, know what the competition can offer in terms of price, quantity and quality. Furthermore, try to use the current fuel infrastructure. The problem is that British Petroleum and Exxon and Shell might not agree with the "newcomer." These have to ask themselves: Who owns the infrastructure, and can we cooperate with them, with or without a government role?

The input-to-output path in Figure I.171 can essentially be broken down into the steps from raw material to output of a firm's value chain (Figure I.7) and involves the scale-up process displayed in Figure I.8. Relating a particular raw material basis to be "harvested" (dry or wet) from a given area of land ("point of collection" of a particular region) to the economic effects of the produced biobased transportation fuel (Figure I.171) provides an expression (Equation I.20) which makes "Land Productivity of Biomass/Waste" proportional to the key performance measures of the agriculture/"farming," biofuels and automotive industries.

Equation I.20:

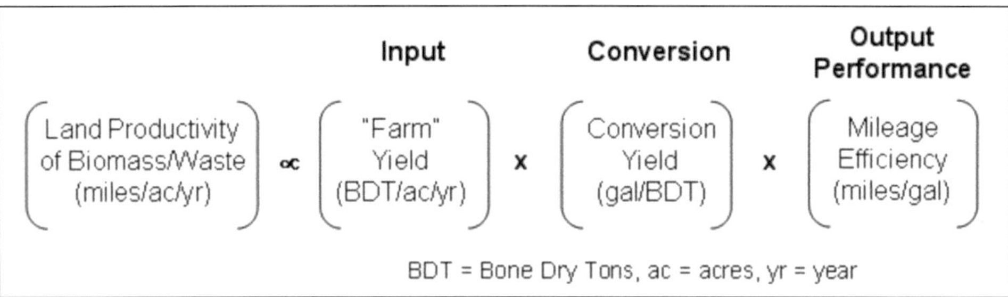

With regard to the overall goals of saving energy and reducing CO2 emission one has also to consider, where both, farm and conversion yields, contribute through these efficiencies to the overall energy and climate balances. The target would be to compete with cost of $85-$100 per barrel mineral oil without public subsidies. Hence, considerations of the following kind may be associated with intentions to start a biofuels firm.

> Our operating costs are lower than more traditional technologies, and as our technology is realizing higher yields and utilizing biomass more efficiently it can economically take biomass from a wider radius and capture economies of scale with larger production capacities. We can compete with $85-$100 mineral oil (unsubsidized cost between $2 and $3 per gallon). We can make that possible through use of 100-ton railroad cars of biomass instead of using trucks because we have a compression technology for biomass to make rail cost effective.

As production and distribution of first-generation biofuels can follow largely established input provision, production processes and other "beaten tracks" (no technical hurdles, "normal" commercial conditions, but acceptance and ethical issues discussed in Box I.1) the value system given in Figure I.171 will only focus on second- and third-generation biofuels with a number of hurdles to be overcome – and on the plenty of technical opportunities and myriad solutions to problems for entrepreneurship and intrapreneurship.

From a general economic point of view biofuels may just represent a major component for a "*biorefinery*" concept embedded in the concept of a "biobased economy" [Runge 2006:565-570, 578-585, 849-873]. In so far, any economic activities related to biofuels can view biofuels as a dedicated segment of transportation fuels or as a part of the broader concept of a biorefinery.

In line with the last "hierarchy" in the US within the biomass program there is a "*Recovery Act – Demonstration of Integrated Biorefinery Operations*" which provides funding opportunities also for biofuels. [97] A number of biofuels startups to be discussed below achieved financing via the biofuels mandate or DOE funded biorefineries (for instance, BlueFire Ethanol, Range Fuels, Mascoma, Verenium Biofuels Corporation).

As a conclusion, *biofuels are proving expensive in terms of upfront capital.* Disregarding end-product competition or substitutive potential, respectively, between types of biofuels among each other (Table I.82) and with respect to petrofuels, the economy of biofuels production for NTBFs and other involved firms is determined by local cost minima along the value system from "raw material" to "output" (Figure I.171, Equation I.20). The *final metrics* is "*capital cost per gallon (liter) capacity.*"

Local minima depend on the following major parameters:

- Feedstock: Type of biomass or waste, respectively (plant- and waste-based versus aquatic biomass like algae), cost of feedstock when cost of input increases due to increased demand for biofuel production;
- Cost of planting, growing, harvesting/collecting the biomass where the location plays a role (national or international; for instance, BioMCN (Table I.87), GBL (Table I.95), BP/Verenium-Vercipia (Table I.84), Amyris (Table I.99);
- Transportation, storage and pre-treatment of the biomass (Bioliq, Figure I.173);
- The fundamental process cost – thermochemical, bioengineering, or biothermal "hybrid" process – with various cost reduction options, such as utilizing known overall processes (Fischer-Tropsch syngas, Figure I.174), utilizing lego-type existing technologies (BlueFire, ZeaChem – Table I.86,Table I.88.), genetically modified microbes or enzymes to increase yield/output or finding the "best" naturally occurring objects for fermentation, particular process step energy efficiency;
- Increasing process efficiencies by various means, such as energy efficiency by processing intermediates and co-products which can be fed back into the process, minimizing water use;

- Separation and purification of end-products which again means issues of energy efficiencies (Cobalt, Table I.96);
- Co-financing biofuels production by simultaneous co-manufacturing marketable products, the "biorefinery model."

Moreover, continuous R&D is needed to improve conversion yields and also engineering has to follow a continuous improvement process to increase process efficiency.

According to Figure I.171 in the end biofuels for transportation addresses the issues of *blending petro-fuels with biofuels or replacing petro-gasoline*. Therefore, it is important to be aware of players and ownerships in the current petroleum value system. The oil industry supply chain (Figure I.172) comprises upstream crude oil exploration and production, and downstream refined products manufacturing.

Refining can be associated with oil firms. However, there are also independent oil refiners. These oil refiners can also serve as blenders, for instance, blending gasoline with additives like octane boosters or with biofuels – bioethanol or biodiesel.

This is followed by Wholesale product distribution from refineries (blenders) to primary distribution terminals and Retail delivery to final customers (end-users) with the automotive industry interfering technically and government politically with the refinery/blending stage. Figure I.172 depicts the sectors of the petroleum industry and shows that the oil industry is largely integrated across the whole value system.

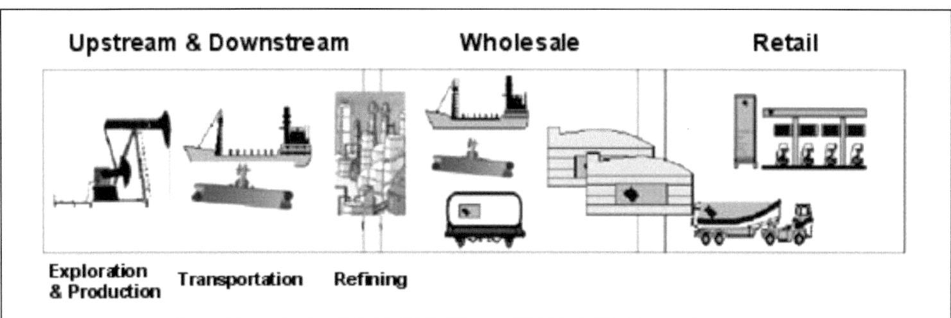

Figure I.172: The petroleum (oil industry) supply chain.

Generic Factors Triggering the Current Biofuels Orientation

The actual biofuel focus and related innovative and entrepreneurial activities occured as an additional peak of a progression cascading for more than one hundred years and triggered by parameters which are induced by comparable events or initiatives, respectively.

The current situation exhibits some sort of "déjà vu" when Germany strove for synthetic fuel during the 1920-1940 period [Runge 2006:271-272]. In comparison to previous ones the current step of the cascade is enforced essentially by four additional factors:

- Societal "green" attitudes in developed societies and climate change issues inducing political actions;
- The "decentralization megatrend" (in particular, for energy);
- The new biotechnological options of genetic modifications of microorganisms (and plants);
- Additional (specific) financing options for technology and entrepreneurship in the biofuels sector by the venture capital industry.

The current main political drivers for the biofuel industry in Europe and particularly the US are:

1. Fight climate change, use environmentally friendly, "renewable" energy, reduce the "carbon footprint," which is the total amount of greenhouse gases produced to directly and indirectly support human activities, usually expressed in equivalent tons of carbon dioxide (CO2).
2. Dramatically reduce, or even end, dependency on foreign petro-oil; in the US the focus is to reduce dependency on imported petro-oil, Germany looks at changing its power mix in favor of "renewable energy sources" (RESs).
3. Spur the creation of a domestic bioindustry emphasizing concepts of a bio-based economy and biorefineries. A special origin of this direction is the widespread belief that petro-oil oil supply will cease "soon."

The new biotechnology options for biofuels occurred in line with scientific/technical progress for the bio-agricultural field. On the other hand, the second and third drivers are interwoven. These drivers initiated corresponding technical innovations in Germany and the US which are still the basis for current biofuel developments. What has changed essentially is the variability of the raw material input.

For Germany, immediately before World War I and then until the WWII for aggression *self-sufficiency ("autarky")* became a characteristics of *national security policy*. Areas of interest were formulated as interests to the national administration and the military, many of them being chemical areas for civilian and military purposes: hydrogenation of coal (*synthetic fuel*), more *efficiency* for use of German coal, fertilizers, artificial fibers, *synthetic rubber, fermentation technology* to release Germany from imports of alcohol etc. Striving for self-sufficiency peaked just before and during the two world wars.

Additionally, over the past 150 years, geologists and other scientists have regularly predicted that the world's oil reserves would run dry within a few years. Particularly, in 1922 the US Geological Survey predicted that the US only has energy oil supply to last 20 years which spurred innovation and cooperation activities in and between German and US firms [Runge 2006: 424, 564-565].

Currently, there is again a public discussion when oil fields will dry up, that global oil production is about to peak ("peak oil") and that there is an absolute end of cheap oil mainly by ever increasing demand from China and India. The bioindustry orientation occurred parallel with the 1920s "oil issue" in terms of the "Chemurgy Movement" in the US, which can be seen as a bottom-up entrepreneurial approach, and the German plant-based "Ersatzstoff"-approach which was largely driven by government [Runge 2006:565-566].

Generally, in Germany (and the US) the emphasis was on the involvement of huge companies with the financial power to master the challenge for synthetic fuel and rubber [Runge 2006:270-272]. Rubber brings in aspects of the current "biorefinery" concept. The German synthetic rubber "BUNA S" was produced in Germany and the US. But, in both countries governmental guarantees for price and sales quantities were needed. In the US additionally governmental financing of related plants were necessary to start large-scale industrial production on BUNA S.

In this context, industry and academia interactions and "technology transfer" in the US for developing the synthetic rubber between DuPont and the University of Notre Dame is notable. Notre Dame's most famous effort in *technology transfer* was Father Julius Nieuwland's groundbreaking work with polymerized-2-chloro-1,3-butadiene, which led to two patents and the development of Neoprene in 1931 by the DuPont chemical company. That particular bit of "intellectual property" was very good fortune for the Notre Dame University – some $2 million when the royalty payments ceased in 1948 [Streb 2002, Runge 2006:272, 692].

The 1973 first oil crisis started when the members of Organization of Arab Petroleum Exporting Countries or the OPEC (consisting of the Arab members of OPEC, plus Egypt, Syria and Tunisia) proclaimed an oil embargo "in response to the U.S. decision to re-supply the Israeli military" during the Yom Kippur War. It lasted until March 1974. For the most part, industrialized economies relied on crude oil, and OPEC was their predominant supplier. With the US actions seen as initiating the oil embargo, the long-term possibility of embargo-related high oil prices, disrupted supply and recession occurred. Correspondingly, in the industrial countries there was a strong movement to become independent from OPEC oil.

> Due to expensive oil the energy crisis led to greater interest in *renewable energy* and *spurred university and other publicly funded research in solar power and wind power*. It also led to greater pressure to exploit North American oil sources, and increased the West's dependence on coal and nuclear power. Notably, already at that time, the Brazilian government implemented a very large project called "Proálcool" (pro-alcohol) that ultimately led to blend gasoline with ethanol for automotive fuel (for FFVs). This project, which produces ethanol from sugar cane, continues and has reduced oil imports and decreased the price of fuel [Seraphim 2009].

The 1973 "oil price shock," along with the 1973–1974 stock market crash, has been regarded as the first event since the Great Depression to have a persistent economic effect. The second (1979) oil price crisis in the US occurred in the wake of the Iranian

Revolution. In 1980, following the Iraqi invasion of Iran, oil production in Iran nearly stopped, and Iraq's oil production was severely cut as well. However, after 1980, oil prices began a six-year decline that culminated with a 46 percent price drop in 1986. This was due to reduced demand and over-production, which caused OPEC to lose its unity.

> In 1979 US President Jimmy Carter outlined his plans to *reduce oil imports* and *improve energy efficiency* in his "Crisis of Confidence" speech. Acting as an example he had already installed solar power panels on the roof of the White House and a wood-burning stove in the living quarters. [98]

Correspondingly, again several governmental research programs, initiatives and pilot projects started worldwide with regard to "renewable energy," adding hydrogen and algae options (see below).

Around 2000 the idea of a "biobased economy" emerged with the "biorefinery" as a central concept [Runge 2006:849-873]. Not only biofuels, but many other CleanTech areas occurred in the spot, such as photovoltaic (PV; solar cells) and solarthermics (solar thermal energy), fuel cells and batteries, and wind turbines (ch. 4.3.5.2). And related innovative and entrepreneurial activities relied considerably on governments and further developments and refinements on several decades old or century old scientific insights and technologies.

A.1.1.3 Intrapreneurship and Entrepreneurship in Biofuels: The Biomass-to-Biofuels Boom

For biofuels for the transportation sector the economic realities for startups show up as an exertion of power and resources as well as the streamlined utilization of financial and infrastructural resources of incumbents. Apart from corporate-internal activities, the theme is *innovation of (mostly) giant companies by means of interrelating to NTBFs and other firms* (Table I.83; Figure I.41, Figure I.51) following rather common New Business Development (NBD) approaches given below [Runge 2006:722-730].

- Corporate sponsorship and funding of external basic and applied research in universities and public research organizations;
- Joint research and/or development alliances (JRAs, JDAs) with startups and NTBFs; special cooperation by which big firm will receive R&D samples from NTBFs for tests;
- Joint ventures with related firms or investing directly or indirectly in startups through "corporate venturing," preferentially in startups/NTBFs in a later stage of development (ch. 1.2.7.2); on the other hand, interrelated NTBFs have access to the resources of the firm's stakeholders, for instance, get help, advice and consulting for process and plant engineering and biofuel analytics (cf. Shell, BP);
- License-in from startups and NTBFs;

- "Cherry picking" (picking "winners") from the masses of early- and late-stage startups in the world fitting their current value chain and long-term strategy rather than doing own research and intrapreneurship;
- Utilizing the firms' capabilities to tap to their advantages into the various financial resources and other aids of policy to fuel inventions and technology developments.

Some of the giant firms had already technical/commercial footsteps in biofuels. For instance, Shell and BP had already large mix-in of ethanol and experiences in processing and plant engineering; Dupont synthesized successfully the alcohol 1,3-propanediol by fermentation.

From Table I.83 and further related text one may extract the following specifics concerning intrapreneurship of giant companies from the oil, chemical and automotive industries in biofuels.

- There is pronounced sponsoring/funding of universities and public research organizations by BP, DuPont and Chevron. BP, for instance, provided $500 million over ten years to establish a dedicated biosciences energy research laboratory attached to a major academic center in the US; the Energy Biosciences Institute (EBI) is led by Berkeley and with Lawrence Berkeley National Laboratory and the University of Illinois at Urbana-Champaign.

There are numerous alliances and cooperations of oil and chemical giant firms with investments in startups and "later-stage" NTBFs to tackle different technologies and/or developments targeting different steps of the value system (Figure I.171).

Shell was said to have over 70 research alliances in biofuels [Kanellos 2009a]. And also large firms that were around 20 years in corn ethanol like privately held US firm Poet LLC set up alliances to enter cellulosic ethanol. Poet had a network of 26 plants in seven US states producing ca. 1.25 billion gallons of ethanol annually (revenues of $4 billion in 2008).

- JVs and other forms of alliances show also up for oil/chemical giants with other large firms.
  For instance, there are the DuPont's connections with Genencor (a unit of Denmark's Danisco, now belonging to DuPont) and Poet LLC emphasizing cellulosic ethanol and also Shell's connections with the Canadian firm Iogen in cellulosic ethanol [Runge 2006:858; Gold 2009].
  Poet's research discovered an enzyme and designed a process that allows converting the starch from corn kernels into sugar and fermenting it without using heat. The process for cellulosic ethanol, which Poet commissioned to Danish industrial biotech giant Novozymes to develop, shall reduce energy consumption and increase its yield of ethanol in the fermented mix [Dolan 2008].

- By 2008 a new emphasis on algae emerged as a raw material for biofuels (biogasoline and biodiesel).
  In particular, its late entry into biofuels suggests that ExxonMobile bets on photosynthetic algae to be a viable, long-term candidate raw material for various types of biofuels. BP seemed to give it a try and also Shell and DuPont. And chemical giant Dow Chemical planned an algae biofuels pilot. The joint project with the firm Algenol (A.1.4, Figure I.179) should test a process to turn CO2 into ethanol [Voith 2009a].
- It seemed that Shell's "biogasoline" ("normal gasoline" and diesel) versus DuPont/BP's cellulosic ethanol and biobutanol (and biodiesel) emerges as the heavyweight fight for the future of the gas tank.
  Recently, French oil giant Total also joined the biobutanol option [Gold 2009a] through an investment in the startup Gevo, Inc. (Table I.99)
- Automotive companies "synchronized" their developments with developments in biofuels through investments in selected NTBFs.
  For instance, not only Shell had stakes in the German firm CHOREN Technologies, but also the German automotive giants VW and Daimler had ones [Runge 2006:254-255; Kempkens 2009]. Correspondingly, Mascoma (Table I.99) snagged $100 million in funding from General Motors (GM), Marathon Oil and other investors plus millions more in government grants, and aimed to produce cellulosic ethanol from wood chips using genetically engineered bacteria [Langreth 2008].

GM and Coskata (Table I.99) said their partnership will enable them to work together on ethanol research and development, as well as to build the infrastructure needed to commercialize the biofuel. GM said it will utilize the fuel from the demonstration facility, and will also provide some of its carbon-based waste, like old tires, as a feedstock for Coskata [Fehrenbacher 2008]. And Virent Energy's investor Honda was testing Virent's fuels in engines [LaMonica 2009].

The JV of Chevron with Weyerhaeuser (Catchlight Energy) was remarkable in that the JV would study "*not only the technology, but also the commercial implications of creating a viable business*. It should *devise a sustainable business model* "from the forest lands to the fuel." That involves harvesting timber, transporting it, breaking technological ground to process it into biofuel, and finding ways of transporting and distributing the fuel [González 2008].

The BP/DuPont partnership, Butamax Advanced Biofuels, should focus on developing a technology program to produce biobutanol from many different types of feedstock and was expected to license the technology to produce biobutanol to other biofuel producers. It would work closely with Kingston Research Ltd., another JV between BP and DuPont. Kingston Research would be constructing a biobutanol demonstration plant in the UK. [Lane 2009h; Lane 2009i]. Biobutanol can be blended with any fuel – gasoline, diesel or ethanol – or can be used as a bio-alternative to chemicals.

Butamax can be characterized (business model, strategy; status 2009) as follows [Anonymus 2008; Lane 2009h]:

- Develop and commercialize biobutanol targeting advanced *metabolic pathways for 1-butanol* as well as other *higher octane biobutanol isomers*;
  develop biocatalysts to produce 1-butanol as well as 2-butanol and isobutanol – the higher octane biobutanol isomers that are of increased interest and utility from a fuels perspective ("high octane biobutanol"),
  develop a genetically-modified microbe, or "ultimate bug," as the catalyst for new technology to significantly improve the conversion ratio in processing feedstock into biobutanol, boosting fuel yield and concentration [Chase 2006];
  have a strong intellectual property position in the butanol areas of greatest interest through patents covering the higher octane isomers as well as the previously announced 1-butanol patents;
- Not only *improve the bio-process to produce commercial volumes* of biobutanol, but also pursue an *integrated commercialization strategy* that incorporates building pilot and commercial scale facilities, a complete fuel evaluation, and a full environmental life cycle analysis,
  work with fuel blenders and distributors globally to introduce biobutanol into the fuels market;
- *License the technology* to produce biobutanol to other biofuel producers;
- Deliver by 2010 a *superior biobutanol manufacturing process* with economics equivalent to ethanol and commercially produce biobutanol in 2013 [Lane 2009h].

Three other (new) ventures, advanced in their routes towards commercial production and working on butanol-based solutions, are in the US, ButylFuel (below text), Cobalt Biofuels (Table I.96) and Gevo (Table I.99) and in the UK Green Biologics (Table I.95).

"*Cellulosic ethanol*," as opposed to sugar or starch-based (corn) ethanol, broadens the choice of feedstock without impacting food supplies. But bioethanol – whether from corn or new sources – runs into something called the "*blending wall*." Right now much of the gasoline in the US (and Germany) contains 10 percent ethanol (E10), which works fine in today's cars and trucks. However, automakers worry that higher levels will damage engine components. So they will void the warranties of most vehicles running on richer ethanol blends. But there is more (Table I.82), if performance of different biofuels is compared [Kiplinger Washington Editors 2007].

Table I.82: Comparing the biofuels bioethanol, biobutanol, biogasoline and their fit with existing infrastructure and vehicle compatibility.

Properties, Features	**Bioethanol (EtOH)**	**Biobutanol (BuOH)**	**Biogasoline**
Energy Efficiency; Mileage Efficiency	70 percent of the mileage of petro-gasoline ("hydro-carbons")	88 percent of the mileage of petro-gasoline; lower vapor pressure – less volatile (than EtOH), more comparable octane number with petro-gasoline, especially if also *isobutanol* is in place	Comparable with petro-gasoline, has ca. 50 percent more BTUs (British thermal units) per gallon than EtOH does
Water content	Attracts water, separating EtOH from water in the production is energy-intense	Does not attract water like EtOH, can be transported in existing pipelines and is less sensitive to colder temperatures.	Negligible
Other Features		Very important chemical solvent and intermediate	
Blending with Petro-Gasoline	Restricted blending with petro-gasoline: "Blending Wall" E10 (max. 10%)	Blending with petro-gasoline (in higher concentration than EtOH) or replacing (100%)	To any extent; replacing petro-gasoline
Transportation, Fuel Station Infrastructure	Need for separate ethanol infrastructure; damage (corrosion) by high water content (trucks, trains, barges and pipelines)	Can be transported in existing pipelines and is less sensitive to colder temperatures.	Compatible with existing petroleum-based infrastructure

Table I.82, continued.

Performance in Vehicles	Needed "flux fuel vehicles" (E85); damages "normal" combustion engines	No need to retrofit vehicles; shown that cars can run on pure biobutanol	Compatible with existing vehicle (combustion engine) operation

The data in this table reveal one central *issues of biofuels legislation*. It is bioethanol and biodiesel oriented. But what is more serious is the fact that usually governmental incentives and tax credits are done on a volume (per-gallon) basis. In essence, in the US ethanol is getting an unusually large credit, considering that it got ca. 20 percent less energy per gallon than biobutanol and more than 50 percent less than biogasoline. Although large volumes of cellulosic ethanol may be used in the coming decade and beyond, its long-term technical feasibility has been questioned because of its low energy density.

To put the level of development of biofuels into perspective with regard to have a full commercial large-scale plant (Figure I.171) consider Iogen Energy partnering with Royal Dutch Shell on a demonstration-scale cellulosic ethanol plant that sold its output in 2009 at a single station in Ottawa, Canada. Emphasizing the *demand* side, the plant produced about 40,000 liters a month (10,560 gallons). By way of comparison, Canada drank 30.2 million gallons of gasoline every day in 2008 and the US guzzled 377.6 million gallons daily. In 2008 in the US ca. 140 billion liters were consumed, in Germany 45 billion liters [Seidler 2009].

In 2009 some new firms already were building – or contemplating building – industrial scale facilities, for instance, CHOREN Industries in Germany (below text) or Verenium (Table I.84) in the US [Gold 2009a].

Scaling-up (Figure I.8) to produce biofuels on a commercial scale is the tricky part for related startups or NTBFs. That has been the hard part for all of the cellulosic ethanol startups. But it was the big oil companies that were perceived to may help these companies eventually reach commercial scale.

Key intrapreneurial and innovative activities of oil and chemical giants in biofuels during the period 2005-2009 are summarized in Table I.83.

One business model for startups for rapidly expanding to commercial-scale operations focuses on collaborations being formed between biofuel startups and "Big Energy" which are comparable to the partnerships formed between biotech startups and big pharmaceutical companies. For US Codexis the Shell deal reflects a desire to apply its biotechnology to markets beyond pharmaceuticals. Alan Shaw (then CEO of Codexis) said Codexis will soon be forming a new business to further its efforts in the bioindustrial field [McCoy 2006].

Shell's deal with microbe/enzyme producer Codexis strikes at the heart of the big challenges standing in the way of biofuel's coming of age: How to economically turn starches into sugars. Once cellulosic material like wheat stalks and corn stovers are broken down, they can be fermented just like corn and turned into ethanol. The problem so far has been finding a way to cheaply, quickly, and massively break down huge amounts of agricultural or municipal waste and having microorganisms that convert efficiently not only C6, but also C5 sugars into ethanol. Basically, converting biomass to biofuels requires breakthrough developments in three areas:

- chemical preparation of the cellulosic biomass (*pre-treatment*) and separation of the cellulose and hemicellulose parts from the lignin,
- conversion of pretreated cellulosic biomass to fermentable sugars (degrading the chemical bonds of the cellulose/hemicellulose) by combinations of enzymes (*saccharification*),
- and the development of novel microorganisms to ferment the sugars to ethanol or other fuels (*fermentation*).

Table I.83: Oil and chemical giants' key activities in biofuels 2005 – 2009.

Company (Remarks)	**Alliance [1] / JV**	**1st Generation Bioethanol**	**2nd Generation Bioethanol**	**Biobutanol**	**Biodiesel**	**Biogasoline**	**"Algae Biofuel"**
British Petrol (BP) for blending in 2005, BP purchased 590 mil. gallons of bioethanol (575 mil. gallons in the US) and 70 mil. gallons of biodiesel 2) BP emphasizes producing cellulosic ethanol through its partnership with Verenium in the US and on sugarcane and ethanol in Brazil, and on biobutanol [McDermott 2009]	a) "*Butamax Advanced Biofuels*" (see below): JV with DuPont b) *Verenium* (formerly Diversa) c} *Verenium* BP has invested additionally $22.5 mil. in Verenium and formed a JV called Vercipia – now BP Biofuels Highlands [Lane 2009a] d) *Universities* BP to spend $500 mil. over the next ten years to establish a dedicated biosciences energy research labora-	a) Retrofitting an ethanol plant to produce biobutanol	c) BP's $112.5 mil. total investment in Verenium is the largest by an oil major in an advanced biofuels company; funding is the basis for a commercial scale cellulosic ethanol plant, will pursue to globally license technology of Verenium as well as developing the commercial scale plant, a team comprised of employees from both BP and Verenium, the 1.4 mil. gallon-per-year (MGY)	a) With DuPont (see also DuPont) first phase of the DuPont-BP venture will consist of using existing technology to convert sugar beets into 30,000 tons, or 9 mil. gallons, of biobutanol annually at British Sugar's facility, second phase of the venture involves developing a genetically-modified microbe, or "ultimate bug," as the catalyst for new technology to significantly improve the con-	e) D1-BP Fuel Crops Limited, to accelerate the planting of Jatropha curcas – a drought resistant, inedible oilseed bearing tree which does not compete with food crops for good agricultural land or adversely impact the rainforest – in order to make more sustainable biodiesel feedstock available on a larger scale, both firms intend to invest around $160 mil. over the next five years [BP 2007]		f) Martek, specializes in engineering fuel and other products from algae, fungi and other microbes [Associated Press 2009a]

	tory attached to a major academic center [BP 2006]; BP funding for the Energy Biosciences Institute (EBI) is led by Berkeley and with Lawrence Berkeley National Laboratory and the University of Illinois at Urbana-Champaign [Birgeneau 2007]; $35 mil. from BP biofuels research center at UC Berkeley (as annual budget) [Langreth 2008] e) *JV with D1 Oils* to develop Jatropha biodiesel feedstock f) *BP and*		cellulosic ethanol plant in Jennings, LA is the first true demonstration-scale plant in US capable of producing ethanol from non-food cellulosic biomass sources, [Lane 2009a] BP to commence cellulosic ethanol production in Brazil by 2013 [Lane 2009c]	version ratio in processing feedtocks into biobutanol, boosting fuel yield and concentration; UK provides fuel station infrastructure and test field, “By getting it out in the pump, letting consumers buy it ... I think that will help government get behind it” [Chase 2006]	D1 Oils will buy out BP a mere £500,000 (ca. $823,000), even though the joint venture was valued at more than £7 mil. ($11.5 mil.) over its two-year lifetime. [McDermott 2009] f) Martek “we believe *sugar-to-diesel* technology has the potential to deliver economic, sustainable and scaleable biodiesel supplies,” fuel will be created from biomass; microbes will convert the biomass into lipids, which will then be turned into		

	Martek Biosciences form algae biodiesel partnership, BP will contribute up to $10 mill. [Associated Press 2009a]				fuel		

1. Company, University or Public Research Institute/Center.
2. http://www.bp.com/genericarticle.do?categoryId=2012968&contentId=7018719 (last access 11/10/2009).
3. http://www.shell.com/home/content/aboutshell/our_business/previous_business_structure/oil_products/fuels/biofuels/biofuels.html (last access 11/22/2009).
4. http://www.virent.com/MeetVirent/our_story.html, http://www.virent.com/News/press/03-26-08_Shell_Virent_Biogasoline_Collaboration.pdf.

Table I.83: continued.

Company (Remarks)	**Alliance** [1] **/ JV**	**1st Generation Bioethanol**	**2nd Generation Bioethanol**	**Biobutanol**	**Biodiesel**	**Biogasoline**	**"Algae Biofuel"**
DuPont with sugar producer Tate & Lyle already produced the di-alcohol propanediol (PDO) for polymers by fermentation [Runge 2006:583]	a) BP and British Sugar; the two companies have applied for more than 60 patents in the areas of biology, fermentation processing, chemistry and end uses for biobutanol, those patents cover the higher octane isomers as well as the previously announced 1-butanol. According to Anonymous, this places the BP/DuPont partnership in a strong intellectual property position [Ebert 2008; Anonymus	c) Converts an existing Broin ethanol plant into a biorefinery (cost of $220 mil.), upgraded facility would operate on both corn and stover instead of on corn alone [Reisch 2006]	a) Together with BP and British Sugar built $400 mil. bioethanol plant [Van Noorden 2008]; b) switchgrass feedstock supplied by Genera Energy and corn cobs supplied by Sun Grant, $140 M funding from parents. 90 scientists from DuPont and Danisco dedicated to project. State of Tennessee has granted $40 mil. towards the demonstration plant [Lane 2009b]	a) Converts the UK's first ethanol fermentation plant to produce biobutanol [Chase 2006], a) by using traditional methods, such as the ABE fermentation process to create biobutanol [ISEE] e) DuPont and BP have developed catalysts to produce 1-butanol (a four-carbon chain with the alcohol group at one end), 2-butanol (a four-carbon chain with the alcohol group bonded to a			d) Produces biobutanol from seaweed (macro-algae, a potentially sustainable and scalable new source of biomass) [Morrissey 2009]

	2008]; b) partnership (JV) with enzyme company Genencor (a unit of Denmark's Danisco, recently acquired by DuPont) [Dolan 2008]; c) with corn ethanol producer POET [Reisch 2006]; d) gets $9.0 mil. for a partnership with the biofuel startup Bio Architecture Lab from the US Department of Energy's Advanced Research Projects Agency-Energy (ARPA-E). [Morrissey 2009]			carbon atom in the middle of the chain), and isobutanol (a branched chain with the alcohol group on one end). Isobutanol and 2-butanol have higher octane ratings, making them better fuels, the companies found that gasoline blends containing 16% high-octane butanols deliver fuel performance similar to blends with 10% ethanol (E10) [EERE 2008]			

Table I.83: continued.

Company (Remarks)	**Alliance [1] / JV**	**1st Generation Bioethanol**	**2nd Generation Bioethanol**	**Biobutanol**	**Biodiesel**	**Biogasoline**	**"Algae Biofuel"**
Royal Dutch Shell Shell Biofuels: Shell currently buys, trades, stores, blends and distributes (essentially Brazilian sugar cane) bioethanol. It is the world's largest distributor – more than 6 bil. liters in 2008 – and continues to build its capability 3) It is said that Shell has five (corn) "ethanol hubs" in the US, providing the distribution infrastructure for 30% of US ethanol [BioFuelWatch 2009]	a) With Canadian company Iogen Corp it has a stake in [Runge 2006:858] b) with the German firm CHOREN Industries it had a stake in [Porretto 2009; Runge 2006:254-255], gave up its minority stake in Nov. 2009; wanted to continue to support CHOREN c) with Codexis it has a stake in, development agreement renewed in 2009 [McCoy 2006; Johnson		a) Produces cellulosic ethanol from wheat straw c) speeds up development of super enzymes that can chew through starchy plants and break them down more quickly, cellulosic material like wheat stalks and corn stovers [Johnson 2009]		b) Develops and produces biodiesel from wood residue ("BtL process") [Runge 2006:254-255]; plans a 200,000 tons per year biodiesel plant costing €800 mio. [Kempkens 2009]; d) will construct with HR BioPetroleum an algae-oil production facility to produce feedstocks for biodiesel ("second-generation biodiesel") [Lane 2007]	e) A thermochemical catalytic process ("Bio-Forming") to convert sugars into hydrocarbons, the chemicals found in petroleum; claimed: sugars are converted directly into gasoline and gasoline blending through a catalytic process, new "biogasoline" molecules have higher energy content than ethanol (or butanol) and deliver better fuel efficiency, Virent's unique cata-	d) To build a pilot facility for growing marine algae and producing algal oils that can, in turn, be used to make biofuels; Cellana, completed construction on the Kona coast of the big island of Hawaii, and the demonstration facility is now operating [Bigelow 2009]; expands the 2.5-hectare (269,000 square foot) pilot project (2007) to a 1,000-hectare facility after two years and later to a full-

Said earlier in 2009 it would scale back large investments in wind and solar in favor of next-generation biofuels [Porretto 2009]; continues to be close to familiar businesses	2009] d) JV (named Cellana) with HR BioPetroleum [Lane 2007] e) with Virent Energy Systems joint research and development effort [EERE 2008; Carey 2009] 4}					lytic process uses a variety of biomass-derived feedstocks to generate biogasoline at competitive costs, sugars can be sourced from non-food sources like corn stover, switch grass, wheat straw and sugarcane pulp, in addition to conventional biofuel feedstock like wheat, corn and sugarcane 4)	scale commercial 20,000-hectare plant, grows only non-modified, marine microalgae species in open-air ponds using proprietary technology. Algae strains used will be indigenous to Hawaii; once the algae are harvested, the vegetable oil will be extracted; expectation: 60 tons of oil per hectare

Table I.83: continued.

Company (Remarks)	**Alliance [1] / JV**	**1st Generation Bioethanol**	**2nd Generation Bioethanol**	**Biobutanol**	**Biodiesel**	**Biogasoline**	**"Algae Biofuel"**
ExxonMobile photosynthetic algae appears to be a viable, long-term candidate for biofuels [Porretto 2009]	*Synthetic Genomics Inc. (SGI);* has so far done early work on algae strains; with $300 mil. designated for research and development at Synthetic Genomics; planned to build a greenhouse and biofuels test facility in San Diego to test different strains of genetically engineered algae and methods of commercial biofuels production [Bigelow 2009a]					Algae-based gasoline/biodiesel [Porretto 2009]	2009: $600 mio. to develop algae-derived biofuels; "commercializing algae-based biofuels will cost billions of dollars more" [McCoy 2009] SGI has plans to build a greenhouse and biofuels test facility to test different strains of genetically engineered algae and methods of commercial biofuels production

Table I.83: continued.

Chevron, has a special emphasis on research and development alliances, studies ways of producing and converting specially grown non-food crops into ethanol and other biofuels, will address the vast range of variables – from genetics to thermo-chemical reactions to economics	a) Biofuels development agreement with the National Renewable Energy Laboratory (NREL) [McCoy 2006] b) agreement with Texas A&M University (four-year period), formed research arrangements with Georgia Tech; the University of California, Davis; and the Colorado Center for Biorefining & Biofuels, which is a consortium of the Energy Department's National Renewable Energy Laboratory, three major Colorado		a) Converts biomass such as forestry and agricultural waste into ethanol and renewable fuel [Reisch 2006] c) converting cellulose and lignin – the compounds plants are made of – into biofuels; the venture will study "not only the technology, but also the commercial implications of creating a viable business there," devises a sustainable business model "from the forest lands to the fuel," Weyerhaeuser is one of the largest producers of pulp		b) Chevron says its large-scale biodiesel plant in Galveston, Texas, is fully operational, has a 22% equity position in the plant [Hess 2007] said it will be able to prove it can make diesel from designer microbes and sugar economically by 2011 [Kanellos 2009]	f) Getting biodiesel from algae	a) Working with NREL on a five-year project to research transportation fuels from algae f) getting biodiesel from algae

	universities, and other private companies [Hess 2007; UC Davis 2006]; c) JV (Catchlight Energy) with timber behemoth Weyerhaeuser [González 2008] d) with Mascoma a feedstock processing and lignin supply agreement [Mascoma Corp. 2009] e) investment in LS9 f) biodiesel feedstock development and testing agreement with Solazyme		and timber in the world. [González 2008]; d) Chevron will supply feedstock that Mascoma will convert to cellulosic ethanol, a by-product of that conversion is energy-rich lignin, which Mascoma will then provide Chevron to evaluate whether it can be turned into transportation fuel.				

The investment of BP in Verenium (Table I.84; itself created from Celulol and Diversa; Table I.99) and finally in their JV, Vercipia Biofuels, emerged as an exemplary bio-engineering-oriented route of a biofuel innovation path of a large (giant) company (Table I.83, Table I.84): Starting with corporate venturing and ending with acquiring the whole ligno-cellulosic biofuels *business* of Verenium which is now run under the name BP Biofuels Highlands. Main features and innovative characteristics of Verenium are given in Table I.84, additional details for Verenium are found with Lane [2009g].

Table I.84: Verenium targeted by BP focusing on its biofuels business. *)

Company (Foundation) Remarks	**Major Funding**	**CEO, Other Executives, Key Researchers; Technology Protection**
Verenium Corporation, Cambridge, MA Formed in 2007 by a merger of Diversa and Celulol (Table I.99). Headquartered in Cambridge, MA, has research and operations facilities in San Diego, CA; Jennings, LA (Verenium Biofuels, LLC) and Gainesville, FL. Organizational Units: Biofuels Business, Specialty Enzyme Business, R&D (complementary components of biofuels). Markets Served: Biofuels, Industrial Processes, Health & Nutrition Annual Revenues and Losses	As of March 31, 2007 had cash, cash-equivalents and short-term investments on hand of ca. $125.5M; together with approximately $20M [Childs 2007]. In 2008, the Jennings Facility was selected for an award under a $240M federal program, operated by the US Department of Energy, to support the development of up to nine small-scale biorefineries in the US. BP's $112.5 million total investment is one of the largest by an oil major in an advanced biofuels company, BP has invested an additional $22.5M in Verenium and formed a joint venture Vercipia [Lane 2009a]. BP put up $90M to	Carlos A. Riva President and CEO; Riva joined Celunol as CEO in 2006, prior to joining Celunol, from 2003 to 2005; Riva served as Executive Director of Amec PLC, a major global construction and engineering company based in UK; from 1995 to 2003; Riva served as CEO of InterGen, a joint venture between Shell and Bechtel that developed more than 18,000 megawatts of electric generating capacity; William H. Baum Executive VP Business Development since 2007 after the merger of Diversa Corporation and Celunol Corporation; joined Diversa in 1997 as VP Sales and Marketing, then Senior VP Business Development and to Executive VP Chemical, Agriculture, and Industrial Enzymes Business; Gregory Powers Executive VP, Research and Development with Verenium since 2008; before joining Verenium, Dr. Powers was VP of Global Engineering at United Technologies Corporation's Carrier Division, (responsible for all engineering activity and strategy development supporting core innovation for the company), held various positions with the General Electric Company; Janet Roemer Executive VP Specialty Enzymes Business Unit; prior to joining Verenium, Ms. Roemer held several positions with BP Group, e.g. chief executive of a $1.7 billion North American chemical business; Has a substantial intellectual property position,

($, mio.): 2008: 57, (26) 2009: 49, (15) 2010: 52, (14) 2011: 61, (6.5) (FORM 10-K (Annual Report of 03/05/2012) For Vercipia: Net losses: $10,353,177 (2009), $6,251,816 (2008) [Vercipia 2010]	develop "low-cost, environmentally sound cellulosic ethanol production facilities in the US." Through a second deal, BP agreed to provide additionally $45 mil. and to form a JV with Verenium for construction of a cellulosic ethanol production plant near Tampa, FL., estimated to cost close to $400M; the JV has sought federal loan guarantees to cover 80 percent of that pricetag [Bigelow 2009a]	including more than 250 issued patents and more than 350 patent applications, as of March 2009 [Lane 2009g], in September 2010, Verenium completed a *sale of its ligno-cellulosic biofuels business to BP*, refocusing the company on its historical strength in enzyme development (FORM 10-K Annual Report of 03/05/2012).
Technology, Goals Business Model: Rely on alliances/JV; Has a strong IP position (patents), and holds exclusive rights to commercialize University of Florida technology for cellulosic ethanol production License-out and technology transfer Biomass from nearby sources Announced plans to build first commercial cellulosic ethanol plant in Highlands County, Florida, with a target	Emphasis on cellulosic ethanol; Verenium is a vertically-integrated firm in the biofuels industry through the combination of assets, technologies and personnel. Verenium claims to be the only company to offer fully integrated, end-to-end capabilities in pre-treatment, novel enzyme development, fermentation, engineering and project development [Childs 2007], "full range of 'field-to-pump' capabilities," has combination of enzyme discovery and enzyme evolution platforms. Enables conversion of nearly all of the sugars found in cellulosic biomass, including both five-carbon and six-carbon sugars, Uses a combination of microorganisms and specialty enzymes to convert up to 95% of available sugars in biomass feedstocks into fuel ethanol. The Jennings 1.4 million gallons-per-year (MGY) demonstration plant will draw on locally grown sugarcane bagasse and specially bred energy cane; in 2009 optimizing its 1.4 million-gallon-per-year demonstration-scale facility. Verenium has established the Jennings site as a permanent cellulosic ethanol "Center of Excellence," where future plant operators will be trained for roles in other commercial sites; the Jennings technology was transferred to BP's plants in Brazil after operations commenced at the US facility [Lane 2009c].	

capacity of up to 36 mil. gallons per year (MGY)	In addition, the Company's process technology has been licensed by Tokyo-based Marubeni Corp. and Tsukishima Kikai Co., Ltd. and has been incorporated into BioEthanol Japan's 1.4 million liter-per-year cellulosic ethanol plant in Osaka, Japan; Verenium and Marubeni are continuing to advance the commercialization of cellulosic ethanol projects utilizing Verenium's proprietary technology in Asia with the opening of a three million-liter-per-year plant in Saraburi, Thailand Verenium's goal: from a cost standpoint to be producing ethanol that is competitive with (today's) grain ethanol (~$2/gal) [Lane 2009g]. *Business Model Biofuels* [Lane 2009g]: Develop integrated solutions for the emerging cellulosic ethanol industry for use in production facilities that the firm owns and operates, individually or jointly with partners, as well as those of third-party licensees; develop novel, high-performance enzymes and to advance technology and process development capabilities, together with BP, at the pilot and demonstration-scale plants in Jennings, Louisiana, and the first planned commercial facility in Highlands County, Florida (36 MGY facility is expected to begin commercial production in 2012); exploit opportunities in the developing market for the production of cellulosic ethanol; incorporate scientific and engineering skills into the production facilities; achieve increased product sales and profit margins to support the future growth and profitability of the firm's portfolio of products sold directly by Verenium and by partners. The BP/Verenium JV Vercipia Biofuels JV planned the first commercial-scale cellulosic ethanol facility in Highlands County, Florida and expected to break ground on that site in 2010. The estimated construction cost for this 36 million gallon-per-year facility is between $250 and $300 million. Production from this plant was expected to begin in 2012 [BP 2009]. Another cost estimate: the project will cost close to $400 million, and the joint venture has sought federal loan guarantees to cover 80 percent of that pricetag [Bigelow 2009a; BP 2009]. Verenium will pursue distribution, research, marketing or production partnerships or alliances

*) From the firm's Web site if not stated otherwise by a reference;
firm's status: early 2010.

Shell's approach to biogasoline and biodiesel for transportation is essentially reflected by the interconnections with the German CHOREN Industries and Virent Energy Systems (Table I.85) and with HR BioPetroleum (Table I.89) focusing on biogasoline (including biodiesel).

Table I.85: Virent Energy Systems as a target of Shell regarding biogasoline and hydrogen. *)

Company (Foundation) Remarks	**Major Funding**	**CEO, Other Executives, Key Researchers; Technology Protection**
Virent Energy Systems, Inc. Madison, WI (2002) Founded by Dr. Randy Cortright and Prof. James Dumesic to commercialize the Aqueous Phase Reforming (APR) process, a technology the two invented and patented while at the University of Wisconsin-Madison. Started as Virent Energy Systems LLC. Employees: 2003: 5 2004: 12 2007: 54 2008: 68 2009: 75. Revenues: 2007: $4M (top line) 2008: $11M [Gillies 2009]. Having received prestigious government and industry recognition, Virent will seek to manufacture, and not just license technology,	2003: wins a competitive Advanced Technology Program (ATP) grant. 2004: grants awarded by the US Department of Energy to further advance Virent's hydrogen research; $2.2M in Federal funding, $550k in State funding, and raised $300k of private equity seed money. 2006: $2M grant from the US Department of Agriculture and Department of Energy for development of converting glycerol (co-product of biodiesel production) into renewable propylene glycol. 2007: $11M in federal funding including a second Advanced Technology Program grant from the National Institute of Standards and Technology. 2007: raised over $30M in equity financing. By 2009: raised about $70M, some	Lee Edwards President & CEO; Edwards brings 25 years of global energy leadership and petroleum industry experience to Virent; was President and CEO of BP Solar, a global solar technology provider; Edwards held a range of executive positions; Eric Apfelbach President and CEO; was replaced in 2008 by a person "likely to have experience in energy markets" (Edwards). Dr. Randy Cortright Founder & Chief Technical Officer; is experienced in the field of catalytic processing of biomass-derived feedstock into chemicals and fuels, is the co-inventor of Aqueous Phase Reforming (APR), the innovative pathway to biofuels and bioproducts used by the BioForming® technology platform. His background includes research and development, process design, start-up, and operations of large scale industrial catalytic processes at UOP LLC, a provider of petroleum and petrochemical process technologies. After leaving UOP, Dr. Cortright earned his PhD in Chemical Engineering, from the University of Wisconsin. In academia, he specialized in catalytic systems for the clean manufacturing of fuels and petrochemicals. He holds seven issued patents. 2008: Virent owns or holds the exclusive rights to 17 pending or issued patents in the US and 41 pending or issued patents in other countries; 6 issued US patents, 2 US patent applications, and 25 foreign patent and patent applications are the subject of exclusive and irrevocable licenses from the Wisconsin Alumni Research Foundation (WARF). Virent is the only source for liquid conversion of sugar-based feedstock into hydrogen and

boils down to "continuous improvement." In March 2010 Virent and Shell started up a demonstration plant at its facilities in Madison, Wisconsin, as part of the development deal that Virent and Shell started in 2008.	$40M from corporate investors and government grants and $30M in venture capital arms of Honda, Cargill and other companies; expected to raise $25 million to $40 million [Gardner 2009, LaMonica 2009]	alkanes. Hydrogen is also a key interest of Shell.
Technology, Goals Technology demarcation point in terms of cost competitiveness: crude oil will remain above the $60 per barrel mark. *Business Model:* Biorefinery orientation for own production; technology licensing; retrofitting existing ethanol plants to Virent's process. Partnering with major energy (Shell) and agricultural (Cargill – supply chain) companies; is building a 10,000 gallon per year plant, to build a 100 million gallon per year plant by 2015 [Gardner 2009].	Started to commercialize the Aqueous Phase Reforming (APR) process by which hydrogen is generated from sugar. Technology has evolved into the BioForming™ process, which enables the production of renewable products: Liquid Biofuels ("biogasoline") Chemicals (e.g. propylene glycol) Fuel Gases (hydrogen – H_2) The BioForming™ process is thermochemical. 2006: expanding the BioForming technology to convert plant sugars into hydrocarbon molecules ("biogasoline"). Virent's BioForming *platform technology* employs *low temperature* aqueous-phase reforming and solid state catalyst (rather than microbes) to convert plant sugars into hydrocarbon molecules. Focuses on catalyst composition, reactor design, and reaction conditions. The BioForming process can economically utilize many types of biomass and carbohydrates from cellulosic and biomass-derived feedstock: Glycerol (also named glycerin; by-product of biodiesel production from vegetable or animal oil); Glucose and Sucrose (from sugar crops); Starches (glucose containing polysaccharides); Long-chained glucose contained in cellulose (plant cell walls); C5 and C6 sugars, such as xylose, arabinose, and glucose contained in hemicellulose (part of the protective covering around cellulose). Is already processing mixed sugar streams. Looked at what is a long-term, sustainable way to get biomass feedstock into the process. Thinks sugarcane and the plant that the sugar cane comes from are going to be the cheapest, most scalable feedstock	

	on a global basis for quite a while. Virent's pitch is explicitly *anti-ethanol*. Around a quarter of existing ethanol production capacity is of the wet mill type which could be converted to low temperature catalytic BioForming technology for biogasoline production at significantly lower capital costs than would be needed for new Virent plants. Hopes to license its fuel process [Gardner 2009]; is building its own full-scale production refinery to reap maximum margin from end products and gain a direct feedback loop for improvement of its process. Finally, in 2007 Shell Hydrogen LLC and Virent set up a five-year joint development agreement to develop further and commercialize Virent's BioForming technology platform for hydrogen production (worldwide market for distributed and centralized hydrogen is estimated at approximately 45 million tons each year) [Shell 2007].

*) From the firm's Web site if not stated otherwise by a reference; firm's status: end of 2009; for more information cf. "Virent-Story-Through-2011.pdf" provided on its Web site.

Virent's business model is not focused on biofuels alone, but emphasizes a "*biorefinery model*" which is directed toward biofuels, biochemicals and biomaterials [Runge 2006:849-866] and adds a specialty (fuel cell grade) hydrogen gas which finds also applications in semiconductor manufacturing, ammonia production or in gas chromatography analytics (as a carrier gas). Correspondingly, Virent has a joint agreement with Shell Hydrogen, a subsidiary of Shell Oil Co., to produce hydrogen from renewable glycerol and sugar-based products [Vanden Plas 2007].

The development of Virent's BioForming® technology platform is supported through strategic partners including Cargill, Coca-Cola, Honda and Shell, as well as 80 employees (by 2013) based in Madison, Wisconsin. Virent recently signed agreements with the Coca-Cola Company to provide the technology and biobased chemical Paraxylene to create 100 percent plant-based, renewable PlantBottle® packaging.

The company has received several grants from the US Departments of Commerce, Energy and Agriculture and has been recognized with many honors, including the World Economic Forum Technology Pioneer award and the EPA's Presidential Green Chemistry Challenge Award.

The fundamental process Virent is relying on is one hundred years old and was developed in Germany. In the early 1900s, research on the extraction of chemicals and energy from coal in Germany was focused on two primary approaches [Runge 2006:424-425].

- Direct liquefaction under pressure by converting coal into liquids with the help of hydrogen and heavy oils (Bergius process; Coal-to-Liquids, CtL), and
- Indirect liquefaction by first gasifying coal and then converting the resulting gas into liquids through the process of the Fischer-Tropsch (FT) synthesis (Gas-to-Liquids (GtL)). The process converts a mixture of carbon monoxide (CO) and hydrogen (H_2) called "synthesis gas" ("syngas") to liquid hydrocarbons using, for instance, iron and cobalt catalysts. A special, currently very economic route of the GtL process uses *natural gas*.

Over time the FT process has been subjected continuously to modifications, became an established technology and is still applied on a large scale. Currently, two companies have commercialized their FT technology, Shell (in Malaysia) and Sasol (several plants in South Africa), using *natural gas* and *coal* as feedstock to produce the syngas, respectively. In South Africa CtL is used on a large scale to produce automotive fuels from coal.

In the US Bioconversion Technology, LLC, (BCT) founded in 2003 and essentially led by Robert E. "Bud" Klepper and Kenneth L. Klepper, started with an emphasis on gasification ("anaerobic thermal conversion"). Bud Klepper was the inventor (and patent holder) of a gasification process capable of processing 25-35 tons per day of coal to *synthesis gas*. Bud Klepper's *engineering company* generated syngas from coal, coal slurry, coal fines, but also other biomass feedstock. The gasification technology is called the Klepper Pyrolytic Steam Reforming Gasifier (PSRG) with a Staged Temperature Reaction Process (STRP). A separate Klepper Ethanol Reactor catalytically converts syngas into ethanol.

According to a comparative evaluation of such systems the Klepper system has the highest energy efficiency of any system and the highest syngas energy content of any thermochemical biomass conversion system that has been developed for biomass inputs of less than 1,000 tons/day [Green Car Congress 2005; Green Car Congress 2007b]. BCT created revenues by licensing its technology globally.

BCT was later transformed into Kergy, Inc. and then Range Fuels by Khosla Ventures, LLC (Table I.99). And the BCT original thermal converter was upgraded to a so-called K2 modular system [Rapier 2006; Schuetzle et al. 2007]. Obviously, Kergy also looked to "optimize an existing and a novel catalyst or catalyst combinations for the conversion of syngas to alcohols." [Reisch 2006]

Gasification of biomass to produce biofuels as a well known approach for more than twenty years is just a capital-intensive process that has the problem of competing against lower cost (but unsustainable) gasification options [NNFCC 2009, Schuetzle et al. 2007; Rapier 2006].

Similarly to the CtL process, and in line with current biofuel efforts, Bergius also succeeded in 1930 in Germany to convert wood into cellulose and treating the biomass with hydrochloric acid to get (biobased) synthetic sugar via saccharification and fermentation. The product was similar to beet sugar in taste and application. And in 1938 building a plant was started that should become operational in 1939 to convert 400,000 m3 of wood per year into sugar [Runge 2006:566].

The Israel-based recent startup HCL CleanTech (now named Verdia, Table I.99) acknowledges Bergius when they wrote on their Web site: "Innovative HCL Recovery Process Revolutionizes the 1930 Bergius Technology for Converting Cellulosic Materials into Fermentable Sugars." HCL CleanTech "has developed a proprietary full HCL (hydrochloric acid) recovery process, which makes an old, industrially-proven German cellulosic to fermentable sugars and ethanol process economically very attractive."

Similarly in the US, Arkenol/BlueFire further developed the acid hydrolysis process (here with concentrated sulfuric acid) [BlueFire Ethanol 2004, Klann 2007] to make it economically viable through the use of new technology like flash fermentation and membrane distillation and purification, modern control methods, and newer materials of construction and focusing on a special type of biomass.

BlueFire Ethanol Fuels (Table I.86) follows a "veterans approach" and biorefinery orientation [Runge 2006:849-866] with a management team of people having 25+ years of experience in project finance, technology commercialization and project development. BlueFire's *biorefineries will be located near markets with high demand for ethanol* and will *use locally available biomass*. It is a cellulose-to-ethanol company with demonstrated production of ethanol from urban trash (post-sorted municipal solid waste – MSW), rice and wheat straws, wood waste and other agricultural residues. By weight, post-sorted MSW is more than 70 percent cellulose.

Bluefire's favorite input is municipal waste, because it can build its refineries on landfills, cutting feedstock transportation costs and using methane emitted from decomposing waste to help the plant generate 70 percent of its own electricity.

The BlueFire process licensed from Arkenol uses naturally-occurring yeast, which has been specifically cultured by a proprietary method to ferment mixed (C6 and C5) sugars (actually, NREL developed *rec. Z. mobilis* (licensed by BlueFire) and *S. cereviscae* yeast) to produce ethanol at 95 percent [BlueFire 2004]).

Arkenol can license its technology to qualified entities for their own project development. However, Arkenol prefers to offer more than just a license. With its team members, Arkenol can provide turnkey engineering, procurement, construction, and operations services. Arkenol will work with developers around the world to license its technology and on an individual project, a corporate or a regional basis. Notably, BlueFire and Arkenol share the same president and CEO (Table I.86).

> Decision-making and action in new biofuels firms is often by a management team with members providing large "science, technology, management and policy experiences and connections."

An entrepreneurial "*veterans approach*" can often be assumed to be founded and run by an "*old boys network*," which is an exclusive informal network linking members of a social class or profession or organization in order to provide connections and information and favors (especially in business or politics), often indicated by current or past affiliations to the same organization.

Table I.86: BlueFire Ethanol Fuels showing an engineering-type approach to innovation and entrepreneurship. *)

Company (Foundation) Remarks	**Major Funding**	**CEO, Other Executives, Key Researchers; Technology Protection and Position**
BlueFire Ethanol Fuels, Inc. Irvine, CA (2006) New name: **BlueFire Renewables, Inc** An over-the-counter (OCT) publicly traded stock company. Revenues (from consulting, DOE Grant/ Reimbursement, in 2009 selling ethanol to Solazyme, Inc.): 2007: $49,000, 2008: $1.075 mil., 2009: $4.32 mil. 2010: $669k [Wikinvest 2009]. Employees 2008: 12 [CI 2009], 2011: 6 full time employees and three part time employees.	2007: Securities Purchase Agreement, Quercus Trust acquired shares of common stock and warrants for total proceeds of $15M, strategic investment (Quercus Trust shares of voting common stock 34.4%); 2007: Department of Energy (DOE) provides grant of $40M for the first of two stages of its second US commercial ethanol production using cellulosic wastes diverted from landfills in Southern California (ca. 17 million gallons per year); has been invited	Experienced Management Team specializing in project finance, technology commercialization and project development: Arnold R. Klann – Chairman - President – CEO; CEO for BlueFire Ethanol, Arkenol, Inc., and ARK Energy; prior to founding ARK Energy, he launched three businesses and managed complex teams for project development and operation; Arkenol is a technology and project development company; ca. 30 years experience in corporate management, project finance, engineering, design, construction, start-up, environmental permitting, driving force behind the research and development effort leading to the commercialization of the Arkenol technology; BS (electrical engineering); (shares of voting common stock 49%); John E. Cuzens – SVP CTO; has been with ARK Energy and Arkenol and is the co-inventor of seven of Arkenol's eight US foundation patents for the conversion of cellulosic materials into fermentable sugar products using a modified strong acid hydrolysis process, experience of 20+ years of project management, experience punctuated frequently with engineering or R&D management assignments; B.S. Chemical Engineering; (shares of voting common stock 6.1%) Necy Sumait – SVP – Director; Senior Vice President for BlueFire Ethanol and for Arkenol, Inc., background in the development of energy projects from inception through financial closing, commissioning, and operations. She has broad experience in siting, regulatory compliance, governmental and community relations and legislative affairs (of federal, state and local agencies); (shares of

Business plan developed in conjunction with Booz Allen Hamilton, in particular, use waste to produce ethanol and put the production near to end users.	to submit a formal application for a DOE loan guarantee to assist in the financing of ethanol production facilities; approved for California Energy Commission Grant of $1M 2009: DOE increased funding to $81.1M for Phase II construction of the cellulosic ethanol biorefinery planned for Fulton, MS, in addition to the previously announced Phase I funding of ca. $7M for development of the Fulton plant (goal 19 mil. gallons) [Wikinvest 2009]; has received a $3.8 mil. reimbursement from the US DOE to be used for preconstruction activities for its second planned biorefinery in Fulton, MS. NOTE: In Dec. 2013 the firm	voting common stock 6.2%) William Davis – VP Project Management; 30+ years of experience; he has served as advisor to the Governor of California for energy conservation and renewable energy policy; additionally he has worked for several Fortune 500 companies managing their energy development activities. Notably: William A. Farone – Technical Advisor for Arkenol, Inc.; 30+ years of technical research experience in the alternative energy, chemical and biotechnology industries, managed the Arkenol Technology Center focusing on the development of new chemical technology and biotechnology, and the optimization of existing chemical processes. Dr. Farone is the chief scientist and technical expert for all equipment application, feedstock processing and product development activities at the Arkenol Acid Hydrolysis Pilot Plant and co-inventor of Arkenol's US foundation patents for the conversion of cellulosic materials into fermentable sugar products. BlueFire Ethanol Inc. has formed a technology development services agreement with William Farone. Farone is president and CEO of Applied Power Concepts Inc. (APC), a producer of higher-value sugar-based chemicals. Under the new agreement, Farone and Applied Power Concepts Inc. will work with BlueFire to continue advancement of the technology. BlueFire will conduct development work at the APC facility [Austin 2009]. *Staff and majority shareholders have been involved in technology development since 1992* as Arkenol. Research and development work completed, patent protections in place, pilot-scale process successfully completed, bioethanol commercial plant projects currently in various stages of development. Bluefire is the *exclusive North American licensee* of "Arkenol Technology," may also utilize certain biorefinery related rights, assets, work-product, intellectual property and other know-how related to nineteen ethanol project opportunities originally developed by ARK Energy; from 1994-2000, a test pilot biorefinery plant was built and operated by Arkenol in Orange (CA) to test the effectiveness of the Arkenol Technology; results fed into another test pilot biorefinery plant in Izumi, Japan, built

<table>
<tr>
<td></td>
<td>got notice from the DOE indicating that the DOE would no longer provide funding under the firm's DOE grant for developing its cellulosic waste facility in Fulton, due to inability to comply with certain deadlines for informing the DOE on future financing the Fulton Project a)</td>
<td>and operated by engineering firm JGC Corporation (Arkenol retained the rights to the Arkenol Technology while the operations of the facility were controlled by JGC);

in 2002 JGC was awarded a contract by the New Energy and Industrial Technology Development Organization (NEDO) of the Japanese Government for the implementation and commercialization of cellulosic ethanol production technology; the Izumi facility enabled to verify Arkenol's technology as being commercially viable via an unrelated third party, setting the stage for the rollout of this technology in the US;

design and engineering of BlueFire's facilities in North America by established engineering firms can rely on use the Arkenol Technology and utilize JGC's operations knowledge [SEC 2009, BlueFire Ethanol 2004]; JGC exclusive licensee for SE Asia</td>
</tr>
<tr>
<td>Technology, Goals, Strategy

Vision:
Build/ develop biorefineries focused initially on bioethanol;

Mission:
Become a world-class producer and supplier of renewable liquid fuels by the production of ethanol from opportunistic sources;

Create a business that develops and owns state-of-the-art ethanol biorefineries that are the lowest cost producers of ethanol.</td>
<td colspan="2">The process [Klann 2007:6] is cellulose/hemicellulose to mixed sugars using Arkenol's concentrated (sulfuric) acid hydrolysis to provide ethanol and open routes to other biobased products, such as biobutanol, to become a biorefinery.
It is an incremental innovation ("improvement") of an old technology developed in Germany in 1930 ("Bergius process") centered on input selection and plant and process engineering.

Claimed: The only cellulose-to-ethanol company worldwide with demonstrated production of ethanol from urban trash (post-sorted MSW), but also rice and wheat straws, wood waste and other agricultural residues;
Bluefire can use the landfill to power its refining process; combined with the use of lignin, a by-product from the Arkenol process; the capture of methane from a nearby landfill allows Bluefire's refineries to be 70% self-powered.

Strategy:
Equity and debt funding for the BlueFire projects will be done on the project level, not the Corporate level, which means little or no dilution to current shareholders [Klann 2007].

Targeting specific geographic areas with the highest demand for ethanol fuels (e.g. California) and available feedstock supply; BlueFire will be project lead and equity owner in projects utilizing their technology;

Bluefire positions its production facilities right on landfills, in order to exploit certain cost advantages: input is from municipal waste cuts feedstock transportation costs and can use methane emitted from decomposing waste to help the plant generate 70% of its own electricity [Wikinvest 2009];
as a broad spectrum of biomass can be used for the process, targeted regions are large urban areas where waste disposal is a problem and landfill</td>
</tr>
</table>

Projections of 20 biorefineries to be in commercial production within the next 7 years in North America.	disposal alternatives are important and areas adjacent to National Forests where there is a pressing (and long-term) need to dispose of dead or diseased vegetation. The plan is, concurrent with the development of its own facilities, to deploy the technology, form associations with a group of companies selected on the basis of well-defined criteria (Joint Venture Development Partnerships with qualified and experienced regional developers throughout the US and Canada); that is the desire to create a portfolio of strategic partnerships. BlueFire's product selling revenue orientation has materialized by an alliance with algal biofuel company Solazyme, Inc. (Table I.90) which purchases and tests BlueFire sugars. Solazyme directly feeds the algae sugar rather than relying on light [Wellsphere 2009].
Risk, Risk Mitigation	Bluefire competes with many other biofuels manufacturers, though its closest competitors are cellulosic (and corn) ethanol companies. Targeting a biorefinery model would allow lifting dependency on just one offering in the future – bioethanol – through downstream processing. BlueFire is well beyond the research and development stage of its business plan and the *technology has been in actual production for over five years* in NEDO's pilot plant in Japan. Without government grants Bluefire's facilities would have no hope of being profitable, illustrating the company's dependence on government aid to be feasible and legislative support to achieve profitability: Federal and state governmental funding to build its refinery helps to offset the hefty installation cost of $5.00 per gallon for a 55 million gallon per year facility. Without such funding, it would be very difficult for Bluefire to make ethanol cost-competitive with other fuels, much less turn a profit [Wikinvest 2009]. According to BlueFire's plan sub $1.00/gal production costs are feasible for facility 2 through 20 [Klann 2007]. Further risk mitigation would require a shift in the auto industry to accommodate for higher blending (cf. the E15 thrust) and/or to build the missing infrastructure for still relatively few FFVs in the US that would allow E85. "Bluefire is Dependent on Legislative Support to Achieve Profitability." [Wikinvest 2009]

*) From the firm's Web site if not stated otherwise by a reference; firm's status: end of 2009;
a) cf. also Jim Lane's article "Being solid and liquid: The screwy, upside-down world of renewable fuels financing" in Biofuels Digest of Oct. 6, 2013.

By the end of November 2011, BlueFire Renewables formed a wholly owned subsidiary, SucreSource, LLC, that will manufacture cellulosic sugars from biomass and will use BlueFire's patented Arkenol Acid Hydrolysis Technology. SucreSource will capitalize on BlueFire's existing process design packages, providing either a 34,000 tons per year or 163,000 tons per year source of cellulosic sugars.

SucreSource was riding a wave shifting the paradigm for energy sources and was created to meet the market's increasing demand for cellulosic sugars not just for biofuels, but also for bioplastic and specialty chemical markets (A.1.1.6). BlueFire claims to be the only cellulose-to-fuel company worldwide with demonstrated production of biofuels from urban trash (post-sorted MSW), rice and wheat straws, wood waste and other agricultural residues. BlueFire received an increase to its Grant totaling $88 million under the American Recovery and Reinvestment Act in December of 2009 [Lane 2011h].

SucreSource is actively pursuing partnership opportunities to deploy the technology. Early in 2012 SucreSource signed agreements with GS Caltex – a joint venture between GS Holdings and Chevron, and a leading Korea-based petroleum company to build a cellulose-to-sugar plant in Korea.

The facility will process 2 tons of construction and demolition debris per day into cellulosic sugar, which will be converted into a high value chemical by GS Caltex's proprietary technology. The facility will be owned and operated by GS Caltex with SucreSource providing the process design package, equipment procurement and technical and engineering support. If the initial facility is validated, SucreSource will work with GS Caltex to develop and build larger commercial scale facilities in Korea and throughout the world [Green Car Congress 2012].

If in the well established Fischer-Tropsch (FT) process syngas is obtained from biomass, the process is referred to as a BtL (Biomass-to-Liquid) which has been developed and was utilized by CHOREN Industries GmbH in Germany to produce SunFuel für Otto engines und SunDiesel® for Diesel engines [Runge 2006:254-255], the last one being used as a blend in Shell's premium "V Power Diesel" fuel in Germany and Austria. CHOREN, founded in 1998, ultimately acted as a holding.

The biomass spectrum for BtL includes wood chips from forest timber or from rapid turnover plantations, straw briquettes, energy, crops or recycled wood (from houses). According to its Web-site CHOREN became a group with several subsidiaries. The number of employees of the CHOREN Group was around 300. Some subsidiaries were already profitable in 2009, but overall CHOREN remained in the red (Capital employed: > €180M with €100M invested in the Beta-Plant).

Revenues for 2007 of more than €4M were reported [Lachmann 2007]. About 20 percent of the Beta-Plant investment was by the federal government and local state government [Wuttke 2008]. As an LLC CHOREN had eight partners including two corporate investors (automotive firms VW and Daimler) with minority stakes and a private businessman with a majority stake [Lachmann 2007]. The key founder Bodo Wolf left CHOREN as a partner in 2008, but stayed connected as a consultant to the firm [Wuttke 2008]. Shell Deutschland Oil GmbH had sold its shares in CHOREN Industries GmbH to all the remaining CHOREN shareholders.

By 2009 CHOREN produced 15,000 tons/year biodiesel (ca. 4.8 million gallons) and planed for a €800 million plant for 200,000 tons/a ("Sigma-Plant") which would correspond to 0.7 percent of the then demand of diesel fuel in Germany where diesel is much more used for cars than in the US. However, the prerequisite for building was that after 2015 in Germany biofuels continue to get a tax exempt [Kempkens 2009].

Simultaneously, there was a European research project OPTFUEL comprising European car manufacturers and led by VW (with a minority stake in CHOREN) to drive large-scale production of 2nd generation biofuels which will be based on CHOREN's Carbo-V®-process. A part of the research funds would be used for the development of economically and ecologically viable concepts for supplying a large-scale plant with biomass (cf. Figure I.173), such as CHOREN's Sigma-Plant.

Apart from own production, product selling and licensing CHOREN's business (revenue) model was based on its proprietary Carbo-V gasification process and on engineering core competence which includes a wide range of services for design, installation and operation for Carbo-V biomass combined heat and power plants in line with the common service offerings of an *engineering company*, specializing in *mechanical engineering and plant engineering* – similar to BlueFire's offerings.

- Concept development for the construction of industrial-scale production units including or integration of the process into existing energy supply configurations
- Advice to and support of companies in organizing biomass supply concepts and the relevant processing technology, or even organization of the complete biomass management, including the securing of long-term raw material provision as main supplier
- Process engineering design for the project
- Permit engineering for the entire plant
- Basic and detail engineering for the Carbo-V section
- Design, construction, delivery and installation of the main process equipment
- Assistance during installation, start-up and commissioning, training of operating personnel
- Technology transfer.

In 2010 CHOREN and the French group CNIM (Constructions Industrielles de la Méditerranée SA) signed an EPC agreement (engineering, procurement and construction). The agreement covered the design and construction of a synthesis gas production facility using biomass feedstocks, whereas CHOREN provides extensive engineering and other technical services [CHOREN 2010].

For CHOREN *about half of the production cost resulted from type and procurement of biomass* (Figure I.171, Figure I.173). A strategic option would be *production in countries where biomass is cheaper*. The other alternative, type of biomass, can be related to the technical bottleneck of the process: The actually used biomass provides too little hydrogen to the syngas of the FT process [Lachmann 2007].

Correspondingly, in November 2009 CHOREN USA LLC reported on a two-year bio-energy cooperative project with the energy crop company Ceres, Inc. which was funded in part by a grant from the US Department of Energy and Agriculture. Ceres would evaluate the *composition* of a broad range of switchgrass and willow plants, and provide biomass samples to CHOREN for thermochemical processing.

The results should be used to identify the most relevant compositional traits and later, to select the plants and traits that improve conversion and maximize fuel yields. Additionally, it was said that CHOREN USA will use the results of the work to help the company selecting its initial US project site. But generally, the C/H proportion from biomass is inferior to that from fossil energy sources. This suggests adding hydrogen from other renewable sources which would give Virent's process a flavor. Further questions for the gasification of biomass were the use of dry versus wet biomass and consistency of the biomass composition.

> Whether it is cosmetics, chemicals or other applications, for the relevant natural product, such as chitosan (cf. Heppe Medical Chitosan GmbH, ch. 2.1.2.4) or here biomass using "biological" ("renewable") stuff as raw material, intermediate or other input for technical processes, the *composition* of the natural, renewable material and also the material's *composition consistency*, which may be sourced from different regions and/or at different seasons, are crucial for technical processes and may interfere with scale-up efforts and quality and purity requirements.

Among other factors, the above issues were also reasons why CHOREN went bankrupt in Germany by mid of 2011. Most of the 290 employees lost their jobs [Rapier 2011; Reuter 2011]. In early 2012 the giant German engineering conglomerate Linde (Linde Engineering Dresden) took over the biomass business of CHOREN saving in Germany 65 jobs [DAPD 2012] and took over the Carbo-V® Technology. More details are given by Rapier [2011]:

In 2009, the author dealing with what happened to CHOREN [Rapier 2011] accepted a job as Chief Technology Officer for the man who was at that time the largest shareholder of CHOREN (which was not his only investment). Both persons shared a belief that oil prices are inevitably headed much higher, and in that case they both believed the CHOREN process would be ultimately economically viable.

In 2009 CHOREN started commissioning the gasification section for its process. That was the beginning of a long process of running for a period of time, and then shutting down and making adjustments. One of the biggest challenges with the gasification was that there was no blueprint; nobody had run a gasifier like this at this scale on a biomass feedstock. Hence, the plant had to work through many new technological challenges, and with a staff of 290 employees, the time it took to work through the issues was costly.

As CHOREN had elected to forgo government funding, private investors have borne the development costs over the past few years. Several of those investors – including Shell – exited at various points due to the time and cost it was taking to work out the technical issues. Ultimately, the largest shareholder was largely funding the ongoing operations of CHOREN from his pocket.

While the plant made good progress, commissioning took far longer than expected. Rapier had in fact warned that it would take at least a year to start up the plant once it was mechanically complete, but it was taking even longer than expected. It finally got to the point that all investors decided to stop funding development, because all of the technical bugs had not been worked out. *It was not that there were any technical show-stoppers, it was just that the timing of how long it would take to work through the issues was uncertain.*

CHOREN Industries and its four engineering-oriented founders, in particular, Bodo Wolf, provide a representative case for a rise of entrepreneurial spirit and private initiative in 1990 almost immediately after the Berlin Wall was teared down and the former German Democratic Republic (GDR) ceased to exist.

The four persons immediately initiated self-employyment by setting up an engineering office [Ahrens 2005] which developed into CHOREN Technologies GmbH. The team started in the biofuels field years before it became a huge wave, with a pilot plant ("Alpha") to test their Carbo-V process in 1998 and a pilot production plant ("Beta") in 2003. Bodo Wolf, after having left CHOREN in 2008, acted as an angel investor for the German startup SunCoal GmbH (B.2) which focuses on BtC – Biomass-to-Coal.

With regard to the FT (GtL) and CtL processes the first to benefit from them was Germany during World War II. Germany, rich in hard and brown coal, built enormous liquefaction plants. After the war, the GDR, to secure its resources, operated the German Fuel Institute (DBI), a center for coal processing in the mining town of Freiberg.

After the German Reunification the DBI left behind a group of highly specialized scientists and engineers who were highly specialized in a field abandoned by the West (after oil-based industries took over coal-based ones) that offered a key to a more efficient way of using biomass. The key persons of the DBI were the engineer Bodo Wolf, Eckard Dinjus (see below) and Bernd Meyer, each of them having their own technological approach to the field.

Dinjus became professor and head of a "BtL Division" (later Institute for Chemical Engineering - Division of Physicochemical Processes) at the Karlruhe Research Center (Forschunszentrum Karlsruhe – FZK, a German National Research Center; now part of the Karlsruhe Institute of Technology – KIT). Meyer became a director at the Technical University of Freiberg. The two professors would rather see more research done on BtL before it makes the transition to general industrial use. Indeed, when they testified as experts at an official hearing, they advised the government against providing CHOREN with loan guarantees [Wüst 2007].

This particular constellation for the BtL-case represents a further example of the key issues of technology transfer, namely that *effective technology transfer of complex industrial processes depends highly on "people transfer," knowledge and experience (know how) bound to people* ("subject matter expert" – SME – providing "tacit technology"), as has been shown by several historic cases.

- Manufacturing of porcelain in the 1700s: after defeating Saxony in a war, winner Prussia transferred many of the Meissen (near Dresden, Saxony) workmen of the porcelain production site to Berlin and Berlin afterwards became a famous competitive manufactory [Runge 2006:405].
- In 1927 US Standard Oil (now ExxonMobil) and German I.G. Farben (see [Runge 2006:271-272]) agreed to cooperate technically in the further development of synthetic oil and fuel by coal liquefaction via the Bergius process [Runge 2006:424]. Standard Oil and I.G. Farben set up the "Joint American Study Company" (JASCO) to work on synthetic oil. JASCO should push commercialization of technologies. Therefore, both partners sent scientific and technical employees as the "carriers of people-based knowledge" into JASCO.
  After WW II captured German scientists continued to work on synthetic fuels in the US.
- The biggest corresponding know-how and technology transfer occurred in the 20th century after WW II when the US and Soviet Union each caught the top 100+ German scientists and engineers who developed the V2 missiles or the world's first jet fighter, the Messerschmidt 262 jet, and moved them to their respective countries. No such technologies existed in the world outside Germany and the German specialists rebuilt the missiles and jets to educate and train Americans or Russians in the technologies [Runge 2006:405]. And German scientists transferred to the USSR also contributed to building the Russian atomic bomb (ch. 2.1.2.8).

Collecting biomass and getting enough of it in one place to make a difference is a key problem in the biomass world. "Trucking costs can become exorbitant. You want to preprocess it at the farm and then ship a high-density, high-energy intermediate to processing plants." [Jonietz 2007]

Focusing particularly on the issues of economical collection, procurement and transportation, the first part of the biofuel value system (Figure I.173), Dinjus of the KIT developed a BtL process called "Bioliq," through stepwise projects together with the engineering firm Lurgi GmbH (now a subsidiary of Air Liquide) as an industry partner. By November 2009 the third of four steps for a pilot plant had been funded with €11 million by the German federal and related state governments. Bioliq aims at the production of biofuels (and industrial chemicals like propylene, ethylene or acrylic acid; Figure I.174) via the FT and methanol/dimethylether (DME) route [Runge 2006:851-852, 858]. In May 2011 there was the topping-out ceremony for the related €60 million Bioliq plant.

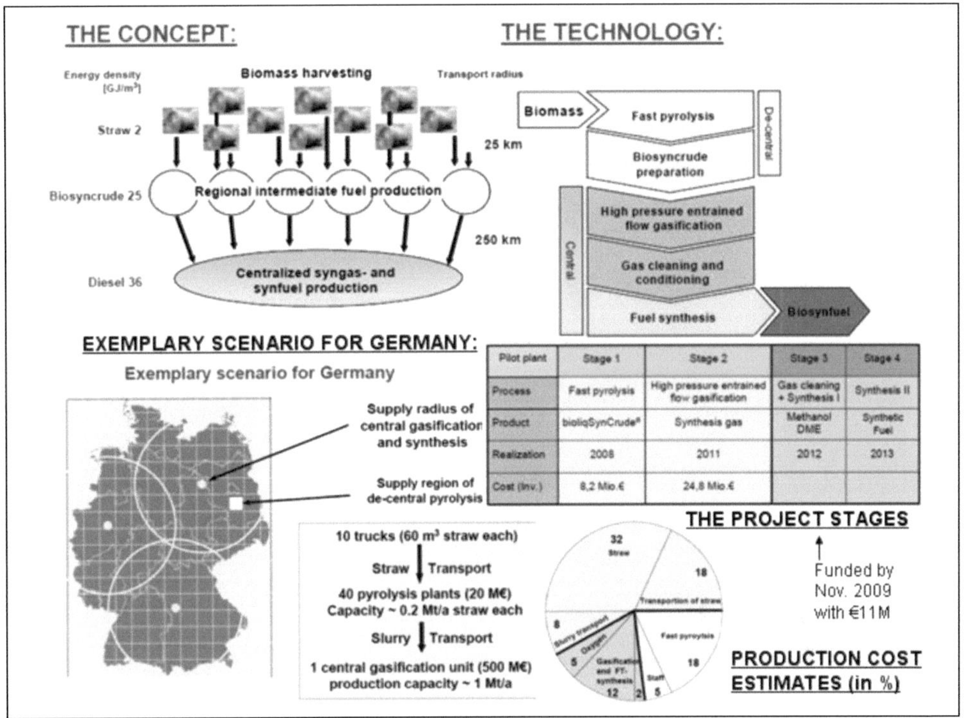

Pilot plant	Stage 1	Stage 2	Stage 3	Stage 4
Process	Fast pyrolysis	High pressure entrained flow gasification	Gas cleaning + Synthesis I	Synthesis II
Product	bioliqSynCrude®	Synthesis gas	Methanol DME	Synthetic Fuel
Realization	2008	2011	2012	2013
Cost (Inv.)	8,2 Mio.€	24,8 Mio.€		

Figure I.173: The KIT – Lurgi Biofuel BtL Project Bioliq emphasizing the input – biomass collection and transportation and proportions of overall production cost.

The overall *systems-oriented* project focuses on the necessary infrastructure for raw material (straw) collection through a *decentralized* stepwise approach. Raw material is collected and transported over short distances (less than 25 km) to mini-plants. Here a "fast pyrolysis" converts the biomass to a mixture of an "oil" and a "coke" which are combined into a suspension.

This "slurry" ("*biosyncrude*") is of such a "high energy density" that its transportation to the actual BtL plant is economically viable. The BtL plant then provides "syngas" which can be further processed [Dinjus et al. 2008]. Transportation of biomass beyond short distances would quickly cost more BTUs (British thermal units) than the biomass fuel would yield. Having or building an infrastructure to collect and store the biomass, hence, is *the* serious cost concern.

Systems thinking concerning biofuels and biomass and "commercially successful" entry into the transportation area brings about further considerations – the *economics will change* upon success of the biofuels firms.

> Rising biofuel production, or the burning of biomass to generate electricity, will drive up demand and prices for the raw material, just as production of corn ethanol helped raise the price of that crop. Biomass is cheap right now because no one wants it. As firms want it, it will become more expensive.
>
> Furthermore, and more important, the laws of supply and demand mean that replacing a significant amount of (petro-)gasoline with biofuels would drastically lower the demand for gasoline. That, in turn, would cause the price of gasoline to plunge, making biofuels less competitive.

The only notable activities concerning cellulosic ethanol in Germany are observed for Süd-Chemie (Box I.3) which was recently acquired by Swiss specialty chemicals firm Clariant. It is a company with €1.225 billion sales (in 2010; employs some 6,500 people) operating on a worldwide scale. Its business units, Functional Materials and Catalysis & Energy, offer products and technical solutions for numerous industrial sectors to facilitate effective use of resources in customer value chains. The Catalysts Division offers solutions for the chemical, petrochemical and refinery industries, for energy storage and hydrogen production, as well as off-gas purification

In 2009 Süd-Chemie opened a pilot plant at the firm's Research Center in Munich supported by the Bavarian Minister of Economics, Infrastructure, Transport and Technology and the EU to produce bioethanol from lignocellulosic biomass. The pilot plant will be using cereal straw to manufacture up to two tons of bioethanol fuel annually.

The process developed by Süd-Chemie and German gases and engineering giant Linde allows biofuels, such as ethanol, to be extracted from biomass, for instance wheat straw or maize (corn) straw, with the aid of enzymes created using biotechnological methods. The partners in this alliance offer complementary competencies. Whereas Süd-Chemie´s expertise lies in the sectors of biocatalysis and bioprocess engineering, Linde's subsidiary, Linde-KCA-Dresden, offers extensive experience in implementing chemical and biotechnological processes on a commercial scale [Süd-Chemie 2009] – and now owns also the rests of CHOREN and its Carbo-V® Technology.

Süd-Chemie´s sunliquid® process uses not only the cellulose contained in plants, but also the so-called hemicellulose. Both can be converted into ethanol, making it possible to increase ethanol production by up to fifty percent compared with conventional technology.

Planned for 2012 a large-scale demonstration plant was started by mid of 2010, located in the immediate vicinity of the new Bavarian BioCampus in Straubing, that will produce up to 2,000 metric tons per year (670,000 US gallons) of bioethanol fuel from agricultural waste, such as wheat straw or maize (corn) straw, bagasse from sugar cane or so-called energy crops [Süd-Chemie 2010].

In July 2012 Clariant inaugurated the new cellulosic ethanol pilot. The plant – the biggest of its kind in Germany – will start producing produce around 1,000 tons/year of cellulosic ethanol, using around 4,500 tons/year of locally sourced agricultural waste as a feedstock. Clariant said studies show Germany potentially has around 22 million tons of straw that could be used for energy production without compromising essential soil regeneration, which would be sufficient to cover around 25 percent of Germany's current gasoline requirements.

Süd-Chemie´s demonstration plant will represent a scaled-down version of the entire integrated manufacturing process. The total project had a volume of altogether €28 million ($35 million) and comprised an investment volume of some €16 million and accompanying research projects amounting to approximately €12 million. These and additional related research projects were subsidized by the Bavarian State Government and Germany's Federal Ministry of Education and Research (BMBF) with approximately €5 million respectively [Süd-Chemie 2010]. The pilot plant represents the interim stage necessary prior to erecting production plants with annual capacities of 50,000–150,000 tons of bioethanol.

The current situation at Süd-Chemie shows that CAPEX (capital expenditures) is radically lower than for other bioethanol plants. Their system is designed to ultimately cost less than $100 million for a 20 million gallons (60,000 ton) plant, and is expected to have OPEX (operating expenditures) that is competitive with first generation (corn) ethanol. The company is on the road transforming the economics of cellulosic ethanol, to compete at parity with gasoline and was expected to commence licensing in 2012 [Lane 2012p].

According to Süd-Chemie and its Unique Selling Proposition (USP) "We are one of the few companies worldwide that have process development and enzyme development under one roof. We are independent from enzyme supply, because we make our own during the process itself, using only a small fraction of pretreated feedstock. We have optimized enzymes for feedstock and operating conditions." "We will deliver the complete technology." "The basic engineering package, also include all biotech software, microorganisms for producing enzymes, downstream processing, for producing the ethanol, and also help with the start up." [Lane 2012p]

Summarizing some aspects of innovation and intrapreneurship in biofuels considered so far the following emerges.

> Research, innovation and commercialization progress in the biofuels area require tremendous capital investments. Startups and NTBFs might have the technology, but they lack the capital to build the (pilot and demonstration)

plants and other infrastructure required to fully prove it. Hence, they need very strong partners and giant oil (and automotive) firms play a key role.

Federal and state governments play a very important role for financial support in terms of tax credits, subsidies, and research support (grants and cooperation with public research organizations; Figure I.34).

But, looking at oil firms as partners, one must be aware of Big Oil's contribution into perspective. For instance, for ExxonMobil, the world's largest publicly traded oil company, the biofuels investment (in a $600 million partnership with biotech company Synthetic Genomics Inc. [Lux Research 2007] over five to six years!) is tiny compared with its spending to find new supplies of crude oil and natural gas [Porretto 2009]. Capital spending of a giant oil company is $20 – $30 billion per year. *Exxon Mobil made $142 million in profit each day of 2008*. Correspondingly, the financial risk of the investments for these firms does not weight very strongly.

It seems that concerning performance data of ethanol (Table I.82), its carbon footprint (Box I.1) and "without a Shift in the Auto Industry, Cellulosic Ethanol is No More Than a Good Idea." [Wikinvest 2009]

In short, assessments of biofuels' roles referring to Shell Chief Executive Officer Peter Voser turned to the statements that *advanced biofuels will not be in widespread use until about 2020* and that it would take "quite a number of years" before there is a commercially proven plant. The company has also been forced to acknowledge that it has been *over-optimistic* about when these ventures will start to pay off [Crooks 2009].

In line with this statement corporate and capital investors were plagued by promises of startups/NTBFs concerning the start date of production by commercial plants which deviate by years from reality. For instance, the following projections were made during the 2007-2009 period on commercial production or generating notable revenues:

- Cobalt Biofuels (Table I.96), founded 2006, planned: a $25 million GPY plant for 2012; for 2015 jumpstart revenue in the chemicals market; claim to first commercial sales of biobutanol in 2011 and "multiple facilities" by 2014; according to its recent fact sheet: Demo-scale plant expected to be operational in 2011; 1.5 million GPY facility operational in 2012;
- LS9, Inc. (Table I.99) founded in 2005, would not reach commercial production levels until 2013;
- Solazyme (Table I.90), founded in 2003, would be at parity with $80 oil by 2012/13.

But often there are associated other concerns with statements about the time of entry into large-scale production, for instance, with regard to Range Fuels [Rapier 2010] and Coskata [Admin 2011a] (Table I.99).

For Range Fuels Rapier [2010] provided firstly common wisdom and advice. Learn to be conservative with claims, because failing to deliver can have far-reaching impacts. Plus, a pattern of over-promising and under-delivering will ultimately destroy your credibility, and thus your ability to get anything done.

He then secondly turned to “Range Fuels: Years of Broken Promises.” And he presented a timeline from October 2006 to February 2010 to show the remarkable evolution of their “progress” that has gone largely unreported and emphasized the “highlights.”

The key point is that in May 2009, while Range Fuels stopped issuing so many press releases, replaced CEO Mitch Mandich was quoted in the New York Times admitting that “the soup’s not quite cooked yet.”

The known amount of money by 2010 that has been poured into this firm (Table I.99) – beyond Khosla and company’s initial investment – is $158 million in VC money, $76 million of DOE money in 2007 to finance the Georgia plant, $80 million from the USDA (a loan guarantee of $80 million, and that allowed the company to secure an $80 million bond in 2010 to fund the plant’s construction in 2010), and $6 million from the state of Georgia. Further, they asked for more DOE money, but were turned down. That turned out to be more than $320 million to build 4 million gallons of methanol capacity.

Rapier [2010] could refer to and cite a US EPA report:

“As for the Range Fuels plant, construction of phase one in Soperton, GA, is about 85% complete, with *start-up planned for mid-2010*. However, there have been some *changes to the scope of the project* that will *limit the amount of cellulosic biofuel that can be produced in 2010*. The *initial capacity has been reduced from 10 to 4 million* gallons per year. In addition, since they plan to *start up the plant using a methanol catalys*t they are *not expected to produce qualifying renewable fuel in 2010*. During phase two of their project, currently slated for mid-2012, Range plans to *expand production at the Soperton plant* and *transition from a methanol to a mixed alcohol catalyst*. This will allow for a greater alcohol production potential as well as a greater cellulosic biofuel production potential.” (Emphases added)

And Rapier concluded: “So taxpayers funded a **40 MGY** wood-based **ethanol** plant and they are instead getting a **4 MGY** wood-based **methanol** plant.”

In line with this, in 2011 Bud Klepper, who is not only Range Fuels’ technical advisor but also the original founder of the company that became Range Fuels, announced that Range Fuels is laying off most of its employees at its plant near Soperton, Ga, after it makes a single batch of ethanol, and the company will shut down the plant while it tackles technical problems and raises more money [Wang 2011].

Some remarkable conclusions of R. Rapier [2010] were as follows.

- Range Fuels' "people had been in the habit of promising the moon to secure ever more funding."
- Investors seem to proceed "how Silicon Valley innovates." "The thing is, the energy industry is full of very smart people who went to the same schools the people in Silicon Valley attended."
- Failure tars an entire renewable industry as being hopelessly unrealistic.

A similar situation in different context can be described for Coskata (Table I.99), founded in 2006. In 2010/2011 biofuel companies, many without revenues or commercial products, continued to shoot for IPOs. Filing for an IPO in the *industrial biotech* boom, which began with a successful listing on the NASDAQ by Codexis in 2010 was followed by IPOs of Amyris, Gevo, KiOR (all in Table I.99) and Solazyme (Table I.90). Then, PetroAlgae, Mascoma (Table I.99), BioAmber and Genomatica (A.1.1.6) have also filed S-1 registrations for proposed IPOs, as did Coskata (Table I.99) in December 2011 when it looked for a $100 million IPO.

In 2010 Coskata lost $28.7 million while recording $250K in revenues and $23.3 million for the nine months ended September 30, 2011. They expected these losses to continue for the foreseeable future. A recent summary of its S-1 registration culminated in a revealing overview entitled "The Risks, Translated from SEC-speak." [Admin 2011] Below are some cited examples of that exercise concerning credibility, execution and delivery.

In SECSpeak:	**In English**
"In place of the plasma gasifier that we used at our Lighthouse facility, we expect to integrate an indirect biomass gasifier with our syngas cleaning technology, which have never been tested together for fuels production. While biomass gasifiers are a proven technology, they have only been used commercially on a limited basis and have experienced operational reliability issues."	Uh, we didn't actually use our proposed gasification machine, a/k/a/ Old Unreliable, because in the demonstration that we did, we decided to demonstrate something else. (a/k/a = also known as)
"We have entered into an MOU with a lender for $87.9 million of debt financing to fund a portion of the cost of constructing Phase I of our planned Flagship facility. We have also received a conditional commitment from the USDA relating to a 90% guarantee of such debt financing…The process for finalizing the definitive documentation with the lender and the USDA may take longer than expected or may not happen at all." (MOU: Memorandum Of Understanding)	Your investment dollars may become, er, marooned (i.e "into the Valley of Death rode the six hundred"), if we don't close this loan.

"A disruption in our supply chain for components of our proprietary nutrient package could materially disrupt or impair our ability to produce renewable fuels and chemicals."	If Rumplestilskin the Magic Coskata Microorganism doesn't get his vitamins, he won't spin our straw into gold.
"Although we currently intend to use the net proceeds from this offering in the manner described in "Use of Proceeds," we will have broad discretion in the application of the net proceeds."	If we spend all this money on, say, golf memberships, the only ethanol we'll see will be at the 19th Hole.

Overall one can say that opinions about the developments and future of biofuels are split reduced to the question "Advanced Biofuels: pipedream or solid investment?" [Bomgardner 2011b].

The industry partly blamed the credit crisis (Great Recession) for its slow pace, but acknowledged that *getting the conversion techniques to work is the biggest problem* [Wald 2009b]:

- "It's certainly turned out to be more complicated technically than people thought it would be," said Brian Foody, the president and chief executive of Iogen (working with Shell).
- BP America also acknowledged slow progress in its company's joint venture with Verenium, in Louisiana. "We aren't seeing fundamental technology issues; it's more a matter of optimizing the engineering of the pots and pans we use to do the cooking, so to speak."

Experiences of the average duration of scale-up from the laboratory to a high-volume commercial plant from the chemical industry could be used as an indicator for estimating that time. Scale-up after laboratory work for the polymers/plastics area (polyethylene, Nylon or Perlon, Kevlar, Biomax/Ecoflex) by thermochemical processes took 6 – 8 years [Runge 2006:655] and probably more than 10 years for bioengineering processes. Evonik Industries (previously Degussa) reported 13 years for an industrial biocatalytic process [Runge 2006:577].

A model for the biofuels industry following the thermochemical path could be the Izumi Biorefinery (a US/Japan cooperation, in operation since 2002 in Japan) which uses Arkenol's concentrated acid hydrolysis technology for the conversion of biomass to ethanol. It took 9 years from completion of base technology developments to deployment, 12 years from ideas and R&D to "demo" deployment [BlueFire Ethanol 2004].

As engineers know, scale-up proceeds in a series of steps, and the scale-up between steps can range from a factor of, say, 10x in the case of a conservative play, to as much as 150x in a single step.

Correspondingly, cellulosic ethanol production has not grown as fast as the industry or the federal US government had hoped it would. An investment firm predicted that cellulosic ethanol companies will be able to supply only 28.5 million gallons by 2010, short of the federal government's 100 million gallon goal [St. John 2009]. The industry acknowledged that it will make a few million gallons of the advanced fuel in 2010, at most, and could fall even further behind the 2011 quota, 250 million gallons [Wald 2009b].

Shell's emphasis on its core competency with regard to the process engineering and particularly the FT-process and the large variety the intermediate syngas provides to produce other commercial industrial products on a large scale referring to existing, proven sub-processes (Figure I.174) represents a strong directional effect for the realities of the future of bioethanol.

Specifically, Shell demonstrated the world's first commercial passenger flight powered by a fuel made from natural gas [Quatar Airlines 2009] after developing and producing jet fuel using a 50-50 blend of synthetic GtL kerosene (*"GTL Jet Fuel.")* and conventional oil-based kerosene fuel.

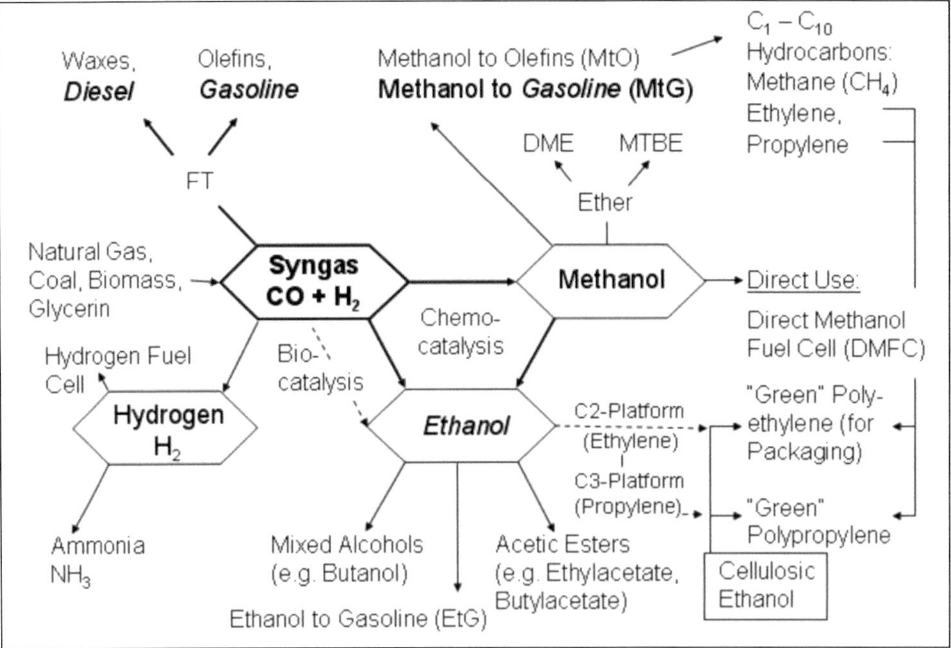

Figure I.174: Syngas routes to selected high value products and applications (DME = dimethylether, MTBE = Methyl-tertiary-butyl-ether, a gasoline additive – "octane-booster").

The flight was the latest step in over two years of scientific work carried out by a *consortium consisting of Airbus, Qatar Airways, Qatar Petroleum, Qatar Science & Technology Park, Rolls-Royce, Shell and WOQOD.* Much of this work was being undertaken at the Qatar Science & Technology Park in Doha. Shell and Quatar Petroleum started building a commercial plant (project Pearl) which should open in 2012 and deliver 1 million tons/a of *GTL Jet Fuel*.

The simplest alcohol (bio)methanol can also be used as a chemical building block for a range of future-oriented biobased products, including bio-MTBE, bio-DME, bio-hydrogen and synthetic biofuels (synthetic hydrocarbons). Furthermore, biomethanol can represent the input for so-called "Direct Methanol Fuel Cells" (DMFCs), as offered successfully by German NTBF Smart Fuel Cell AG (now SFC Energy AG) [Runge 2006:328-335, 623].

Europe and particularly Germany, has a very long history of synthetic methanol. Currently it is produced, for instance, through pyrolysis via syngas (Figure I.174). The first patent to produce methanol via syngas was granted to the chemical giant BASF in 1913; its first large-scale commercial plant started production in 1923 [99].

For centuries, methanol was first produced from pyrolysis of wood, leading to its common English name of wood alcohol or wood spirit (in German Holzspiritus or Holzgeist). Today, China is the largest producer and consumer of methanol in the world. And in 2008 it utilized 3 million tons of methanol (of a total of 45 million tons consumed worldwide) as a fuel blend. [126] Methanol usage in China for other products is: formaldehyde (38 percent), MTBE (20 percent), acetic acid (11percent), fuel use (4 percent), other uses (27 percent) [Engeler 2008].

In China (contrary to the US or Germany) methanol has been approved in 2009 for use as a motor vehicle fuel and as a renewable fuel for transportation. Like ethanol in Brazil and the US, China now permits methanol to be added to pure gasoline so that it can make up to 85 percent of the mixture for use in flex fuel cars (FFVs). As the US has E85, China has M85. However, among alcohols as biofuels methanol has the lowest energy content and stoichiometric air fuel ratio meaning that fuel consumption (on a volume or mass basis) will be higher than other alcohol or hydrocarbon fuels (Table I.82).

Similar to the US, to become (more) independent from crude oil imports, China promotes biomethanol as a biofuel – and also biobutanol (cf. Green Biologics, Table I.95) [Blanco 2009]. In the 1980s in Germany there was a large field test of M15 and M85 fuels sponsored by a German ministry which involved more than 1,000 vehicles and all German automotive firms, the mineral oil industry and many research institutes. [126] But there was no significant follow-up. [126]

More importantly, for biomethanol entrepreneurs, methanol plays a key role as an alternative to petroleum feedstock for the petrochemical industry (MtO, Figure I.174), which brings in the names of chemical giants like German BASF and US Dow Chemical. In particular, there is a strong emphasis on methanol-to-ethylene (MtE) and methanol-to-propylene (MtP) (Figure I.174) processes. Also here China plays a leading role [Heathcote and Fryer 2008]. And recently already the concept of a "methanol economy" has been put forward [Bullis 2006].

Based on gasification of largely municipal waste to syngas the primary focus of the Canadian firm Enerkem (headquartered in Montreal) is the commercial production of cellulosic ethanol. However, its three step process first requires the production of methanol as a chemical building block for the production of ethanol (Figure I.174). Enerkem can sell its methanol as an end-product, or use it as a key intermediate to produce other renewable chemicals.

Enerkem announced in 2012 the initial production of cellulosic ethanol from waste materials at its demonstration facility in Westbury, Québec. Its technology has been developed and tested during the past 11 years. Enerkem has already produced cellulosic ethanol at its smaller scale pilot laboratory facility. The newly installed equipment for the conversion of Enerkem's methanol into cellulosic ethanol is now used in combination with the larger methanol equipment already in operation at Westbury [SpecialChem 2012].

With Europe's strong orientation towards biodiesel as a biofuel and its by-product glycerin as a cheap raw material it was almost natural that entrepreneurial ideas are generated around glycerin as a basis for biofuels and other biobased feedstock materials following a biorefinery approach as a commercialization model.

And, indeed, in the Netherlands Bio-Methanol Chemie Nederland BV (BioMCN) introduced a *glycerin-to-syngas* process to produce biomethanol (Table I.87). It is a typical engineering, biorefinery approach to entrepreneurship of founders with profound experience in business and technology ("veterans approach") taking advantage from pre-work of an industrial partnership.

Table I.87: BioMCN showing an engineering-type approach to innovation and entrepreneurship. *)

Company (Foundation) Remarks	**Major Funding**	**CEO, Other Executives, Foundation, Key Researchers; Technology Protection and Position**
Bio Methanol Chemie Nederland BV (BioMCN), Delfzijl, The Netherlands (Nov. 2006) BioMCN claimed to be the first company in the	Main shareholder: the European private equity firm Waterland; remaining shares owned by management and some of the original founders, the Japa-	Rob Voncken CEO; before joining BioMCN fulfilled several positions at Dutch chemical firm DSM in the areas of business development, marketing & sales, general management; was responsible for the management of several outsourcing, divestment and acquisition projects; Siebolt Doorn, co-founder; inventor of the BioMCN technology and co-founder of BioMCN [Ecofys 2006]

<table>
<tr>
<td>world to produce and sell industrial quantities of bio-methanol and the largest 2nd generation biofuels producer in the world.

Employees:
2009: 80

Sales (in €):
2007: 101 mio. (net profit €5.7 mio.),
2008: 85 mio. (Net loss 12.9 mio.);

Capex 2007: €3.4 mio.
Capex 2008: €26 mil., majority on the construction of the commercial bio-methanol plant [BioMCN 2008; BioMCN 2009].

Since March 2008 producing biomethanol in a 20,000 tons pilot plant;

in July 2009 successful start-up of a 200,000 tons per year bio-methanol plant.

October 2008: winner of the European Responsible Care Award</td>
<td>nese chemical firm Teijin and Dutch investment company NOM [BioMCN 2009];

in 2009 received from Waterland €39M to fund the expansion of the facility to double production capacity to 400,000 tons by 2010 [Lopez 2009].

EOS Demo (Energie Onderzoek Subsidies - Energy Research Subsidy) and IBB-subsidies (Innovatieve BioBrandstoffen) granted total investment subsidies for an amount of €.8.2M; first amount of €0,7M received in 2008;

in October 2007 extra financing facility of in total €30M on top of existing subordinated loan (€10M) already granted in 2006.</td>
<td>Paul Hamm, co-founder;
co-initiator of the consortium and temporary CEO of BioMethanol Chemie Holding
[Ecofys 2006; Van Zanten 2007].

Foundation approach (Figure):
technology veterans sharing previous firm affiliations and serial entrepreneurs;
a quasi management buyout (MBO);
team acquired a natural gas-to-methanol plant with two lines of 330,000 tons per year capacity jointly owned by Dutch firms DSM, Akzo Nobel, and Dynea (Figure) and retrofitted the existing lines;
November 2006 – The new owner of the methanol plant is BioMethanol Chemie Holding (BV), a consortium of Econcern, NOM, Oakinvest,. S. Doorn and P. Hamm.

Fundamental European patent (EP1897851) represented an improvement of a known glycerin-to-syngas process, where liquid glycerol droplets – in a stream of inert gas or steam – are introduced into a catalyst bed; leads to significant carbon deposition on the catalyst and catalyst deactivation;
the improvement is that glycerin is not introduced into the catalyst bed as a liquid, but is fed to the catalyst bed as a vapor, together with the necessary amount of steam this results in significantly diminished carbon deposition on the catalyst.

Figure: BioMCN foundation and founders' interconnections.

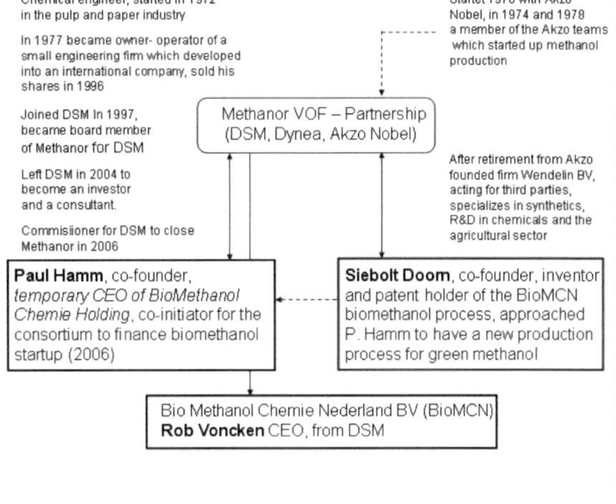
</td>
</tr>
</table>

Technology, Goals, Strategy	The innovative and patent-protected process converts crude glycerin, a residue from biodiesel production, into biomethanol via syngas ("glycerin-to-syngas"). The key steps of the process: purifying crude glycerin and then converting it into bio-syngas. Syngas is sent to a methanol converter to produce crude biomethanol, the crude biomethanol is purified by distillation. Continuous improvement of the process on lab scale and pilot plant scale, e.g. regeneration procedure to remove carbon from the catalyst. By using biomass instead of natural gas as a feedstock means supplying the market with a "green" product that has a higher market value. Flex fuel cars can run on any mixture from 100% gasoline up to 85% bio-methanol (M85). Actually, it was a *retrofitting* approach: BioMCN acquired in 2006 a natural gas-to-methanol plant. Methanol to be used as a feedstock in the sense of a *biorefinery* approach for downstream producing, for instance, formaldehyde or acetic acid. Even though Delfzijl is located in the very North of The Netherlands, it can be considered an ideal location due to excellent road, rail and water links. Also, 15% of Dutch chemicals are produced in Delfzijl. The global methanol market was believed to be around 35 – 38.000 kt of which some 20 – 25% is sold in Europe [BioMCN 2009]. Business focus was mainly on Europe as the primary market for BioMCN. In Europe BioMCN claims to have a market share of 5% in the existing methanol markets. Commercialization/revenue model: owner-operator, BioMCN will spend several years concentrating on the production of bio-methanol from glycerin "We are aiming to expand into biorefining but that is a long-term concept which will not be realized for 10 years." [Headline News 2009] It is assumed that most of the revenues currently come from selling biomethanol as an input for the petro-gasoline octane booster MTBE (to oil companies for blending with petro-gasoline).

*) From the firm's Web site if not stated otherwise by a reference; firm's status: early 2010.

BioMCN showed healthy further development. In 2011 BioMCN and the Investment and Development Agency of North Netherlands (NOM) have, together with Visser & Smit Hanab and German giants Linde and Siemens, formed a consortium to build a large scale biomass refinery. This refinery will be built next to the biofuel plant that BioMCN opened in Delfzijl in 2010. This consortium asked the Ministry of Economic Affairs, Agriculture and Innovation (EL&I) for a subsidy. The proposed refinery will be the largest of its kind. The biomass refinery can process approximately 1.5 million tons of residual wood, which will yield more than 500 million liters of second generation biomethanol [BioMCN 2011].

In line with the historic production of methanol from pyrolysis from wood (via syngas) also acetic acid (as a key component of "wood vinegar" – Holzessig in German) can be obtained from wood pyrolysis. [99, 126] This leads almost directly to ZeaChem, Inc.'s biofuel process.

Whereas the bioengineering route usually converts (C6 sugars or C6 and C5 sugars) to bioethanol or biobutanol, Zeachem's bioengineering patented process relies on lignin separation and converts the fermentable sugars into acetate (ethyl acetate, which is the ethanol ester of acetic acid; CH3CO-OC2H5) and then gasifies the remainder, tough lignin and all other, into hydrogen before mixing the two streams in a reaction called "hydrogenolysis" (a chemical reaction whereby a carbon-carbon or carbon-heteroatom single bond is cleaved or undergoes "lysis" by hydrogen) to produce ethanol.

There are two remarkable points: 1) a big refinery rather than oil company appears as an investor; and 2) ZeaChem Inc. did not invent anything. "There's no new bugs, no new equipment. We're taking things that already exist." [Verser 2009] A Lego-type process is followed; only *commercial-off-the-shelf (COTS) components* are used, no unproven technology.

The lessons from ZeaChem entrepreneurship is the focus on people (team), building on experience (what you can do best), timing and persistence. There is a very structured engineering-type approach to financing and risk mitigation. And, as in case of BioMCN, the human component comprises a people's network of shared company affiliations which brings in trust and stability.

Table I.88: Zeachem showing an engineering-type approach to innovation and entrepreneurship using only proven technology. *)

Company (Foundation) Remarks	**Major Funding**	**CEO, Other Executives, Key Researchers; Technology Protection**
ZeaChem, Inc. Lakewood, CO (2002); Incorporated in 2002, founded 1998 – "Two guys in a pickup" [Verser 2009]. Headquarter in Lakewood, (CO)	Funding together with profit from service contracts and grant funding $34M Series B (2008), several VCs and Valero Energy Corp., the largest pe-	According to its EPC services Burns & McDonnell ZeaChem has 30 employees, half of whom work in the ZeaChem R&D facility in Menlo Park, Calif., where scientists and engineers conduct laboratory-scale research to prove the theoretical chemical viability of each step of manufacturing. **Figure 1**: ZeaChem Management Team – experienced persons ("veterans"} with interpersonal relationships (and co-working).

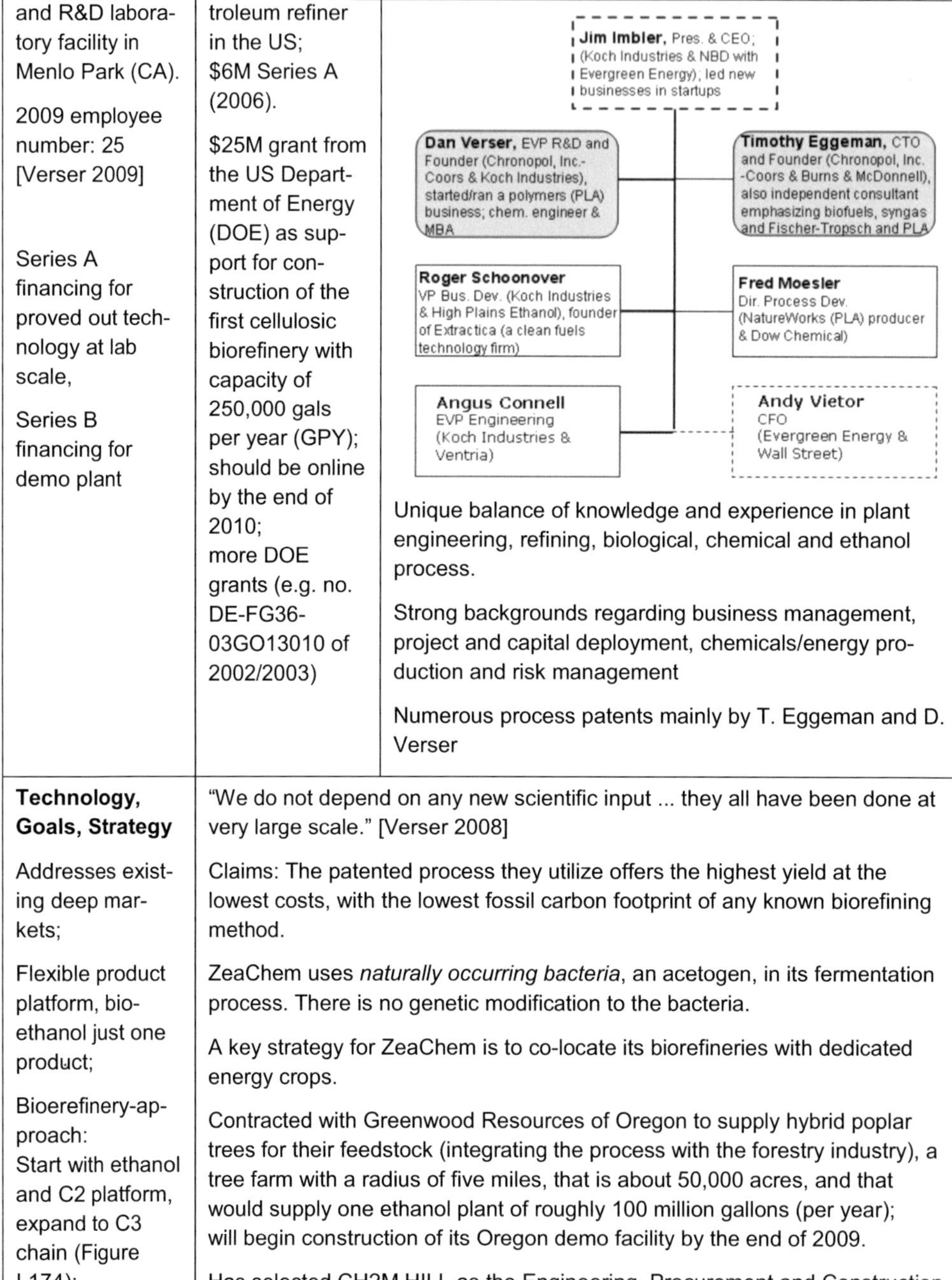

and R&D laboratory facility in Menlo Park (CA). 2009 employee number: 25 [Verser 2009] Series A financing for proved out technology at lab scale, Series B financing for demo plant	troleum refiner in the US; $6M Series A (2006). $25M grant from the US Department of Energy (DOE) as support for construction of the first cellulosic biorefinery with capacity of 250,000 gals per year (GPY); should be online by the end of 2010; more DOE grants (e.g. no. DE-FG36-03GO13010 of 2002/2003)	Unique balance of knowledge and experience in plant engineering, refining, biological, chemical and ethanol process. Strong backgrounds regarding business management, project and capital deployment, chemicals/energy production and risk management Numerous process patents mainly by T. Eggeman and D. Verser
Technology, Goals, Strategy Addresses existing deep markets; Flexible product platform, bioethanol just one product; Bioerefinery-approach: Start with ethanol and C2 platform, expand to C3 chain (Figure I.174); Use known microbes, equip-	"We do not depend on any new scientific input ... they all have been done at very large scale." [Verser 2008] Claims: The patented process they utilize offers the highest yield at the lowest costs, with the lowest fossil carbon footprint of any known biorefining method. ZeaChem uses *naturally occurring bacteria*, an acetogen, in its fermentation process. There is no genetic modification to the bacteria. A key strategy for ZeaChem is to co-locate its biorefineries with dedicated energy crops. Contracted with Greenwood Resources of Oregon to supply hybrid poplar trees for their feedstock (integrating the process with the forestry industry), a tree farm with a radius of five miles, that is about 50,000 acres, and that would supply one ethanol plant of roughly 100 million gallons (per year); will begin construction of its Oregon demo facility by the end of 2009. Has selected CH2M HILL as the Engineering, Procurement and Construction (EPC) contractor for its first biorefinery. ZeaChem's technology is *a parallel hybrid system* of gasification and	

ment, processes; Focus on management team with execution on technical and commercial level; Have strategic support; Recurrency: Expand technology into other products and repeat; Maintain the option of awarding direct licenses to qualified parties. ZeaChem has a number of potential business segments that will necessarily involve strategic partners	fermentation: a Lego-type "hybrid" fermentation esterification hydrogenolysis process. ("indirect" ethanol fermentation and chemically high efficient, well established hydrogenolysis) – current products and future C2-platform products. A differentiator: ZeaChem's ability to produce a range of cellulosic biobased products to serve a variety of market sectors; produce many chemicals and fuels within various carbon chain product groups; production facilities will be capable of producing the products that will yield the best margin – should market conditions change, a ZeaChem facility will have the option of changing the products produced. **Figure 2:** The ZeaChem Proceess. 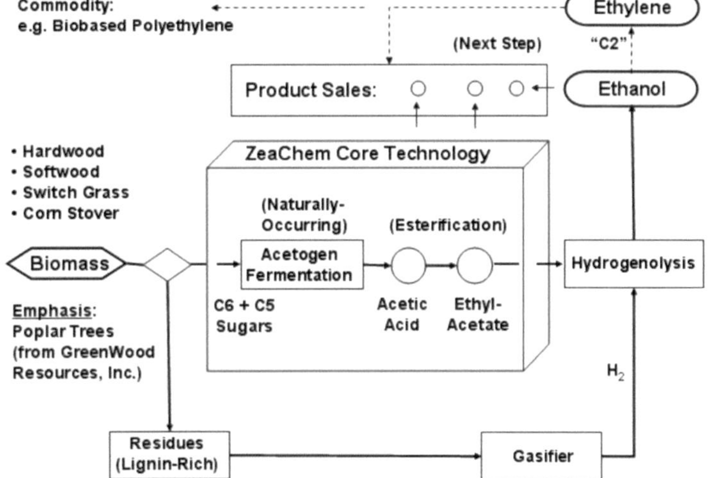 Revenue Model: 1. Able to produce ethanol at <$1/gal 2. License technology for early plants, 3. Sell products, and monetize markets and geographies. Moving forward by combinations: 1. Strategic investors. 2. Government support Partnerships: Strategic – Valero Energy Corp. (acts also as investor), Feedstock: Greenwood, VC: such as MDV. Reduce risk by alliances along entire value chain Feedstock supply, Technology (successful demo plant for core technology, warrantees from

	non-core technology vendors)	
Risk Mitigation: [Imbler 2009] Bottom-Line: Have more money than you need Once technology established, utilize traditional project finance	**Risk**	**Mitigation Strategy**
	Demo. Capital Cost	Freeze design; provide explicit project definition
	Scale-up	Establish clear performance goals for each level of scale-up; design for process flexibility
	Integration	Use Lego block deployment if possible to minimize potential of integration issues
	Technology	Keep plant #1 based on known technology and processes where possible (remember risks are multiplicative)
	Engineering	Use experienced knowledgeable EPC, but maintain tight supervision on daily basis
	Operations	Evaluate partnering with experienced operator
	Economics	Fix feedstock costs and sellout production from plant to 3rd party credit worthy entity
	Funding	Ensure business has adequate funding accessing a variety of different sources

*) From the firm's Web site if not stated otherwise by a reference; firm's status: early 2010.

Zeachem's "indirect" ethanol fermentation which combines a thermochemical and a bioengineering (fermentation) approach to bioethanol corresponds to a "*hybrid*" process based on syngas as schematically outlined in Figure I.175.

By the end of 2011 ZeaChem launched its 250,000 gallon per year core process, at their Boardman, Oregon, biorefinery. "'We came in on schedule and significantly under budget," reflected ZeaChem CEO Jim Imbler. "We had a guaranteed maximum price, but in this case we received a rebate check. One of our long-term VCs said to me I've seen a lot of things, but I have never seen a check come back." "It came down, we think, to the choice we made to use known processes, known vendors."

ZeaChem integrates feedstock from a portion of GTFF's residual fiber with local agricultural residue suppliers to achieve feedstock costs 50 percent less compared to Brazilian sugar cane and 80 percent less, compared to corn based processes [Lane 2012l].

In Colorado, Zeachem has received a conditional commitment for a $232.5 million loan guarantee from the US Department of Agriculture's 9003 Biorefinery Assistance Program [Biofuels Digest 2012]. Development of the first commercial biorefinery is already underway. The facility is expected to have capacity of 25 million or more GPY, and is expected to be competitive, upon completion, with $50 oil, with a targeted $1.96 operating cost per gallon. Its first commercial plant will be located adjacent to

ZeaChem's 250,000 GPY integrated demonstration biorefinery in Boardman, which is a logistics and transportation hub for the Columbia River system through the Port of Morrow and is projected to be operational by late 2014. Under the conditional commitment, ZeaChem must meet specified conditions before the 60 percent loan guarantee can be completed, and must also source the loan. Silicon Valley Bank was the bank of record for the project.

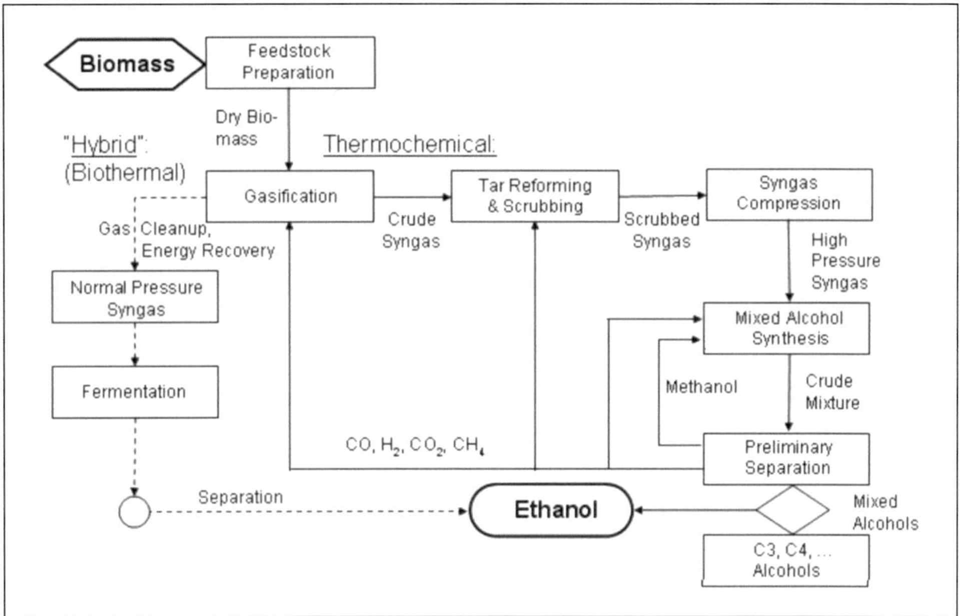

Figure I.175: Flow diagram of the thermochemical and a "linear hybrid" approach to bioalcohols.

The total project cost for the 25 million gallon per year biorefinery is estimated to be $390.5 million, and the remainder of the project cost shall be covered through equity contributions by ZeaChem and its investment group [Biofuels Digest 2012].

The news of ZeaChem's startup came at a time of significant blow-back for the cellulosic biofuels movement and sector, after the US EPA waived down the scheduled 500 million gallon mandate for cellulosic ethanol, first proposed back in 2007 when the current Renewable Fuel Standard (RFS) was developed, down to 10.45 million gallons for 2012.

The overall mandates for 2012, under the Renewable Fuel Standard, are as follows expressed in ethanol-equivalent gallons (actual volumes in parentheses)

- Biomass-based diesel: 1.5 billion gallons (1.0 bil. gal)
- Advanced biofuels: 2.0 billion gallons (1.3–1.5 bil. gal)
- Cellulosic biofuels: 10.45 million gallons (8.65 mio. gal)
- Renewable fuels: 15.2 billion gallons (14.5–14.7 bil. gal).

A.1.1.4 The Special Algae-to-Biofuels Boom

As for the biomass-to-biofuel movement the same occurs for the third-generation biofuels. For the emerged boom of algae-to-biofuels, "*everything new was old again.*" In the US the National Renewable Energy Lab (NREL) in Golden (CO) led a $25 million study of algae from 1978 to 1996, before money dried up and government research shifted to ethanol [Gold 2009b; DOE 2009]. DOE canceled the program in 1996, saying *the process could not be made cost-competitive with petroleum refining* [Voith 2009b]. Similar research programs and experiments were persued in Germany, also with no promising results and corresponding endings.

Furthermore, all kinds of research avenues were explored, but when the funding shriveled during later years, knowledge, experts and know-how were lost. Programs that started during the late 1970s and early 1980s were stopped in the years of low energy prices that followed. In the US, now, there was all this biofuel work going on, and they are all going back to that public domain research. In the context of the biofuel boom interest in algae-based biofuels also exploded and venture capital and corporate money flowed also into this field. It is estimated that over 75 percent of the companies who had algal aspirations in the1980s and 1990s no longer exist [Rapier 2009c].

A notable difference to bioalcohols and biogasoline, however, is that past efforts with algae focused essentially on research rather than development and demonstration [Pienkos 2008; Pienkos and Darzins 2009]. The 18 years of research at the National Renewable Energy Laboratory (NREL) yielded a lot of knowledge, but it resulted in nothing resembling a commercial product or process. In the best-case scenario, when all is said and done, algal biofuel could cost $50 per barrel. But that will not happen anytime soon, and it could take a decade [Madrigal 2009; 2007].

There are about 30,000 species of algae. "100 are well known and between 15 and 20 are used for production." [Robinson 2009]. Hence, there is plenty of scope for competition. But the diversity of algae also meant that there is scope to produce niche algae for different conditions. And it is unlikely that there will be a single type of algae that work well in the cold climates, the tropics, salt and fresh water.

Furthermore, there is no parallel agricultural enterprise equivalent for cultivating algae at a similar scale. "We don't have the infrastructure yet to grow literally thousands of acres, maybe millions of acres of algae," NREL's researcher Darzins said. In short, the science of algae cultivation (algaeculture), agronomy-for-algae, does not exist.

"Algae growers" are still learning how to protect their fragile crop from predators and invasive species.

It was thus clear that a significant basic science and applied engineering R&D effort including a rigorous techno-economic and life-cycle analysis (LCA) will be required to fully realize the vision and potential of algae [DOE 2009]. In this regard, a recent publication [Lardon 2009] reported algal biofuels would not have a positive energy balance; in other words, you had have to put more energy in than you would get out.

Algae constitute single-celled or simple multi-cellular photosynthetic organisms. They produce their own food by using energy from sunlight to synthesize complex molecules from carbon dioxide and water – both in sea and fresh water. Algae range in size from microscopic organisms to giant seaweeds some hundred meters in length (micro- versus macro-algae). They contain chlorophyll and other pigments which give them a variety of colors. They manufacture their food by photosynthesis (Figure I.177). Inside their mushy cells, algae contain up to 50 percent vegetable-oil-like lipids by dry cell weight. Genetic engineering claims it can tailor the oil composition, productivity, and other traits of the algae.

> Algae are very interesting: they created the oxygen (Figure I.178) in our atmosphere, and also oil, both essential as a basis of our existence.

The close connections of algae to petro-oil are as follows. "Geologists view crude oil and natural gas as the product of compression and heating of ancient organic materials (i.e. kerogen) over geological time. Today's oil formed from the preserved remains of prehistoric zooplankton and algae, which had settled to a sea or lake bottom in large quantities under anoxic conditions. Over geological time the organic matter mixed with mud, and was buried under heavy layers of sediment resulting in high levels of heat and pressure (known as diagenesis). This caused the organic matter to chemically change, first into a waxy material known as kerogen which is found in various oil shales around the world, and then with more heat into liquid and gaseous hydrocarbons in a process known as catagenesis." [100]

In principle, for biofuels algae offer many advantages over traditional oilseed crops, such as corn, soybeans or rapeseed.

- Better Yield:
  Algae yield far more oil than traditional oil seeds. Up to 50 percent of an algae's body weight is comprised of oil, whereas oil-palm trees – currently the largest producer of oil to make biofuels – yields approximately 20 percent of their weight in oil. Although many different parts of plants may yield oil, in actual commercial practice oil is extracted primarily from the seeds of oilseed plants.
  The draw is that algae have the potential to produce up to ten times more oil per acre than traditional biofuel crops such as oil palm [Waltz 2009b].

- Rapid Growth:
  Algae grow up to 15 times faster than oilseed crops grown on land. According to a rule of thumb, bacteria will divide once an hour and algae once every day.
- Better Use of Land:
  Algae can be grown in marginal lands in places away from the farmlands and forests, thus minimizing potential stresses to our food chain and ecosystems.
- Reduced Pollution:
  Algae can reduce pollution by utilizing via photosynthesis large amounts of potentially harmful CO2 from industrial emissions to grow rapidly.
- Frequent Harvests:
  Daily harvesting diminishes the risk of crop failures in comparison to terrestrial plants.
- Algae may produce directly bioethanol (sugars to ethanol, Figure I.178) or other fuels by fermentation.

Algae provide carbohydrates, proteins, and lipid oils, essentially triglycerides (Figure I.178) which may be converted to fatty acid methyl esters (FAMEs) and biodiesel. Considerable research has focused on using algae for the production of hydrogen gas (H_2, Table I.92). As a target for commercialization and production of biofuels the algae focus so far was on the following:

- Food and feed applications: after extracting the algae oil the remaining protein-rich algae biomass can used for food and feed;
- Nutritional additives and fine chemicals: fine chemicals for pharmaceuticals and cosmetics or, for instance, astaxanthin pigments as adding color to the flesh of farmed fish or as antioxidants. Fatty output that makes algae a cosmetic ingredient can also be used for biodiesel. Nutritional and health care additives are produced as capsules, tablets and correlative;
  food- and probably also cosmetics-related products may encounter acceptance issues in various societies if genetically modified objects are used
- Sequestration of carbon dioxide (CO2): CO2 is absorbed by algae, CO2 is the feed for algae (Figure I.177)
- Raw material for biogas plants: algae biomass can serve as a raw material for power generation in combined heat and power (CHP) plants; anaerobic digestion is a series of processes in which microorganisms break down biodegradable material in the absence of oxygen, here to biogas.

For most of these applications rather than open cultivation ponds (or algae enclosed in plastic tubes) that are exposed to the sun closed photobioreactor systems with various layouts and shapes are used. This prevents any exposure to heavy metal contamination or dust, sand or microbiological contaminations (voracious microbes that feed on algae "like a pack of jackals at a buffet") and guarantees consistent and reliable supply of algae. Their use in health care requires the products to be free from pesticides and herbicides.

Bioreactors allow controlling the growth of algae through feeding and control of algae exposure to radiation.

The below Figure I.176 illustrates qualitatively the market value of the various algae products in relation to their market volumes or anticipated volumes, respectively, for a state of commercial scale offerings of biofuels. However, it must be considered that increasing value is or may be associated with increasing concern about genetically modified objects as components of the products. For the highest value offerings it will be the consumer (of a particular society) who decides upon the purchase. High value products can fetch high prices, but some of these markets already have strong players.

Cultivating algae in open ponds or other open systems may encounter similar problems as agriculture do, namely the weather in terms of variations in temperature and sunshine as well as wind or storm which may bring dirt, dust and pollution to the algae.

With regard to biofuels, how to grow algae cheaply on a large scale is one of the biggest challenges facing the industry. When cultivating algae difficulties may be encountered when trying to *scale-up*. In large numbers, the organisms sometimes *crowd* one another out and emit toxic waste that halts the production process. That means, even if you can do it in a test tube, getting the same kind of quality on a large scale could be an issue.

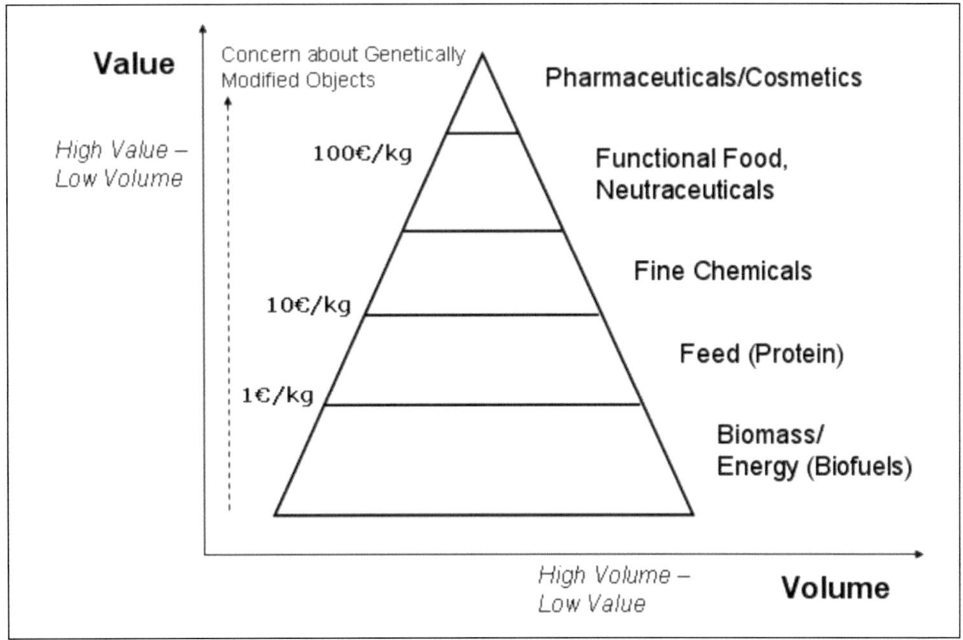

Figure I.176: The algae value ladder for different offerings.

Also related to crowding in open ponds there is the issue of the necessary *radiation exposure* (from sun light above): In a crowded algal assembly only the layers nearest to the surface will get sufficient light ("self-shading") which limits productivity.

Finally, one approach for NTBFs targeting biofuels from algae would be to focus also on the high value-low volume part of algae (Figure I.176) as co-products that can be an additional source of revenue and can improve the overall economics of a biofuel process.

To overcome the problems of scale, cost and price competitiveness with petro-oil companies will be working with a different combination of inputs, conversion methods, extraction techniques, and outputs (Figure I.177). Algae have been sampled from local sources, extreme environments, and genetics labs. They are grown in sunlight and in the dark, in high-tech tanks (bioreactors) and low-tech ponds.

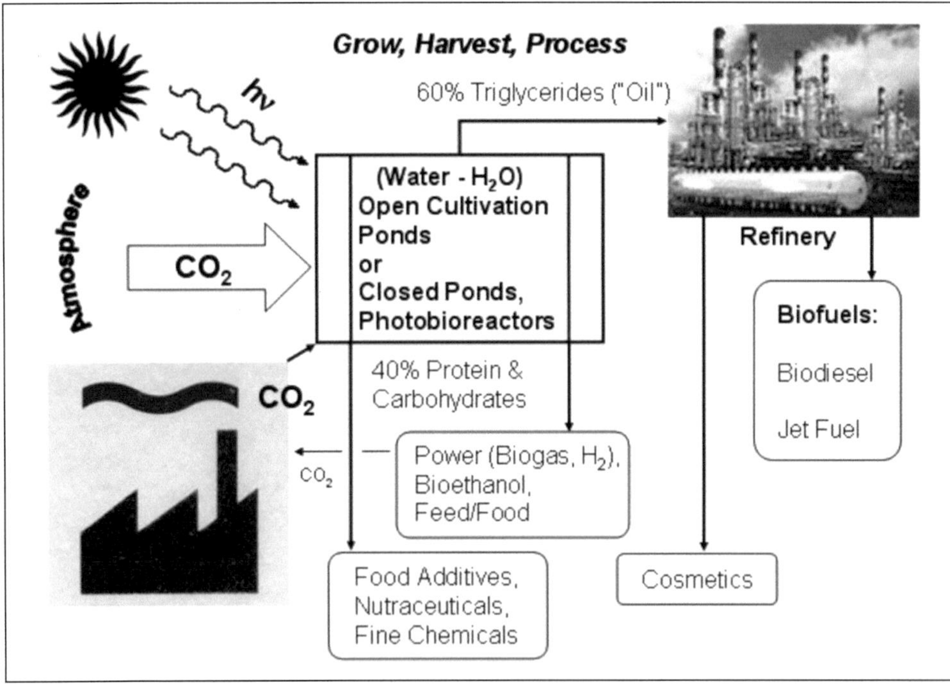

Figure I.177: Algal biomass output streams based on (sun) light and carbon dioxide (CO2) from the atmosphere or a nearby CO2-emitting plant as algae feed and major products and applications.

Today, there is no commercially viable algae approach to biofuels. Production costs need to come down by about a factor of 10. For instance, when the algae first must be harvested, the open-pond method gets expensive and a cost-effective dewatering process must be put into place. On the other hand there is a capital cost ladder associated with lowest cost for open ponds, then closed ponds and highest capital costs with closed photobioreactors.

It was estimated in 2009 that the best process and strategy will take three to five years to reach commercialization [Voith 2009b]. According to NREL researcher Darzin it will take at least five or ten years before anyone finds a way to produce commercial quantities [Wheeler 2009].

Algae as a biomass input exhibits generally a broad versatility depending upon whether algae oil or whole algae is used for further processing. It is also possible to use algae as input for a gasification route to biofuels (Figure I.175). Biogas produced from algae contains essentially methane (CH4) and carbon dioxide. And when methane is used for power generation a closed loop for CO2 is generated (Figure I.177).

Algae grow in freshwater or seawater (salt water). During its growth they produce freshwater (2 gallons of seawater to make 1 gallon fresh water) according to Figure I.178.

In the context of bioethanol, it is interesting to note that cyanobacteria, formerly called blue-green algae, also produce ethanol which is the basis for the entrepreneurial operations of Algenol Biofuels (Table I.91). Genetic manipulation transformed each cell into a small "ethanol factory.

There are notable firms, one being an NTBF, which are successfully active in the high value nutritional additives and fine chemicals algae business. In the US (in Kailua Kona, Hawaii) there is Cyanotech Corp., incorporated in Nevada in 1983, with Common Stock trading on the NASDAQ Capital Market. Cyanotech (revenues $13.9 mio. in 2009, $17.0 mio. in 2011) claims to be the world's largest producer of natural Astaxanthin for human consumption.

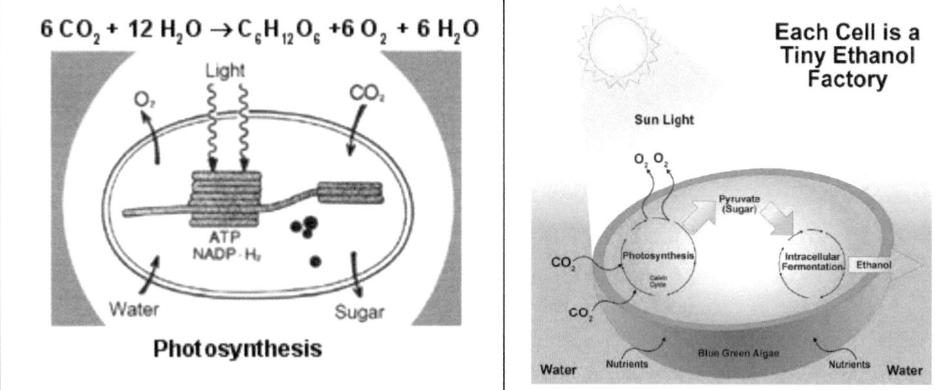

Figure I.178: How algae produce directly bioethanol: Direct To Ethanol® technology (Source – Right: [Woods 2009a; Ahlm 2012:2]).

In Germany, BlueBioTech GmbH [Runge 2006:577], founded in 2000, is rather successful in commercializing algae as food additives and for functional food and cosmetics. It specializes in research, development, production and sales of microalgae as well as natural dietary supplements, feed additives and aquaculture feed. It also provides custom manufacturing services. Further directions target co-operation with industrial partners to develop and produce microalgae and/or extracts, for instance, as active pharmaceutical ingredients (APIs). It sells branded food additives (BluBio® brand) to consumers in capsules, tablets and correlative directly via its Web shop or through common distribution channels.

BlueBioTech GmbH has a group structure. In 2002 BlueBioTech International GmbH was founded as a hub for marketing and distribution activities with partners around the globe. It has several locations in Germany. Particularly, R&D is located in Germany as well as plants for processing algae powders.

BlueBioTech appeared as the sole manufacturer, with German expertise and technology, of microalgae in China. The capacities of the facilities in Germany were not large enough to fulfill the great market demand. Therefore, BlueBioTech International produces larger amounts of powders of *Spirulina* and *Chlorella* (the names of the most important microalgae species) through a joint venture in China on a Chinese sun island. Here there is no farming or industry. "The climatic conditions at the certified algae farm are excellent. The water, in which the algae grow, comes from a depth of 120 meters. It can't be any purer!" Currently, ca. 400 metric tons are produced per year.

As a private firm (LLC) BlueBioTech keeps a low level about their revenues. It is known that the firm expected revenues of €5 million in 2003 and the founder confirmed €6 million. Hence, one would estimate revenues of €10 to €15 million in sales for 2009.

Concerning algae for biofuels and the role of startups and Big Oil a situation emerged similar to that for biomass-to-biofuels: We see alliances of startups or NTBFs with giant oil companies, such as Shell, ExxonMobile or Chevron, or chemical giants DuPont (Table I.83) and Dow Chemical or with other large companies. Others active in the algae-biofuels industry said *ExxonMobile's investment validated the sector.*

The oil industry's view is reflected by ExxonMobile's view of the area, as expressed by a vice president at ExxonMobil Research and Engineering Co. (Emil Jacobs, italics added):

> "Growing algae does not rely on fresh water and arable land otherwise used for food production. And lastly, *algae have the potential to produce large volumes of oils* that can be *processed in existing refineries* to manufacture fuels that are *compatible with existing transportation technology and infrastructure*."

"This is not going to be easy, and *there are no guarantees of success.*" The project (with SGI; Table I.83) involves *three critical steps*: identifying algae strains that can produce suitable types of oil quickly and at low costs, determining the best way to grow the algae and developing systems to harvest enough for commercial purposes [Porretto 2009].

"We *pulled together a pretty high-powered team* ("think tank approach") to look at alternative energy sources and we *looked at all biofuels.*" After examining for ability to scale-up "meaningfully," technical challenges, environmental impact and economics, ExxonMobil arrived at algae, Jacobs said [Lemos-Stein 2009].

Also Chevron seemed to think that algae will provide the biofuel of the future [Scheffler 2009]. Similar to ExxonMobile, also Dow Chemical made excessive due diligence [Woods 2009a] before it announced plans to build together with Algenol Biofuels a $50 million pilot plant at Dow's huge industry complex in Freeport that will test Algenol's technology on a large scale.

The project was focused with regard to several important implications. It could point the way to a more sustainable path for making ethanol. It also could help determine the feasibility of using biofuels not just to power cars, but to produce common building block chemicals (Figure I.174) currently derived from fossil fuels. This means it emphasizes the biorefinery concept (Table I.91).

Royal Dutch Shell established a joint venture in 2007 with HR BioPetroleum to build a pilot facility for growing marine algae and producing algal oils that can, in turn, be used to make biofuels. HR BioPetroleum Inc., incorporated in the State of Delaware and headquartered in the State of Hawaii, is a developer of large-scale microalgae production technology. It is a University of Hawaii, School of Ocean and Earth Science and Technology based company. [102] It offers algae products, such as algae oil, biodiesel, and animal feed proteins; carbohydrates for the production of ethanol and petroleum-based products; and military jet fuel. The joint venture was called Cellana LLC and Shell took the majority share.

Cellana sees itself possibly best positioned in the race to algae biofuels. *As for cellulosic alcohols, to be of any significance, algae technology needs to be scaled-up. Any aspiring algae company needs to find a route to mass markets. One route is to do this with a global partner, big enough to handle the technology risk.* For an algae company to be attractive for a large partner (or, to attract funding from any other source), it needs to have the right competences, and have structured and solid programs as part of its business plan (Table I.89).

Cellana should construct an algae-oil production facility to produce feedstock for biodiesel. The 3 MGY pilot plant was next to the Maui Electric power plant at Maalaea

(Hawaii) on the Kona coast of Hawaii Island. It would make the Maui Electric plant's CO2 emissions as feedstock for the algae [Lane 2008].

Assuming everything occurs successfully as planned the first phase of the commercial facility was envisioned to be in operation by 2011. The site, leased from the Natural Energy Laboratory of Hawaii Authority (NELHA) [101], is near existing commercial algae enterprises, primarily serving the pharmaceutical and nutrition industries. The facility would *grow only non-modified, marine microalgae species in open-air ponds using proprietary technology.*

An academic research program supported the project, screening natural microalgae species to determine which ones produce the highest yields and the most vegetable oil. The program included scientists from the Universities of Hawaii, Southern Mississippi and Dalhousie, in Nova Scotia, Canada. This demonstration plant would be an important test of the technology and critically of commercial viability. [102]

Table I.89: Innovation architecture and business plan components for the Shell – HR Biopetroleum joint venture Cellana, LLC. *)

Professional, experienced management ("veterans approach")	Edward T. Shonsey, CEO and Director, HR BioPetroleum, Inc. served e.g. as Executive VP, Internal Development of Verenium Corp. and as its Interim CEO and also CFO; also President and CEO of Syngenta Seeds Inc. C. Barry Raleigh co-founded HR Biopetroleum, Inc. in 2004; served as its Chairman and President; he is an experienced manager of large research organizations.
Top (experienced) algae scientists ("veterans approach") Technology has been validated in production of algae oil and antioxidants/carotenoids such as astaxanthan, at pilot operation located in Kona. Over a period of several years, Dr. Mark Huntley, CSO and a co-founder of HR Biopetroleum has utilized a proprietary two-stage algae cultivation system at this site; he overcame a number of key challenges – namely, open-pond contamination and low productivity.	Mark Huntley Chief Science Officer, co-founded HR Biopetroleum, Inc. in 2004, is a thought-leader in marine biological sciences generally and algae cultivation technologies specifically; has been active in algae-related research and development for more than 20 years; held research faculty positions at both the Scripps Institution of Oceanography and the University of Hawaii. Huntley held also a senior management role for 10 years in a special program; as an entrepreneurial CEO he took Aquasearch Inc., a marine biotechnology company focused on the production of astaxanthan and other value-added products from cultivated marine microalgae, from startup to a public market valuation of $200 million in four years; is skilled at organizing interdisciplinary R&D programs.

Algae production experience The key advancement made by Huntley and his co-inventor, Redalje, was to couple the continuous, large-scale production of a pure culture of algae in the sterile, controlled photo-bioreactors with the larger-capacity open ponds used for large-scale production	Dr. Donald Redalje co-founder and Member of Scientific & Technical Team, co-founded HR BioPetroleum Inc. and served as a Member of its Scientific & Technical Team; is a leading expert in microalgae physiology and biochemistry; He holds numerous patents in the area of marine biotechnology and is a co-inventor of the ALDUO™ process.
Structured programs	Strain selection, cultivation development, extraction, scale-up, product development
Strong partners: a) Scientific and b) commercial	a) Cellana has an academic research program including scientists from the Universities of Hawaii, Southern Mississippi and Dalhousie, in Nova Scotia, Canada b) capacity to take technology risk, professional execution, professional culture (Shell)
Location, pilot facility, Proximity to scientific and commercial partners	Will make the nearby Maui Electric Powerplant's CO2 emissions as feedstock for the algae; the "Kona Demonstration Facility" (KDF) on the Kona coast of Hawaii Island was leased by Cellana from the Natural Energy Laboratory of Hawaii Authority (NELHA); was also near existing commercial algae enterprises, primarily serving the pharmaceutical and nutrition Industries.
CO2 feed for algae	Access and permits to operate, piping etc. from Maui Electric plant to the adjacent algae facility; non-modified, marine microalgae species in open-air ponds
Sales contracts	Vegetable oil and protein/carbohydrates
Commercial roll-out plan	?

*) From the firm's Web site if not stated otherwise by a reference.

However, as observed for German CHOREN, in 2011 Shell withdrew from the project set up according to best practice and endowed with many resources (cf. also the approach of venture capitalist Vinod Khosla; Table I.98). H R BioPetroleum acquired Shell's shareholding in Cellana in January 2011; HRBP became the sole owner of Cellana, including its six-acre demonstration facility in Kona, Hawaii. Shell provided an undisclosed amount of short-term funding for Cellana during the transition. In 2011, it was one of the most advanced operational demonstration facilities among algae-to-biofuel organizations and companies in the US.

The envisioned first phase of the commercial facility to be operational by 2011 did not materialize. Still in the research and development stage, the company had been working toward developing a commercial facility on Maui by 2013. Obviously, there again was the issue of scaling-up. But, the focus on breeding and growing only non-modified, marine microalgae species and strain selection introduced another obstacle.

HRBP commented: "We will continue to operate Cellana's Kona demonstration facility and to continuously improve the economics for growing marine algae using HRBP's patented process." "Based on HRBP's and Cellana's results to date, we believe this technology holds great potential for the economical production of algae and algae-derived products for applications within the aquaculture and animal feed markets, as well as for the production of algal oil for conversion into biofuels." [Cocke 2011; Algae Industry Magazine 2011]

Biofuels Entrepreneurship Related to Algae in the US and Germany

In 2009 in the US there were about 100 startups [Voith 2009b] and worldwide there were an estimated 200 algae companies [Waltz 2009]. In this technologically highly risky field the American GreenFuel Technologies was the first high profile algal concern to go under, but it would not be the last. The prominent startup GreenFuel, which grew out of Harvard University and MIT research (founded in 2001, staff of 50 in early 2009 [Kanellos 2009b]), went bust early in 2009 after blowing through $70 million [Madrigal 2009].

Greenfuel raised millions of dollars for R&D, had a bioreactor development arrangement with the German firm IGV (Institut für Getreideverarbeitung) [Schibilsky 2008] and landed a high-profile deal in Spain to erect test facilities for an algae farm project. The multi-year deal in Spain was worth $92 million to build greenhouses that grow algae, which can be harvested for vegetable oil to make biodiesel or to make animal feed. The project developer was Spain's Aurantia; the algae would be fed sunlight and carbon dioxide from the Holcim cement plant near Jerez, Spain.

Processing was in vertical thin-film algae-solar bioreactors. Getting the whole thing to run smoothly, though, was tougher than expected. The company also found its system would cost more than twice its target [Kanellos 2009b].

The week GreenFuel folded, the DOE awarded an Arizona utility $70.6 million to scale-up the firm's technology [Waltz 2009]. Additionally, Mark Edwards cited by Rapier [2009c] also argued that GreenFuel made "some serious mistakes in executing strategy." "We are a victim of the economy," said (a representative of) a VC company which invested in Greenfuel.

However, the company had also been chronically saddled with delays and technical problems. In 2007, a project to grow algae in an Arizona greenhouse went awry when the algae grew faster than they could be harvested and died off [Rapier 2009b]. In retrospect GreenFuel's claims were viewed so overblown that they "became a joke" [Waltz 2009a].

And generally, there was *much suspicion about the claims of algae startups*, with regard to their technological development state, prediction of time when fully commercializing their technology and particularly with regard to numbers of quantities of biofuel (gal/ac/yr) to be obtained from land. And it was also questioned in how far data obtained on the lab scale will also be valid for large scale production.

Finally, several analyses pointed out algae firms' claims of productivity to violate various physical laws. Near-term technologies may allow algae to produce up to 6,000 gallons of oil per acre per year. "If you really push the limits, then maybe 10,000 gallons per acre," commented a researcher at a National Laboratory. "This figure could improve with advances in cultivation, species selection, breeding and genetic modification, but only to a certain extent. The laws of thermodynamics and the limits of photosynthetic efficiencies just won't allow it." "When you see 20,000 or beyond – that's total bologna." Yet there are companies claiming they can make up to 100,000 gal/ac/yr, and raking in tens of millions in investment based on those promises.

There was a broad opinion that "only a handful of companies are really serious." [Rapier 2009; Waltz 2009b] Wesoff [2009] presented the citations: 1) "As soon as I see an article touting algae's production of oil per unit area over terrestrial plants – I know the author(s) are clueless about the financial economics of algae fuel processing."; 2) "Bottom line – in our opinion the reality of economically viable algae fuel production is still quite a few years in the future – unless someone finds a truly novel short cut through the Laws of Thermodynamics and basic economics."

Correspondingly, there is a serious *credibility issue* for most of the numerous algae startups entering the scene and adding to the "algae bubble." "Most algae-to-fuel companies refuse to reveal much information about their technologies, which has led to more skepticism." [Waltz 2009b]

> Hence, one of *the basic prerequisite to start an algae firm is to get credibility*, for instance, by corporate venturing of big firms in the startup or research or development alliance agreements with big firms, public research institutes or universities, and creating an advisory board with scientists with high reputation.

After a proof-of-concept in the (research) laboratory in many cases the development status of algal startups is demonstration on the level of a Feasibility Assessment Unit (FAU).

In the US Solazyme which established an alliance with Chevron (Table I.83) was the first NTBF entered the algae biofuel field in 2003, after which the algae field has attracted many entrants. As Virent Energy Systems (Table I.85) Solazyme has its technological roots in hydrogen (H_2) generation. This occurred while the "Zeitgeist" of "The Hydrogen Economy" [Runge 2006:325] was in full swing.

With its growing investor population and grant acquisitions, strategic partnerships, and a varied product line (not wholly focused on biofuels) based on its platform technology, Solazyme was assumed to be able to weather future storms. Solazyme not only created biofuels for the transportation industry, but was experimenting with ways to tailor its processes and products for the cosmetic and food and feed industries (Figure I.176).

By the end of 2009, Solazyme has produced only limited quantities of biofuels, including several hundred thousand liters to the US military, which is seeking to promote alternatives to conventional fuels and it focused on the high margin areas of the algae business (Table I.90).

Table I.90: Solazyme's approach to biofuels and additionally high value products using algae. *)

Company (Foundation) Remarks	**Major Funding**	**CEO, Other Executives, Foundation, Key Researchers; Technology Protection and Position**
Solazyme, Inc. South San Francisco (CA) (2003) A renewable oil and bioproducts company utilizing innovative algal biotechnology Employees: 2006: >20, [Wilson 2007], 2007: 33 [ZoomInfo 2008], 2008: 45 (late '08 60) [Melendez 2009], 2009: 65 [Melendez 2009].	[CrunchBase 2009b] Latest round $7M Series C, brings the total funds to $76M; will be used for commercialization of its technology; Solazyme added $12M in an interim round standing at $57M [Lemos-Stein 2009] $45.4M, 8/27/08 (?) Total Investment:	Jonathan Wolfson CEO and co-founder; overseeing the management team and strategic direction of the company; held a variety of positions in finance, business, and law; was a co-founder and President and Chief Operating Officer of InvestorTree, a financial software and ASP services firm; also worked as an investment banker for Morgan Stanley, in the M&A department of Fried, Frank, Harris, Shriver & Jacobson, and also as a business/legal analyst; he holds J.D. (Juris Doctor) and MBA degrees. Harrison Dillon is the President, CTO and co-founder; overseeing technology strategy, intellectual property and legal affairs; he is trained in the field of microbial genetics, formerly managed the biotechnology patent portfolio of the University of Utah in the University's Technology Transfer Office. Dr. Dillon received a J.D. Arthur Grossman is the Chief of Genetics at Solazyme. Dr. Grossman is a world renowned scientist and has spent over twenty-five years in microalgal research.

Revenues: 2007: $2.6M [ZoomInfo 2008]. When doubled the size of the firm (from 25 to 52 employees) in 7 months developed a human capital management plan and changed organization was necessary ("10 – 25 - 150" rule, Table I.72) [ZoomInfo 2008; Yatedo].	$11.6M Series D (06/2009), $45.4M Series C (8/2008), $10M Series B (03/2007) (cf. also [Crunch Base 2009b] $5M Debt funding (09/2007) 2004: Harris & Harris Group, Inc. invested $310,000 as part of a larger funding round [Harris & Harris 2005]. 2009: received a $21.8M federal grant from US DOE to build its first integrated biorefinery in rural Riverside (PA); located on the site of Cherokee Pharmaceuticals' existing commercial biomanufacturing facility (make diesel from sawdust and cooking oil from food-factory waste; on an industrial scale), Solazyme will add $3.9M; goal: to prove this (refining) can be an in-	Currently, he shares his time between Solazyme and the Carnegie Institution/Stanford University where he is a Senior Staff Scientist at Carnegie and a Professor at Stanford; overseeing of current R&D projects and the development of strategies for initiating and implementing new projects. Prior to becoming a member of the management team, Dr. Grossman was the Chairman of Solazyme's Scientific Advisory Board and has been a key part of the R&D team since 2004. Jurgen Dominik is the SVP of Process Development; most recently, Mr. Dominik was senior vice president of global operations for CP Kelco focusing on manufacturing, logistics, capital spending, and process research. Mr. Dominik brings over 30 years of experience in technology, manufacturing, research and development, and international management. His duties included the management of design, construction, startup, and operation of large-scale fermentation facilities and natural product extraction manufacturing plants. Peter Licari is the Senior Vice President of Research and Development at Solazyme. Dr. Licari has over 20 years experience in biochemical engineering and bioprocessing, most recently as a Senior Vice President at Kosan Biosciences, Inc. where his responsibilities included development and manufacturing operations; he served also as Senior Scientist at BASF Bioresearch Corporation, responsible for fermentation process development. In 2005 he obtained an MBA degree. David Brinkmann VP of Manufacturing, has over 30 years of technical, operations, and leadership experience in the bioprocess industry. Mr. Brinkmann came to Solazyme from CP Kelco, where he spent ten years managing all operational aspects of a large biotechnology pilot plant and semiworks that provided process R&D and manufacturing support programs, such as productivity improvement, cost reducetion, and new product development. David Isaacs SVP of Government Relations of Solazyme, is responsible for formulating and executing the company's government strategy at the federal and state level. In this capacity, he is working to advance the company's

	dustrial method, not just an experiment [DiStefano 2009]; two contracts with the US Navy, one worth $200,000 to supply 1,500 gallons of jet fuel in 2010; another separate contract worth $8.5 mil. to research, develop and deliver 20,000 gallons algae derived fuel for use in Navy ships [DiStefano 2009; Nagappan 2009]. Both grants used for learning to scale-up 2007: received a $2M grant from the National Institute of Standards and Technology to develop a substitute for crude oil based on algae [Bullis 2008a].	priorities on funding opportunities, legislation, regulatory matters, and policy, particularly with regard to the Congress, the Departments of Defense and Energy, the Environmental Protection Agency, and key states (cf. also BluefFire, Table I.86). Solazyme believes in intellectual property and in-licensed technology; has issued patents and more than 20 (published) patent applications. The founders Dillon and Wolfson met on their first day of college back in the 1980s; wanted to start a company together but had only a vague notion of what to do. Dillon went into biotechnology and ultimately also became a patent attorney. Wolfson got a law degree and an MBA and went into finance [Kanellos 2008]. Foundation of the company in Dillon's garage in 2003 [DiStefano 2009]. Originally focused on hydrogen (H_2) production by GM algae [Kanellos 2008] (Dillon granted patent US 7135290 "Methods and compositions for evolving hydrogenase genes"). When the two first sought funds, most venture capital firms were intrigued by their idea, but did not know how to position the company in their portfolio. Ultimately, investors shuttled the two to the partners who handled pharma deals [Kanellos 2008]. "We're developing a production platform (and) are looking for commercialization partners." It doubled its lab and office space to 7,000 square feet and increased head count from 20. Wolfson said Solazyme had narrowed its initial algal library from "the low thousands" to five or six strains [Wilson 2007]. Patented process: maximize triglyceride production through fermentation (US Patent 2008/0124756 A1) [Waltz 2009b].
Technology, Goals, Strategy	Solazyme uses synthetic biology [Waltz 2009b] for the renewable production of biofuels, industrial oleochemicals, and health and wellness ingredients. Self-Description: It modifies microalgae to produce tailored triglyceride oils that can be re-	

	fined into biodiesel in the same facilities that refine petroleum. They produce crude oil, which can be turned into anything that is made from oil. They genetically engineer the cells' ability to handle different feedstock, as well as the structure of the oil produced. It is not photosynthetic. *Some algae naturally produce oil more effectively when fed biomass in the dark* – an adaptive mechanism that allows them to survive in the event that sunlight is blocked for extended periods. Solazyme enhances this ability. It feeds its algae various cellulosic and other waste materials rather than CO2 and sunlight. Some algae produce polysaccharides from biomass, instead of oils. Solazyme uses synthetic biology also to modify these, as well, and some may be commercialized. The company uses different strains of algae to produce different types of oil. The process also has significant advantages. First, keeping the algae in the dark causes them to produce more oil than they do in the light. That is because while their photosynthetic processes are inactive, other metabolic processes that convert sugar into oil become active. Just as important, feeding algae sugar makes it possible to grow them in concentrations that are orders of magnitude higher than when they are grown in ponds. Solazyme uses traditional industrial fermentation equipment [Voith 2009b]. Solazyme is the only microbial biofuel company to produce an oil-based fuel, Soladiesel®. Solazyme said it produced thousands of gallons of fuel from algae that was tested to meet strict ASTM international standards for jet fuel. Solazyme has already road-tested its diesel fuel for thousands of miles in unmodified cars. Co-founder Dillon said the company is about 24 to 36 months away from hitting its target manufacturing cost of \$2 to \$3 a gallon, or \$40 to \$80 a barrel [AP 2009b]. Feed wood chips, switchgrass, waste glycerin (cf. BioMCN, Table I.87) to algae in a process where the algae will convert that biomass into crude oil, which "can be used to make" diesel fuel, jet fuel, high-nutrition edible oil like olive oil, or plastics. As Solazyme can convert sugars directly into oils without photosynthesis this allows the organisms to be grown in fermentation tanks, which reduces the costs of a still-expensive process. Solazyme is not only focused on scale-up, production and road testing of a variety of advanced biofuels, but simultaneously diversifies its platform into other products and markets. It develops products across distinct market segments, leveraging algae's unique oil and material production capabilities: - Fuels - Chemicals,

	- Nutritionals (Human + Animal Nutrition), e.g. selling the non-oil biomass for animal feed, - Health Sciences (Cosmetics + Nutraceuticals). Solazyme is interested in developing collaborative R&D and commercialization relationships, working with partners to develop and commercialize new or improved processes and products in its target markets (e.g. Chevron, BlueFire, Imperium Renewables, Inc.) The US Navy agreed to pay for 22,000 gallons of Solazyme jet fuel and ship fuel for delivery in 2010 [DiStefano 2009]. B&D Nutritional Ingredients (B&D) has formed a strategic partnership with Solazyme Health Sciences whereby B&D will promote and distribute Solazyme Golden Chlorella {algae} products for the functional foods and dietary supplement markets [Neutraceuticals World 2009]. In 2010, the Company launched its products, the Golden Chlorella line of dietary supplements. In March 2011, the Company launched its Algenist brand for the luxury skin care market through marketing and distribution arrangements with Sephora S.A. (Sephora International), Sephora USA, Inc. and QVC, Inc. In 2010, Solazyme and Roquette, the global starch and starch-derivatives company headquartered in France, formed Solazyme Roquette Nutritionals, a joint venture bringing to market an entirely new line of microalgae-based food ingredients. [103]

*) From the firm's Web site if not stated otherwise by a reference; firm's status: early 2010.

Indeed, Solazyme has positioned itself first and foremost as a "renewable oils" company, as a producer of sustainable *triglyceride oils* that can be used to replace or enhance oils derived from petroleum, animal fats, and plants. According to Figure I.176 fuels actually stand at the bottom of the company's target markets when considering the concept of profit margins.

Correspondingly, in April 2012 Solazyme announced an agreement with Bunge Global Innovation to build a factory in Brazil that would make triglyceride oils for both chemical and fuel products. Under the joint venture, whose financial terms were not disclosed, the factory would rise next to Bunge's Moema *sugarcane* mill and have an annual capacity of 100,000 metric tons of oil. It would start production in the second half of 2013, making oils for fuel as well as additives for soaps, detergents and plastics [Cardwell 2012].

In February 2011, Solazyme entered into a joint development agreement with The Dow Chemical Company to jointly develop and commercialize non-vegetable, microbe-based oils and related products like those for dielectric insulating fluids and other industrial applications. In March 2011, the company entered into an agreement to purchase a development and commercial production facility with multiple 128,000-

liter fermenters, and an annual oil production capacity of over 2,000,000 liters (1,820 metric tons) located in Peoria, Illinois for $11.5 million [Riddell 2012].

In March 2011 Solazyme filed for a $100 million IPO. Solazyme had found little support over the last year when it comes to Wall Street. The company, which priced its initially well-received IPO at $18/share, managed to raise $227 million that year. In doing so, it raised a sum of capital that has thus far been projected to be adequate to sustain the company until it becomes cash flow positive. But, by May 23, 2012 the closing price was $9.72, a 46 percent discount to its IPO price (ca. $11 end of 2013). The market and its analysts continued to express doubts over the advanced biofuels industry [Bomgardner 2011d; Quon 2012].

The company had raised $128 million in venture capital since its founding in 2002. In its Annual Report 2011 (US SEC Form 10K) Solazyme reported revenues of $38 million for 2010 and a net loss of $16.3 million and $39 million for 2011 and a net loss of $53.9 million. However, a breakdown of total revenues (2011) reveals that there are only $12 million in reveues from selling products or licenses:

Research and development programs	$26.793 mio.
Product revenue	$7.173 mio.
License fees	$5.000 mio.

This is – after almost nine years of existence – an amount obtained also by the algae firms Cyanotech Corp in the US and German BlueBiotech GmbH mentioned above. Particularly with regard to nutritional and cosmetics aspects one can expect Solazyme to encounter difficulties in selling its products in certain areas of the world, particularly Europe, due to the use of genetically modified objects for production.

The promise of using algae to make biofuels – a dream scientists have chased for decades – might have seemed particularly welcome in a time of stubbornly high oil and gasoline prices. But the path to commercial-scale production has been circuitous.

Algenol Biofuels (Table I.91) with its proprietary Direct to Ethanol™ process is the only bioethanol producing algae company. Furthermore, rather than harvesting the algae as such for post-processing of algae's "biomass deliverables" the "algae" are utilized "like getting the milk rather than killing the cow" [Woods 2009a; Woods 2009b].

It actually uses *cyanobacteria* (formerly known as blue-green algae). Cyanobacteria have been isolated from various habitats both from freshwater and marine systems. Since cyanobacteria show an impressive variation in physiological properties a collection of cyanobacteria from various habitats also provides an excellent platform for experimentally selecting suitable strains.

The organisms produce some ethanol naturally, but Algenol has patents to selectively breed and genetically manipulate them to pump out more. In essence, each cell acts as a tiny ethanol factory (Figure I.178).The Algenol process consumes CO2, which means biologic sequestration and generates photolytically sugar which is fermented intracellularly into ethanol. The outstanding feature of this technology is that it fundamentally uses carbon dioxide as its source to make ethanol. However, some types of cyanobacteria are responsible for so-called "algae blooms" through explosive growths and some cyanobacterial blooms are toxic. Some common bloom-forming species produce potent toxins that can even be lethal.

Algenol is *privately funded* and was not seeking outside investment during the stages of development and primary commercialization. Algenol's ongoing financing needs continued to be financed privately through the initial commercialization phase but Algenol continued to investigate the optimal *funding opportunities* including licensing fees and royalty payments, partnering arrangements, government financial support and government land access for facilities. In the US, in particular, Algenol planned to seek federal, state and local assistance to bring US facilities online [Gelsi 2008].

Algenol works directly and indirectly with a number of collaborators and (small to large) companies in the US and Germany to further develop and commercialize its existing technology. It reported to have 100 employees and consultants (40 Ph.D.'s) in early 2010 and biological laboratories in Baltimore, Maryland and Germany [Woods 2009b]. This makes Algenol's *entrepreneurial constellation rather complex* concerning other involved companies, intellectual property rights (IPRs), scientific advisors and R&D arrangements, credibility build-up as well as strategic partnerships with large firms and financing through federal and county agencies.

Apart from its own research efforts in Baltimore and cooperation in the US Algenol organized massive R&D support in Germany (Figure I.179; Box I.25). Algenol was getting help from experts at the Johns Hopkins University and the University of Maryland Biotechnology Institute. Frank Robb, a professor at UMBI's Center for Marine Biotechnology, had been contracted by the company to help with its research. Joseph Katz, a professor of mechanical engineering at Hopkins, signed on as a consultant, in part to help design the "bioreactors" in which the blue-green algae are to grow [Wheeler 2009].

As outlined for BioMCN (Table I.87) and ZeaChem (Table I.88) Figure I.179 exhibits *Relationship Mapping*. This is a method of visualizing, describing, and analyzing all the individual and organizational relationships of an existing firm, may be also prospective business partners, by establishing a *dynamic* "map." This map provides background on the target company by showing direct, indirect, and business and social relations among individuals within the organization and of the company as a whole.

Networking of NTBFs discussed so far, hence, reflects a special relational mapping restricted to organizational relationships for a particular time span.

Figure I.179 is an illustration of what is described in detail in Table I.91 and Box I.25.

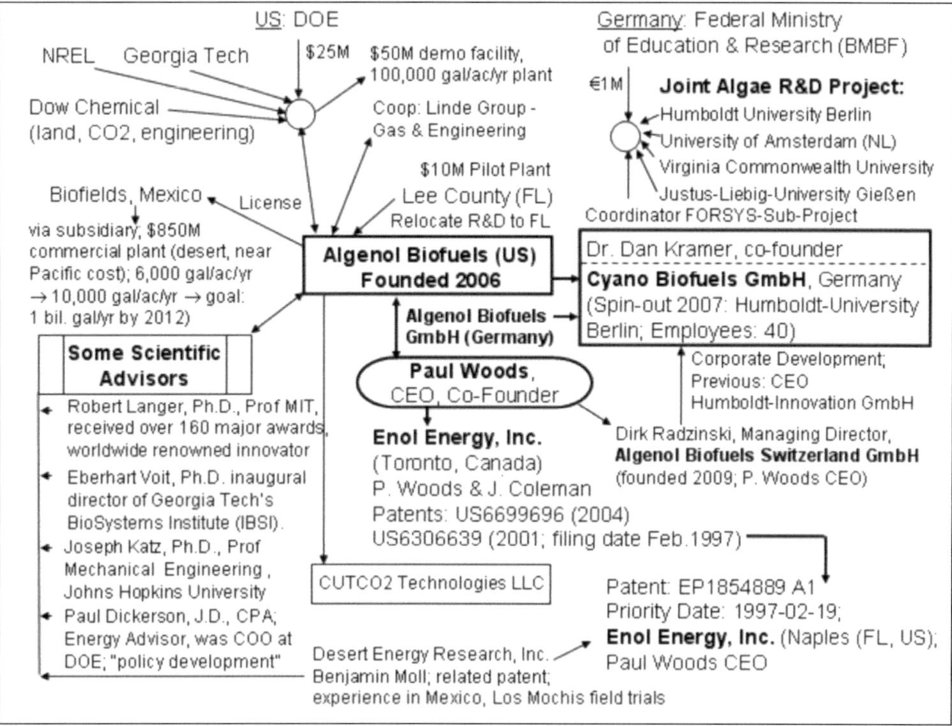

Figure I.179: Algenol's encapsulated web of firms, R&D and financial resources, partnerships as well as persons (by 2009/2010).

Table I.91: Algenol Biofuels and its path to bioethanol directly from algae. *)

Company (Foundation) Remarks	**CEO, Other Executives, Foundation, Key Researchers; Technology Protection and Position**
Algenol Biofuels, Inc. Bonita Springs (FL) (2006) Administrative offices in Bonita Springs, but a laboratory in Baltimore (MD) with ca. 15 company scientists and technicians	Paul Woods CEO, co-founder and co-inventor of Algenol's basic technology; when working as a student in Canada in genetics at Western Ontario University in 1984, Paul Woods discovered a method for producing ethanol from cyanobacteria (blue-green algae); related patents with Prof. John Coleman (now Algenol's CSO) as a co-inventor are assigned to the firm Enol Energy, Inc. (Figure I.179); The split of IPR between a firm owned by a person to just exploit these in another startup is also observed for BlueFire (Table I.86). Wood's invention did not appear to have much commercial promise until the early 2000s. Meanwhile, he started a couple natural gas businesses in Canada and US. He started his business career in 1989 at Alliance Gas Management, which completed an IPO in 1997. He built the firm, raised

[Wheeler 2009]. **Major Funding:** Some $70M in private backing [Gelsi 2008; Wesoff 2009], invested by P. Wood personally and a few partners [Wheeler 2009] 2010: Lee County (FL) approved a contract with Algenol, a $10M grant to build a facility during the next two years [CNN 2010] Algenol got a Department of Energy (DOE) grant for up to $25M, or no more than half the cost of a $50 million facility at a Dow Chemical plant.	$80 million in capital, and sold the business in 1999. In 1997 he founded United Gas Management Inc. in the US. But the natural gas marketing company he launched wound up in bankruptcy and was sold in 2000. He said he retired after that, until launching Algenol [Wheeler 2009]. In March 2006 Woods formed Algenol along with Craig Smith and Ed Legere armed with patents, several test facilities around the world and, according to the literature, some $70 million in private backing [Gelsi 2008]. He and his partners started Algenol Biofuels Inc. to commercialize the process "algae-to-ethanol" on an industrial scale. More on the background of Algenol's foundation and the associated foundation of Cyano Biofuels GmbH (Figure I.179) in Germany is described in Box I.25. Craig Smith, MD Executive VP COO and co-founder; from 1993 to 2004 Dr. Smith served as Chairman, President and CEO of Guilford Pharmaceuticals, Inc., a publicly held biopharmaceutical company that he co-founded in 1993; from 1988 to 1992 Dr. Smith was a VP and Senior VP of another publicly held biotechnology company; from 1975 to 1988 he served on the faculty of the Department of Medicine at The Johns Hopkins University School of Medicine; has served on several corporate and charitable boards and is still a member of the Johns Hopkins Alliance for Science. Edward Legere, MBA Executive VP and CFO and co-founder; he has over 18 years experience in the biotechnology industry as a consultant and active business manager and has over ten years of public company experience in the role of member of the Board of Directors; he has also served as the President and CEO of the publicly traded biotechnology company Peregrine Pharmaceuticals, Inc. John Coleman, Ph.D. CSO and co-inventor, was the Vice-Principal (Research and Graduate Studies) at the University of Toronto Scarborough, and a Professor in the Department of Cell and Systems on the St. George campus of the University of Toronto prior to joining Algenol. He was Chair of the Department of Botany from 1998 to 2004 and played a primary role in the formation of the new Department of Cell and Systems Biology in 2006. Dr. Coleman's research interests and experience lie in the fields of molecular biology and biochemistry of photosynthetic carbon metabolism in higher plants and cyanobacteria. Initial proof of science for Algenol's technology was generated by Dr. Coleman at the University of Toronto between 1989 and 1999. Since then, the process has been refined to allow algae to tolerate high heat, high salinity, and the alcohol levels present in ethanol production. The key basic patents for Algenol's technology are those assigned to Woods and Coleman (and a third inventor) given in Figure I.179.

Technology, Goals, Strategy Business Model: Owner-operator and partner for bioalcohol production from algae; Expand into "green carbon" building block monomers (bio-refinery model), such as ethylene or propylene (from ethanol); Look also into biobutanol production [Woods 2009c] Licensing business **Revenue** (2010): $3.1 mio. (according to Wikipedia) **Employees**: 120	Using a patented metabolically enhanced cyanobacteria, or blue-green algae, to directly synthesize ethanol – one of the few, if only, companies working with direct production of this algae; blue-green algae, since they also use photosynthesis – sunlight – to convert nutrients and carbon dioxide into fuel. Algenol estimated that its technology can produce 10,000–12,000 gal/ac/yr of ethanol in the near term. This has been questioned based on a simple calculation which would make it hard to believe those yields will make Algenol's biorreactors economical [Waltz 2009b]. So far, Algenol claims its test facilities have yielded 6,000 gallons of ethanol per acre per year, with yields expected to grow to 10,000 gallons of ethanol per year. The metabolically enhanced algae are resistant to high temperature, high salinity, and high ethanol levels; they do not produce human toxins. Algenol's "Direct-to-Ethanol" process gathers ethanol produced by algae without destroying the algae and without the necessity of refining oil into biodiesel. This method, if viable and scalable, seems to have huge potential cost and embedded-energy advantages. "Ethanol is almost infinitely mobile in a cell, and essentially leaks out into the bioreactor after synthesis," Coleman said. "Through some various condensation steps we collect it." [Hamilton 2009] The algae strains are genetically modified – and that might be a hard sell in the US [Wesoff 2009] and elswhere. The production plants will need vast tracts of land for row upon row of algae-filled bioreactors (filled with sea water), but the company is targeting desert or arid lands; so no usable farmland will be taken out of cultivation. The ethanol-making process will yield fresh water as a by-product (Figure I.178) which could be used to irrigate nearby lands. [Wheeler 2009] Algenol had licensed its technology to Biofields of Lomas de Chapultepec, Mexico (for more than $100 million [Donner 2011]). Biofields said it has committed $850 million to building the industrial-scale ethanol facility on 102,000 acres in the Sonora Desert. As the algae grow, Algenol will tap into carbon dioxide from a nearby power plant and funnel it into the tanks. By 2012 Biofields' subsidiary Sonora Fields S.A.P.I. de C.V. intended to produce a whopping one billion gallons of ethanol per year. Much of the ethanol shall be transported by ship to Mexican oil refineries nearby to be blended into gasoline [Waltz 2009b; Gelsi 2008]. The Mexican company secured an exclusive license for the Algenol technology until 2013 when the company expected to reach its 250 MGY target. It will be run by a former Mobil Oil senior construction executive. According to CNN for expansion, by the end of 2009, Biofields had invested $30 million in the project, which is reporting yields of 6,900 gallons per acre at the Sonora site [Lane 2010b]. As displayed in Figure I.179, by December 2009 Algenol got a Department

	of Energy (DOE) \$25M grant (approximate half of the estimated cost of a \$50 million facility) to install a photobioreactor-based algae-to-ethanol demonstration plant at a Dow Chemical site in Freeport, Texas [Voith 2009; Wesoff 2009; Hamilton 2009]. The rest of the capital would be provided by Algenol, which would also own and operate the plant. Dow would contribute 25 acres of land, the CO2 supply and technical expertise. Dow's chemists and engineers would help design a process that can scale-up for commercialization [Voith 2009a]. In Algenol's case, the photobioreactors are simply plastic covered troughs housing a mixture of saltwater, algae, nutrients, and CO2; plastic material will be supplied by Dow [Wesoff 2009]. The initial target production was up to 140 gallons of algae fuel per day, or 51,000 gallons per year at a yield of 2,120 gallons per acre [Lane 2010n]. Dow would also help develop advanced plastic films for covering the bioreactors. But fuel-quality ethanol must be distilled from the bioreactor condensate, which is a major focus for the pilot plant. For this process Algenol will use advanced membrane technology and separations, a Dow competency, that are more energy efficient [Voith 2009a]. Dow would also assist with process engineering [Hamilton 2009]. The \$25 million Energy Department grant to help fund the Algenol Dow plant designed to produce 100,000 gallons of ethanol a year at a target cost of \$1-\$1.25 per gallon will be associated by continued research by the National Renewable Energy Laboratory (NREL) and Georgia Tech [Clanton 2009]. The Algenol-Dow experiment could have several important implications. It points the way to a more sustainable path for making ethanol as a biofuel, but also producing common building block chemicals now derived from fossil fuels (such as ethylene). With regard to process engineering Algenol followed also a further route (outside Dow). By the end of 2009 Algenol and the German Linde Group, an internationally leading gases and engineering company, have agreed to collaborate in a joint development project in order to identify the optimum management of carbon dioxide (CO2) and oxygen (O2) for Algenol's algae and photobioreactor technology. The cooperation will aim the companies to join forces to develop cost-efficient technologies that capture, store, transport and supply CO2 for Algenol's proprietary process for the production of biofuels. In 2008 Algenol Biofuels and Codon Devices announced that the companies have entered into a multi-year partnership utilizing Codon's proprietary BioLOGIC^{TM} protein engineering platform. A key strength of Algenol is: "Exceptional group of partners from industry and academia, some are announced publicly, some surprises are in store…" [Lane 2009c]. Some of this is discussed Box I.25.

	As Algenol sought to seek federal, state and local assistance to bring US facilities on line [Gelsi 2008], Algenol accompanied technical advances with Current Policy Initiatives [Woods 2009b].

*) From the firm's Web site if not stated otherwise by a reference; firm's status: end of 2009.

It is interesting to note that the other US chemical giant, DuPont, concentrates on macro-algae to produce biobutanol (Table I.83).

Box I.25: Algenol Biofuels as a catalyst for helping to found and use Cyano Biofuels GmbH in Germany as an "external" R&D resource.

When Algenol founder Paul Wood re-assed the commercial promise of his early algae-to-ethanol invention during the emergence of the "algae boom" of the early 2000s he turned to Germany, in particular to Cyano Biotech GmbH co-founded by Dr. Dan Kramer (with two additional co-founders) in 2004 in Berlin, Germany. As he has been occupied by research of natural products delivered by bacteria the foundation idea of Dan Kramer was to investigate the little studied potential of cyanobacteria to produce natural products, particularly those that have application in or for pharmaceuticals, utilizing genetic optimization.

Cyano Biotech GmbH was created as a spin-out from the Institute of Biology of Humboldt-University, Berlin, with the aim to commercially exploit the results of 20 years of R&D in the field of cyanobacteria generated by the group of Prof. Dr. Thomas Börner (Dept. of Genetics), in particular, by metabolic engineering. On the basis of "incubation options" ("Gründerlabor") and special "seed fundings" and grants by German ministries and Berlin authorities during the first three years Cyano Biotech utilized the university laboratories and facilities [BMWi]. By the end of 2007 Cyano Biotech GmbH moved to own facilities in the Technology Park Adlershof/Berlin.

During the pre-startup phase, prior to foundation, Dan Kramer performed a feasibility study [122], including a proof-of-concept, in cooperation with an industrial partner in Potsdam adjacent to Berlin with respect to the commercial opportunity of cyanobacteria as a source for biologically active compounds [BMWi]. There is high evidence that this firm in Potsdam is AnalytiCon Discovery GmbH (an R&D spin-out of AnalytiCon AG) and is currently a cooperation partner of Cyano Biotech. With currently 60 employees it is one of the leading companies in research and development of products made from natural materials for the pharmaceutical, food, and cosmetics industries. AnalytiCon Discovery claims to be leading worldwide with libraries (MEGAbolite®, NatDiverse™) of natural product small molecules as screening compounds to accelerate natural product based drug research.

After two years the feasibility study was completed and also a business plan finalized. Dan Kramer acting as managing director looked for two technology-oriented co-founders, one being a specialist in natural products research, the other in cyanobacteria and bio-active compounds. After the "seed phase" financing of Cyano Biotech was by a special governmental support program comprising a public subsidy and equity stake by a governmental equity investment organization for NTBFs (tbg – Technologie Beteiligungsgesellschaft). Based on his experience as project leader Dan Kramer grasped the essentials of "professional management" by a "learning-by-doing" approach; accounting and controlling activities for the firm were outsourced [Bowie].

Cyano Biotech, (with 8+ employees in 2008 [Bowie]), is one of the few firms, if not the only one, in the world that identifies and characterizes natural products by screening cyanobacteria ("drug discovery"). It optimizes derived related bioactive compounds regarding pharmacological needs employing combinatorial biosynthesis to generate novel lead compounds (or "active pharmaceutical ingredients") for the pharmaceutical industry. A part of the natural products was sold as extracts from bacteria. But another firm was planned to commercialize the isolated natural products itself [Technology Park Adlershof 2008]. In 2010 Cyano Biotech had revenue of €276,000 (according to the German Creditreform – Firmenwissen Database).

Cyano Biotech's business is aligned with the diverse microalgae-based products and applications in the pharma, agrochemical, food and cosmetics industry as well as in regard to the sustainable production of raw materials, CO2 sequestration and waste water sanitation.

Equally important for the assessment of the entrepreneurship is the fact that, within ca. 80 miles, Humboldt-University in Berlin is at a center of various cyanobacteria R&D (Figure I.179) and commercialization activities for more than 20 years which covers other research universities, public research institutes and private firms involved in cyanobacteria commercialization. This constellation appears formally as a part of the general biotechnology "Competence Cluster Berlin-Brandenburg" which has additional links to other related areas in Germany and Austria. For instance, a scientific advisor from the Austrian Academy of Science runs simultaneously a cyanobacteria project with Cyano Biotech.

By subject and by (spatial) proximity Cyano Biotech's leaders can be assumed to have had links to people in a related "*joint project*" (in German Verbundprojekt) of the cluster. A "Verbundprojekt" is a special German approach to technical innovation (ch. 1.2.6). It is a systemic interconnection tied together by a common explicit goal (and achievable result) through coordination and control and assigning different contributing sub-projects to different elements (organizations) of a related value system.

This project was the "Hydrogen from Microalgae: With Cell and Reactor Design to Economic Production" project (HydroMicPro; Table I.92) funded by a €2.1M grant of the Federal Ministry of Education and Research (BMBF) and conducted jointly by universities, research institutions and enterprises. The notion microalgae comprises essentially also cyanobacteria.

HydroMicPro targeted developing an inexpensive, highly efficient production process with optimized biology and process technology, which is suited for the mass production of hydrogen. It focused on the photobioreactor, gas separation by membrane processes, biological sensor technology for cellular oxygen, biotechnological optimization of algae, and systems integration and was led and coordinated by the Karlsruhe Institute of Technology (KIT). It included also practical field tests [Knuber-Knost 2009].

A tempting offer from America (Paul Woods to Dan Kramer of Cyano Biotech) marked the birth and catalysis for further development of Cyano Biofuels GmbH. Decades of research and commercialization activities and competence in and around Berlin concentrating on cyanobacteria obviously attracted Paul Woods [Technology Park Adlershof 2009].

Paul Wood's idea to further develop and commercialize his discovered method for producing ethanol from cyanobacteria led him look for partners to assess his approach for feasibility (cf. below Perkin case; A.1.2). And linked to the available competence the order for the feasibility study was addressed to Berlin.

According to Dan Kramer the promising results initiated foundation of a second firm in April 2007 ("Unsere Untersuchungen haben so gute Ergebnisse geliefert, dass wir im April 2007 eine zweite Firma gegründet haben," said Kramer) [Viering 2009] Together with a colleague (Dr. Heike Enke) Dan Kramer founded Cyano Biofuels GmbH, again as a spin-out of Humboldt University. Cyano Biofuels with about 40 employees in 2009 [Seidler 2009, Martin 2009] and more than 30 scientists [Humboldt Innovation 2009] succeeded in making the cyanobacteria produce preferentially ethanol rather than sugar by a proprietary process.

The formal setup of Cyano Biofuels as an LLC (GmbH) involved legally several partners (Figure I.179) – from Cyano Biotech and from Algenol Biofuels GmbH. According to an official legal documentation ("Elektronischer Bundesanzeiger") Algenol Biofuels GmbH was a one-man-firm of Paul Woods without any employees. Algenol Biofuels Inc, Bonita Springs, FL, USA is the full owner of Algenol Biofuels GmbH. Additionally Algenol Biofuels Switzerland played a role (Figure I.179).

Concerning genetic engineering of cyanobacteria, the rod-shaped bacterium *Zymomonas mobilis* for making Mexican agave tequila turned out to be three to four times as efficient as local beer yeasts when applied to the plants. Correspondingly Cyano Biofuels took the gene for the enzyme pyruvate decarboxylase (Figure I.178) out of the tequila tribe and implanted it into the genome of cyanobacteria [Donner 2011].

In the US using corn for biofuels reached 3,700 liters per hectare per year. Under *ideal laboratory conditions* researchers of Cyano Biofuels in Berlin claimed to have achieved 112,000 liters. And in November 2011, a pilot plant with 3,000 tubes as bioreactors was announced to be built in Texas [Donner 2011].

In essence Cyano Biofuels is a research and development firm for biofuels and building block chemicals utilizing optimized cyanobacteria to be channeled into industrial scale production via a direct process [Humboldt Innovation 2009]. The vision and the mission of Cyano Biofuels embraces knowledge and know-how transfer from a university into business, development and optimization of cyanobacteria for the production of biofuels (currently bioethanol).

Some key challenges on the route to commercialization of algae-to-ethanol not tackled by Cyano Biofuels include design of massive pipeline systems to bring seawater to the algae, cleanliness for the large scale growing and processing units to avoid other bacteria to overcome the cyanobacteria in the bioreactors and separation of ethanol [Seidler 2009].

In 2010 Cyano Biofuels was acquired by Algenol Biofuels and is now a member of the Algenol Group. Algenol LLC, Bonita Springs, Florida, held previously a minority position in Cyano Biofuels. According to an official legal document ("Elektronischer Bundesanzeiger") the minority stake was 40 percent which was transferred to Algenol Biofuels Switzerland GmbH.

Growth of Cyano Biofuels in terms of number of employees was:

2011 ca. 50 employees [Donner 2011], 2010 ca. 40 employees, 2009 33 employees, 2008 24 employees (the last figures from "Elektronischer Bundesanzeiger").

The Institute of Biology of Humboldt-University, Berlin (Dept. of Genetics), is not only the connection between the startups Cyano Biotech and Cyano Biofuels (via the co-founder Dan Kramer). It has two working groups dealing with the production of biofuels or hydrogen by cyanobacteria providing simultaneously the connection to the joint project HydroMicPro (Table I.92).

Furthermore, it participates in the FORSYS (*Research Units for Systems Biology in Germany*) research project funded Germany-wide by the Federal Ministry of Education and Research (BMBF) with € 45 millions until the end of 2011. The co-founder of Cyano Biofuels acts here as the project coordinator (Figure I.179) for the FORSYS sub-project "Systems Biology of Cyanobacterial Biofuel Production." This sub-project is supported by a €1M grant from the BMBF [Kramer 2007; Glocalist 2008].

When Cyano Biofuels was taken over by Algenol Biofuels Paul Wood characterized the situation as follows (Humboldt Innovation, Mar. 23, 2011):

"The Greater Berlin is known for its diverse and high-profile research on microalgae and a perfect place to attract research talent for Algenol. We want to support Cyano Biofuels' networking with German universities and further believe that generate new ideas and technologies emerge."

An overall position of a joint project or "joint R&D project", respectively in the German industry development approach, such as HydroMicPro, is shown in Figure I.40 in relation to a structural value system. Furthermore, there is a planned connection to the Enertrag AG hybrid power plant project combining wind, hydrogen and biogas (Figure I.104) located also in the German competence region for algae.

Table I.92: Example of a "Verbundprojekt" in Germany ("Hydrogen from Microalgae" – HydroMicPro) to illustrate its systemic character.

Organization, Partners	Sub-Project Goal
Professor Clemens Posten KIT Institute of Life Science Engineering (Karlsruhe)	*Coordination*: Expected result: prototype reactor that allows an economically efficient hydrogen production by microalgae. Next steps for large plants: Automation of the plant, optimization of service life, and mass production of materials (microstructurization of membranes and coating of transparent materials).
KIT Microalgae Working Group; KIT's Engler-Bunte Institute (from Chemical and Process Engineering Department)	Develop an optically structured photobioreactor (first enlarge the inner surfaces of the reactor); Photobioreactor: First step, a high amount of biomass shall be produced. Such high amounts will also be needed for producing other valuable materials from algae in the future. Second step, the system will be optimized for hydrogen production.
University of Bielefeld	Identification and biotechnological optimization of the algae for biomass and H2 production in the photobioreactors.
Organization, Partners	**Sub-Project Goal**
Max Planck Institute for Molecular Plant Physiology (MPI), Potsdam (adjacent to Berlin)	Focus on the regulation and control of cell-internal oxygen concentration and develop a sensor based on transgenic and physiological/biochemical processes
University of Potsdam (adjacent to Berlin)	Develop a method to measure cell-internal oxygen concentration.
Ehrfeld Mikrotechnik BTS GmbH (EMB), Wendelsheim	Provide experience gained from using a micro photobioreactor and elaborate the concept of the production plant.
IGV (Institut für Getreideverarbeitung) GmbH, Nuthetal (near Berlin)	Contribute experience from the use of a thin-film reactor; assess the new process under close-to-practice field conditions. (IGV will be described later)
OHB-System AG, Bremen (ch. 2.1.2.4; [Bläske and Kiani-Kreß 2010])	Connect the bioreactor system for hydrogen production to energy-transforming systems; in addition, evaluate adaptation of the system to space applications.

Table I.92, continued.

The KIT Institute for Technology Assessment and Systems Analysis (ITAS)	Evaluate the potential contributions of HydroMicPro technology to sustainable hydrogen supply; identify the ecological and socioeconomic "hot spots" of the technology; feedback: based on the results, processes for hydrogen production from microalgae will be improved and appropriate applications in the energy system identified.

Another algae startup in the US, Sapphire Energy, Inc. (Table I.93), is not only notable with regard to the huge amount of invested capital, but how it positioned itself: "We're an energy company," "and, really, who we are competing with is big oil and gas." [Schwartz 2010]. The founders express competitive antagonism (Table I.32).

During the process of setting up Sapphire founder and serial entrepreneur Jason Pyle was also in contact with Big Oil: "We've had conversations with all six of the largest oil companies in the world." [Bigelow 2008] Morrison [2008] characterized this situation from a different perspective: One of the distinguishing factors of algae startups is that they tend to dream, and talk, rather large, and Sapphire is no exception. Like other companies, however, its algae have yet to be proven at commercial scales.

The origins of Sapphire began in 2006 as a handful of venture capital leaders began looking for the right technology. Typically, the innovator who develops a new technology looks for the right venture capital firms to provide funding for the idea. Pyle said his discussions began with Kristina Burow, a chemist-turned-partner at Arch Venture Partners, biotech CEO Nathaniel David and scientist Mike Mendez. "We started *analyzing different kinds of biofuel deals and technologies* and asking ourselves what's great about this and what's not," Pyle said (emphases added).

After determining that their best prospect was to become a producer of gasoline and diesel fuels Pyle said they set out to identify the best green technologies for making it. They found what they were looking for in the research of Stephen Mayfield, an algae biologist at The Scripps Research Institute in La Jolla, and Steven Briggs, a professor of cell and developmental biology at UC San Diego. The founders and their scientific collaborators officially launched Sapphire in May 2007 [Bigelow 2008].

Table I.93: Sapphire's approach of serial entrepreneurs to algae-based biofuels following opportunity identification and assessing technical and financing options. *)

Company (Foundation) Remarks	**CEO, Other Executives, Foundation, Key Researchers; Technology Protection and Position**
Sapphire Energy, Inc. San Diego (CA) (2007) Headquarter and primary research labs in San Diego, CA, engineering and project management, Orange County (CA), a research and development complex in Las Cruces, New Mexico; to build a 300-acre full size open-pond algae farm demonstration project in Luna County, New Mexico by the end of 2010 (Intgrated Algal Bio-Refinery). Employees: 2008: ca. 80 [Bigelow 2008], 2009 ca. 120 [Bigelow 2009b]. **Major Funding:** $100M million in a second venture round [Bigelow 2008]	Jason Pyle, MD, PhD CEO and co-founder, also on the Board of Directors; was formerly CTO and co-founder of Epoc, Inc., a privately held medical engineering company. Dr. Pyle holds an appointment as adjunct professor of bioengineering at Vanderbilt University where he has worked to develop cross-disciplinary programs of biological and engineering research. As the co-founder and CTO of Pria Diagnostics, Dr. Pyle was named Innovator of the Year (2006) by Frost and Sullivan. Dr. Pyle holds numerous pending and issued patents in the engineering and biological sciences. In addition to his broad technical abilities, Dr. Pyle has established numerous corporate partnerships between small technical companies and some of the world's largest corporations. Cynthia J. Warner President; brings more than 27 years of experience in the energy, refining and transportation industries. A chemical engineer by training and one of the very few senior women in the oil and gas industry, Ms. Warner served as an executive with energy industry giants British Petroleum, Amoco Oil Company and UOP. Warner left her post as Group Vice President of Global Refining for BP. At Sapphire Energy, Ms. Warner is tasked with driving the company's initiative to transition technology trials and research into commercial-scale crude oil operations. She is a featured leader in the 2008 book "Becoming a Resonant Leader: Develop Your Emotional Intelligence" (Harvard Business School Press). Mike Mendez VP Technology; has held a number of top industry positions at the forefront of the molecular biology revolution. In addition to serving as Director of Bioengineering at GenWay, Mr. Mendez was also associate director of Exploratory Research at Syrrx, Inc. (presently Takeda Pharmaceuticals). There he established a new department that focused on novel platforms for over-expression, purification, and crystallization of membrane proteins. Mr. Mendez co-founded and led the technical program at MemRx, a structural biology company. He has served as a genetic consultant and scientific adviser for numerous biotech and academic institutions; he is also the founder and principal scientist of Gryffin Consulting, Inc., a genetic engineering consulting firm specializing in the areas of gene therapy and antibody and membrane protein production. Tim Zenk VP Corporate Affairs; has spent much of his career shaping public policy – in helping leaders become better leaders and the public become more educated about key issues impacting the nation and the globe. He is known nationally for his political acumen, particularly regarding his

$50M first large venture funding. (5/28/2008) [CrunchBase 2009c]; for construction of the Algal Bio-Refinery in New Mexico. Sapphire received a $50M demonstration-scale grant from the DOE and a $54.5M DOE Loan Guarantee [Lane 2009k]	work on key campaigns ranging from gubernatorial (governor-related) to congressional to presidential. His global work for the Clinton/Gore administration has left him with professional and life experiences that will last forever. His passion for legacy energy solutions is top on his agenda. Tim Zenk had worked on projects with some of the Sapphire investors previously [Schwartz 2010]. Stephen Mayfield, PhD co-founder, Chairman Scientific Advisory Board; is Director of the San Diego Center for Algae Biotechnology (SDCAB), and the John Dove Isaacs Professor of Natural Philosophy at the University of California San Diego. Formerly a Professor of Cell Biology, and Associate Dean of the graduate school at The Scripps Research Institute (TSRI), Dr. Mayfield has worked on the molecular genetics of green algae for over 25 years. His research focuses on understanding gene expression in the green algae; he is a leading expert on the genetics of algae. He set also the stage for the use of algae as a platform for therapeutic protein production, including the expression of human monoclonal antibodies. These studies resulted in the founding of Rincon Pharmaceutical. Nathaniel David, PhD co-founder, ARCH Ventures; teaming with ARCH in 2009, Dr. David was building new companies that create disruptive technologies to address global-scale problems. Co-founder of Sapphire Energy and formerly CSO and co-founder of Kythera, he is also co-founder of Syrrx (acquired by Takeda for $270 million in 2005) and Achaogen. Dr. David has demonstrated experience creating and growing innovative biotechnology companies. He was named one of the Top 100 innovators in the world under 35 (2002) by the MIT Technology Review. He holds numerous pending and issued patents in fields such as nanovolume crystallography, antibiotic resistance, and aesthetic medicine. Kristina Burow co-founder, a Principal, ARCH Venture Partners, joining the firm in 2004; is primarily focused on companies in the life sciences and materials sciences. Ms. Burow joined ARCH from the Novartis BioVenture Fund in San Diego where she was involved in numerous investments in the life science sector. She was a co-inventor of key technology platforms that formed the core of Kalypsys, a GNF spin-off company. Ms. Burow holds an MBA. from the University of Chicago, an MA in Chemistry from Columbia University, and a BS in Chemistry from the University of California, Berkeley. Sapphire Energy originated from a debate between three friends, Jason Pyle; Kristina Burow and Nathaniel David: “Why is the biofuel industry spending so much time and energy to manufacture ethanol – a fundamentally inferior fuel?” In the end a biofuel company was envisioned with the goal to be the world’s leading producer of renewable “petrochemical” products. “To produce a droplet of gasoline” from algae to show their possibilities to investors launched a cascade of funds which has allowed the founders, at

	latest count, upwards of $200 million of running room to get their algae biofuel approach up to scale. Backers include Bill Gates' venture capital wing, Cascade Investments, and the US Department of Energy. Sapphire has invested more than any other private or public entity ever in the business of turning algae into an industrial crop and something that can be considered a true drop-in replacement fuel [Schwartz 2010]. As part of the American Recovery and Reinvestment Act and through the biorefinery assistance program in the 2008 Farm Bill, Sapphire has partnered with both the US DOE and the USDA to build a next-generation algal biorefinery. Sapphire is also collaborating with scientists from the Department of Energy's Joint Genome Institute; University of California, San Diego; The Scripps Research Institute; University of Tulsa; and San Diego Center for Algal Biotechnology. The ambition of the founders of Sapphire is: not just to replace a small fraction of the oil use in the US, but the algae that Sapphire is working on could replace all of it. And founder Pyle believed that there will be many winners in the algal biofuel space. "In a trillion dollar market, it's hard to believe in a winner take all strategy." [Morrison 2008] Sapphire has over 230 patents or applications spanning the entire algae-to-fuel process [Lane 2009k] – from genetically engineering algae to maximize the production of biological oils to extracting the oils, which constitute the "*green crude*" that can be refined into gasoline, diesel, and jet fuel. Part of the IP stems from Rincon Pharmaceuticals founded by Mayfield to begin commercializing his research on algae as vehicles in which to produce biotech drugs. Rincon was acquired by Sapphire in 2008 [Gellene 2009] Sapphire Energy is supported by a syndicate of investors led by co-founder ARCH Venture Partners; along with The Wellcome Trust; Microsoft founder Bill Gates' Cascade Investment, LLC; and Venrock, the fund of the Rockefeller family.
Technology, Goals, Strategy Business Model: Develop technology and operate along the entire pond-to-pump value system – except refining. Focus on manu-	Sapphire produces so-called "*green crude*" (that exhibits many of the same molecules that are in petro-crudes from the ground) which can be refined into "normal" fuels – gasoline, diesel and jet fuel. These meet ASTM standards and are compatible with the existing petroleum infrastructure. Sapphire said its technology is "carbon neutral" because its algae absorbs as much carbon dioxide as a car releases when its fueled by renewable gasoline. When on its Web site referring to "Green Crude Production," Sapphire tells us that "the world needs a radical new solution." Sapphire works with multiple strains, based on geography, climate, what is available naturally,

| facturing in-frastructure compliant *green crude* that fits with the fuel transport and distribution systems we use today.

In line with a bio-refinery approach focus on concen-trating on the best co-products to produce for sales.

Claim: not to use genetically modi-fied organisms. | and what can be manipulated. Acccording to Sapphire it does not develop genetically modified organisms, but selective breeding. Using high throughput screening (HTS) Sapphire said it looks at 8,000 strains every single day and is just entering the pre-commercial demonstration phase [Schwartz 2010].

Sapphire plans to use non-potable water like agricultural runoff and salt water and locate its biorefinery in the desert. As algea grow in brackish water scientists are evaluating how different species of algae react to variations in salinity, pH, temperature, humidity, and other factors. HTS helps accelerated identification of the strains that are best-suited to produce lipids under any given condition.

Stephen Mayfield (a Sapphire co-founder and scientific advisor) would require pumping CO2 into the desert. Just how this would work without carbon dioxide escaping into the atmosphere is not clear, but Sapphire officials say it is *one of many issues the company must address* as it develops its 100-acre pilot facility near Las Cruces, NM [Bigelow 2009b].

But, in 2008 the founder Pyle said "we use genetic engineering, directed evolution, synthetic biology and (agricultural) breeding" and specifically that does not include fermentation. And with regard to the process in which algae "directly converts sunlight and carbon dioxide into hydrocar-bon products," Pyle said "all of our systems are photosynthetic." [Bigelow 2008]
According to Waltz [2009b] experts said some of their organisms are ge-netically engineered, but the company has not yet publicly confirmed this.

Sapphire is also collaborating with scientists from the Department of Energy's Joint Genome Institute; University of California, San Diego; The Scripps Research Institute; University of Tulsa; and San Diego Center for Algal Biotechnology.

Sapphire claims: "We don't have any questions about whether the tech-nology works. The only question is about the cost of production." [Bigelow 2008] Sapphire's concept calls for creating enormous "algae farms" throughout the desert lands of the southwestern United States [Bigelow 2008].

Sapphire is focused on the entire "pond to pump" value system. It will do everything but refining which will be done by a partner, Dynamic Fuels, in Louisiana. The algae and processes developed are field tested at a New Mexico research and development center where all the processes – from biology to cultivation to harvest and extraction – can be performed at a pilot scale. [Lane 2009k; Schwartz 2010].

Sapphire's position: "We're an energy company," "and, "really, who we are competing with is big oil and gas." There might be a distribution deal with Big Oil (e.g. Shell), but not being acquired by Big Oil.

There are large amounts of biomass left over from the process. In line with |

	its "*Algal Bio-Refinery*" approach Sapphire studied to figure out what are the best co-products to produce, and it will probably deliver these to others that take the material and do something with it [Schwartz 2010]. During 2008/2009 Sapphire's products were used for first commercial airline test flights using algae-based, drop-in replacement fuel and a first vehicle to cross the US fueled by a blend of algae-based gasoline in an unmodified engine. Within 3 years Sapphire Energy expected to be nearing completion of a demonstration and test facility and well on its way to producing 1 million gallons of diesel and jet fuel per year over the next 5 years. By 2018, Sapphire expected to grow this to 100 million and by 2025 1 billion gallons of diesel and jet fuel per year and to be able to produce green crude at \$60 – \$80 per barrel [Lane 2009k]. Sapphire is pro-actively lobbying; it wants to cooperate on policy, wants to cooperate making sure there is a playing field that allows everybody to compete. As a prerequisite for the development of the (transportation) biofuel industry, Sapphire assumes that there has to be a price on carbon and to have cap and trade. In this line Sapphire is also heading up the task force appointed by New Mexico Governor Bill Richardson, a former Energy Secretary in the Clinton Cabinet, who has visions of New Mexico's leadership in renewable energy [Schwartz 2010].

*) From the firm's Web site if not stated otherwise by a reference; firm's status: end of 2009.

Sapphire had been moving quickly to build a 300-acre algae farm as a large-scale demonstration of its process for making algae oils which was planned to be completed by the end of 2010 (Table I.93). In 2012 the US government supplied over \$100 million of the investments, including a \$50 million Recovery Act grant designed in part to spur job creation. Sapphire is a major beneficiary of the US government – in line with its strategy concerning public policy and execution by implementation of a VP Corporate Affairs (Table I.93).

Sapphire's rapid expansion raised the question of whether it is scaling-up its technology too soon. Some of its ideas for reducing the cost of algae fuels appeared at too early a stage to be implemented at the new farm. The new funding will allow Sapphire to finish building its algae farm near the small town of Columbus, New Mexico; a 100-acre segment of the farm has already been finished. When the whole project is complete, planned by 2014, Sapphire would have the capacity to produce about 1.5 million gallons of algae crude oil, which can be shipped to refineries to make chemicals and fuels such as diesel and gasoline [Bullis 2012a].

According to its Web site in April 2012 Sapphire Energy said co-founder Jason Pyle has stepped down as chief executive officer of the startup company that has raised then close to \$350 million to develop algae as a viable biofuel alternative to crude oil to become a member of the Board. Founding CEOs do not often walk away from startups that have amassed a $1 billion valuation and that have drawn nationwide attention for developing potentially transformational technology. As usual there is an official reason for such a step and a hidden one.

Bullis [2012a] assessed the current situation of Sapphire rather detailed. Knowing when to move technologies out of the lab and into large-scale demonstrations is a perennial challenge for energy startups. According to some experts, Range Fuels (ch. A.1.1.3; Table I.99), founded to produce ethanol from wood chips, foundered because it built a large-scale plant too soon, before the bugs had been worked out of its technology at a smaller scale. As a result, the plant did not work well enough to be economical (cf. also CHOREN Technologies, A.1.1.3).

"Sapphire hopes to lower the cost of producing algae fuels by changing every part of the production process. That includes increasing the quality and the amount of oil produced, reducing the cost of building ponds, and developing low-cost ways to harvest the oil." Sapphire is working with Munich-based Linde Group to develop a low-cost way to supply the algae with carbon dioxide, which is a key to high productivity. Linde has developed systems for supplying greenhouses with carbon dioxide from a refinery. "The company aims to have a product that is competitive with oil priced at $85 per barrel, and it expects to meet this goal once it reaches full-scale production in about six years."

When complete, the new 300-acre algae farm project is expected to produce about 100 barrels of algae crude per day, or 35,000 a year. Sapphire Energy's vice president of corporate affairs Tim Zenk said the process will not be commercially viable without the economies of scale that will come with much, much bigger farms – 1,000 to 5,000 acres.

> Achieving these cost targets will require significant innovation. In 2011 studies from the National Renewable Energy Laboratory (NREL) concluded that *algae-based diesel made by scaling up existing algae technologies would cost several times as much as conventional diesel.* According to one of the studies, it would cost about $9.84 per gallon to make algae diesel, as opposed to $2.60 per gallon for petro-diesel, at January 2011 costs.

Phil Pienkos, a research scientist at NREL, said that Sapphire is doing a number of good things to reduce costs. Yet he said making algae fuels competitive will be a challenge. "It takes a certain amount of faith that there is going to be a business there," he said.

Furthermore, experts agree that non-fuel markets can be profitable for Solazyme and other algae firms, but they warn that *investors will be impatient* to access the multi-billion-dollar fuel market. *That may set the industry up for failure*, because it will be many years – if ever – before algae can be cost-competitive with petroleum [Bomgardner 2011d].

One aspect with all the above discussed algae startups, Cellana/HR Biopetroleum, Solazyme, Algenol Biofuels and Sapphire Energy, is the strong focus on policy initiatives and lobbying for governmental financial and legislative support, specifically through politically experienced firm representatives. In case of algae it is not surprising that, apart from the US Department of Energy (DOE) and Department of Agriculture (USDA), in the US also the Department of Defense is a target for seeking support for activities in algae, particularly, with regard to the algae-to-military jet-fuel production process [Gaithwaite 2009].

> For entrepreneurs in biofuels, the potential payoff is obviously big enough that it is worth hiring appropriate persons and spending time and money away from firms' sites to take the risk that the administration and authorities will not pick their technologies. Success in the capital simultaneously increases the firms' credibilities and facilitates to win funds from venture capital.

Whereas innovative and entrepreneurial activities have the same key targets in the US and Germany regarding plant-based biomass or waste, biofuels, reducing carbon dioxide emissions and lifting/reducing petro-oil dependencies, things are different for aquatic biomass (algae).

Basically, the US "owns sunshine" for certain areas or states, respectively, as a natural resource, as are coal, petroleum or forests/wood [Runge 2006:287]. This means, national advantages in natural resources and traditional industries can be fused with related competencies in broad technological fields thus providing the basis for technological directions and advantages in new product fields and often new and strong and innovative companies.

> In Germany activities with algae are generally held back to a certain degree by the all over on average low duration (and strength) of sunshine, at least with regard to open pond or outdoor settings. Hence, related innovation and entrepreneurial activities which are also supported by governmental programs emphasize closed photoreactors, greenhouses and algae feeding, light, temperature control and mass transfer (algae; CO2 in, oxygen (O2) out) and an *engineering approach* to optimize relevant process parameters.
> Furthermore, due to stronger societal attitudes against GMOs, in Germany, more than in the US, bioengineering routes are only rarely followed by startups, but the focus is on breeding and cultivating naturally occurring algae.

There are some visions in Germany to install huge bioreactors with sea water and algae at the Mediterranium, in which CO2 from power plants are converted to biomass [Anonymus 2007b] similar to the constellation now being established by Sapphire in New Mexico or Algenol and Biofields in Mexico (Table I.91).

Disregarding the above described exceptional startup Cyano Biofuels, rather than following strongly the fuels/energy route, in Germany algae activities including those of NTBFs, focus largely on *higher value products from algae* (Figure I.176) and *carbon dioxide sequestration.* Due to "cap and trade" legislation in Germany (as also requested for the US, for instance, by Sapphire (Table I.93)) carbon sequestration works as an incentive for firms because trading and selling from created CO2 emission permits will lead to financial revenues.

The current general view is that algae are only an option as an energy source in combination with high value algae products. In Germany's largest algae farm (near Klötze) with an annual production of 60 tons biomass per hectare (2.47 acres) utilizing the ingredients is the priority. And everything that cannot be extracted and sold at a high price will be converted to animal or fish feed and sold at competitive prices. Also experts from large German power companies, such as RWE, share the general view: "currently we exclude large energetic exploitation {of algae}" [Müller-Jung 2010].

Correspondingly, compared to the US level, there are relatively little entrepreneurial activities with algae in Germany. On the other hand, its giant power suppliers E.ON, RWE and EnBW or their respective subsidiaries run or have run several pilot projects targeting carbon dioxide sequestration requirements by transferring flue gas containing CO2 of their power plants into algae and to algae fuel producers – for instance, in cooperation with the NTBFs BlueBioTech GmbH (mentioned above), Novagreen Projektmanagement GmbH or Subitec GmbH and federal or state financial support (Table I.94).

There is also considerable research in Germany with regard to producing biogas, consisting essentially of methane (ca. 65 percent) and carbon dioxide (ca. 30 percent) from algae as a biofuel for cars, trucks and busses. A conversion of algae into biogas that then will be burned in a *closed loop process* (Figure I.177) would capture CO2 permanently from the air!

Table I.94: Algal pilot projects of German power suppliers concerning carbon dioxide sequestration and producing high value algae products involving NTBFs.

Power Supplier and Project Partners	**Details**
RWE Power [RWE 2009a; RWE 2009b] With: (Federal) Research Center Jülich, Jacobs University Bremen , *NTBF* Phytolutions GmbH (a spin-out of the Jacobs University Bremen). at: RWE Power's Coal Innovation Center, at its Niederaussem power plant site. Partial financing is envisioned via selling CO2 emission permits. Operational since 2008,	Goal: optimize the whole process from algae cultivation to end product; produce 60 – 100 tons of biomass per hectare and year Tests: different types of algae and reactors for energy efficiency, photobioreators (outdoor) without greenhouse, other applications, such as biofuel and building material; inquire into *hydrothermal carbonization* (HTC), also for opening a new field of

project end 2011. The pilot plant covers 600 m2 (148 acres) (can be extended to 1,000 m2) Process details: A suspension of microalgae from sea water is mixed with flue gas and transferred into a photobioreactor in a greenhouse. Currently patented vertical column photobioreactors (transparent plastic hoses fixed in V-form) from Novagreen are used; shall be replaced by "endless hoses."	chemistry and development of new materials. Results so far: The plant produces up to 6,000 kg (dry) algae mass and binds 12,000 kg CO_2; a high value product has been obtained.
RWE Power [RWE 2010] With: German biotech firm Brain AG at: RWE Power's Coal Innovation Center, at its Niederaussem power plant site Biotech firm Brain provides innovative enzymes and synthesis routes and pathways. Its comprehensive "natural toolbox" shall allow synthetic biology to produce innovative microorganisms that are able to capture more CO_2 from a lignite-fired power station	Goal: convert carbon dioxide into microbial biomass or biomolecules – joint research alliance between RWE Power and BRAIN AG using "designer microorganisms" in search of biotechnological solutions to CO_2 conversion and developing further intelligent uses. Applications to be explored include building and isolation materials and the production of fine and specialty chemicals.
EnBW Energie Baden-Württemberg AG [EnBW 2008]. With: *NTBF* Subitec GmbH (founded in 2000; a spin-out of the Fraunhofer Institute for Interfacial Engineering and Biotechnology (IGB) in Stuttgart at: Eutingen, a biogas plant; running and cultivation a system for microalgae with CO_2 feeding from a (block) combined heat and power (CHP) plant EnBW acts as a project client, Subitec as the owner and operator of the pilot plant	Goal: carbon dioxide sequestration by algae; focus on efficiency of CO_2 binding. Subitec's business model: provide its patented thin channel airlift and forced flow photobioreactors and develop and operate pilot plants of various dimensions and develop concepts for utilizing and post-processing algae related biomass [Ripplinger 2009]; the pilot plants will sometimes be constructed at the company's own expense, and sometimes on behalf of customers or cooperation partners.

Table I.94, continued.

E.ON Hanse [E.ON 2008; Anonymus (2007b] At: a natural gas storage facility at Hamburg-Reitbrook; provides flue gas from a CHP Sponsors together with the city of Hamburg run the project TERM (Technology for the Exploitation of the Resource Macroalgae) of several universities from northern Germany and NTBF Subitec. E.ON adds a researcher and several technicians; who research together with Subitec as a partner; E.ON provides technology, logistics, infrastructure and research services and an area of 1 hectare to TERM and Subitec. SSC Strategic Science Consult GmbH coordinates and leads the project.	Goal: planning, construction and technical assistance for the operation of a cultivation system for microalgae; E.ON and Subitec add reactors for cultivating algae and develop these further; Operation is in an open air outdoor (no greenhouse); Project cost: €2.2M; Hamburg's contribution €0.5M [Anonymus 2007b] R&D focus: ▪ Optimizing bioreactor technology for optimal growth and high efficiency for converting primary energy (day light) into biomass, ▪ Optimize microalgae on the species level, biomass composition and physiology, ▪ Automate the plant, to facilitate industrial production.
E.ON Ruhrgas [Böttcher, C. 2006] With: IUB (International University of Bremen) and BlueBioTech GmbH is working on research and development A project "Biofixing of Greenhouse Gases with Microalgae Biotechnology" (2005-2007) was funded by E.ON	Goal: produce biodiesel and animal feed from flue gas and marine microalgae using a small greenhouse with a 150 l photobioreactor.
Using carbon dioxide emissions from an E.ON Ruhrgas 350 MW coal-fired power plant in Bremen supported by the state (city) of Bremen was a basis of the Greenhouse Gas Mitigation Project (GGMP) [Golon 2006] 4 micro-photobioreactors were connected to the flue gas from the power plant. Project partner: Jacobs University Bremen BlueBioTech GmbH Novagreen.	Goals: Experimental micro-facility to test the tolerance of microalgae to the flue gas ▪ Test facility, establishing proof of principle ▪ Planning and construction of pilot plant with a capacity of 500 tons CO2/year ▪ Planning and construction of plant to treat 15,000 tons CO2/year Status/results: microalgae removed CO2 (and NOx) from flue gases and recycled it in form of

With sales of the CO_2 emission permits cost of the plant and fuel production are envisioned to be covered	biomass and derived products; no negative effect of flue gas on the production of microalgae; concentrations of pollutants were significantly below the values accepted for animal feed

The innovative/entrepreneurial approaches in Table I.92 as well as Table I.94 reveal that *governmental/public funding and grants in Germany are provided preferentially to defined technical projects with partners from industry (large firms and NTBFs) and public research organizations* (universities and/or federal/state research centers) with a systemic orientation whereas in the US the focus is often on individual firms (large or small).

Another topic of interest in the context of algae as biomass (RWE Power) is "*hydrothermal carbonization*" (HTC) which was developed by the Max Planck Institute of Colloids and Interfaces in Potsdam/Germany. It is a new variation of biomass conversion.

In contrast to other biomass carbonization techniques that require dry biomass, the HTC process is a highly efficient "wet"" process that avoids complicated drying schemes and costly isolation procedures (biomass + water +catalyst+ pressure in the absence of air) to produce carbonaceous materials ("biocoal"). This would also apply to wet biomass from algae. The method – "pressure cooking biomass till it boils dry" – is relatively inexpensive, widely applicable and quickly scalable and can produce clean energy in the form of gas or oil along with a "*biocoal*" powder. And it opens a new field of chemistry and development of new materials, such as nanotubes. [104]

A German NTBF founded in 2007 to exploit HTC technology is Suncoal Industries GmbH (B.2).

Apart from BlueBioTech GmbH in Germany there are two algae-oriented NTBFs particularly notable, NOVA green Projektmanagement GmbH (also written as Novagreen; Table I.94) and IGV GmbH (Institut für Getreideverarbeitung; The Institute for Cereal Processing Ltd.; Table I.92). Novagreen has its root in horticultural engineering (project management!); IGV originated with food and nutrition engineering, but then also added algae-related biotechnology as a key area of activities.

Novagreen was founded in 2004 by an engineer (Rudolf Cordes) and Dr. Theodor Fahrendorf, with more than twenty years of (industrial) experiences in plant physiology, biochemistry and biotechnology. Novagreen originated with the firm Agrinova Projektmanagement GmbH, which was established by Cordes in 1997. Agrinova is active in cultivating and breeding fruit and vegetables. Agrinova focuses on "secondary plant ingredients" which are marketed as food additive with protection capacity against cancer. Hence, Novagreen can be seen as a spin-off complementing horticulture by algae-oriented aquaculture.

Agrinova does not only breed backcrossings to still more primitives to get more substantial varieties which are closer to the wild-type related varieties. It develops also advanced harvesting technology and processing know-how. The company isolates and extracts raw and finished materials and supplies raw materials and pharmaceutical intermediates for the manufacturing industry. Its mechanical engineering branch develops prototypes and machines to harvest, for instance, cabbage more softly directly in the field. Agrinova sells its product directly via its Web online ("Agrinova-Shop") and supports its marketing by public presentations about the health supporting roles of their products.

T. Fahrendorf joined Agrinova in 2005 to lead the science direction of the firm (CSO). Cordes and Fahrendorf became partners and managing directors of Novagreen GmbH (LLC) which focuses on algae and could rely on development systems set up already in 1998 by Cordes which were subjected to continuous improvements. [Agrinova].

According to its Web site Novagreen is a developer and provider of novel bioreactors for the production of microalgae in a closed environment. Simultaneously Novagreen is a producer of selected bulk microalgae and ingredients using these bioreactors for the food, cosmetics, and pharmaceutical industry.

The bioreactors are adapted to producing different types of algae – and for freshwater or sea water. Novagreen's patented production platform using a unique three-layer film "tubing system" ("hoses") can be implemented in almost any standard greenhouse facility worldwide to fit existing infrastructure. It offers also a power (heat) concept for connecting to an existing biogas plant utilizing the waste heat of a combined heat and power (CHP) plant, to reduce carbon dioxide (Figure I.177).

Hence, Novagreen may provide new streams of revenue for the agro industry, particularly horticulture. Production takes place in a closed and controlled environment that guarantees the high quality distributors and consumers are looking for (high quality certified microalgal products). Novagreen *uses pre-existing, well established horticultural production systems, distribution and marketing channels*. Novagreen is introducing production of microalgae as an integral part of modern horticulture.

Novagreen pursues R&D activities, usually in cooperation with other firms (Table I.94) via projects subsidized by policy. For instance, for optimization, in its laboratory all the species of algae are matched to the carbon dioxide sources as well as their intended use.

Microalgae can also be used for production of recombinant proteins. Novagreen is planning the production of antibodies and vaccines for the veterinary market. For 2012 it planned to establish a production system for heterologous antibodies and vaccines in transgenic microalgae. Since 2008, preliminary tests were run in cooperation with European and American institutions [Daniel Meier Medienteam 2010].

According to the Creditreform/Firmenwissen database Novagreen had revenues of €310,000 in 2010. It is not clear whether this corresponds to sales of offerings. One can assume that there is additionally a large hidden contribution of project capital attributed to Novagreen. Furthermore, apart from the two founders, the GmbH (LLC) has a third partner who may contribute equity. Moreover, it is not obvious in how far Novagreen and Agrinova are run as financially independent firms.

In the sense of an engineering-type firm technical offerings by Novagreen include [Fahrendorf 2008]:

- Developing algae-bioreactors and production plants for algae up to 100 hectares
- Power concepts for biogas plants
- V- and H-reactors
- Foil tunnels
- Harvesting and processing technology (pumps, driers)
- Production of algae for various purposes
- Precursors for pharmaceuticals
- Reference proteins
- Transgenic algae
- Commercialization (marketing) of the customers' algae.

Services covers:

- Consulting, concepts for utilizing heat, assessments
- Tailored solution that guarantee an optimal utilization of heat
- Construction of plants
- Management, central laboratory and analytics
- Approval and commercialization of the algae.

In 2010 Novagreen's demonstration greenhouse with production of algae was coupled for the first time to a biogas plant. With its self-developed, proprietary cascade system it was tested under which conditions algae show best growth and bind the most carbon dioxide of the biogas. The greenhouse is covered with photovoltaic glass panels and will produce die 630 Kilowatt-Peak (KWp) electricity per hour. The light transmitted to the algae in the photobioreactors suffices to generate 80 metric tons of biomass per hectare. Additionally the demonstration setup was used for trial with algae that produce essential omega-3-fatty acids which is interesting for the animal feed industry as a fishmeal replacement [Daniel Meier Medienteam 2010].

DTB – Deutsche Biogas AG, a German NTBF in renewable energy – produces and sells electricity and heat from biogas plants, which it designs, builds, and operates in partnership with farmers. In 2011 it started am algae program together with Novagreen. The aim of the common project was to test Novagreen technology, and further develop and optimize the technology. Here, the waste heat from the biogas plant is used for the operation of the greenhouse and thus integrated into the value chain. With the help of the algal and special cultures carbon dioxide from the exhaust gases of the related CHP will be "stripped," so that the production of biomethane is not only CO2-neutral, but has reduced CO2 [DTB 2011].

Previous discussions related the German IGV GmbH (Institut für Getreideverarbeitung; The Institute for Cereal Processing Ltd.) to the US startup Greenfuel Technologies founded in 2001 as a cooperation partner regarding biofuel from algae (ch. 2.1.2.8; A.1.1.3). However, IGV represents also one prototypical case of *necessity entrepreneurship* enforced by the re-organization of the industry and the science and technology system of the former socialistic German Democratic Republic (GDR – in German DDR – occupied by the USSR) after the German Re-Unification around 1990.

Due to the devastation of Germany after World War II and particularly the lack of resources in its Eastern part and lack of money for imports the GDR followed often the technological paths of Nazi Germany to become self-sufficient (autarky) concentrating on optimizing what is available and the focusing on the notion "Ersatz" which is "substitute," such as "Ersatz-Holz-Zucker" (synthetic sugar from wood) [Runge 2006:270-272, 566] (cf. above CHOREN Industries and Bioliq – Figure I.173).

IGV addresses necessity entrepreneurship of employees of a state-owned research institute in the former GDR, located in the German state Brandenburg rather close to Berlin. The privatization of IGV corresponded formally and regarding the result to a management buyout (MBO; ch. 2.1.2.4), but there was a legally complicated process behind it as for a public research institute or firm the "normal" approaches to an MBO did not apply in the capitalistic Federal Republic of Germany.

IGV was founded in GDR in 1960 in Brandenburg which is a state with a dominant food industry. Ca. 50 percent of Brandenburg's area was and is used for agriculture. And also after the German Re-Unification, the food industry with a turnover of €2,402.5 million (in 2004) remained the industry showing the highest overall revenues. It was set up as a *practice-oriented research institute for the milling, bakery and food industry*.

Furthermore, its location in Nuthetal near Brandenburg's capital (and Berlin) provided an environment with universities and non-university research institutes for food science and nutrition. And for decades the IGV was a leader of the food industry and in processing of vegetable raw materials in the GDR.

When the IGV added biotechnology to its main fields of research Prof. Dr. Otto Pulz became the leader of biotechnology in 1975. Since 1981 he was engaged in biotechnology of algae, particularly focusing on design and construction of photobioreactors, on active ingredients and raw materials for cosmetics [succidia]. That means he started during the first "algae wave," in the late 1970s and early 1980s.

Another important person of IGV was Peter Kretschmer who is a passionate scientist ("Ich bin eben ein leidenschaftlicher Wissenschaftler") [Steyer 2007]. In 2010 the engineer P. Kretschmer was promoted with a doctoral degree by the Technical University of Berlin at the age of 72 [Meuser 2011]. He was seen as “the Gyro Gearloose of the GDR-food industry” (“Peter Kretschmer war der Daniel Düsentrieb der DDR-Lebensmittelindustrie.”)

Dipl.-Ing. Peter Kretschmer (in 2012 74 years old and still the managing director of the new IGV) had a research focus on bread and bakery and he became a world renowned expert in this field. Also apprenticeship for people in the bakery industry was established and provided by the IGV under his supervision.

Basically, Kretschmer had to respond to the command of the GDR government: “Reduce the time of our women to stay in the kitchen” (“Verkürzen Sie unseren Frauen die Zeit in der Küche.”). Even if this aimed officially at improving the life balance of women between family, children and profession and career, actually the GDR was is heavy need of more people to have its industry grow [Steyer 2007].

As described by Meuser [2011], after the German Re-Unification, “The German Treuhand Agency” was leading the re-organization of the socialistic economy, industry and research organizations of the former GDR. In 1990 IGV got the legal status of a limited liability company (LLC – GmbH) and the Treuhand became its single owner. The Treuhand wanted to eliminate the existing IGV and laying off ca. 180 employees.

As a consequence, there was a massive layoff of employees and in 1990 Peter Kretschmer became managing director supposed to be able to lead IGV into the market environment. All IGV employees were aware that the IGV could only survive focusing on its core competencies. This orientation was appreciated by the German Science Board which acts as a consultant for the Federal Government.

In 1991, after a careful assessment, the German Science Board suggested to keep the IGV with all its research priorities. Many options were explored to transfer IGV ownership of the Treuhand into a State custody. However, after all the options had turned out to be negative, only privatization remained as a solution.

Since an additional objective of IGV’s future was the preservation of the autonomy and independence, in 1994 an MBO occurred with retaining the LLC-structure in which Dipl.-Ing. Peter Kretschmer, Dr. Helmut Barnitzke and Prof. Dr. Otto Pulz were partners [Meuser 2011]. Later H. Barnitzke changed into the Supervisory Board of the firm, and P. Kretschmer became managing director and Prof. Pulz a representative.

IGV now is a private and independent applied research institute with key competencies in food processing, biotechnology and processing of biomass materials with currently about 100 employees. The proportion of people with a scientific or engineering education and technical employees (masters, chemically-technical assistants and laboratory technicians) is about 4 to 1.

The focus of the competence and research spectrum is on production and process innovations, their efficient development, commercialization and technology transfer to small and medium-sized enterprises in food processing and related areas. The scientific and technical services include the operation of an accredited testing laboratory.

Sales and distribution of IGV's offerings take place worldwide [BMWi 2011]:

- Ca. 30 percent of the orders are captured in the state of Brandenburg or Berlin.
- Ca. 60 percent of the orders are requested by medium-sized or large firms in Germany or the EU.

Since 2002 there is a close cooperation of IGV with the University of Applied Sciences of Lausitz (in German Fachhochschule) concerning phototrophic biotechnology whose major theme is algae. In 2006 this cooperation was formalized and contractually extended. In particular, it included establishing phototrophic biotechnology as a field of education and research and having Prof. Pulz of IGV as a visiting professor to hold lectures [Witzmann 2006].

Common research fields focused on

- Building a unique collection of microalgae species (after four years the collection had more than 250 originals – an invaluable genetic potential);
- Search for species-specific active substances (an example is to develop a special line of cosmetics);
- CO2 sequestration and climate protection (Table I.92).

A similar approach of necessity entrepreneurship out of a large research institute of the Academy of Science in Berlin-Adlershof (in the GDR-part of Berlin) induced by the German Re-Unification is observed for ASCA GmbH (Angewandte Synthesechemie Adlershof GmbH), active in the area of fine chemicals and active pharmaceutical ingredients. It acts as a private research institute focusing essentially on contract research. As the IGV also ASCA financed its early life and survival by special grants which were established by German state and federal governments and the EU to support the transition of the socialistic system of the former GDR into the Nippon-Rhineland capitalism (ch. 1.2.4) of the Federal Republic of Germany.

> And looking also at the below discussed case of "Bioprodukte Prof. Steinberg Produktions- und Vertriebs GmbH & Co KG" one realizes that independence and determining one's own destiny (Table I.39) are very strong drivers for entrepreneurship demonstrated by people who were living in a society where both aspects were suppressed and became founders when the restrictions were lifted.

IGV now exhibits healthy growth. Total revenues of IGV amounted to €6.52 million in 2010, €4.70 million were achieved by selling to the market and €1.83 million were obtained by grants related to projects financed by the public (German state and federal governments as well as the European Union). Currently growth in revenues is almost determined by sales to industry. While the proportion at 2002 was 50:50 (each ca.

€2.15 million), the contribution from grants stayed almost constant at ca. €1.8 million. For instance, the proportion for 2007 was €4 million to €1.4 million [IGV 2011].

Looking at algae IGV developed plants for breeding and harvesting algae in *closed systems focusing on photobioreactors*. Since the middle of the last century microalgae have been produced in open ponds, mainly in South East Asia and the US. These ponds are about 15 cm deep and are stirred at one or more points. In 1995 a technique was developed in Germany for cultivating microalgae which centered on a *closed system of glass tubes*, rather than a "continuous aquarium," to expose the algae to maximum light.

Cultivation of algae in glass tubes has considerable advantages over cultivation in ponds, particularly for high value products (Figure I.176):

1. Light can reach the algae from all sides through the glass tube; they are exposed to maximum light and can grow well. There are no areas which are deficient in light such as those occurring in the deeper layers of ponds.
2. All the environmental factors which are important for healthy growth of algae, such as pH, temperature and carbon dioxide supply, can be controlled and adjusted to the optimal setting.
3. External uncontrollable influences, such as rainwater, dust, insects, waterfowl and their droppings, blue-green algae etc., can be eliminated. And this difference in quality can be measured, for example, in terms of heavy metal and toxin contamination.
4. There is no evaporation of water over a large area.
5. Only a fraction of the space taken up by a pond system is required.

Over the years IGV built up much experience in using algae for food and nutrition, pharmaceuticals, cosmetics and animal feed. But, for a long time, Pulz had also the idea that one could use the microalgae for the production of biofuels using particularly the type *Chlorella vulgaris* and to use carbon dioxide generated by power plants or exhaust gases generated by particular manufacturing plants. Already in 1996 Otto Pilz cultivated *Chlorella* in plate reactors for biodiesel.

The related techniques were implemented at a lime burning plant in cooperation with the German conglomerate Preussag AG which at that time (until 2000) was focusing on exploitation of mineral resources and their processing, for instance, mining of coal, metal ores, potash and rock salt and limestone. Back in 1996/1997 Prof. Karl-Hermann Steinberg of Preussag AG (1995 – 1999 Director of Innovation, Preussag AG Hannover responsible for microalgal research, etc.) was also considering the question of how carbon dioxide emissions from power stations, for example, could be put to good use.

Together with the holder of a patent for algae-breeding IGV developed the Preussag algae manufacturing plant. That is, the production plant and the method of cultivation were protected by patents. Construction of a microalgae production plant began in

Klötze in 1999 (in the German state Sachsen-Anhalt; see below). One year later the first algae of the *Chlorella vulgaris* species were already being cultivated in this unique facility, the first of its kind in the world. However, the low petro-oil price forced Preussag as the owner/operator of the algae production plant to give up after four years.

Prof. Pulz continued to complete Europe's largest microalgae production plant. It produced 130 metric tons of algae per year. The system of production halls contained 500 km of arm-width glass tubes filled with green water and meandering through the halls. The water is mixed with a starter culture; then the tiny organisms multiply with the help of light and carbon dioxide. At the end of the plant a large centrifuge concentrates the liquid to a thick green grits, which is dried. The result is purest microalgae – packed into bags as a fine greenish powder. The powder is used as an additive to food or cosmetics [Schürmann 2007; Schibilsky 2008].

In the meantime, IGV's researchers developed a so-called 3D-matrix system in which to grow two to three times as many microalgae as in the conventional, essentially linear photobioreactors made of glass tubes. Here, microalgae grow in geometric structures. The novelty was that the distribution of algae is in three dimensions, as the light is distributed, and thus light and algae are constantly in contact. Thus, one achieves a better photosynthetic performance and highest yield. If sunlight is not sufficient energy-saving light bulbs will support providing necessary light intensity [Schibilsky 2008].

US startup Greenfuel Technologies took notice of this new type of reactor which led to development cooperation with IGV. For instance, a stepwise process of feeding the algae was found to be important. At first, the algae are still supplied with everything needed for rapid growth. Then the food intake is reduced. And the algae react to the sudden shortage by converting up to 70 percent of their weight into oil, which can be processed into biodiesel [Schürmann 2007; Schibilsky 2008]. However, not related to technical shortcomings, a number of other factors ultimately led to the bankruptcy of Greenfuel as described in previous chapters (ch. 2.1.2.8; A.1.1.3).

One lesson learned concerning algae is that so far technology has not reached its goal and "under the present price situation a large-area production of microalgae for energy under Central European climatic conditions is not realistic." [Schibilsky 2008]

The IGV/Preussag cooperation in Klötze gave rise to a further technology entrepreneurship case of a former citizen of the GDR involving the Preussag partner of IGV, in particular, Prof. Karl-Hermann Steinberg, and the marketing of algae under the Algomed® brand. When Preussag stepped out of the algae project in 1999 Prof. Steinberg founded the firm "Bioprodukte Prof. Steinberg GmbH" (BPS) to organize distribution and selling of algae products [Zentner 2004].

Also in 1999 the "algae-patent" was licensed to a startup "Ökologische Produkte Altmark GmbH" (ÖPA) and the algal production plant was finalized in Klötze. It was claimed that nowhere else in the world algae are produced with such high standards of purity and quality. But in 2001 ÖPA went bankrupt. [105] Steinberg attributed commercial difficulties to the termination of production after less than one and a half years in charge. Due to excess capacity and a more expensive building the company became uneconomical [Voigt 2008]

Prof. Dr. Karl-Hermann Steinberg (born in 1941), growing up in the former GDR, graduated in chemistry (Dr.) at the Merseburg Technical University and became full professor of chemical engineering, Leipzig University in 1991. Further steps in his life included [106]

1989 – 1990: Deputy Minister of heavy industry of the GDR, environmental protection department;
1990: Minister for environmental protection, nature conservation, energy and reactor safety in the "de Maizière government" of the former GDR;
1991: Joined Noell GmbH, Würzburg (Germany), in essence a subsidiary of Preussag AG and active in various fields, such as systems and mechanical engineering, steel construction and machinery, energy and environmental technology, process engineering and services;
1995 – 1999 Director of Innovation, Preussag AG Hannover.

In 2004 insolvency assets of ÖPA were taken over by Prof. Steinberg's firm assisted by a group of private investors to form "Bioprodukte Prof. Steinberg Produktions- und Vertriebs GmbH & Co KG." The facility was partially rebuilt, renovated and modernized and production of microalgae was launched within a new constellation. The next year the laboratory was extended, an extensive collection of algal strains was built up and a scale-up line established to produce and sell ALGOMED® products, for instance, also via its online shop or pharmacies [Zentner 2004; Voigt 2008] [105].

"We had to invest nearly €4.5 million to build technology anew and eliminate the damage caused by the long shutdown," said Steinberg [Voigt 2008] For his investments Prof. Steinberg could get a loan guarantee of the State of Sachsen-Anhalt for €1.2 million [Lieske 2004] – and it can be assumed that his political experiences and networking had paid off.

With 17 employees in 2008 sales of the business went up. In 2006, revenues from sold algae in many different forms were almost €1 million; one year later sales climbed to €1.2 million euros and for 2008 Steinberg expected to reach €1.5 million. 60 percent of the products were exported. Customers from Switzerland, France and even Malaysia were on the list [Voigt 2008].

But, since January 2008, the company belongs to the French group Roquette Frères, which is Europe's biggest starch producer [Voigt 2008] – and since October 2008 Steinberg worked as an external consultant. [106] The new firm's name is Roquette Klötze GmbH & Co. KG.

In line with the German firms Novagreen, "Bioprodukte Prof. Steinberg Produktions- und Vertriebs GmbH & Co KG" and the private research institute IGV the US firm Solix Biofuels™ (now Solix BioSystems) with its AGS™ Technology (Algae Growth System) cultivates oil-rich microalgae in a *controlled* environment. It is involved in the production of a "biocrude" (algae oil), "green" diesel (jet fuel and biodiesel), methane, chemical intermediates, feed and other important products. Actually Solix, headquartered in Ft. Collins, Colorado and founded in 2006, aims to build a commercially viable alternative to petroleum-based fuels and chemicals. It is a university spin-out transformed into a *VC-based startup* with currently a *"veterans" management team.*

According to research done by the Laboratory for Algae Research & Biotechnology, closed system photobioreactors, like Solix's AGS™ Technology, have seven times the biomass productivity of open pond systems.

Solix Biofuels was founded by private entrepreneurs Jim Sears and Doug Henston (then CEO), Colorado State professor Bryan Wilson (then CTO), and Colorado State University (CSU) itself. Working to refine and scale Sears' original bioreactor design, the group has called on the resources of CSU's Engine and Energy Conversion Laboratory in constructing a working prototype of a closed-tank bioreactor [Madrigal 2007].

By August 2006 a first generation prototype had been built, tested, and analyzed, and a second generation prototype was launched. Furthermore, Solix housed at Colorado State University, has spent a year sorting through 40 strains of algae collected from around the world. The startup sought for the best strains and best environment for the organisms [Procter 2008; Narvaes Wilmsen 2006].

Now Solix Biofuels' AGS algal production system is designed to enable the industrialization of algae at scales suitable for large volumetric production of biocrude in volumes. At the center of the AGS™ Technology is Solix Biofuels' proprietary photobioreactors. The photobioreactor contains closed chambers rather than tube systems as used by its German counterparts. The technology circulates algae within the chamber using controlled turbulence in order to maximize exposure of algae to light and thus algae growth by photosynthesis. More details of the technology are given in The Energy Blog [2006].

Solix provides essential technology for industrial algae production. Apart from acting as a producer it offers also its system as an integrated, flexible algae growth system addressing various capacities utilizing Solix's proprietary, floating photobioreactor panels (Lumian™ panels) to provide a high productivity growth environment for the outdoor cultivation and evaluation of algae species.

Funding steps according to CrunchBase were:

- $16M in Series B funding (3/28/2011)
- $2M in Venture Round funding (1/4/2010)

- $2.5M in Venture Round funding (11/23/2009)
- $500k in Venture Round funding (11/6/2009)
- $6.3M in Series A funding (7/2/2009)
- $10.5M in Series A funding (11/12/2008).

Solix converts the algae biomass it produces at its Demonstration Plant with peak production capacity of 3,000 gallons per acre per year of algal oil into biocrude oil using its proprietary extraction process that removes the triglycerides from the biomass (algae oil, Figure I.177) [Ritch 2010]. Solix is also offering the residual biomass that remains after the triglycerides have been extracted. This biomass is rich in protein and carbohydrates and has potential as a source for various food ingredients or as an animal feed or for aquaculture. It is also rich in other products including amino acids, carotenes and antioxidants.

Solix Biofuels' demonstration facility is located at Coyote Gulch in southwestern Colorado. This large-scale facility complements the pilot facility in Fort Collins, Colorado. The demonstration facility is located on land provided by its partner, the Southern Ute Alternative Energy Fund. It is using waste water generated during coal-bed methane production thus reducing the need for fresh water.

In 2010 Solix was in talks with potential partners interested in building, owning and operating plants using Solix's technology to produce algal oil and downstream products [Ritch 2010]. In the same year Solix signed an agreement to investigate the use of algae for the German firm BASF, the world's largest chemical company. For BASF the emphasis is on commitment to generate growth from industrial biotechnology and on algae representing an addition to BASF's technology portfolio as they offer the potential to produce a number of specialty chemicals and products.

A.1.1.5 Structuring Entrepreneurship in Biofuels

Previous sub-chapters have largely focused on the *large firms' intrapreneurial approach to biofuels involving NTBFs* in the field or a technical/*engineering approach* following an engineering, procurement and construction (EPC) project for scaling-up to commercialization (Figure I.180). Without much own research the last approach concentrates on process and plant engineering. Additionally, it focuses on

- Improving and *modifying existing, proven technologies* and exploiting whatever is available or purchasable, such as using Lego-type commercial-off-the-shelf (COTS) components, retrofitting existing plants etc.;
- Managing and execution by "*technology veterans*" with large experience in the (technical) field and a related *management team with long and broad managerial and project leading experiences* in the oil industry or related industries and connected by previous common affiliations and presently common interests; moreover, the veterans approach has emerged often from "old boys networks";

- Having managers or persons, respectively, with experience in government relationships (grants, funding, permits etc.) to play an important role in the management team.

In an RD&D scale-up process as depicted in Figure I.180 this approach steps in essentially at the "Initial System Prototype" phase.

On the other hand, there are startups (often RBSUs) which have to do much research and start from scratch in the process ("Research") which later have to add engineering competencies for scale-up ("Demonstration & Development"). This characterizes the proceedings as a *scientific/engineering approach*.

In particular, with regard to the "veterans" aspect it is to be noted that new firms' foundation was done often by *serial entrepreneurs*. On the other hand, the examples of Sapphire (Table I.93) (and later Coskata, Table I.99) exhibit a pro-active role of venture capital for foundation.

Given the myriad of technological options, innovation approaches and financing options by private and public organizations the "acid test," the rigorous or crucial appraisal to start a biofuel firm, is the comparison of the startup's production cost versus the petro-fuel production cost or oil price, respectively. Startups usually flag their (competitive) position by the price (in $ or € per gallon/liter) they would have to charge for their biofuel.

For RBSUs considering a scale-up approach from research and lab to a large-scale commercial plant one issue is whether a corresponding calculation done for a lab or pilot plant setting is also valid for the large-scale manufacturing setting. Still, most biofuel NTBFs will have to prove they can deliver on those low prices at full-scale commercial production.

However, the basis of the individual calculations is mostly not compatible due to intransparency of the underlying components and assumptions. For instance, the price per gallon may be calculated as *cost of production* minus revenue from co-products, and assuming particular feedstock cost (type of feedstock, dry or wet?) and their process energy cost which may refer totally to natural gas cost or even may contain energy delivery through a side-chain of their biofuel process.

Comparing the RD&D process for biofuels (Figure I.180) for a policy-driven market with the common value chain (Figure I.7) of technology-based firms one observes that marketing plays a minor or even negligible role.

Right now there are many products and technologies out there and no dominant design, and startups are trying to choose from these and make decisions about. But one cannot be sure that the best technology will win. It may be that the best management team or the best marketing team will make it or the realities of the market will decide otherwise.

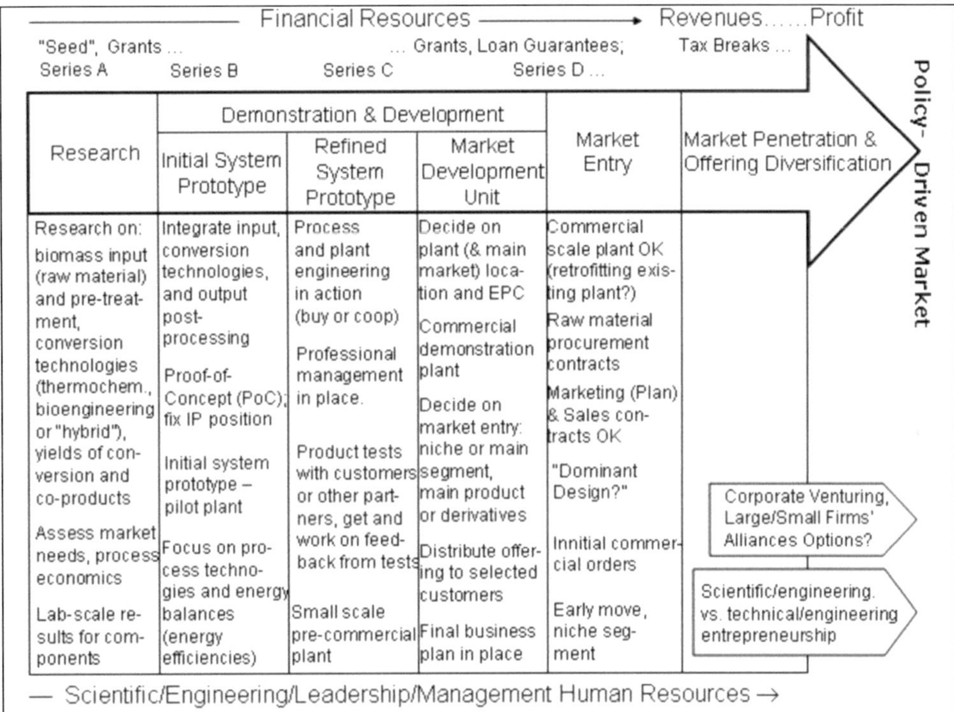

Figure I.180: The RD&D staged path of innovation for biofuels.

Pamela Contag [2008a] of Cobalt Biofuels (Table I.96) and focusing on biobutanol describes entrepreneurship in biofuels in the US as follows (author's emphases and additions in braces): "If public financial support does not dry {in the US and also Germany}, there is a good startup funding base (via grants) for new biofuel firms. However, the plant-and waste-based base biofuel business is essentially *squeezed between* two commodities: the biomass feedstock of the *agriculture community* (or municipal communities) and the *oil industry*." "Although this represents an opportunity for a lot of people, it is a very unpredictable place to be, and it is very high risk."

"The price of oil determines how investors think about the adoption rate of biofuels. Technology development around biofuels resides in the hands of people with a relatively short-term view. Biofuel projects take huge amounts of capital to get to commercialization. There is a hurdle, because my first commercial-scale plant will be first in kind. It's very difficult to put a finance package around a first-in-kind technology."

And Rick Wilson of Cobalt Biofuels added [NewNet]: "If you develop a technology, you have to build your first plant to prove to the world you can do it and for that you require an investor with deep pockets, which are definitely lacking in this space. The venture investors like to develop the technology, while "project finance" investors enjoy funding after the technology is proven commercially, but there remains a grey zone in the middle where someone is needed to step up and provide the capital for the first plant, and it's a big number."

Concerning the industry, for the *biofuel race* in the world ca. 350 companies, from startups to oil and chemical giants, are developing second- and third generation biofuels using a *bewildering array of technologies* [ISEE] on the laboratory scale and very many pilot and demonstration plants are operating or are under construction.

Every year Biofuels Digest publishes (on the Web) a list with the "50 Hottest Companies in Bioenergy." However, replacing mineral oil with biofuels is a tough business. *The competition is intense.* It is a multimillion-dollar question of how to translate a beaker of success to global scale. It is estimated that, even as the industry develops, many of the companies – probably most – will not survive.

Basically, converting biomass to biofuels requires breakthrough developments in the production orientation in any type of process (Figure I.171):

- Thermochemical,
- Bioengineering/biotechnological or
- "Hybrid" ("biothermal") approaches relying and combining thermochemical and bioengineering sub-processes.

A research roadmap for biofuels of the US Department of Energy [DOE 2006] may be a guideline for entrepreneurs (and venture capitalists) to put their corresponding ideas and opportunity identifications into perspective, whether they are from the US or Europe. Though this roadmap acknowledges the validity and public support options for other technologies and type of biofuel, there is a strong focus on the bioengineering approach and particularly (bio)ethanol. This means, there is an implicit effect that may direct entrepreneurial ideas toward the explicitly expressed goals and related routes.

Indeed, in the US the majority of biofuel startups follow the bioengineering and biotechnology route (Figure I.171) using microbes/bacteria, yeast or enzymes to convert (various forms of) sugar into alcohols (bioethanol or biobutanol). But when biomass is broken down into sugars, it still contains substances such as lignin that can poison other microorganisms. In most processes, lignin has to be separated from the sugars to keep the microorganisms healthy. On the other hand, the tolerance of the algae to lignin, however, makes it possible to skip this step, which can reduce costs [Bullis 2008a].

For a bioengineering or biothermal process *naturally occurring* species can be used or *genetically modified objects* (GMOs) – "*designer microbes*" – focusing on several approaches, methods and processes (Figure I.181).

Being proteins *enzymes* participate in cellular metabolic processes with the ability to enhance the rate of reaction between biomolecules. They thus represent biocatalysts (for "industrial biotechnology") which may perform the same functions as "chemocatalysts [Runge 2006:165-169, 571-583]. Both are related as *generic technologies* (Table I.12).

Yeast is a microorganism of the type of single-celled fungus and used as an agent in baking and in brewing beer and is responsible for the conversion of sugars in must to alcohol ("alcoholic fermentation"). Usually chemical process and plant engineers of the world love chemocatalytic processes and, as you can control the yield and scale, make a plant or refinery easier from different (and the least expensive) feedstock.

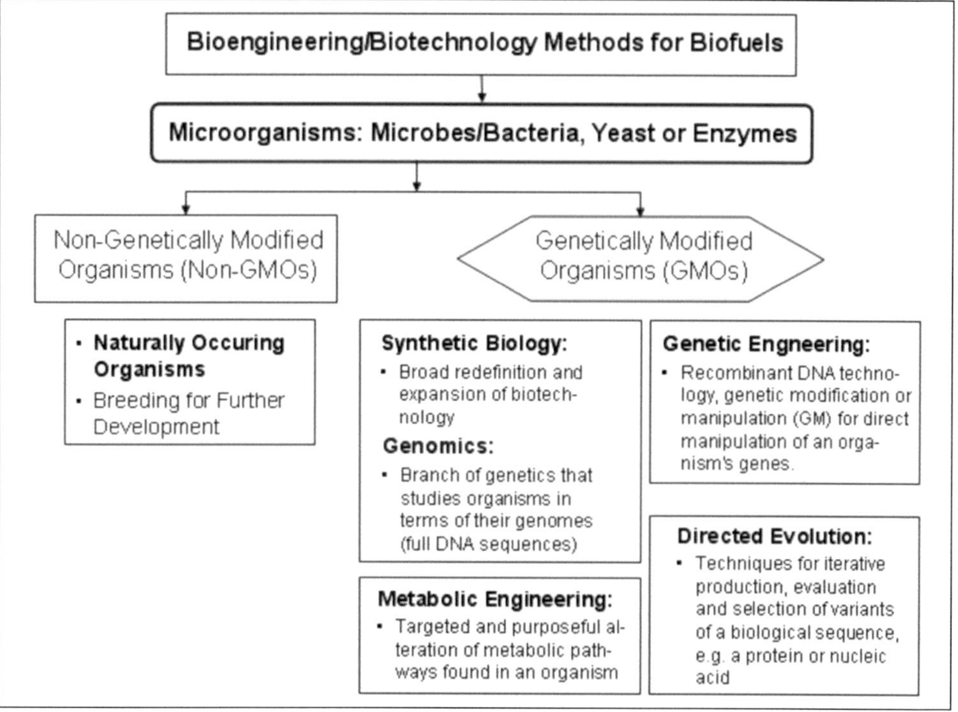

Figure I.181: Bioengineering approaches to convert biomass including algae to biofuels.

Irrespective of the particular bioengineering approach each company has a fundamental decision to make: whether to engineer a biofuel-producing capability into a well-known, robust industrial organism or to engineer industrial fitness and other necessary attributes into an organism that is a natural producer of the molecule of interest.

Generally, with regard to cost minimization an issue of production economy is in how far production can use proven, purchasable and integratable sub-processes or components. Correspondingly, a question for a bioengineering approach is whether to develop microbes/enzymes in house or purchasing them (for instance, the US firm KL Energy purchases enzymes from the big Danish enzyme company Novozymes).

A further idea is to replace the need for enzymes, which are often expensive, with a *mixed culture of bacteria*. The availability of cost-effective enzymes for breaking down cellulose will be critical for the success of the second generation biofuels field. But hydrolysis of cellulose and hemicellulose requires about 20 distinct enzymes that are normally provided by commercial suppliers, such as Novozymes (Denmark) or the Genencor subsidiary of Danish Danisco (no owned by DuPont). But the hydrolysis process can also result in the production of by-products, including acids, ketones and aldehydes, which can inhibit the growth of cells as well as the secreted enzymes.

Working and experimenting with microbes or enzymes always opens a way for *serendipity*. For instance, Mark Emalfarb, founder of Dyadic International Inc, just wanted a better enzyme to soften blue jeans. The search led him to a new fungus from Russia, and then to a *serendipitous mutation* that turned the organism into a biofactory capable of churning out vast amounts of enzymes that can give denim (the fabric for jeans) a prized lived-in look. "By accident, we came by the world's most prolific fungus," he said [Anonymus 2006a]. Actually, after the fall of the Berlin Wall, he hired Russian scientists and took the fungus to the US. [108]

With technical hurdles abound there may be also societal, attitudinal (acceptance) hurdles with bioengineered biofuels. Genetically modified crops have met generally with stiffer resistance from a public in Europe that has labeled such crops as "Franken foods" [Mandaro 2007] and a similar attitude may extend into the biofuels area. In particular, in the US, "the food versus fuel" debate was bad press for biofuels and the "Frankenalgae" debate would be even worse." [Wesoff 2009]

More on Bioalcohols: Rerun of Biobutanol

Using naturally occurring microbes is not only practiced since ages to ferment sugar to ethanol. Since 1916, it is known that microbes, such as *Clostridium acetobutylicum*, can ferment sugar to produce a *mixture* of acetone, butanol, and ethanol in large volumes – by the '"ABE process" exploited mainly for its acetone during the First World War. The butanol was a by-product of this fermentation (twice as much butanol was produced). Yet microbial breweries were discarded by the 1980s in favor of a cheaper petrochemical route, via the reaction of carbon monoxide and hydrogen (from syngas) with propylene. Currently, there have emerged several startups following the ABE route [Kiplinger Washington Editors 2007].

For instance, over years, David Ramey in the US, founder of the engineering and consulting firm Environmental Energy, Inc., then ButylFuels, LLC, has *further developed and patented the original ABE process* that makes the fermentation process *more economically viable* and competitive (by a continuous two stage anaerobic fermentation process without significant amounts of acetone or ethanol). In particular, he demonstrated to the public that there is an alcohol made from corn (butanol) that can replace petro-gasoline totally (Table I.82). In 2005 he ran a conventional unmodified

"2 Buick Park Avenue" car with no modifications across the US with 24 miles per gallon on butanol.

ButylFuel was supported by several federal and state grants ($0.6 mio. by US Department of Energy Small Business Technology Transfer Program [Wilder 2004]) [109] Then, for collaboration with Dr. S.T. Yang at the Ohio State University, he obtained a $1 million dollar grant through the SBA's Small Business Innovation Research (SBIR) program to research, develop and commercialize butanol fermentation.

In particular, the project was to *develop novel engineered Clostridia strains* for fermentation to economically produce butanol as a biofuel from sugars derived from starchy and lignocellulosic biomass [Ramey 2007]. ButylFuel was planning to market its *biobutanol as a solvent* first, and then market it as a fuel in the future. Generally, it is assumed that *existing bioethanol plants can cost-effectively be retrofitted to biobutanol production.*

David Ramey (of ButylFuel) is a *veteran in biobutanol.* He started around 1990 when he asked himself "Why Not Butanol in the 1970s." He noted that (in the US) "people are surprised to learn that it hasn't been firmly on the radar screen as an alternative fuel. On the other hand, butanol was on the alternative fuels map three decades ago. We had a choice to subsidize either ethanol or butanol and we went with ethanol."

Then, Ramey's butanol was produced by his own patented process, and for his pioneering efforts to bring this organically derived fuel to market, he was recognized as the "1996 Technologist of the Year" by the Ohio Academy of Science [Ramey 2007].

> Though by all criteria biobutanol is a much better biofuel than bioethanol (Table I.82) policy driven by the agricultural corn-lobby attributed the lion's share of support to bioethanol.

But by 2012 ethanol producers began switching to biobutanol and chemicals [Admin 2012; Bevill 2012]. Longer term, butanol is a superior "drop in" biofuel and can directly replace gasoline as a fuel. It is a superior blend stock as well, and can be blended with diesel as well as other biofuels, such as biodiesel, ethanol and isobutanol. The blend stock opportunity for butanol exceeds $80 billion per year. Butanol also has the potential to be upgraded to aviation jet fuel, a $50 billion market driven by increasing global interest in reduction of carbon emissions.

For ethanol producers, it is the path of least resistance in getting around the ethanol blend wall. For the high priests developing the new technologies and magic bugs, it is an opportunity to partner with companies that have feedstock, infrastructure, 90 percent of the required steel in the ground, and existing markets for co-products. [Admin 2012]. In monetary terms it was estimated that butanol has a high value of £900 (ca.$1,500 in 2012) per ton compared with the £300 per ton price of ethanol.

The production principles of butanol according to an ABE-process from agricultural residues is the same as that of cellulosic ethanol. It involves four steps: 1) pretreatment, which opens the cell wall structure and removes lignin; 2) hydrolysis of hemicellulose and cellulose into simple hexose (C6) and pentose (C5) sugars using enzymes; 3) fermentation of simple sugars into butanol using a microbe; and 4) recovery of the butanol.

However, there is an inherent paradox in the microbial fermentation of butanol: Butanol-producing bacteria produce the enzymes that convert simple sugars into the alcohol, but butanol itself is toxic to those same bugs. This butanol inhibition (once its concentration rises above about 2 percent) results in a lower alcohol concentration in the fermentation broth, which leads to lower yields of butanol and higher recovery costs. These are challenges that surface when even highly pure feedstock is used [Ebert 2008; Van Noorden 2008].

To grasp the related opportunity, butanol production is generally in search for input substrates, which are not only economical on the lab level, but also on the production level. The major barrier to butanol production has been the high cost of the conventional starch fermentation process. US ButylFuel was already on the route to improve butanol yield through ABE fermentation by genetically manipulating related microbes.

But, in 2012, ButylFuel with 4,000 ft² lab and office space in Ohio and 40 employees, many of whom have advanced degrees in microbiology, biochemistry or biochemical engineering, merged with the UK startup Green Biologics Ltd. (GBL; Table I.95).

GBL's strengths in biobutanol technology were seen to complement ButylFuel's strengths in the design, build and operation of large scale bioprocessing facilities, particularly in the US market. Post-merger, the combined entity is claimed to be a global leader in biobutanol and other C4 chemicals, with skills and assets spanning microbiology and metabolic engineering through advanced fermentation and commercial production scale [GBL 2012].

The key butanol players divide neatly a pair of producers pursuing essentially isobutanol – Gevo (Table I.99 and Butamax (Table I.83, a BP-DuPont JV) – and two pursuing n-butanol, Green Biologics and Cobalt Biofuels (Table I.96).

In the UK since 2003 the startup Green Biologics Ltd. (GBL) pursued optimization and "*re-commercialization*" of the n-butanol fermentation process aiming for a two- to three-fold reduction in cost (Table I.95). GBL focuses on *thermophilic microbes and thermostable enzymes*. These are robust, faster, more effective and cheaper than conventional microbes operating at ambient temperatures. GBL also *looked into input options (feedstock)* and directed their *market orientation towards India and particularly China* taking the same view as BioMCN for biomethanol (cf. Table I.87) – and recently also towards the US.

Green Biologics is pursuing a model in which it and an ethanol producer will co-invest in a project, and both earn off the increased revenue flow from the sale of biobutanol into higher-value markets. Green Biologics offers the sales and marketing for n-butanol. Payback was expected to be within three years for an ethanol plant partner [Admin 2012].

Table I.95: Pursuing biobutanol as a biofuel and C4 chemicals and derivatives by Green Biologics. *)

Company (Foundation) Remarks	**Major Funding**	**CEO, Other Executives, Foundation, Key Researchers; Technology Protection**
Green Biologics, Ltd. – GBL Abingdon, UK (2003) Vision: Become the world's leading supplier of advanced fermentation techniques for conversion of lignocellulosic plant material to renewable biofuels and chemicals. Employees: 2003: 4, 2004: 6, plus access to Georgia University scientists 2007: 13 [VentureBeat], 2008: 20 [Guardian 2008], 2009: 25. Revenues: 2003: £160,000 [Koenig 2005] 2008: £700,000 – forecasted	By 2012 GBL has raised over $15 million in equity financing from angel investors and venture capital firms. Launched a £6.5M (ca. $10M) round (Series B) to close in April 2010 [Lux Research 2009] 2009: completed round with Hong Kong investment group Morningside (its Dr Gerald Chan joining GBL's board). 2008: £3.5M (ca. $6.33M) fundraising round 2007: £1.58M (ca. $3.2M) completed Awarded £560,000 (ca.	Sean Sutcliffe CEO since 2008, came from Biofuels Corporation Trading Ltd, which operates one of Europe's largest biodiesel plants, where he has been CEO since 2005; worked for BG Group plc for 14 years in a variety of roles spanning operations, business development and strategy, most recently as Executive Vice President with responsibility for Corporate Development and New Businesses; he is a Chartered Mechanical Engineer with an Engineering degree from Cambridge University. Dr. Edward Green founder and CSO, gained PhD in Biochemical Engineering in 1993 from the University of Manchester Institute of Science and Technology (UMIST); after 5 years in academia in 1998 joined Agrol Ltd., a UK startup where he established a multi-disciplinary team that developed a high temperature ethanol process; has delivered technical improvements in microbial fermentation processes for biofuel production over the past 17 years contributing to numerous scientific publications and patents. Fergal O'Brien VP Commercial Operations; a Biochemist with 25 years experience in the Biotechnology/Fine Chemical Industry; held positions in R&D, operations, business development and senior management for a number of UK-based companies including Celltech, Enzymatix, Chiroscience, Chirotech and Dow Pharmaceuticals; was CEO of Warwick Effect Polymers Ltd from 2004-6 and joined GBL from his own Business Development Consultancy where he has focused on assisting SME's. GBL utilizes closely connected "Advisors": Robert Rickman, Feedstock Advisor Steve Vaux, Feedstock Advisor Dr. Martin, Comberbach Bioprocess Consultant

[pipeline 2008]. Estimated $150,000 annual burn per employee [Lux Research 2009]. Funding mode: equity, grant aid from central and local government, contract services.	$1.1M), with £250,000 (ca. $500,000) from the Department of Trade and Industry-led Technology Program and £310,000 ($610,000) from shareholder investors and business angels £250,000 grant awarded to GBL and EKB 2005: First funding round, £63,000 (primarily business angel investors in the community)	Professor, David Jones Scientific Advisor. GBL Technology Strengths and/or Differentiator: A unique collection of thermophilic microorganisms; a comprehensive and searchable database for the culture collection (library of organisms includes over 120 *Clostridia* strains and over 800 thermophilic organisms used for high temperature processes); unique access to large-scale fermentation at the University of Georgia; advanced separation process. In-house IP covers microbe and fermentation processes as well as solvent recovery, but some overlap exists with other fermentation technologies; their patented biobutanol: Butafuel™. GBL has wide ranging portfolio of proprietary technology relating to ABE fermentation using *Clostridia* organisms as biocatalysts.
Technology, Goals, Strategy Goal: to produce a wide range of C4 chemicals and derivatives, including C4 biofuels The technology and IP estate includes lignocellulosic processing which allows utilizing both C5 and C6 sugars to extract much higher energy content than processing sugar and starch alone.	The company has a biorefinery approach, but focuses on the production of only n-butanol (not isobutanol) on the basis of ABE fermentation; follows genetic manipulation of microbes to *improve butanol yield*; they optimize and "re-commercialize" the butanol fermentation; although *cheaper feedstock* decreases major costs, energy expenditure required for solvent recovery is a major challenge that has to be resolved; GBL aims to retrofit (grain) ethanol plants; considers also co-products (acetone and ethanol of ABE process); process (Figure) aiming for a two- to three-fold reduction in cost. To circumvent "butanol toxicity" claims to have developed "solvent tolerant" strains; has a microbial platform technology based on a unique and proprietary collection of heat resistant microorganisms (thermophiles) and thermostable enzymes that operate at higher temperatures than other industrial microorganisms; metabolic engineering is used to generate a second generation of Industrial thermophiles (Figure). GBL's microbial platform technology provides flexibility across a "range of different feedstock" due to its options to use different microbial species and strains for specific feedstock; microbial strains and cocktails are tailored for wide variety of feedstock	

Business model: License technology (focus on "upgrading" biobutanol plants); provide consultancy, contract services and contract research; solution provider (integrated technology provision participation in production assets); focus on China and India as key markets (US added recently) Services: High throughput screens for thermophiles and thermostable enzymes; microbial expression systems in robust hosts; metabolic pathway engineering to improve product yield and other strain characteristics; fermentation process development to improve both yield and productivity.	fermentation; yet process remains unproven at large scale. Using its library of thermophiles and thermostable enzymes GBL has isolated a cocktail of thermophilic microorganisms for the rapid enzymatic hydrolysis and release of fermentable sugars from biomass. The company planned to integrate this patented hydrolysis technology with a proprietary butanol fermentation process. Partners with EKB Technology (specialist in process technology) to develop the advanced fermentation process for butanol with improved yields. EKB aims to create new and highly configurable platform technologies that combine advances in bioreactor technologies and previously separate downstream process technologies into a single step; also suited to chemical syntheses that utilize fermentation technologies and biocatalysts. Ultimate GBL process: continuous fermentation, advanced separation process. Develops also advanced and renewable fermentation technologies for conversion of biomass to higher value chemicals and biofuels; solving problems in both existing and emerging markets for fuel and bulk chemical manufacture as well as environmental waste treatment. Equipped at GBL's Milton Park headquarters with a 300 liter pilot plant and a 140 liter pilot plant allowed producing small amounts for thermophile fermentation testing and new patent developments, provides also confidence to GBL's feedstock partners. The most significant emerging opportunities for GBL are in China. "China is a key market for Green Biologics." The focus is introducing its improved technology to their plants to radically reduce the cost of local biobutanol production. Green Biologics works directly with commercial butanol producers at scale (300 m3 to 400 m3 fermentation volume) by licensing its microbial technology and process solutions; working with several butanol producers in China, due to the accessibility to existing plants and large market for butanol. It expects to reach commercial production there, with an estimated capacity of 30,000 tons/year; this serves simultaneously to rapidly demonstrate the GBL technology to potential clients around the world. it is also working with sugar and ethanol producers in India, due to accessibility there of cheaper feedstock, and converting sugar ethanol plants into butanol plants. Ties to China are broadly developed, from the Hong Kong investor Morningside to appointing Intelligent Sensor Systems (ISS) as its commercial representative in China to support GBL in exploiting the rapidly expanding commercial opportunities for its technology and services to Chinese biochemical and biofuel producers and appointing Professor Zhihao Sun as a Scientific Advisor.

In 2008 GBL and the Energy Research Institute, Shandong Academy of Sciences and Green Biologics (SDERI) of Jinan, China. established a technical collaboration agreement on the production of biobutanol; involved transfer of GBL's technology into a purpose built pilot facility at SDERI's research center; serves GBL as a local commercial demonstrator and supporter of marketing of GBL technology to Chinese biobutanol producers;

also in 2008 an agreement between GBL and Laxmi Organic Industries to develop and construct a commercial scale demonstrator for biobutanol in India; the demonstrator plant was expected to produce 1,000 tons of butanol a year starting in 2010, the biobutanol plant will run on molasses produced by the Indian sugarcane industry.

The £3.5 million ($6.33 million) fund raising was intended to roll out GBL's renewable chemicals technology. In 2005 GBL developed, for instance, a novel solvent system to remove chewing gum waste from pavements.

The company estimated a total fuel butanol market size of £3 billion growing at 4% per year.

Figure: GBL's bioengineering approaches to biobutanol.

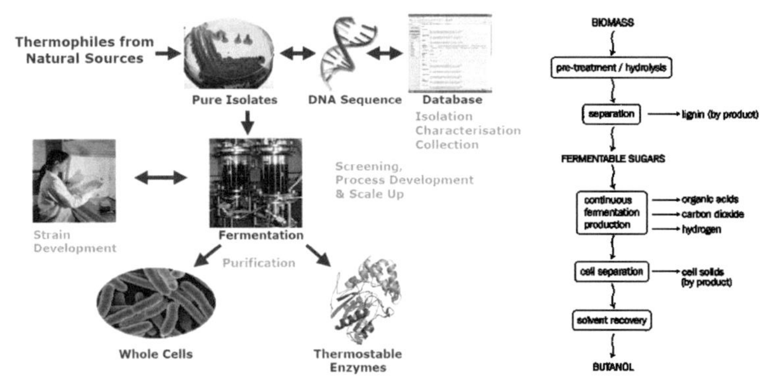

*) From the firm's Web site if not stated otherwise by a reference;
firm's status: early 2010 and selected relevant additions till 2012.

In the US one of GBL's close competitors is Cobalt Biofuels (Table I.96) which addresses *cost reduction* through three processes or areas, respectively: 1) strain development, 2) reaction management and 3) vapor compression distillation (VCD) as a separations technology that removes alcohol from the fermentation steep. Cobalt brought down the overall cost of production through a systemic approach involving the three key sub-processes or areas rather than focusing on just one or another sub-process. In contrast to GBL, which already created revenues in 2008 by contract services and consulting, Cobalt Biofuels did not seem to generate revenues (that far).

Table I.96: Pursuing biobutanol as a biofuel and for C4 chemicals and derivatives by Cobalt Biofuels *)

Company (Foundation) Remarks	**CEO, Other Executives, Foundation, Key Researchers; Technology Protection**
Cobalt Biofuels, Inc. Mountain View, (CA) (2006)	Dr. Pamela Contag CTO and founder (President and CEO Cobalt 2005-2008); serial entrepreneur, prior to founding Cobalt Biofuels, founded Xenogen Corp. in 1995 and served as President and concurrently as CEO of Xenogen Biosciences. Xenogen Corp. went public in 2004; sold it as it merged with CaliperLS in 2006.
Employees: 2008: 25 (Oct. 20, 2008) [Contag 2008b];	Pamela Contag is a representative of a "*stage-oriented entrepreneur*" founding and leading a new firm to a particular state of development and then handing over to professional management (ch. 2.1.2.6).
2010: 40 [Stroud 2010].	With more than 25 years of microbiology research experience, Contag has widely published in the field of non-invasive molecular and cellular imaging. She received her PhD in Microbiology at the University of Minnesota Medical School in 1989; since December 2008 she has been a director of Delcath Systems.
Biobutanol Scale-Up Plan: 35,000 gallons per year pilot in 2009, 2.5 mil. GPY pre-commercial in 2010 25 mil. GPY plant in 2012	"I generally invent and develop technology and then take on investors who ultimately direct the company. I put all my energy into the demonstration of the technology and business model," Pamela Contag said [Ainsworth 2008]. Hence, after having raised \$25M she withdrew to CTO to hand over commercialization to investors and a professional management. Dr. Rick Wilson CEO; over twenty years of global energy commercial and technology experience, recently including VP of British Petroleum's Global Derivatives Chemicals business unit, background in process engineering and broad executive experience in the fuels and chemical industries; received an MBA from the University of Chicago Graduate School of Business and a PhD in Chemical Engineering from Lehigh University, where he focused on energy efficiency. During his seventeen years at BP/Amoco, Dr. Wilson held a variety of technical, trading and executive positions and was also responsible for BP's \$3.5 billion petrochemical business, ultimately spun out as Ineos.
2015: jumpstart revenue in the chemicals market [Contag 2008b]. **Major Funding** In 2011 raised a new \$20 million Series D venture round \$25M (10/2008) Series C, to	Mark Dinello SVP Engineering; extensive experience in the chemicals and refining industry; during his 30-year career with BP/Amoco Dinello held a variety of senior positions in engineering, procurement, construction, and management, for chemical and refining operations worldwide. He earned a BS Degree in Chemical Engineering from the Pennsylvania State University and an MBA from The University of Chicago. Prior to joining Cobalt, Mark founded and served as President and Senior Consultant of Plan B Consulting Inc. At Cobalt, he will be responsible for capital projects, operations, engineering, and company health, safety and environmental policy and practice.

expand from lab scale production to pilot facility – capacity of 35,000 gallons of fuel per year, 2007: $12M Series B; 2006: $1M Series A. Opened up its first pilot plant in 2010 [Fehrenbacher 2010].	David Walther, PhD: Director Engineering; has over ten years of experience directing research teams focused on developing and implementing microsystems in the areas of biosensing, power and energy; he was granted Exceptional PI status for several US Government, State of California and Industry Sponsored Research Grants. Hendrik Meerman, PhD: Director Bioprocessing; is an acknowledged expert in fermentation technology and bioprocess development, possessing a keen understanding of microbial physiology. Prior to joining Cobalt, Dr. Meerman served for over ten years in process development at Genencor International, where he was responsible for developing and transferring several scalable processes that quickly moved product concepts from the bench to commercial manufacture. Cobalt and, in particular Pamela Contag and her husband Christopher as inventors, and their university and Xenogen, have proprietary technologies in microbial physiology, strain development and fermentation. It is interesting to note that Cobalt sought professional services to establish its communication and media strategy. Cobalt Biofuels asked Ecofusion, a strategic communications and media company, to build an *identity, a story, and a public relations platform* to spread the word about biobutanol and the range of solutions the company will develop in the future. After constructing and delivering a new biofuel brand and message systems Ecofusion implemented a third phase of the communications strategy; planning and managing a press and media rollout for major milestones in the company's early development [Ecofusion]
Technology, Goals, Strategy Business model: Very low cost producer/owner incl. low cost feedstock and process efficiency; Be determined by a project-by-project basis; Sell co-products into the chemical solvent market, then butanol (primarily as a substitute for	Cobalt follows the ABE-process to produce normal butanol (not isobutanol) with modified *Clostridium* microorganisms thus increasing the amount of butanol produced to decrease cost. Generally some other factors are also seen as important (Figure): mainly the consumption of energy and the consumption of water. To reduce energy the company has licensed a new technology, called vapor compression distillation (VCD), for separating the butanol and water; traditional butanol separation (distillation) accounts for 40-70% of total production energy [Contag 2008b]. VCD removes alcohol from the fermentation steep using one-half the energy required for typical separation techniques [Fehrenbacher 2010]. To reduce water use, the company has turned to proprietary water purification and recycling systems; the company has further increased butanol production by engineering a bioreactor [Bullis 2008b]. Additionally, it is said [Lane 2010m] that residual "lignin is passed to the onsite boiler and generates sufficient power to serve the needs of the biorefinery, with significant excess power exported to the grid." One of Cobalt Biofuels' key advances is a technique for genetically engineer-

gasoline); See to become a licensor; Assess how to grasp opportunities overseas.	ing strains of *Clostridium* so that they produce a luminescent protein whenever they produce butanol; "When the *Clostridium* are happy and producing butanol, they're also producing light," Contag said. When they are paired with light detectors, the company can quickly sort through new strains of the bacteria, as well as tailor their environment, to increase production [Bullis 2008b]. Their patented reaction management technology – production monitoring technology – maintains their continuous fermentation process at peak production rates and an optimal concentration of butanol in the steep, for extended periods of time. It is this bioreactor technology that forms the basis of the production process. Cobalt does not do genetic engineering but accelerates the evolution of the bugs to produce more product at higher concentrations by conditioning them to adapt; is focused on putting waste biomass to good use [NewNet] and can tailor its microbes for different regionally available feedstock, optimizing its process for deployment anywhere [Fehrenbacher 2010]. But it can also use more traditional feedstock including corn and sorghum; this means Cobalt can site their facilities in a wide range of geographies and use the feedstock available locally. The focus is on DOE-favored, low-cost feedstock with high hemicelluloses content, - Wood pulp, - Sugar beets and beet processing by-products, - Energy crops, forage or sweet sorghum [Contac 2008b], in particular, forest waste and mill residues. In 2010 Cobalt announced a breakthrough in producing biobutanol from beetle-killed lodgepole pine feedstock. They claimed to be able to scale-up to a commercial facility within the next two years, (first commercial sales of biobutanol in 2011) and "multiple facilities" by 2014 [Fehrenbacher 2010]. Has a strategic partnership with EPC firm Fluor Corp., with a strategy of designing low-cost plants [Lane 2010m]; engineering and construction giant Fluor Corporation should bring its technology to commercial scale. Fluor would provide engineering consulting, advise Cobalt on how to scale up, put together a design package for Cobalt's future plant, and execute the construction part of the project [Stroud 2010]. Cobalt intends to position butanol as a high value chemical or fuel additive; the plan is to sell into the chemical solvent market at pre-commercial stage [Contag 2008b]; can offer also a small amount of acetone;

and as Cobalt scales up production, it plans to sell the butanol as a substitute for gasoline [Bullis 2008b].

The interrelation how Cobalt brought down the cost through three processes or areas, respectively, is illustrated in the Figure.

Figure: Cobalt's systemic approach to three key processes for biobutanol.

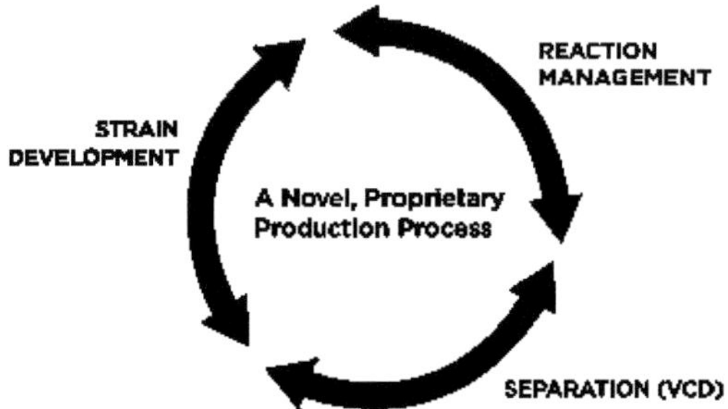

Strain Development: developed proprietary, high-throughput processes for identifying and engineering the optimal microbial strains for converting a given plant material (a range of feedstocks).
Reaction Management: the patented reaction management technology poises the continuous fermentation process at peak production rates for extended periods of time. This increases productivity and ensures optimum feedstock utilization (creating sensor production strains using bioluminescence; real-time monitoring allows avoiding poising of fermentation).
Vapor Compression Distillation (VCD): patented fluid separations technology removes alcohol from the fermentation steep using approximately one-half the energy required compared to typical separation techniques; has the additional advantage of drastically reducing water usage (recycling the VCD-purified water back into the production process).

Cobalt was looking early for opportunities overseas, though the company had not been ready to discuss them [Wang 2008].

*) From the firm's Web site if not stated otherwise by a reference; firm's state: early 2010.

By the end of 2011 Cobalt Technologies appointed as chairman and CEO Bob Mayer replacing Rick Wilson. Most recently Bob Mayer was CEO of Genencor [Admin 2011b]. And currently there are additionally some new faces in Cobalt's executive team. That means, after five years of existence Cobalt's founder/leadership team has been almost completely changed.

Over the 2011/2012 period a lot of demonstrations occurred by Cobalt regarding its various sub-processes. In 2012 Cobalt has successfully demonstrated one of its advanced biocatalysts in partnership with the National Renewable Energy Laboratory (NREL). It completed multiple fermentation campaigns in a 9,000 liter fermenter, exceeding the target yield and other performance metrics for a commercial scale facility.

The demonstration showed the biocatalyst's ability to convert non-food based substrates into renewable n-butanol and resulted in high sugar conversion and high yields of butanol. "Ultimately, we're showing performance is achievable at commercial scale across our technology platform," said Bob Mayer, CEO of Cobalt Technologies.

The advanced (non-GMO) biocatalyst fermentation demonstration confirmed that the Cobalt process to produce renewable butanol could be 40-60 percent less expensive than production of petroleum-based butanol using the traditional oxo-alcohol process.

While Cobalt's technology is claimed to have the ability to perform on a continuous basis, this testing was conducted using batch processes to fully demonstrate the flexibility of the technology to meet the needs of potential customers and partners. The butanol produced during this demonstration will be sent to several customers for product certification (cf. this approach with that of Perkin, Table I.100; A.1.2).

Concerning sub-processes on the road to commercialization, as reported on its Web site as news, in March 2012 Cobalt's dilute acid hydrolysis *pretreatment* process (Figure I.171, Figure I.185), which extracts sugars from lignocellulosic biomass, was validated on woody biomass, bagasse and agricultural residues.

Cobalt conducted the testing in the Andritz pulp and paper mill demonstration facility. Andritz is a supplier of technologies, equipment and plants for the pulp and paper industry. The test runs processed up to 20 *bone-dry* tons of biomass per day. This milestone also marked the first phase of Cobalt's partnership with specialty chemical company Rhodia in Brazil to develop bio n-butanol refineries throughout Latin America utilizing bagasse as a feedstock.

Cobalt and Rhodia intend in the medium term the construction of multiple biorefineries co-located with sugar mills, firstly in Brazil to demonstrate Cobalt's technology on local and competitive feedstock. Subsequently, the proven technology shall be extended to other Latin American countries.

Furthermore, the US Naval Air Warfare Center Weapons Division (NAWCWD) awarded a manufacturing contract to global specialty chemical firm and catalyst supplier Albemarle Corp. (in February 2012) to complete its first biojet fuel production run based on biobased n-butanol provided by Cobalt Technologies. For this production run, Albemarle will utilize NAWCWD alcohol-to-jet (ATJ) fuel technology to convert Cobalt's biobased n-butanol into biojet fuel at its Baton Rouge, La. processing facility.

The resulting jet fuel should be tested by the NAWCWD as a continuing process for military certification through the Department of Defense. According to Cobalt this underpins two main objectives set out by the firm. "First, it basically helps us scale up and derisk the catalyst." "The second thing this relationship really does for us is that it allows us an avenue and platform to actually generate quantities of fuel required for certification, both military and commercial."

And there was another important aspect of the collaboration with Albemarle that extends beyond exclusively jet fuel production; it provides exploiting other chemical derivatives of n-butanol. NAWCWD's ATJ technology is capable of converting n-butanol into other high-value platform chemicals like 1-butene. For example, the process allows Cobalt to produce butadiene using the 1-butene pathway from n-butanol as the starting point. Butadiene is a valuable industrial chemical used typically for the production of synthetic rubber (A.1.1.6; Box I.26). With regard to 1-butene, Cobalt addresses directly what Gevo (Table I.99) is doing on the basis of bio-isobutanol.

Cobalt Technologies as well as Gevo reveal a shift of biobutanol producers putting the emphasis on a biorefinery aspect with the main stream of revenues being generated by biobased chemicals (cf. A.1.1.6):

> Biofuels (butanol) + chemical co-products (butanol compounds) →
> Chemical products (butanol derivatives + chemical intermediates for platform chemicals, plastics, and rubber) + biofuel (butanol for blending or butanol as a replacement for petro-gasoline)

The challenge and constraint for entering the biofuels race is delivering technology, which enables *cost-efficient production*. Cost efficiency means biofuel that can compete with oil at around $80/barrel.

Many industry participants appear to be focused on large volume production facilities ("owner/producer model") that are highly dependent on significant quantities of biomass feedstock gathered (and transported costly) from long distances. *People have to be broad-minded about what is out and what is in* (technology intelligence!). Structuring technical hurdles according to the biofuels value system (Figure I.171) simultaneous provides startup opportunities in terms of solving associated problems.

The very fundamental problem is growing enough green plant material or harvesting or having enough cellulosic residuals or usable waste. The first hurdle for everyone in the game may be characterized by the questions: *how are we going to grow and/or collect, transport, store, and pretreat, if applicable, the biomass for processing; which kind of biomass; is it wet or dry?*

Furthermore, an inherent issue of processing biomass is *consistency*. The fluctuation in feed material composition and quality that ensures success in the real world is usually far from laboratory-controlled conditions. Processes that work well in the lab often run into problems when scaled up to commercial size. For instance, Iogen (Table I.83) found that enzymes that effectively convert pure wheat straw to sugars fail when faced with 1,000-pound bales laced with dirt, soil, dead mice, and stones [Carey 2009].

In general, it has often turned out that when a cheaper biomass substrate is used for a bioengineering process, additional microbial inhibitors are generated during the pretreatment process. Raw material for second-generation bioalcohols and biogasoline are listed in Table I.97.

Table I.97: Selected raw material for biofuels

- Agricultural residues: like corn stover or cobs or cornstalks (stalks that remain after the corn has been harvested), straw (from cereals), bagasse (the fibrous residue remaining after sugarcane or sorghum stalks are crushed to extract their juice);
- Municipal waste: paper trash and pulp, other municipal garbage, municipal solid waste (MSW);
- Residues from forestry, wood processing or "recycled" wood (for instance, from houses), construction and demolition wood waste, timber harvesting residues, wood chips (for instance, KL Energy Corp. using Ponderosa pines), sawdust, Cobalt Technologies also tried beetle-killed lodgepole pine feedstock);
- Highly productive ("energy-rich") existing or cultivated grasses and trees: switchgrass, eucalyptus and hybrid poplar;
- Carbon-based waste, like old tires.

This raw material for second-generation bioalcohols and biogasoline is associated with a number of issues concerning the most efficient raw material or by-products generated by a particular process. For instance, for its two-step thermochemical process the NTBF Range Fuels (Table I.99) reported that "over 10,000 hours of testing has been completed on over 30 different non-food feedstocks with varying moisture contents and sizes, including wood waste, olive pits, and more." [107]

Sugars found in wood in the form of lignocelluloses are not naturally well digested by microorganisms which convert biomass into usable raw materials. First these complex sugars have to be released and broken down into digestible units. This process often gives rise to harmful by-products, including furans, which can have a strong inhibiting effect on the fermentation process.

However, reminding ourselves of the heavy localization of corn ethanol plants in the US "Corn Belt" and the role of lobbyists from the related states in pushing special focus on corn ethanol in legislation, it may be that the technology development for second-generation bioalcohols and biogasoline will exhibit considerable "*path dependency*" in terms of corn alcohol versus lignocellulosic alcohol or raw materials' availability for input rather than performance (and perhaps even price).

And thinking of bioengineering methods there are still some doubts: "It is not yet clear whether a fully synthetic genome will ever be deployed in a live production environment. A fully synthetic microorganism may not have the robustness which is needed for large-scale industrial bioprocesses." [Sheridan 2009]

Tackling the issue of procurement of biomass for processing and production (input for conversion) is approached by startups and NTBFs in various fashions depending on whether the emphasis is on biofuels for transportation or biorefinery: When we are talking about full-scale production and biofuels we speak of thousands of tons of biomass per day, per biorefinery. On the other hand there would be a sub-1,000 ton per day input for biobased chemicals or the smaller scales for fine chemicals or nutraceuticals.

There are a number of factors which may drive input selection. Criteria may refer to

- *Offering destination*: where to sell the products, and probably co-products, most profitable (product localization. market localization; California represents ca. 20 percent of the US ethanol consumption)
- *Input localization*: where input is plentiful and cheap and in sufficient proximity to the production plant(s) to minimize transportation and storage cost, addressed for Brazil by BP/Verenium JV Vercipia (Table I.83, Table I.84), Cobalt/Rhodia (Table I.96), and Amyris (Table I.99)
- *Input-output efficiency*: localization selection for the production plant, which means focusing on biomass input where it is cheap and where there is a huge existing market for the biofuel, such as ethanol in Brazil, many autos in California; FFVs in Brazil.
- *Legislation localization*: China, for GBL E85 of biobutanol (Table I.95), biomethanol for BioMCN (Table I.87).

Issues of localization of input have been tackled by the German Bioliq approach (Figure I.173). In this sense, for instance, Range Fuels' model (Table I.99) required to "bring systems to sources where biomass is most plentiful, instead of having to transport biomass to a central processing site. This reduces transportation costs and related transportation fuel consumption." [107]

Coskata's model (Table I.99) based on sugarcane bagasse biomass was explicit in this regard. "To generate 100 million gallons in this model, Coskata will need 1 million tons, or 900,000 tons of biomass. That will require 15,000 hectares, or 37,000 acres. That's 51 square miles, or the area within 4 miles of a 100 million gallon refinery. A mighty plantation, but not long hauling distances." [Lane 2009o].

Venture Capital for Biofuels

Rather than looking at corporate venturing and/or aliiances with large firms the alternative option for entrepreneurship in biofuels in the US was venture capital firms which played a very aggressive role for CleanTech and especially biofuels (VC- based startups). Fundamental assumptions and rationales of VCs in that area are:

> There were lots of efforts underway in the race to find a better biofuel, and that it is possible that one particular technology will take a dominating position ("dominant design"). Hence, put money into a host of promising new technologies (startups) and push them out into the market. Vinod Khosla (Khosla Ventures), more than anyone, was investing in many different efforts, on the idea that one of them will pan out, and he said. "If you back a lot of horses, it's more likely you're going to win." [Marshall 2006]

Getting involved in some project and provide some project equity actually means to really prove the technology at the commercial scale.

However, in CleanTech and especially biofuels venture capital had filled also roles that were previously occupied by project finance (venture) capital. In biofuels, venture capitalists were foregoing their customary role as just technology investors. They contributed to the cost of demonstration plants and were putting their money into infrastructure as well. This emphasis might stem from the perceived big biofuel opportunities that let venture capitalists fund biofuel infrastructure to push themselves toward a potentially bigger payoff [Barron 2007].

But there was another change of approach of VC-backing of biofuel startups, essentially driven by billionaire Vinod Khosla who founded Khosla Ventures in 2004. Khosla, an electrical/biomedical engineer and MBA by education, himself operated as a very successful entrepreneur and was founding Chief Executive Officer of Sun Microsystems before he turned in 1986 to the venture capital firm Kleiner Perkins Caufield & Byers (KPCB) as an Experienced Team Member focusing rather early on biofuels [Khosla 2006]. By 2009/2010, funding CleanTech startups was about two thirds of Khosla's existing portfolio [Schonfeld 2009b, Khosla 2009].

Khosla's approach of a "*science project to a company*" in biofuels was driven by his "own passion – green investing" and the related opportunities he envisioned for himself. His answer to the challenges was an unprecedented coordination of capital, intellect, and pragmatism associated with the confidence to succeed. And he referred to an *innovation architecture* which he calls an "innovation ecosystem at work," *solving large problems* by harnessing the power of ideas fueled by entrepreneurial energy of scientists, technologists, and entrepreneurs – very bright people working on solving a problem.

This is actually a "*VC-based and managed" multiple-stage entrepreneurial process* being close to an innovation or new business development process of corporate entrepreneurship (intrapreneurship). The innovation project related approach relies on three "legs" [Khosla 2008b]:

- Bright academics from various scientific/engineering disciplines whose competencies complement each others (founders and advisory board members);
- Talented entrepreneurship-minded people who want to make money – and make a difference – based on profound professional management experience (15-35 years) in related industries (oil industry or biotechnology or bioengineering area) – essentially industry veterans with track records ("entrepreneurs and executives");
- Intelligent capital and financial resources through committed people stepping up to the challenge and experience towards building businesses for the long run.

Khosla Ventures did seed, A and B and C investments and "don't mind larger technology risks especially in the smaller seed fund" [Schonfeld 2009a]. However, Khosla's basic attitude underlying his approach is not different from that of other venture capitalists: "If we did not have corn ethanol priming the pump, it would be too risky for me to invest in cellulosic ethanol," he said [Anonymus 2006a].

Khosla's focus on people with significant experience in the related or closely related field is in line with the above described innovation/entrepreneurship architectures for NTBFs with "subject veteran" founders, such as Bodo Wolf of German CHOREN Industries, David Ramey of ButylFuel, Arnold R. Klann of BlueFire Ethanol, Charles Wyman and Lee Lynd of Mascoma Corp. or Paul Woods of Algenol Biofuels.

Khosla's innovation architecture is often characterized as a *VC-backed spin-out* (RBSUs) *with professional and experienced managers* from almost the point of firm foundation. Khosla himself was also often heavily involved in firm foundation. For instance, one company was originally formed casually in response to a bet.

In a conversation with Professor Frances Arnold of the California Institute of Technology, Khosla suggested that "You can't do that with synthetic biology economically yet." She disagreed, and argued that a couple of graduate students working in the area could design bugs to make fuel – economically. Along with Matthew Peters and Peter Meinhold, also of Caltech, Gevo (Table I.99) was formed [Khosla 2008b].

Another example was the transformation of Kergy, Inc., a Silicon Valley engineering-type startup in alternative energy. It is a Menlo Park startup, which raised $3.3 million in a first round of funding from Khosla Ventures to become later Range Fuels and Khosla appeared as one of the founders (cf. also KiOR; all in Table I.99).

Khosla's process is structurally related to particular innovation approaches in large and giant firms:

- The leadership sponsor process of innovation (Table I.98) as described by Runge [2006:748,749] and
- Utilizing features of a standard "Stage-Gate®" innovation or New Product Development (NPD) processes of large firms [Runge 2006: 653-654].

Khosla was investing heavily in biofuels as he could see a financially viable path to the future. "The *risk profile* has to work for investors," Khosla said: "They see that this works in the marketplace over two years, not 20, and then they take the next step. It's stair-step, incremental investment." [Oneal 2006].

As in staged innovation processes (Stage-Gate, PhaseGate) Khosla also applies the "*failing fast*" principle [Runge 2006:787] as he said: "The ones we have cut off, we cut off relatively early." [Rapier 2009a] But additionally, one key risk, Khosla said, is the power of the oil lobby [Oneal 2006], which in the US already played a "negative role" in the 1920s when there was a strong movement towards a "biobased chemistry" [Runge 2006:565-566].

Table I.98: Comparing leadership activities of the innovation sponsor process and Khosla's innovation "ecosystem process.

Sponsor Process (Sponsor's Job)	**Khosla's "Innovation Ecosystem"**
Set the context – communicate a clear vision	Have a clearly communicated vision and mission (defining a related portfolio): "My mission now is to put the fossil in fossil fuels."
Choose projects to sponsor	Initiate a startup or find portfolio-driven startups/NTBFs to be invested in
Find and select innovators – bet on people, not just plans	Focus on bright scientific/technical and managerial innovators ("top-down team building")
Form cross-functional project teams – strive for functionally complete teams	Build multidisciplinary teams with complementary competencies
Support the team – provide resources and a "one stop shop" for decisions that will stick	Organize and lead equity financing for the "startup-project"
Guide the team – set milestones, ask the right questions, know when to redirect the team's efforts	Organize (and lead) staged financing referring to milestones ("gates"; series A, series B,...) and influence decision-making
Reward the team – keep them on track	dto.

In Figure I.182 Khosla's CleanTech Portfolio [Khosla 2009a] is presented. Related biofuels startups or NTBFs are further described and characterized in Table I.99. This set shall simultaneously be used to discuss entrepreneurship options in biofuels, related technologies and their hurdles, and selected business models of the new founded firms referring to the value system in Figure I.171.

> As is found often in previously discussed cases the majority of NTBF leaders of the set of companies Khosla has invested in have a very strong technical background and management experience by doing and executive management training doing a job rather than by higher education.
> Concerning the appropriateness of his portfolio "Khosla says his previous investments in CleanTech companies have generated nearly $1 billion in profits." [Reisch 2011b]

Notably, Khosla Ventures has not invested in algae firms so far. Generally, he has five criteria for investing in cutting edge energy technologies. Algae meet four of those criteria but fail on the last: He seems to believe that the engineering problems of growing and harvesting algae are manageable. And he obviously believes algae will have manageable startup costs and a quick innovation cycle.

But, Khosla does not believe algae (exploiting algae biomass, not necessarily ethanol from algae) will be able to compete unsubsidized with petroleum and other alternatives unsubsidized in the next five to seven years – after looking at "maybe two dozen." "The economics of algae don't seem to work," he said [Kho 2009]

And skepticism has not only increased with the VC community. The sentiment is: "We just don't believe in the economic" and one is not sure that "algae is going to come down the cost curve." This view is also shared by BP. BP, which has invested in algae startups Martek Biosciences (Table I.83), questioned the viability of different types of algae technology, and more specifically the kind that ExxonMobil recently invested $600 million in Synthetic Genomics [Phuong Le 2008; Oilgae Blog 2008; Kho 2009].

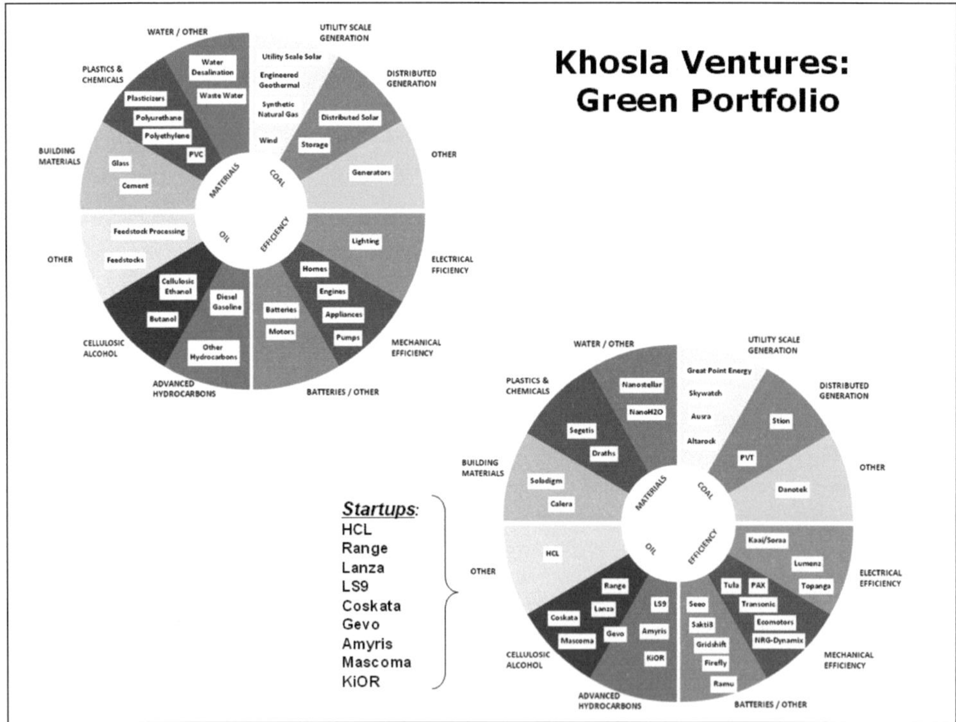

Figure I.182: Khosla Ventures Green Portfolio by industry, industry segments and applications/materials as well as technologies and equity-backed biofuels startups [Khosla 2009b].

When in 2008 the UK's Carbon Trust Investments, which invested in biobutanol firm GBL (Table I.95), set out to fund algal biofuels research, it was confronted with a mélange of overzealous claims coming from the industry. Companies were projecting biofuel yields ten times what is theoretically possible and proposing techniques that are not now and may never be economical.

A year later, after wading through the claims and gathering opinions from a network of more than 300 experts, the agency announced the creation of the Algae Biofuel Challenge, a £16 (ca. $24) million fund that would support the development and large-scale production of algal oil. The Carbon Trust's experience navigating algae excitement is one that generally funding groups and investors in the biofuels industry increasingly face [Waltz 2009b].

The firms in Table I.99 provide developments of startups/NTBFs, usually until the end of 2009, focusing on various types of biofuels backed by Khosla.

Table I.99: Selected startups backed by Khosla Ventures reflecting a venture capitalists' portfolio approach to CleanTech and specifically biofuels. *)

Company (Foundation) Additions and Remarks	**Funding: Khosla Ventures and Several Other VC Firms**	**Technology**	**CEO, Other Executives and Key People**
Altra Biofuels, Inc. Los Angeles, Calif. (2004)	Over $415M; in 2006 the company secured $63.5M	Leverages various kinds of biofuel manufacturing processes. 2009: A half-built $220M, 110M-gallon (416M-liter) per year corn ethanol plant developed by Khosla-backed Altra in Nebraska has been sold piece-by-piece. Altra began building the dry mill ethanol facility in Carleton in late 2006 but halted development in November 2007 after completing approximately 50% of the project because the company was unable to secure additional financing	CEO Larry Gross
Cilion Goshen, CA (2006) As California represents 20% of the US ethanol consumption, Cilion has distinguished itself by its *destination-based business model* and *energy efficient* ethanol	$200M, Cilion raised $105M in debt financing (2007), for the construction of two ethanol plants in California; $170M (09/2005)	Converts corn into ethanol to power cars and trucks; rather than building a central plant to supply the whole country, Cilion was putting up multiple plants near population centers and livestock markets, including three plants in California and two in New York. Apr. 2009: A leak in the tank at the Cilion Ethanol plant in Stanislaus County caused it to collapse into itself last month. The facility re-	Mark L. Noetzel, President and CEO, entered 2007 from BP; previously Group Vice President of BP PLC

facilities. The destination model creates *added value by distillers grain* which is a co-product of ethanol production. Distillers grain is an economic, high protein dairy and cattle feed – localization in producer area or country, not near (steel) industry	Series B	mained closed, ceased operations in March. Other factors behind the demise of the facilities, industry experts say, include *too-rapid growth*. Originally, said Harrigfeld, "they were to be down three to four weeks. Now, because of the market, they are not sure when, or if, they are going to reopen. (Sonya Harrigfeld, director of the Stanislaus Department of Environmental Resources) [Anderson and Moran 2009]. "When Cilion was formed in 2006, they announced they would have 8 plants in operation by 2008 and achieve an energy return of better than twice that of gasoline. Here in 2009 they have zero plants in operation." [Rapier 2009]	
Hawai'i BioEnergy Honolulu (2006) Partnering also with Hawaiian Electric Co. on testing and implementing such clean technologies as solar power; Hawaiian Electric is partnering in an algae production project with Maui landowner Alexander & Baldwin Inc. and startup HR BioPetroleum Inc. (Table I.89). The plan was to create a commercial-scale algae facility adjacent to the Ma'alaea Power Plant [Moresco 2008]	Hawaii BioEnergy <$1M [Rapier 2009]	Researching the development of ethanol plants on Hawaiian islands; Hawai'i BioEnergy (HBE) is a corporation established by three of Hawai'i's largest landowners. Mission is to reduce Hawai'i's energy costs, green house gas emissions, and dependence on imported fossil fuels through the research and development of local renewable bio-energy projects; a variety of energy crops, including but not limited to sugarcane, woody biomass, and algae. Hawaiian Electric is also a partner in the proposed BlueEarth Biofuels LLC 40 MMGYbiodiesel processing plant on Maui, which was expected to be operational in early 2010. The goal was to use locally grown oil feedstocks such as algae, jatropha or palm.	Paul S. Zorner President and CEO; joined Hawai'i BioEnergy in 2008; was the CSO and Executive Director of Principle Energy Limited, a venture oriented to establish sugarcane conversion to ethanol and power in Mozambique; also served as the Chairman of the Board of Directors for Kuehnle AgroSystems, a Hawaiian company specializing in the research and development of algae as a source of renewable fuels

Mascoma Corporation Cambridge, Mass. (2006) Founders of Mascoma Corp.: Charles Wyman and Lee Lynd; Charles Wyman's interest in alternative fuels propelled his career during the late '70s and early '80s when he served as the Director of the Biotechnology Center for Fuels and Chemicals at the National Renewable Energy Laboratory (NREL). Wyman (doctoral degree in Chemical Engineering from Princeton University in 1971) became an authority in the field of cellulosic ethanol in 1996 when he published the "Handbook on Bioethanol." Alliance partner of General Motors (GM), Chevron, Marathon Oil	$61M (5/2008), Series C, {$30M (11/2006), Unattributed} Raised about $100M in equity investments and ca. $100M in state and federal grants by 2009. That included the $61M in a third round of funding, with GM and Marathon Oil. Secured $26M from the Department of Energy and $23.5M from the State of Michigan to build commercial plant in Michigan [St. John 2009]	Producer of biofuels from lignocellulosic biomass using microorganisms and enzymes; unique technology developed by Mascoma *uses yeast and bacteria that are engineered* to produce large quantities of the enzymes necessary to break down the cellulose and ferment the resulting sugars into ethanol. Claim: combining the two steps (enzymatic digestion and fermentation) significantly reduces costs by eliminating the need for enzyme produced in a separate refinery; process, called Consolidated Bioprocessing or "CBP"; started producing cellulosic ethanol from wood chips at a demonstration-scale plant (1,000-5,000 gallons scale) in Rome, NY. Said its microbes can convert plant material like wood chips, tall grasses, corn stalks and sugar cane bagasse into sugar; still feedstock testing: coop with Chevron and GM; Stepped down CEO Jamerson would become chairman of Mascoma and CEO of the company's Frontier Renewable Resources subsidiary in partnership with timber and mining company JM Longyear, that meant developing a 20-40 MGY cellulosic ethanol plant in Kinross, Michigan for 2012 (sought funding for $250 million to $300 million).	Jim Flatt, PhD, Acting President - Executive Vice President, Research & Development / Operations; served as Sr. VP of Research for Martek Biosciences Corporation (Columbia, MD), while the company searched for a new CEO; in 2009 CEO Bruce Jamerson stepped down. Co-founder Lee Lynd, was working on a farm where he noticed heat energy emanating from a compost pile. The observation of microbes producing energy from biomass sparked Lynd to speculate on the possibility of using biomass as a fuel source. He followed this idea with tremendous passion for decades; Masters and PhD degrees in engineering from Dartmouth College where in 1987 he joined the Dartmouth faculty.
Range Fuels Broomfield, CO (2006) Started as Kergy, Inc. and Bioconversion Technology (BCT), LLC, founded in 2003 in Colorado Replacement of Range Fuels founder and CEO	Undisclosed round from Khosla, plus a $75M grant from DOE; dreamt of IPO already for 2008; $100M Series B, $28.2M Series C (4/2008)	Using modular facilities to bring the conversion process to the biomass source, thereby reducing the energy expended with supplying the facility with feedstock; will grow as more biomass becomes available; Claims it can account for fluctuations in feed material in terms of type, consistency, moisture content, quality; Over 10,000 hours of testing has been completed on over 30 different	David Aldous CEO and Director; experience in the energy, oil and petrochemical industries, was Executive VP Strategy and Portfolio for Royal Dutch Shell; Replaced former CEO Mitch Mandich Key persons from Kergy, Inc. connected to Range Fuels:

Mitch Mandich, who led the company through a $100M capital raise and secured $80 million funding for cellulosic ethanol demonstration from the DOE [Lane 2009e]; Business Model: Designing, building, and operating its plants; Be first to market with commercially produced cellulosic biofuels; Thermochemical syngas approach allows production of various biofuels (e.g. bioethanol, biomethanol) and chemicals); Rapidly gain market share by capturing the best plant locations (is independent from type of biomass). Will need to capture a healthy percentage of the fuel ethanol market and be cost competitive to earn a reasonable return on the already substantial investment.	Range Fuels' Soperton Plant was supported by over $250M in support from public and private sources; huge loan guarantees. Its first commercial cellulosic biofuels plant under construction and was scheduled to begin production in the second quarter of 2010. Pilot plant (25 tons per day scale) was at its Development Center in Denver, CO (here 25 employees) [Schuetzle et al. 2007] Has raised the necessary capital to begin construction of a commercial-scale cellulosic biofuels plant.	non-food feedstocks. Originated with Bioconversion Technology's process of Bud Klepper (as described in the text). Two-step thermochemical process; converted into syngas, cleaned syngas is passed over a proprietary catalyst and transformed into cellulosic biofuels; to low carbon biofuels, such as cellulosic ethanol and methanol etc. (Figure I.174); Working on a 100 million-gallon-per-year facility by 2011 [Fehrenbacher 2008]. Had some delays; said it had raised $100 million to build a commercial scale plant in Soperton, GA to make ethanol from wood waste, and said the plant's first phase of 20 million gallons per year would be complete in 2009 [St. John 2009]; Despite some delays, it said it is about half done with construction of the full-scale plant in Soperton, GA [Wald 2009b]; the plant shall produce about 40 million gallons of ethanol per year and 9 million gallons per year of methanol; about 1,200 tons per day of wood chips and forest waste feedstock are expected to be processed at full operating capacity [Schuetzle et al. 2007]. (Cf. text in A.1.1.3 on "Range Fuels: Years of Broken Promises.") In December 2011 Range Fuels, one of the first companies in a wave of startups that promised cheap biofuels made from sources such as wood chips rather than corn, shut its doors for good and was forced to auction off its assets!	Arie Geertsema (here Senior VP), Mitch Mandich CEO, Robert ("Bud") Klepper (inventor of the Range Fuels technology). Robert "Bud" Klepper acted as advisor and Chief Technical Specialist and Inventor; brought many years of process equipment design and fabrication experience to the team. Arie Geertsema on the Scientific Advisory Board; was Managing Director of Corporate Research and Development and managed a team of over 400 R&D staff at Sasol, the largest and most successful gasification company specializing in coal gasification and gas-to-liquids production; was also a Director of the University of Kentucky's Center for Applied Energy Research and a tenured associate professor in chemical engineering (with doctorate in chemical engineering and an MBA).

Coskata Warrenville, IL (2006) Incorporated by GreatPoint Ventures; was the culmination of research of Aaron Mandell and Andrew Perlman, two GreatPoint Ventures partners, had been doing since 2001 into alternative concepts for low cost cellulosic ethanol production. Mandell began following ethanol research taking place at Oklahoma State University and the University of Oklahoma with the help of founding scientist Rathin Datta. When the University's scientific teams identified a potent set of anaerobic microorganisms for the conversion of synthesis gas to ethanol, Mandell secured rights to license the technology and began to formulate the development strategy. The team initiated experimental work at Argonne National Laboratories, and started to advance the organism and build a top-tier biofermentation technology team.	\$40M (11/2008), Series D; Raised \$19.5M in a second round of funding that should be used towards construction costs of its first 100 MGY cellulosic ethanol plant \$10M (06/2006), Series A	Vision: be the global leader in the syngas to biofuels platform, beginning with ethanol. Hybrid technology: Gasification of raw material is released into a bioreactor where microbes convert the gas into ethanol. The integrated biorefinery – utilizes Westinghouse Plasma Gasification on the front end and Coskata's syngas-to-biofuels conversion process on the back end – delivered to General Motors for early testing of bioethanol. The company can co-locate with steel mills to convert CO into gasoline. Ethanol commercial-scale plant should produce from either biomass (like wood biomass, agricultural waste, energy crops, switch grass) or municipal solid waste or other recycled materials (like old tires – one reason GM is interested); the emphasis was on sugarcane bagasse. Claims a yield of 100 gallons per ton of feedstock at a cost of less than \$1 per gallon. Differentiator: secret sauce of microorganisms + microreactor for syngas conversion in a "hybrid" process [Fehrenbacher 2008] Pilot-scale facility opened in Warrenville, IL. first for hybrid process; about to build a 40,000 gallon demonstration plant for cellulosic ethanol in Madison, PA., Uses filter – membrane for separations. Business model: Biorefinery orientation (bioethanol, co-products); Technology licensor (to feedstock suppliers, chemical manufacturers, petroleum companies, ethanol distributors/blenders,	William Roe President and CEO; prior to Coskata, a 29 year career with Nalco, the world's largest provider of industrial water-treatment chemicals and process additives, served as COO; Dr. Rathin Datta CSO; more than 32 years experience in developing and commercializing process and product technologies for both established and emerging companies. Rathin founded Vertec Biosolvents, a technology, manufacturing and marketing company dedicated to providing biologically-derived renewable resource alternatives to petroleum-based solvents (ch. A.1.1.6). Prior to Vertec Biosolvents, Rathin was the VP of Research for the Michigan Biotechnology Institute, where he led the commercial development of lactic acid/polymer technology and the commercial development and successful implementation of a fluidized bed reactor for specific waste treatment technology; Richard E. Tobey VP R&D, spent 28 years developing and commercializing new products and processes for The Dow Chemical Company within their Ion Exchange, Anti-Microbial,

In the spring of 2007, Coskata moved out of Argonne. (Cf. text in A.1.1.3 "The Risks, Translated from SEC-speak.")		project developers); hopes that licensees enable rapid scale-up; owner and operator (of currently) a demonstration-scale facility; orientation: flex ethanol [Roe 2009]	Pharmaceutical and Agricultural business units.
Gevo, Inc. **Pasadena, CA** (2005) Gevo Development, LLC is the development arm; seeking opportunities to access ethanol production assets to make biobutanol by retrofitting ethanol plants at a low capital cost. Spin-out: three co-founders, renowned researchers Frances Arnold (Prof.), Matthew Peters and Peter Meinhold of the California Institute of Technology, plus co-founders James C. Liao (Prof.) and Christopher Ryan. In 2007 acquired an *exclusive license* to use UCLA's method for modifying *E.coli* bacteria for use in biofuel development; has exclusive rights to integrate Cargill's world class microorganisms (yeast strains) into Gevo's Integrated Fermen-	Reported $17M Series C; (05/2009) $10M as Series B; backed also by French oil giant Total SA. Awarded $1.8M from the US Departments of Energy and Agriculture's Biomass Research & Development Initiative to help fund ongoing development of its yeast strain to produce biobutanol [Pruitt 2009a]	Technology: an enzyme process, developed at CalTech, that converts biomass (corn stover, switchgrass, forest residues, and other sustainable feedstock) to next-generation biofuels like butanol, which can be used in the existing petroleum supply chain; looks specifically into isobutanol; fermentation of all sugars including mixed (C6, C5) sugars. Business model includes selling products and licenses; raise capital to acquire assets either through direct acquisition, joint venture or tolling arrangements; retrofitting of existing ethanol plants; focus on biorefinery approach – products include biobutanol, isobutanol, biodiesel, jet fuel and biobased plastics [Lane 2009d]. Differentiates through three pieces from other firms: Gevo's "veteran team of research scientists" has developed a proprietary process" based on "Protein Engineering of Biocatalysts" (to convert agricultural waste products into different types of renewable, alcohol-based, liquid fuels), "veteran leadership team" almost entirely from polylactic acid plastics firm NatureWorks, LLC; Metabolic Engineering of Suitable Host Organisms (engineering suitable host organisms that utilize carbon and energy efficiently for fuel production; strains to exhibit increased yield and productivity to be	Highly experienced management team with roots in biobased chemicals and polymers. Patrick Gruber, CEO; several general management positions in technology and business development for Cargill Inc., one of the founders of NatureWorks, LLC (formerly Cargill Dow, LLC) focused on polylactic acid (PLA) where he was the VP of technology and operations and the CTO from 1997 until 2005 [Runge 2006:130, 245]; Christopher Ryan, EVP; served as COO and CTO for NatureWorks, LLC; David Glassner, Executive VP Technology; led the development of novel yeast biocatalysts for the production of lactic acid and ethanol at NatureWorks, LLC; also during this time he led the development of cellulosic processing technology and economic models for PLA manufacture; Jack Huttner Executive VP Commercial & Public Affairs, came from DuPont

tation Technology (GIFT®) process for the production of butanols from cellulosic sugars that are derived from biomass; Gevo and engineering firm ICM have entered into a strategic alliance for the commercial development of Gevo's technology GIFT® process for production of biobutanol and hydrocarbons from retrofitted ethanol plants.		sufficient to produce commodity chemicals – "green chemicals" – and fuels on a large scale); Process Engineering (developed a proprietary process technology to enhance productivity and lower product separation costs). Has announced (9/30/09) the start up of its 1 million gallon per year demonstration plant (through deploying its technology by retrofitting existing ethanol plants to produce biobutanol). Claims successful retrofit completed in less than 3 months [Pruitt 2009a]; for expansion, Gevo planned to acquire three to five ethanol plants over the next 12 to 18 months [Gevo 2009].	Danisco Cellulosic Ethanol (DDCE); prior to joining DDCE, he was VP of biorefinery business development at Genencor; Brett Lund VP & General Counsil, served as chairman of the legal, IP, and licensing group for Syngenta's biofuels business; Glenn Johnston, VP Regulatory Affairs, prior to joining Gevo he was director of regulatory affairs with NatureWorks, LLC.
LS9, Inc. **San Carlos, CA** (2005) How the idea of the technology and firm foundation came across is described by Svoboda [2008]. In 2008 made about 5,000 liters of biofuel in a pilot fermenter at its headquarter [St. John 2008]. Already in 2008 wanted to raise $75M to $100M to build a demonstration plant (2.5 mil. gallons per year) By the end of 2010 began engineering work on a full-scale commercial plant (up to 100 mil. gals.); could be up and running by 2012	$30M (12/20/2010) Series D; $25M (10/2009), Series C; $15M (10/9/07) Series B; $5M (3/1/2007) Series A; $20M from Lightspeed Venture Partners, Flagship Ventures and Khosla Ventures (latest Khosla investment, $5M). In 2009 applied for a multi-million dollar Government Integrated Biorefinery grant which would cover 80% of the re-	Uses *synthetic biology* to develop biofuels (gasoline, diesel, and jet fuel) from *traditional* feedstock that contain more energy than current biofuels; require less energy to produce and can be distributed through the existing petroleum infrastructure; commercializing and scaling-up DesignerBiofuels™ products; Basically, a technology platform with designer microbes converting renewable materials directly (one step to ultra-clean diesel) into transportation fuels and chemicals (strategic partnerships) [Del Cardayre 2009]. Metabolic engineering replaces whole swaths of genes inside microbes to turn them into tiny chemical factories; have engineered a strain of *e. coli* with a genome that can convert sugars into a fatty acid methyl ester (FAME) which is chemically equivalent to California Clean diesel; LS9's *1-step technology*, compared to competitors' multi-step processing technology, seemed to be highly cost competitive; Does not have to kill its microbes to	The company was bringing together leaders in synthetic biology and industrial biotechnology. George Church is Professor of Genetics at Harvard Medical School and a co-founder of LS9; directed one of the first funded genome technology centers since 1987 – now a DOE GTL systems biology center focused on photosynthesis and biofuels. Chris Somerville is Director of the Energy Biosciences Institute (BP funded) and a professor of plant and microbial biology at the University of California Berkeley, and a co-founder of LS9. Bill Haywood CEO; was Senior VP Manufacturing for Tesoro Petroleum, where he was responsible for the company's seven refineries.

[St. John 2008]. In 2009 revealed that it has a promising opportunity to purchase an already existing plant at a reduced cost but would not say where as of yet; Expected to have commercial plant by 2013 in Brazil as the cost of sugarcane is the lowest in Brazil [Stromeyer 2009]. Business model: Single platform: multiple products (biofuels, specialty chemicals); Low-cost producer; Proprietary single-step technology; Feedstock agnostic technology; Capital efficient scale-up (retrofit); Localization-oriented [Del Cardayre 2009]	trofit and operating cost associated with a demonstration plant. Biorefinery emphasis; two partners, with P&G for sustainable chemicals and with Chevron; equity investment from Chevron; P&G has invested tens of millions of dollars in LS9 to produce "green surfactants" [Stromeyer 2009].	get the oil; they secrete it naturally and then can live to feed, digest and excrete more dollops of oil; has a similar microbe that can make fatty alcohols [Kanellos 2009; Del Cardayre 2009]. Said that LS9, along with its competitors, will have to prove it can deliver on those low prices at full-scale production; LS9's goal was to be able to show that it could produce synthetic diesel for \$45 to \$50 a barrel by mid-2011 [Kanellos 2009]. Would not reach commercial production levels until 2013 [Lane 2009f, Del Cardayre 2009]. Indications for the issues of early promises: LS9 went through two leaders in the three years since its founding; Haywood took the helm of LS9 from Robert Walsh, former president and a 26-year veteran of Royal Dutch Shell. Walsh replaced LS9's first acting CEO, Doug Cameron, who was also chief scientific advisor for Khosla Ventures, in July 2007.	Stephen del Cardayre VP R&D, biochemist by education, spent 9 years at Codexis and Maxygen, was directly involved in the development, application, and commercialization of technologies for the engineering of biocatalytic processes for the pharmaceutical and chemical industry. Wei Huang VP Process Development and Engineering, over 17 years of industrial bioprocess experience, including process scale-up, facility and equipment design, process simulation, construction support, facility start-up, operation support, as well as process development and research. NOTE: By Jan. 2014 LS9 was acquired by Renewable Energy Group, Inc. (REG) with most of the LS9 team incuding .the whole R&D leadership group.
Amyris Biotechnologies Emeryville, CA (2003) Did not start as a fuel company in 2003; it started with \$40M in funds from the Gates Foundation to develop Artemisinin, used for the treatment of malaria; transforming itself additionally into a next generation biofuels company. Amyris engineered yeast to produce a	Raised over \$130M from the sale of equity from 2/2008 to present to support scale-up operations and initial commercial plant work [Lane 2009j]. \$41.8M (10/2009), Series C, \$70M (09/2007), Series B, \$20M (10/2006),	Uses synthetic biology to create biofuels that can replace gas, diesel and jet fuel and chemicals; is engineering microbes specifically for that purpose. Make fuel from any kind of fermentable sugar, start with sugarcane via relationships with producers in Brazil. Amyris' portfolio of patents includes renewable diesel, renewable jet fuel, renewable gasoline, and renewable lubricants. Building relationships with feedstock producers around the world; can retrofit existing ethanol plants. JV with Brazil's second-largest sugarcane grower, a demonstration	John Melo CEO, before joining Amyris, Melo was President of US fuels operations for BP. Paul Adams Senior VP Fuels, spent 25 years with BP in its supply and trading business, where he was instrumental in building internal processes and policies to successfully maximize BP's profitability in the supply chain. Jack D. Newman co-founder and Senior VP of Research; over a decade of experience

malaria drug now being developed by Sanofi-Aventis; its genetic processes to deliver the low cost anti-malarial drug could be exploited for producing biofuels (since around 2006). Products: Renewable Fuels, Chemicals, Malaria treatment (Artemisinin-based). As a producer of biofuels covering the whole value system. Amyris Fuels, LLC. (wholly owned subsidiary), formed to develop a robust network for supplying and distributing renewable fuels. The company is growing its footprint by sourcing current generation of biofuels – such as ethanol – from US and international producers and bringing them to market at the lowest possible cost. Formed wholly-owned subsidiary Amyris Brasil tapping into one of the most economical and sustainable energy sources – sugarcane. Fuels Industry Experience: Amyris Fuels has understanding and	Series A. In 2009 received $25M through the US government's advanced biorefinery project stimulus award. Wanted to generate diesel fuel from sugarcane; Amyris wanted to use its $25M award for a pilot plant that will produce a diesel substitute by fermenting sweet sorghum and other petrochemical substitutes [Riddell 2009]	facility was located amid Brazil's sugarcane fields; Brazil is the No. 1 exporter of ethanol and is moving into biodiesel production. Planned to pump out a billion gallons within the next five years; planned to develop renewably sourced gasoline and jet fuel – but diesel was an ideal place to start. "Diesel fuel is what drives industry." Amyris Brasil announced that it has entered into letter of intent agreements with three sugar and ethanol producers in Brazil, Bunge Ltd., Cosan and Açúcar Guarani, with the purpose of partnering for the production of high value renewable specialty chemicals and fuels. These products should be distributed by Amyris. In December 2009, Amyris announced it had entered into an agreement with the São Martinho Group to acquire a 40% stake in the Boa Vista mill; the parties would convert this mill to achieve the first production of Amyris products; wanted to invest up to $200 million in the project [Lane 2009j]. Agreements are key steps toward building-out a fully integrated renewable products company – a company that encompasses the technology, industrial-scale manufacturing and product distribution capabilities. Also investigate the feasibility of developing an optimal economical model using Amyris technology to produce cane-derived diesel fuel from molasses rather than from traditional sugarcane juice. New "capital light" model: Amyris will partner with a mill and provide its technology through an off take agreement – not a licensing agreement. Amyris Brasil will provide mill owners	researching bacterial physiology and genetics, co-authored the groundbreaking work underlying the technology of microbial terpene production. Neil Renninger co-founder and CTO; has a cross-disciplinary understanding of both the micro-world of strain engineering and the macro-world of chemical engineering; received a doctorate in chemical engineering from the University of California, Berkeley, studying the metabolic engineering of bacterial cells for chemical transformations. Jeff Lievense Senior Vice President of Process Development and Manufacturing; 25 years of industrial experience in bioprocess engineering and a proven track record developing, scaling, and commercializing advanced fermentation processes. Dr. Lievense served also as VP of technology and process development for the R&D organization of Tate & Lyle, where he led the fermentation R&D program, commercializing three large-scale industrial fermentation products and pioneering a production process for Bio-PDO^{TM} (1,3-propanediol).

<table>
<tr>
<td>seeks to deepen experience in the fuels industry;

Has developed advanced capabilities in fuels marketing, supply, distribution, blending, systems development, accounting, and risk management [Lane 2009j].

GreenLane®:
a typical “crossover strategy involving renewables” – prepare to commercialize renewables by learning from operation with non-renewables [Runge 2006:584, 855]

In the US Amyris currently transports, stores, and markets ethanol from US domestic and overseas sources through Amyris Fuels, LLC. – will allow quick and reliable distribution of biofuels in the US; building strong customer relationships throughout the world and credibility as a reliable current generation fuel supplier.</td>
<td></td>
<td>with yeast strains, production processes and engineering design to produce Amyris products.

The mill owner will provide capital to convert mill to produce Amyris products. Amyris Brasil will then purchase Amyris products from mill owners at contracted price and distribute product directly to customers.

To achieve planned 2011 commercialization, has engaged a leading engineering, procurement and construction management (EPC) firm for final design and construction of commercial production facilities.

Stages to commercialization:
Emeryville Pilot Plant – Designed to mimic the full-scale fuel manufacturing process;
Campinas Pilot Plant – second pilot plant, in Campinas, Brazil, is strategically located at the doorstep of Brazil’s sugarcane industry, similar to the Emeryville plant will validate technology for use in Brazilian production conditions

Q2 2009 Amyris Renewable Products Demonstration Plant – also in Campanis (14,000-square-foot facility) to conduct in-country scale-up, demonstration and optimization of all Amyris fuels and chemicals manufacturing processes;
the production of more than 10,000 gallons of Amyris products under conditions representing full-scale manufacturing.</td>
<td></td>
</tr>
</table>

Celunol Corporation (now Verenium Corp.) Cambridge, MA (1994) Founded 1994 as BC International; traded on the NASDAQ, bought by Diversa for $150M+, became **Verenium** (Table I.84) in 2007 [Childs 2007]. Business Units: Biofuels, Specialty Enzymes	Reportedly raised more than $60M from Khosla Ventures, Braemar Energy Ventures, Charles River, Rho Ventures	Produces enzyme-based (bioengineering) cellulosic ethanol; has proprietary biotechnology processes and project development know-how. Biomass used: sugarcane bagasse, corn stover, rice and wheat straws, wood waste, energy crops. Celunol's microorganisms consume C6 and C5 sugars (not commercially fermentable by yeast as C6); emphasis on cellulosic ethanol from agricultural waste left over from processing sugarcane. Exclusive licensee of key cellulosic ethanol (CEtOH) technology developed at the University of Florida; R&D facilities in Gainesville, FL; Operating plant in Jennings, LA; Pilot facility (1st CEtOH in US) operational Nov.2006, demonstration-scale CEtOH facility entering construction in 2007 when merger was already announced [Howe 2006] CEtOH facility in Osaka, Japan (wood waste) developed by Celunol licensee Marubeni Corp; Licenses technology domestically and internationally.	Diversa Enzymes prospects in hot springs, ocean beds, soda lakes, and on the Arctic tundra for genes potentially useful in industry. Verenium achieved growing portfolio of specialty enzyme products and "unique technical and operational capabilities." Verenium claimed to be the only company to offer fully integrated, end-to-end capabilities in pre-treatment, novel enzyme development, fermentation, engineering and project development [Childs 2007].
HCL CleanTech Ltd. Israel/US (2007) In 2012 re- named to **Virdia** Virdia is headquartered in Redwood City, but has a technology center in Danville, Virginia, and a research center in Tel Aviv, Israel. Improved a freely available technology, uses a ca. 80	$5.5M (06/2009), Series A In 2012 Virdia closed its latest round of financing, raised over $20 mil. from insiders, Khosla Ventures, Burrill & Company and Tamar Ventures; in addition, the company received $10 mil. in a venture debt	Virdia has developed the CASE™ (old acid solvent extraction) process, which converts cellulosic biomass to high quality fermentable sugars and lignin, and is based on a series of patented and patent-pending technologies. Use of fuming hydrochloric acid to catalyze the hydrolysis of cellulose to glucose and, generally, all polycarbohydrates to their constituent monomers. Hydrolysis yields of the sugar fraction are over 95-97% (a significant improvement compared to traditional enzymatic processes) and lignin solids are recovered practically intact.	Eran Baniel – Founder and CEO, serial entrepreneur. Robert Janse, Head of engineering, 33 years of experience in corn, wheat, sugar processing and fermentation at Tate & Lyle. Paul McWilliams (USA) – US Engineering, 31 years at Cargill working on wet milling plants. In 2012 Virdia got a new CEO Philippe Lavielle, a veteran of the industrial biotech sector. Lavielle

years old, industrially-proven German cellulosic to fermentable sugars and ethanol process, the Bergius process (named after its Nobel Prize winning developer) – technically superb, but associated with operating high cost. Virdia focuses on fermentable C6 and C5 sugars and lignin from biomass for industrial uses. Strategy: Virdia's cellulosic sugars and lignin are intermediate products in supply chains that can lead into biochemicals, biofuels, plastics and carbon fibers, as well as nutritional supplements for food and feed. Verdia is looking for partners to offtake those quantities of sugars and firms interested in the conversion of the sugars.	deal with Triple Point Capital.	Virdia's use of fuming hydrochloric acid (HCl) allows a large variety of feedstocks to be used with minimal change of configuration. Using the proprietary technologies developed in house, it has improved the recovery of the acid, as well as the recovery of valuable by-products, such as high quality lignin and high quality tall oils. Its technology for the recovery of HCl from aqueous solutions and industrial processes can also provide complete acid recovery solutions to HCl dependent industries (such as the PVC industry). In 2012 Virdia together with Virent (Table I.85) debut drop-in aviation biofuels made from *drop-in cellulosic pine tree sugars.* In 2012 Verdia announced a deal with the Mississippi Development Authority to build a plant to derive sugar from wood chips, a plentiful by-product of the state's forestry industry. That deal included $75 million in low-interest loans and up to $155 million in tax incentives over a 10-yr period. The first plant, due to start up in late 2014 or early 2015, will have capacity for 150,000 tons (300 million lb) of sugars per year. Virdia eventually aims to build plants for 500,000 tons (1 billion lb) per yr.	replaced co-founder Eran Baniel as CEO, who now serves as Vice President of Business Development. Before joining Virdia, Lavielle was a member of the executive management at Genencor. The company is now led by a new management team with decades of industrial-scale manufacturing experience ("veterans") in industrial biotech, chemicals and sugar production.
KiOR Inc. US / KiOR BV ,The Netherlands (2007) In 2007, KiOR was founded by Khosla Ventures and a group of catalyst scientists who shared a vision of making renewable fuels from cellulosic biomass through a	$12.9M (06/2008), Series B, $1.4M (11/2007) Series A	KiOR targets the fuel markets only. The technology originally pioneered by Bioecon in 2006. Converts biomass, particularly the recalcitrant polymeric biomass residue, to valuable molecules which can be utilized by the chemical and fuels industry. Biomass catalytic cracking process – a thermochemical process that produces biocrude from grass, wood and plant waste that can then be refined – has significantly lower	The Company was incorporated and commenced operations in July 2007 as a joint venture between Khosla Ventures and BIOeCON BV. KiOR BV, a Netherlands company, was formed on March 4, 2008 and commenced a process of liquidation in March 2010. As of December

one-step catalytic process. Biofuel JV of Khosla Ventures and Dutch biofuel startup BIOeCON Number of employees (around 2008): 15 Khosla Ventures' involvement with what would become KiOR started with a call from an engineer from Holland to Vinod Khosla. In contrast to the inventor's interest in a licensing business model, Khosla envisioned that the KiOR process could lead to an oil exploration and production company. Has adopted a build, own and operate strategy; Hopes also to license the technology to customers like oil refineries and feedstock owners.		capital costs compared with other biomass conversion technologies (claim); Develops and commercializes Biomass Catalytic Cracking (BCC) technology. BCC technology converts lignocellulosic biomass into a "biocrude" which is suitable for upgrading to transportation fuels; "Biocrude": a mixture of small hydrocarbon molecules that can be processed into fuels, such as gasoline or diesel in existing oil refineries. Renewable biocrude oil can be refined in a conventional hydrotreater into light refined products (gasoline and diesel blendstocks). Technology produces hydrocarbon blendstocks that will "*drop in*" to the existing transportation fuels infrastructure for use in vehicles on the road today. Technology platform combines proprietary catalyst systems with well-established fluid catalytic cracking, or FCC, processes that have been used in crude oil refineries to produce gasoline for over 60 years. Constructed a pilot unit outside of Houston, Texas to continue developing and validating the technology; this pilot unit has amassed over 9,000 hours of operation and evaluated more than 250 catalyst systems [Admin 2011d] In 2010 was producing 15 barrels per day of biocrude (229,000 gallons per year) using its fast pyrolysis technology and a proprietary catalyst	31, 2010, all of the operations of KiOR BV were combined into the operations of KiOR, Inc. Khosla targeted the hiring of mission critical technologists to the company, ultimately leading to the hiring of Fred Cannon as KiOR's President (in 2008), and later CEO (2010). Prior to KiOR, Cannon was president of AkzoNobel Catalysts LLC from 1997 until the divestment of the business in August 2004. KiOR Columbus, LLC, a wholly owned subsidiary of the Company ("KiOR Inc."), was formed on October 6, 2010.
LanzaTech Ltd. Auckland, New Zealand Founded in 2005 in New Zealand and now headquartered in Roselle, Illinois In 2010 LanzaTech	Between foundation in 2005 and 2010 LanzaTech has raised $30 mio. in venture capital and $10 mil. from the New Zealand gov-	The LanzaTech Process captures gas (CO) as a resource; innovation lies in using a bacterium to produce ethanol not from a carbohydrate, but from a gas (cf. Coskata) A "hybrid" ethanol production process that can be retrofitted to industrial facilities, generates ethanol from	Notably, David C. Aldous, was a member of the LanzaTech Board since October 2008; a former Executive VP of Strategy and Portfolio for Shell – was also Range Fuels' CEO and

announced that it had engineered a microorganism that can produce 2,3-butanediol, a chemical precursor that can be used to make the solvent methyl ethyl ketone (MEK), which is used in dry erase markers and in the manufacture of plastics and textiles. The same chemical can produce butanes and butadiene, which can then be used to make a variety of plastics and hydrocarbon fuels. While the chemical market is smaller than the fuel market, it can be more profitable, since chemicals such as MEK sell for more than twice the price of ethanol.	ernment. Throughout 2005 and 2006, the company raised funding through New Zealand-based angel investors and secured grants. $3.5M (4/2007), Series A (for pilot plant) Series A investment was from a consortium led by Khosla Ventures; the Series B financing was led by Qiming Ventures. In 2012 it closed Series C investment led by the Malaysian Life Sciences Capital Fund ($56 mio.). New investors included Petronas Technology Ventures Sdn Bhd, the venture arm of Petronas, the national oil company of Malaysia.	the carbon monoxide (CO) of waste flue gases (little or no hydrogen as is in syngas; e.g. steel and other industries); can also use thermochemical syngas based on any biomass resource (municipal waste, organic industrial waste (tires), waste wood); CO used as a food source for proprietary LanzaTech microbes during the biofermentation process (hybrid process), non-genetically modified, non-pathogenic bacteria, isolated from natural environments [Lanza Web]. The carbon monoxide containing gases are scrubbed, cooled and sent to a bioreactor. The carbon component is used as a food source for the proprietary LanzaTech microbes during the biofermentation process. The microbes use this energy to produce ethanol. With its hybrid process LanzaTech claims to become the lowest cost, highest volume producers of fuel ethanol. Claimed major advantage over existing gas to liquid conversion technologies: Able to virtually eliminate capital cost associated with gas conditioning. Rapidly growing patent portfolio, adopted a stage-gated critical path through process piloting to commercialization [LanzaTech Media Release Aug. 18 2009]. Claims to have one of the world's largest collections of industrial fuel and chemical production microbes. The fermentation suite comprises more than 20 bench-top gas fermentation reactors and a test-bay allowing the development and demonstration of several prototype reactor designs in parallel and at scale.	Director; Currently, LanzaTech is led by a multinational Board of Directors and Management Team with offices in New Zealand, China and the US. Dr. Jennifer Holmgren is the Chief Executive Officer. Jennifer has over 20 years of experience in the energy sector including a proven track record in the development and commercialization of fuels and chemicals technologies. Prior to joining LanzaTech, she was Vice President and General Manager of the Renewable Energy and Chemicals business unit at UOP LLC, a Honeywell Company. Dr. Sean Simpson is the Chief Scientific Officer and co-founder of LanzaTech. He spent the first 12 years of his life living in various countries around the world, including Mauritius, Zambia and Gibraltar, before his family returned to England. He now lives with his family in Auckland.

*) From the firm's Web site if not stated otherwise by a reference;
Firms' states: usually end of 2009, if not stated otherwise.

The biofuels investments of Vinod Khosla in startups (Table I.99) show various fates so far, from bankruptcies to good financial returns by IPOs, from promising developments to recent change of development directions away from biofuels and covering a very broad spectrum of technical approaches – and cooperation (LS9 and HCL Cleantech/Verdia) or interconnections ("networking effects") via technology or people (for instance, Range Fuels and LanzaTech).

For instance, in January 2012 LanzaTech purchased a facility in Soperton, Ga. previously owned by bankrupt Range Fuels at auction for $5.1 million. Range Fuels' lender took control of the facility for non-payment and held the auction to recoup some of a $38 million loan, which had been guaranteed by the Department of Agriculture. Range was also awarded a $43 million grant from the Department of Energy to help construct this facility. LanzaTech acquired the facility for its location and access to cheap feedstocks from local timber operations.

LanzaTech's plan was to leverage some of the existing technology at the facility alongside own proprietary technology to produce renewable and domestic fuels and chemicals from the bountiful waste biomass in the region [Bomgardner 2012b]. The Soperton site, already renamed Freedom Pines Biorefinery, will be LanzaTech's first production facility. The firm is currently working to launch a demonstration facility in Shanghai that will use waste gases from a steel mill operated by China's Shougang Group.

The fact that Range Fuels and LanzaTech share a lead investor – Khosla Ventures – has raised eyebrows because LanzaTech bought the Soperton site for a fraction of the amount spent developing the facility.

As a summary, VC-backed startups, at least in biofuels, can be differentiated by

- Financial backing of early stages ("technology investors");
- Backing specifically industrial scale-up of proven R&D ("project investors," late-stage funding, also corporate venturing);
- VCs being pro-activley involved in firm foundation;
- Initiating firm foundation and proceeding along a defined development path.

Furthermore, entrepreneurship in biofuels (and also other CleanTech areas) exhibits a number of founders who were riding the first wave of renewables in the late 1970s and 1980s.

As observed with corporate venture oriented startups with oil companies in the venture capital community impatience was also rising with the pace of commercialization. Generally, it is a rather normal process when the R&D/pilot phase changes into the engineering/scale-up phase to change the management team, in particular, the CEO who will have to emphasize commercialization rather than technology. But an indicator of problems with biofuels was successively fast change in NTBFs' leadership teams.

For many new biofuels firms the proof-of-concept in the lab is not being translated quickly enough into production results at the plant and they still have to prove they can deliver on those low prices at full-scale production. Pressure related to the promises and postponed milestones into commercialization have built up. Examples of CEO or founders, respectively, stepping down or being replaced include LS9, Range Fuels and Mascoma [Lane 2009e].

The Great Recession 2009/2010 had also a great influence on financing (biofuels) startups' further developments by VC firms. For instance, biotech and CleanTech fundraising from venture capital declined in the third quarter of 2010 [Voith 2010a; Voith 2010b]

On the other hand, policy continued to step in. There have been established a number of programs by the US Federal Government to support technology entrepreneurship. The US Department of Agriculture (USDA) and Department of Energy (DOE) were investing $47 million over three years in eight pilot-scale R&D projects to make biofuels and other products from various biomass sources [Mukhopadhyay 2011c].

For a bargain price of $1,000, start-up companies can get up to three of the thousands of unlicensed patents in the Department of Energy's portfolio. The aim is to double the number of startup companies emerging from DOE's 17 national laboratories, which hold more than 15,000 patents. Only 10 percent of federal patents are currently licensed to be commercialized, according to the agency. By simply submitting a business plan and signing a generic agreement, available as a template on the DOE Web site, interested startups can apply to license up to three patents from a single laboratory at the reduced $1,000 fee [Mukhopadhyay 2011b].

Also the US military will play an important role here. The US military consumes more energy than is used by two-thirds of all nations worldwide! DOD needs cheaper and more abundant energy sources to power its global operations.

Collaboration between the Departments of Defense and Energy was established to reduce the US's dependence on oil. DOE is the nation's largest funder of the physical sciences and DOD will act as a test bed for innovative technology. There are three areas that will benefit from the partnership: batteries; fuel cells; and alternative fuels derived from sources such as biomass, natural gas, and algae [Mukhopadhyay 2011d].

In 2011 many of the firms listed in Table I.99 targeted an IPO for financing further developments. Coskata's approach when it looked in December 2011 for an $100 million IPO has already been discussed (A.1.1.3).

Basically, revenue streams of these NTBFs include

- Grant revenue,
- Licensing revenue
- Biofuel sales and sales of related products.

Aspects of partnering in biofuels include

- Feedstock sourcing and testing;
- Product development and testing;
- Final end use product formulations;
- Commercialization of developed technologies;
- Equity investments and development support.

Mascoma (Table I.99) also filed an S-1 registration for a proposed IPO. For Masoma's IPO several red flags were raised [Fehrenbacher 2011; Admin 2011c], for instance, expressed by "The Risks, translated from SEC-speak" [Admin 2011c].

Wanting to raise $100 million, Mascoma generated $34.5 million in revenue along the way, primarily government funding for R&D. They have not yet commercialized their corn ethanol technology or the hardwood process. The accumulated deficit as of June 30, 2011 was $118.722 million. The net losses were $30.4 million, $38.3 million and $25.7 million for the years ended December 31, 2008, 2009 and 2010, respectively, and $14.8 million for the six months ended June 30, 2011 [Admin 2011c]. Mascoma said for its revenues in 2010, government grants constituted "86 percent of our revenue" while "product sales and other service agreements constituted 14 percent of our revenue." [Fehrenbacher 2011]

"It's not a pretty prospectus: Mascoma's auditors have questioned its ability to remain a going concern, and its debt carries interest rates as high as 11%. This is all to make cellulosic ethanol a fuel whose commercial viability many experts question. The trouble is, Mascoma's plant was supposed to be virtually built by now. The company's plan in early to mid-2008 was to leverage a grant from the state of Michigan to get debt financing, but that never did happen because the markets for energy-project finance closed after Lehman Bros. collapsed. More disturbing, perhaps, is the silence about Mascoma that has emanated from venture-capital firm Khosla Ventures, the company's second-biggest shareholder." [Mullaney 2011]

IPOs of other firms in which Vinod Khosla invested include Amyris, Gevo, and KiOR (all in Table I.99). But here, for instance, Amyris raised around $363 million through its IPO and the IPO translated Khosla Ventures' $15.59 million investment into a worth of $65.36 million meaning a 4X return at that time [Wesoff 2010].

According to McDonald [2011] "Much advanced biofuel development is a combination of hype and science fiction, but these companies have practical business plans, near-term commercial objectives and the financial resources to get across or near the goal line." Cellulosic ethanol is the biggest disappointment, and so *now attention is likely to switch to drop-in (road-ready) biofuels* like renewable gasoline and diesel or jet fuel.

For instance, for policy in Germany, in 2012, one year after its introduction and aiming to become the major type of fuel for Otto-engines, E10-gasoline flopped. E10 achieved just 13 percent of all transportation fuels. Users did not accept E10 – not because of concerns it might damage the current engines, but the population has serious doubts whether such biofuels show benefits for the environment and the climate [Eicher 2012].

Regarding companies, for instance, early in 2012 *Shell* announced that it has built the next generation biofuels pilot plant at Shell's Westhollow Technology Center in Houston, TX, to *produce drop-in biofuels rather than ethanol*. It uses a thermocatalytic process technology licensed from its commercial partner Virent (Table I.85), which is similar to the process being used at the Virent pilot plant in Madison, Wisconsin [McDermott 2012].

The benefits of drop-in biofuels from the perspective of being able to use existing fuel infrastructure without modification should not be overlooked. Drop-in biofuels have the same properties as conventional fuels. This eliminates the need for additional blending and storage infrastructure as well as engine modifications (Table I.82) that may be required for the use of more ethanol in blends with conventional fuels. That means there are a lot of sunk costs there financially.

In early 2011 Khosla-backed *KiOR* focusing on *drop-in "biocrude*" (Table I.99) filed for an IPO. At that time KiOR signed an offtake agreement with Hunt Refining of Tennessee for biofuels produced at the facility KiOR is developing in Columbus. The company was planning to invest $500 million in three wood chip-to-biofuel plants in Mississippi and the Columbus plant was expected to be online in 2012. The state's development authority was granting KiOR $75 million based on the deal with Hunt [Admin 2012d]

"We are a development stage company with a limited operating history, and we have not yet commercialized our cellulosic gasoline and diesel nor have we generated any revenue. Until recently, we have focused our efforts on research and development, and we have yet to generate revenue." "As a result, we had generated $108.7 million of operating losses and an accumulated deficit of $130.4 million from our inception through December 31, 2011. We expect to continue to incur operating losses through at least 2013 as we continue into the commercialization stage of our business." [KiOR 2012]

In raising $150 million by its IPO, the Khosla Ventures-backed KiOR raised 50 percent more than expected at the time of its initial filing, but well short of the $241 million potential the company had tipped in filings over the past month. Through the IPO Khosla Ventures retained up to 70 percent voting control through the structure outlined by the company. That means, Khosla Ventures controls a majority of the outstanding common stock and will continue to control a majority of KiOR's common stock after the IPO. As a result, KiOR is a "controlled company" [Admin 2011d; Admin 2011e].

"In November 2007, Khosla Ventures bought in to the old BioeCON technology (valued at the time at $2.6 million) for $4.4 million at $0.36 per share. By June 2008, the company sold another $10 million in shares to Khosla at $0.97 per share – and another $15 million came in via a promissory note. The company came back in spring 2010 with a $95M capital raise at $9.80 per share – that's a lot of added value based on the $30 million invested." [Admin 2011e]

KiOR said it has completed construction of its first commercial-scale facility on budget and ahead of schedule. The company was commissioning the plant in Columbus, Miss., and expected to begin production in summer 2012. Once it is fully operational, the facility will have an annual capacity of 11 million gal. The feedstock will be local southern yellow pine [Bomgardner 2012a].

> Furthermore, many promising biofuels startups changed their focus softly, declared as "initial orientation," from biofuels to biobased specialty chemicals.

Amyris Biotechnologies (Table I.99) now intends to become a leading provider of renewable specialty chemicals and fuels worldwide. In 2010 it went public [SEC 2010], as mentioned above. Its initial focus was on farnesene, a sesquiterpene that exists as a variety of isomers and stereoisomers. It intended to convert farnesene into diesel and jet fuel and materials for detergents, cosmetics, perfumes, and lubricants. Capital injection was $244 million in funding since its inception [Tullo 2010]. The Amyris strategy: commercialize farnesene on a contract manufacturing basis, then turn to farnesane, produced by adding hydrogen to farnesene. Farnesane is the company's showcase diesel molecule.

Amyris had, as its primary post-IPO challenge, to tackle the proof that it can replicate its lab and pilot results at scale. Apart from *scale-up* Amyris' further Achilles heel is the *dependence for the near term on Brazilian sugarcane resources for its sugar feedstocks*. Sugar is half the price of farnesene in the Amyris equation – near as any analyst has been able to decipher, and with the price of Brazilian sugar doubling in the 2008-10 time frame before retreating in 2011 – the dependency will not only be on Brazilian harvests, but India's (which are more subject to variance) [Administrator 2010].

According to Amyris' annual report 2011 [Amyris 2012] total revenues and net loss, respectively, developed as follows:

Revenues ($ mio.)	**2011**	**2010**	**2009**	**2008**	**2007**
Product sales	129.8	68.7	61.7	10.7	-
Grants and collaboration revenues	17.2	11.7	2.9	3.2	6.2
Total revenues	147.0	80.3	64.6	13.9	6.2
Net loss ($ mio.)	(179.5)	(82.8)	(64.8)	(42.3)	(11.8)

In 2011 and 2012, Amyris leveraged contract *manufacturing capabilities* to begin producing Biofene®, Amyris's brand of farnesene, at three sites in three countries around the world: Biomin (Piracicaba, Brazil), Tate & Lyle (Decatur, Illinois), and Antibioticós (Leon, Spain).

With regard to *execution* its strategy Amyris established a number of partnerships, co-operation, JVs with established large companies addressing their anticipated applications of farnesene. In 2010 Amyris announced a series of agreements with The Procter & Gamble Company (P&G). The agreements focused on the use of farnesene in certain specialty chemical applications within P&G's products.

Also in 2011 Amyris Inc. and French energy giant Total have expanded a partnership to produce renewable diesel products through a joint venture, to which Total will contribute an additional $105 million on top of the $180 million the companies already have committed, according to financial filings (approximately 17 percent equity interest in Amyris). The 50-50 joint venture has the exclusive rights to produce and market renewable diesel and jet fuel worldwide. It also has a non-exclusive agreement to develop and market other non-fuel products [Riddell 2011].

In a partnership, French tire maker Michelin has joined with Amyris to develop renewable isoprene. Amyris said it will use technology similar to its process for making farnesene [Bomgardner 2011e]. Under the agreement, Amyris and Michelin will partner to contribute funding and technical resources to develop Amyris's technology to produce isoprene from renewable feedstocks. Amyris expected to begin commercializing this isoprene in 2015 for use in tire and other specialty chemical applications. Michelin is committed to off-take volumes on a ten-year basis. In addition, Amyris retains the right to market its renewable isoprene to other customers.

Furthermore, in 2011 Amyris signed a collaboration agreement with the Japanse firm Kuraray Co., Ltd. to develop innovative polymers from Biofene. Under the agreement, Kuraray will use Biofene to replace petroleum-derived feedstock such as butadiene and isoprene in the production of specified classes of high-performing polymers.

And, as part of an effort to solve production problems, in 2012 Amyris made a $59 million private placement of common stock and issued $25 million in convertible bonds. Most of the new capital came from existing investors. The firm, will use the new funds to pay for the scale-up of its commercial operations [Bomgardner 2012c].

However, *in February 2012 Amyris said it is giving up making fuels*. Instead, it will focus on higher value products, such as moisturizers for cosmetics. The company learned firsthand just how difficult it is to achieve the kind of yields seen in lab tests in large-scale production. Range Fuels, one of the first of the current crop of companies, recently went out of business. Others were giving up on making biofuels too, also hoping to break into markets for higher value chemicals.

Amyris's technology may still be used to make renewable fuels, but this will happen not at Amyris, but under joint ventures established with Total and Cosan. These ventures will need to build up their own production capacity. Amyris had said that in 2012 it would produce 40 to 50 million liters of farnesene, basically a fragrant oil.

Amyris also said it is indefinitely delaying plans for one of two large production facilities it was to have built this year [Bullis 2012b].

In 2012 Neil Renninger, a co-founder of Amyris Technologies (Table I.99), stepped down as CTO. Concerning science versus business the visions of the CEO and CTO for the company proved incompatible. CEO John Melo came from a big-company culture at odds with Amyris's freewheeling researchers. The scientists balked when he tried to apply big-company rigor and measure employees' contributions and their performance [Grushkin 2012].

A related execution of its strategy is observed for Gevo which also went public. And Gevo seemed to hope to become profitable by turning corn into chemicals. According to its prospectus Gevo showed a net loss of $78.579 million from June 9, 2005 (date of inception) through September 30, 2010 [NASDAQ 2011]. According to Gevo's annual report 2011 total revenues and net loss, respectively, developed as follows:

Revenues ($ mio.)	**2011**	**2010**	**2009**	**2008**	**2007**
Bioalcohol sales and related products	63.74	14.77			
Licensing revenue		0.14			
Grant, research and development program revenue	0.81	1.49	0.66	0.21	0.28
Total revenues	64.55	16.40	0.66	0.21	0.28
Net loss ($ mio.)	(48.21)	(40.11)	(19.89)	(14.54)	(7.23)

According to Gevo's IPO prospectus, venture capitalists own 60 percent of the company (Khosla Ventures, 26.8%; Virgin Green Fund, 10.5%; Total Energy Ventures International, 9.2%; Burrill & Company Life Sciences, 7.1%; Malaysia Life Sciences, 6.3%) and LANXESS Corporation owned 4.7 percent. The German firm LANXESS is the world's largest synthetic rubber producer. But, after the IPO, LANXESS increased its position to 9.1 percent. Total Energy Ventures is the VC arm of French oil multinational Total SA. Gevo management and directors own 22.7 percent.

In and after 2011 Gevo received important foundational patents for its operations, but it was also involved in a heavy litigation with Butamax Advanced Biofuels LLC (Table I.83).

In 2012 Gevo was awarded US Patent No. 8,071,358, covering additional "Methods of Increasing Dihydroxy Acid Dehydratase (DHAD) Activity to Improve Production of Fuels, Chemicals, and Amino Acids." This invention further details and protects the innovations contained in the *Gevo yeast organism* to turn an industrial yeast strain into a highly efficient cell factory to produce isobutanol. Also in 2012 the USPTO granted

US Patent No. 8,153,415 entitled "Reduced By-Product Accumulation for Improved Production of Isobutanol." The '415 Patent" covers technology which *eliminates two pathways* that compete for isobutanol pathway intermediates in yeast.

Gevo was awarded US Patent No. 8,101,808, "Recovery of Higher Alcohols From Dilute Aqueous Solutions." This patent addresses the *separation technology* used to produce propanols, butanols, pentanols, and hexanols. The claims also address *how ethanol plants can be retrofitted* to produce higher alcohols. It solves the long-standing problem for the practical production of higher alcohols, specifically how to separate these alcohol products from fermentation broth and achieve economic concentrations.

This is the technology, along with Gevo's proprietary yeast, being implemented at the Luverne, Minnesota plant, which Gevo acquired in September 2010. The Gevo Integrated Fermentation Technology (GIFT® system) permits the *continuous removal of isobutanol as it is formed.*

In 2011 Gevo was awarded US Patent No. 8,097,440 "Engineered Microorganisms Capable of Producing Target Compounds Under Anaerobic Conditions." It refers to Gevo's yeast technology to enable the low-cost, high-yield production of biobased isobutanol. Gevo has been awarded a *patent for an anaerobic yeast utilizing a novel enzymatic structure*. Gevo believes the most efficient and economical way to make isobutanol through fermentation is to use yeast that is anaerobic, or does not need oxygen.

In 2011 Gevo filed a lawsuit against Butamax™ Advanced Biofuels, LLC and its affiliate DuPont. Butamax has publicly disclosed its use of Gevo's claimed technology in several later-filed patent applications. Butamax has attempted to reach commercial-scale production of isobutanol for several years (Table I.83). To produce commercially relevant levels of isobutanol, however, one must use the technology covered by Gevo's 8,153,415 Patent. This patent illustrates the importance of eliminating these pathways before Geve's competitors do.

In March 2012 the USPTO rejected all patent claims of Butamax covering isobutanol-producing yeast in US Patent No. 7,851,188 which is currently being asserted against Gevo. Gevo also successfully petitioned the USPTO to reexamine Butamax's claims in US Patent No. 7,993,889 covering a method of producing isobutanol using a recombinant yeast microorganism. The USPTO actions in the '188 and '889 patents also reinforced Gevo's position that the technologies and process steps claimed by Butamax were known in the field, published in numerous scientific journals or invented by others, including Gevo, before Butamax applied for its patents.

The recent emphasis of Gevo turned to biobased chemicals and intermediates, in particular, biobased solvents (isobutanol and n-butanol) and synthetic rubber and plastics, based on isobutene derived from isobutanol. Concerning biofuels the focus turned to biojet fuel. Further promising developments with Gevo depend essentially on achieving commercial scale production of isobutanol. To drive developments Gevo is working directly with its important potential customers.

Gevo, Inc. through its wholly owned subsidiary, Gevo Development, LLC, has entered into a joint venture transaction ("JV") with Redfield Energy, LLC of Redfield, SD, to retrofit Redfield's existing ethanol plant into an isobutanol plant with an expected production capacity of approximately 38 million gallons per year (MGPY). The retrofit commenced by year end 2011, and Gevo expected to begin commercial production of isobutanol at the facility in the fourth quarter of 2012.

In July 2012 Gevo succeeded to ferment isobutanol in large (250,000 gallon) commercial fermenters, isolate the product and get it into tanks and railcars. The learnings gained in achieving this milestone are viewed as enormous and further derisk Gevo technology.

In 2012 Gevo established itself as a company producing bio-based solvents, meeting industry standard specifications for all current isobutanol and n-butanol applications, particularly derived solvents for the coatings market.

Also in 2012 Gevo received a USDA $5 million grant for development of jet fuel from woody biomass and forest residues. The award is a portion of a $40 million grant presented to the Northwest Advanced Renewables Alliance (NARA), a consortium led by Washington State University (WSU). Other NARA members include the firms Weyerhaeuser, Catchlight Energy and Oregon State University, Pennsylvania State University, and the University of Minnesota.

The airline industry and the US Department of Defense are eagerly looking for near-term alternatives to petroleum-based jet fuel. Gevo previously announced its progress to airline engine testing using starch derived isobutanol to jet fuel. Gevo expects to receive full fuel certification by 2013 from the American Society for Testing and Materials (ASTM) for its biojet fuel.

In 2011 Gevo successfully completed the construction and commissioning of the world's largest ATJ biofuel demonstration plant at South Hampton Resources' facility near Houston, TX. The facility has begun operations and is delivering test volumes of ATJ biofuel to Gevo's initial customers. Gevo was awarded a contract by the Defense Logistics Agency to supply up to 11,000 gallons of ATJ based biojet fuel to the US Air Force. Also the German airline Lufthansa evaluates Gevo's renewable jet fuel.

The German chemical firm LANXESS plays a key role for Gevo. It is not only the world's largest synthetic rubber producer. Butyl rubber represents 25 percent of LANXESS' sales. It is the world's largest purchaser of isobutene and, for instance, in the long term, biobased isobutene will account for half of LANXESS synthetic rubber production at its plant in Sarnia, Canada. LANXESS strengthened its commitment to produce premium synthetic rubber from biobased raw materials and, as part of this commitment, it increased its minority shareholding in Gevo, Inc. in early 2011 amounting to 9.1 percent after having invested $17 million in Gevo's IPO. LANXESS initially invested $10 million in Gevo as part of a private placement in May 2010.

The LANXESS-Gevo partnership is working on a unique method that may hold the key to the sustainable production of isobutene and created a breakthrough dehydration process that converts isobutanol into isobutene. The dehydration process has not only proven to be successful in the laboratory by the end of 2011, but has also undergone several months of practical testing in a small-scale reactor at LANXESS' site in Leverkusen, Germany

In 2011 Gevo announced a groundbreaking agreement with The Coca-Cola Company (Coca-Cola) to create renewable para-xylene from plant-based isobutanol. Gevo will work to develop an integrated commercial-scale system to produce renewable para-xylene, a key building block towards reaching Coca-Cola's goal of leading the beverage industry away from fossil-fuel based packaging by offering an alternative made completely from renewable resources (PET plastic packaging). The global market for PET is 54 million metric tons and has a value of $100 billion, with approximately 30 percent used for plastic bottles.

Isobutanol that can be converted into para-xylene using known chemical processes is a key raw material in PET production. Gevo has previously set up a cooperation and is supplying the Japanese chemical giant Toray with lab-scale quantities of renewable para-xylene. Toray has successfully converted Gevo's para-xylene into PET films and fibers. Toray employed its existing technology and new technology jointly developed with Gevo and used Gevo's para-xylene and commercially available renewable mono ethylene glycol (MEG) to produce fully renewable PET (all of the carbon in this PET is renewable).

Box I.26: Drivers for Synthetic Rubber from Biobased Intermediates [Bomgardner 2011e].

The common automobile tire contains rubber that is extracted from latex-bearing trees and rubber that is synthesized from petroleum feedstock. Industrial biotechnology companies such as Amyris, Gevo, and Genencor (belonging to Danisco which is now owned by DuPont) want to give tire manufacturers a third option: biobased rubber intermediates. Microbial fermentation targets three renewable rubber intermediates: isoprene, isobutene, and butadiene. Five-carbon isoprene is used to make synthetic latex similar to that of the rubber tree. Isobutene and butadiene are four-carbon intermediates used to make butyl rubber and styrene-butadiene rubber.

Two leading tire makers – Goodyear and Michelin – along with synthetic rubber manufacturer LANXESS have entered into partnerships with industrial biotech firms to advance the commercial production of biobased rubber intermediates. But, it is still assumed that new renewable sources will not be commercially available for another three to five years.

For instance, Goodyear partnering with Danisco's Genencor confirmed that biobased isoprene meets specifications for the catalysts it uses in rubber manufacturing and has even made concept tires with it. But, the project originally targeted 2013 for commercialization has been pushed back a few years. DuPont acquired Genencor's parent company, Danisco, and now the new owner is weighing in.

But it is not just that tire makers want to ride a "green wave." They are also motivated by tightening supplies of both natural and synthetic rubber, driven in recent years by strong global demand, especially from emerging economies, by soaring postrecession demand and constraints on the expansion of rubber plantation acreage. The cost of a common grade of natural rubber shot up.

Furthermore, today, the chemical intermediates come from the cracking of liquid feedstocks in ethylene plants. But as petrochemical makers switch to lighter natural gas feedstocks, production of C4 and C5 chemicals is drying up. Hence, tire makers want something to help them *control volatile raw material costs*.

It is expected that overall global demand for both synthetic and natural rubber will grow to 35.9 million metric tons by 2020, from 25.7 million metric tons in 2011. Demand will be met roughly equally by synthetic and natural rubber.

Basically, it is assumed that for renewable isoprene, isobutene, and butadiene the volumes produced in the next five to ten years will remain quite small. And regarding their suppliers "it very much remains to be proven if they can produce on a cost-competitive basis {compared} with more traditional petrochemical pathways." Despite these facts industry representatives maintain that renewable feedstocks will be valuable to cushion swings in raw material costs.

Though also emphasizing jet fuel another biofuel NTBF that changed its business direction and strategy is Virdia (Table I.99). It focuses on high-quality C6 and C5 cellulosic sugars which can replace corn, beet, and cane sugars (!) and dry solid lignin.

These sugars that do not compete with sugars for food consumption and lignin are ready for fermentations or chemical conversions as products and intermediates for industrial uses. That means Virdia offers products that are directly usable by industries with well established conversion processes for sugars (Figure I.184). Virdia can act as a raw material (input) supplier (Figure I.183). In this way, Virdia can address the following markets:

- Diesel, jet fuel, gasoline, butanol (sugars);
- Surfactants, lubricants, plastics, synthetic rubber (sugars);
- Lignin as an energy source (for Virdia plants (cf. Table I.88) and other manufacturing plants);
- Lignin-based complex carbon fibers to incorporate into composite materials for a large number of industries.

Early in 2012 Virdia announced major company milestones, including a new brand and CEO, and a $75 million deal with the Mississippi Development Authority to build manufacturing plants in the state. The agreement includes an incentive package with $75 million in low-interest loans, as well as up to $155 million in various tax incentives over a 10-year period.

Furthermore, to fund its piloting activities and engineering plans, Virdia recently closed its latest round of financing, raising over $20 million from insiders, Khosla Ventures, Burrill & Company and Tamar Ventures. In addition, the company closed a $10 million venture debt deal with Triple Point Capital.

The company will build, own and operate its first plant, and after that the company will pursue other business models including licensing its technology There are a number of competitors in the rush for sugar as a raw material, including Renmatix, Inc. (A.1.1.6) and BlueFire's SucreSource (Table I.86) [Lane 2012r].

Emphasizing diesel and surfactants and lubricants Virdia has interfaces to LS9 (Table I.99). As other "biofuels startups" LS9 has shifted its emphasis more towards the chemicals side of its portfolio [De Guzman 2012a]. LS9 now positions itself as a supplier and licensor of technologies to the fuels industry, but will entertain direct project participation on a case-by-case basis. In its chemicals orientation it envisions more of a partner-venture model or potentially select LS9-only investments.

LS9's target is to provide drop-in chemicals and fuels. Its chemical products are the building blocks for many functional materials, such as surfactants, lubricants, emollients, and functional fluids. Drop-in biofuels are fuels containing hydrocarbons identical to those in petroleum-based gasoline.

By the end of 2011 LS9 scaled its technology to the 20,000 liter scale, demonstrating continued progress in the scale-up and commercialization of its biobased chemicals and fuels technology platform. From initial production of 1,000 liters at the Company's pilot plant in South San Francisco, California, LS9 has utilized a 20-fold step-up process to produce approximately one ton of a specific chemical for its strategic partner, Proctor & Gamble (P&G).

LS9's initial products in the chemicals arena were sugar-based fatty alcohols (C10-C18) and specialty esters, such as biodiesel fatty acid methyl esters (FAMEs) and fatty acid ethyl esters (FAEEs) under the banner of UltraClean Diesel, which can be directly blended into current petroleum-based diesel. And the company said it has already shipped a ton of fatty alcohol from its pilot facility (and headquarters) in San Francisco to P&G for sampling into surfactant products [De Guzman 2012a]. These products are directed against natural oils as a raw material (Figure I.184).

Biobased feedstock means essentially traditional feedstock. These include traditional feedstocks, such as sugarcane and corn syrups, waste products such as molasses and glycerin, and emerging feedstocks such as sweet sorghum syrup and the hydrolysates of plant biomass (for instance, from Virdia). Addressing all of these, LS9 avoids the many issues with the procurement and pretreatment processes in biomass/waste to biofuels conversions (Figure I.184).

By mid of 2011 LS9 in conjunction with partner Virdia (HCL CleanTech) was awarded a $9 million grant from the Department of Energy (DOE) to improve and demonstrate an integrated process to convert biomass feedstocks into fermentable sugars (Virdia part) and then into diesel and other fuel and chemical products (LS9 part). Virdia addresses surfactants and lubricants explicitly as its markets. LS9 is also looking to build a brownfield 10,000-25,000 tons/year facility initially in Brazil to produce sugarcane-based chemicals [De Guzman 2012a].

For June 2012 LS9 planned the opening of its scaled-up demonstration facility. LS9 uses genetically engineered microbes to convert biobased feedstock to diesel and chemical intermediates in Florida. The Okeechobee plant, which will start making biodiesel, contains a 135,000-l fermentation vessel, a jump from earlier production of 50,000-l quantities. The output shall provide commercial samples for testing by prospective customers [Bomgardner 2012a].

LS9 is hoping to hit their commercialization target by the end of 2012 (obviously they were about 85 percent). Its 135,000 liter fermentation vessel is a key: “At that scale, we are close to world-scale fermentation, which is about 3-4x away. We are well along the pathway towards de-risking our technology processing. The Florida facility has four (each at 700,000 liter) world-scale fermentation capability.” “The company is looking for a strategic round of funding this year to go to commercial-scale up by the end of 2014.” [De Guzman 2012a]

Since around 2010/2011 bio-lubricants are in the spotlight of NTBFs like Virdia or LS9 which originally targeted biofuels. And also Amyris formed a joint venture with distributor US Venture to produce, market and distribute finished lubricants for the North American market using Amyris’ farnesene-derived base oils. Amyris is said to be working on the production of a complete line of renewable lubricants, including hydraulic, compressor, turbine and gear oil and grease, as well as 2-cycle and 4-cycle engine oil [De Guzman 2011a].

Typically lubricants contain 90 percent “base oil” and less than 10 percent additives. Base oils are mostly derived from a mixture of fractions of the crude petroleum oil refining process. Natural (vegetable) oils (Figure I.184) are also used as base oils and there are already a lot of biobased lubricants in the market especially derived from vegetable oil.

For instance, bio-based hydraulic fluids are estimated to grow 5-10 percent per year worldwide and now represent 2-4 percent (US) and 3-7 percent (EU) of the hydraulics markets mostly because of advancement in performance, cost and its “green” factors. Amyris’s modified yeast converts the cane syrup to farnesene (Biofene™) which then has to be finished chemically to create base oils [De Guzman 2011a].

Structuring Complexity of Biofuels for Entrepreneurship and Intrapreneurship

With all the outlined empirical observations concerning involved firms (players), their technologies, types of offerings, financing and leadership/management approaches, development and innovation processes (Figure I.180) and hindsight, one can take a more fundamentally structured view of biofuels and their promise to be available, affordable, and clean.

Taken all the input and process variables together biofuels production and commercialization represents a combinatorical complex problem. Systemic complexity of entrepreneurship in biofuels referring to the input-conversion-output block (Figure I.5) may be approached by Equation I.21 where co-product variety may induce the consideration whether the co-products shall become significant contributions to the firm's revenue stream or fed back into the production process, for instance, for energy (steam) generation.

Equation I.21 and the implicitly associated hurdles (Figure I.171) represent simultaneously the space of business opportunities, the *opportunity landscape*, in biofuels. But, there are almost too many options to choose from, particularly if the chosen option shall give a sufficiently reliable assessment regarding 1) the overall energy efficiency (energy input; Equation I.20:), 2) overall Greenhouse gas emissions (by products and processes) and 3) whether the cost to produce in relation to the mineral oil price calculated for laboratory or pilot plant arrangements will also materialize in large-scale commercial plants.

Equation I.21:

Biofuels Input-Output Complexity →
Input type variety (types of biomass/algae) ⊗ Input location/transportation variety ⊗
Conversion sub-process variety (thermochemical, bioengineering, "hybrid") ⊗
Microbe/bacteria/yeast/enzyme/microorganism variety ⊗ Scale-up approaches ⊗
Output/product variety (type of biofuel) ⊗ co-product variety ⊗ by-product variety ...

As the space provided by the biofuels value system is so large and complex, one sees a myriad of diversity in the range of business models – and entrepreneurial risks. Major risks and development hurdles are listed below. A very detailed description of risks and hurdles can be found, for instance, in the offering prospectus of Amyris Biotechnologies [SEC 2010].

- Contextual legislative/regulatory risks (changes of biofuels related laws and programs, for instance, blending mandates, requirements of certificates; "policy and politics");
- Special Regulatory Risks (plant permissions);

- Funding Risks (by VCs, large firms, government, funding priorities among different types of biofuels or input); research approach highly dependent on grants: change of government – change of programs and financial support;
- Technological Risks (scale-up, cost)
- Environmental Risks (societal attitudes; products are not so "green"; GMOs);
- Supply Risks (type of biomass input; cost of biomass procurement and logistics, biomass price developments);
- Financial Risks (CAPEX, OPEX; profit, exit in VC-backed NTBFs);
- Operational & Execution Risks (management, partnerships with other firms);
- Market Risks (adoption; oil price, natural gas price for thermochemical processes, acceptance of GMOs for bioengineering/hybrid processes; special situation of policy-driven markets);
- Infrastructural Risks (Big Oil's role for the transportation fuels industry).

When producing next generation biofuels from renewable, non-food feedstocks like wood-chips, roughly 25 percent of the output is lignin. Firms can burn the lignin and thus transform a "by-product" into an energy contribution to drive their processes.

However, lignin and the chemical properties and functional attributes of a wide range of lignin derivatives, for instance, lignins in carbon fiber, open also new opportunities for entrepreneurship: High purity lignin extractives (and their subsequent derivatives) which can be engineered to meet the chemical properties and functional requirements of a range of industrial applications that until now has not been possible with traditional lignin by-products generated from other processes.

On the input and output level one can differentiate "plant-based biomass" and appropriate waste versus aquatic biomass like algae and output in terms of bioalcohols (biomethanol, bioethanol, or biobutanol), biogasoline and biodiesel and jet fuel.

For most business plans of biofuels startups addressing a policy-driven market was central. And often the business models for many biofuel companies were predicated on a much higher price of crude oil, making biofuels more attractive.

Concerning output the fundamental "pure biofuel versus biorefinery model" was actually often reduced to a "biorefinery light" model providing just small volumes of selected biobased chemicals for various industries originally encountered as co-products.

The actual complexity expressed in Equation I.21 can be visualized by a number of graphics. In Figure I.183 key components for business models and the financing models are summarized. There is a broad diversity in the range of business models. Input, conversion technologies and output options are largely specified in Figure I.184.

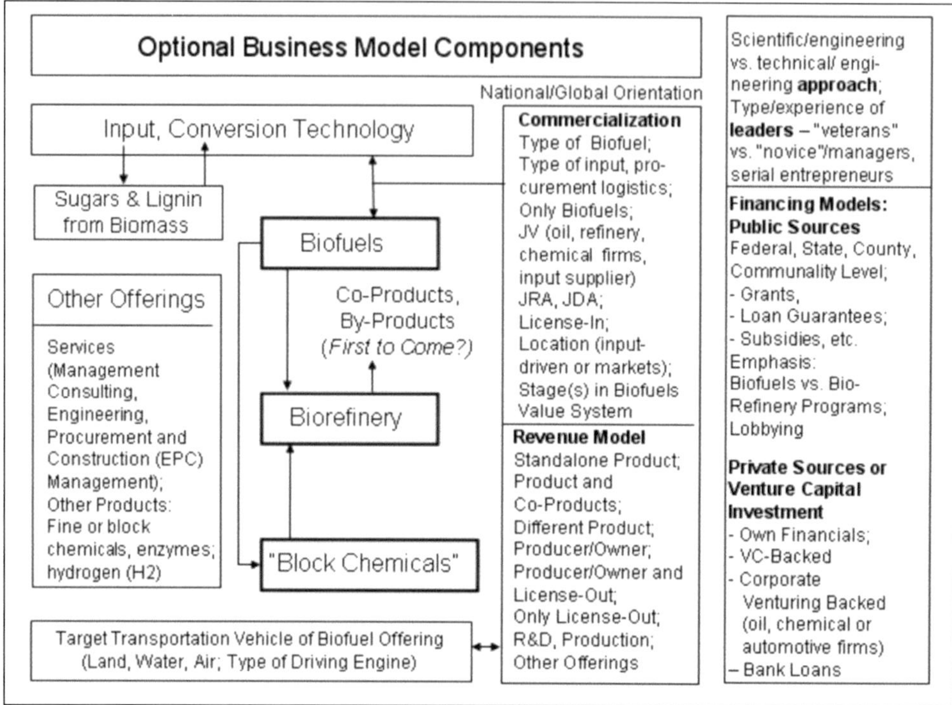

Figure I.183: Optional components and financing options for business models of biofuels NTBFs.

One important model for these CleanTech companies is to make strategic alliances with big or giant companies. For example, they can trade sales and marketing rights for a capital investment. Or they can sell the licensing rights for a product in exchange for an investment. Or they establishe JRAs or JDAs.

There was a clear focus of the cooperative model on "*Big Oil*" and "*Big Chemistry*" and sometimes also the automotive and tire industry. As many NTBFs emphasized a producer/owner and licensor approach, as they stumble, big companies will be able to snap up technologies on the cheap, when and where they need them [Carey 2009].

Finally, an important business model addresses input localization, establishing production (subsidiaries or JVs) where input is plentiful and cheap and in sufficient proximity to the plant(s) to minimize transportation and storage cost. In this regard, for instance, BP/Verenium JV Vercipia (Table I.83, Table I.84), Cobalt/Rhodia (Table I.96), Amyris and LS9 (Table I.99) addressed Brazil. On the micro-level one also finds localization near steel or power plants for flue gas (LanzaTech and some algae firms) or land fills (waste).

For to-be entrepreneurs in biofuels this does not only mean to know critically these competitive technologies, but also to know the patent landscape around the anticipated technology to be used and the related existing intellectual property rights.

Figure I.184 summarizes essentially the technologies discussed so far for generating second and third order biofuels.

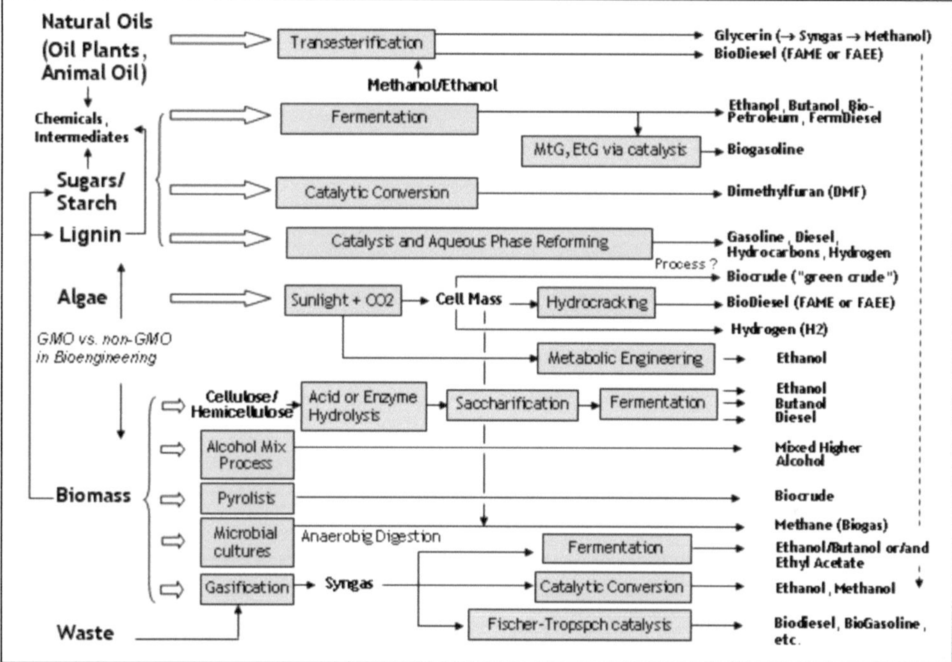

Figure I.184: Biomass conversions: many feedstocks, many conversion options, many products; different economics and energy balances (adapted from Khosla 2009a]).

Figure I.184 does not address explicitly the pretreatment and hydrolysis sub-processes leading to fermentable sugars and correspondingly whether the subsequent fermentation proceeds with C6 or C5 sugars or C6 and C5 sugars. The pretreatment process of biomass has several options, for instance,

- Water-based (hot water or steam explosion, combination of both)
- Chemistry-oriented (acids, alkaline bases, ammonia or oxidative processes);

- Solvent-based (alcohols, esters or both); especially “supercritical water” (A supercritical fluid is any substance at a temperature and pressure above its critical point, where distinct liquid and gas phases do not exist (A.1.1.6). Supercritical water is suitable as a substitute for organic solvents. Supercritical water oxidation is a process that can be used to advantage in the destruction of hazardous wastes.)
- Biological (enzymes or microbes; enzyme activity and cost?).

Which process will be adopted by around 2020 as the leading technology is still hard to predict because process economics are key and industrial scale production are just emerging.

Figure I.185 illustrates rather than completely reflects the issues and complexity of cost and energy efficiency consideration for manufacturing second-order bioalcohols.

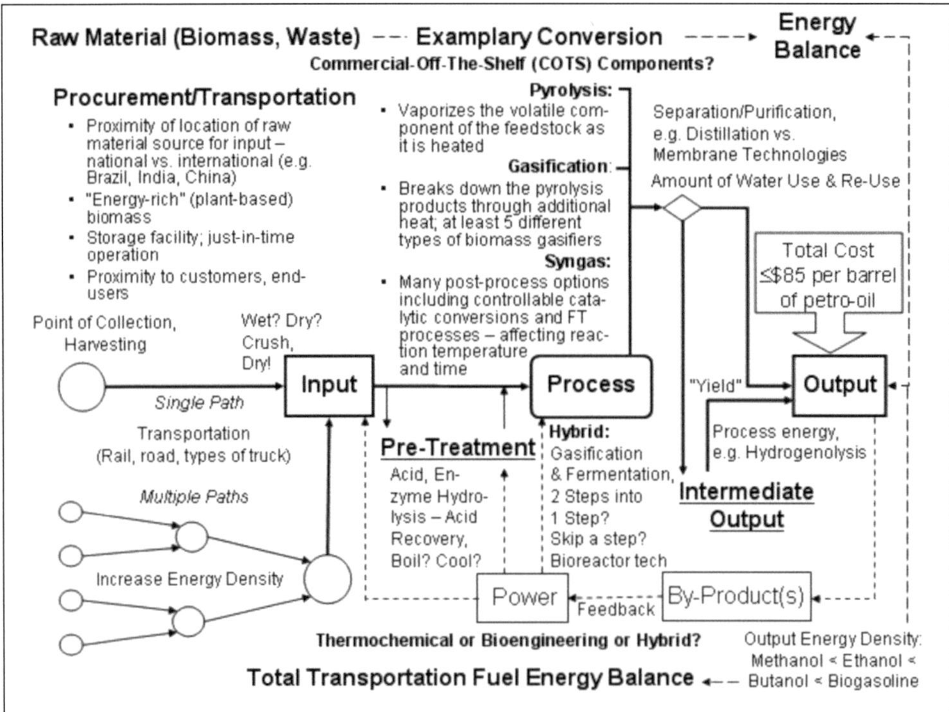

Figure I.185: Some biofuels energy and cost efficiency options for input and conversion emphasizing second-generation biofuels.

One key issue of biofuels economics is find feedstock (Table I.97) at a good relatively stable price over a mid-term period. Basically, it seems that sugarcane and its bagasse and the plant that the sugarcane comes from is going to be the cheapest, most scalable feedstock on a global basis for quite a while. Obviously, sugarcane is going to be best within some distance from the equator. And biofuels startups seek localization of subsidiaries or JVs to be established in Brazil.

Correspondingly, firms, such as BASF and Monsanto as described above (A.1.1.1), are genetically engineering sugarcane to produce more sugar per acre and take less fertilizer and less water.

In different climates other (non-waste) feedstock exhibit also good potential. For instance, sugar beets, sorghum, cassava, and things like this have very high sugar content.

That means economically, any biofuels startup whose process requires to first converting biomass into sugar(s) is squeezed between two benchmarks:

- Output cost has a ceiling related to the price of petro-oil (currently ca. $100 per barrel rather than $85 as given in Figure I.185 for 2008/2009).
- If fermentable sugar is the key input cost will have to meet the price of Brazilian sugar derived from sugarcane.

By mid of 2011 one would have spotted the probable winners to be those with deep pockets and patience, such as Royal Dutch Shell, BP, chemical giant DuPont, agriculture giants Archer Daniels Midland (ADM) or Cargill or the rare startups with significant revenues from businesses other than biofuels, such as making biobased chemicals (specialty or fine chemicals or building block monomers or drugs) or serving the health and nutrition segments as in case of algae.

But, what are then options of biofuels startups lacking the above features as a preferred exit strategy, acquisition or IPO? Obviously, going public is pretty much difficult. If there is a very sound prove that the startup has economical and scalable biofuels, and it is IP protected, then there is a chance to become a public company. But the reality is that most firms that are venture-backed actually will sell. Many people think if they can get a good valuation on a merger and acquisition transaction, they should go ahead and do that instead of get in a critical public market.

All these difficulties do not mean advanced biofuels are not coming, or that they will not play a role in fighting climate change. But everything will happen more slowly than many venture capitalists and government expect.

Indicative of this situation, in 2012 BASF and Archer Daniel Midlands (ADM) made opposite announcements within few weeks around the same overall subject. BASF invested in a company which proposes a singular technology to address a "sugar platform" to transform biomass into fermentable sugars (A.1.1.6) while ADM at the very end of the same biochemical value chain decided to exit from its five-year-old venture bioplastic's JV Telles with Metabolix [Molitor 2012, Runge 2006:871].

Telles produces Mirel, Metabolix' biobased biodegradable plastic at a fermentation facility in Clinton, Iowa, that has capacity to produce 50,000 tons per year of Mirel, a polyhydroxyalkanoate, or PHA, plastic. The facility is adjacent to an ADM corn-processing plant and was built and owned by ADM. With ADM exiting the venture Metabolix will lose access to the facility and its corn sugar feedstocks. The intellectual property related to Mirel will revert back to Metabolix. "ADM says financial returns from the five-year-old venture, called Telles, were too uncertain." [Bomgardner 2012d] Already in January 2010 it has been reported that the "PHA project between Metabolix and ADM has suffered delays and cost overruns." [Taylor 2010].

The ADM decision puts a spotlight to the entire family of polyesters and their monomers (for instance, including the 3-hydroxypropionic acid – 3-HP [Runge 2006: 867-868]).

Beyond the production of bioplastics several companies intend to convert the 3-hydroxypropionic acid into acrylic acid (for instance, OPX Technologies; A.1.1.6). "No doubt that the ADM decision will have an impact on future IPO's or investment rounds of start-ups whose businesses have to pass the so-called 'valley of death' when development cost unexpectedly increase and the perceived risks and uncertainty." [Molitor 2012]

"Both decisions point the current perceived risks and uncertainty in the white biotechnology field: on one hand biorefinery concept as an alternative to a petrochemical feedstock and on the other hand the scale-up of fermentation processes to economic viable and attractive margins." [Molitor 2012]

The mid-2012 economic developments indicated the possibility of another downturn – with consequences for the biofuels field and one is reminded of what has happened before. In the early 1980s, higher-mileage cars and an economic downturn sent petroleum prices swooning, killing off many renewable energy efforts, including those supported by Big Oil.

Remarks Concerning the Role and Effects of Policy on Technology Entrepreneurship and Innovation

With all the issues and problems of biofuels and Cleantech in general "thank goodness cleantech has the government as a customer." [Bormgardner 2011c]

Policy as an "investor" in capital-intensive innovation and entrepreneurship is not new, particularly not in Germany (ch. 1.2.6; Box I.2, Box I.3; A.1.1.2). But also in the US much more than one hundred and fifty years ago this was not unusual, for instance, with regard to Samuel Morse (1791 – 1872), the co-inventor of the telegraph code and American contributor to the invention of a single-wire telegraph system based on European telegraphs. Morse combined *marketing and political skills to secure state funding for development work,* and to spread the concept of communication over vast distances on the continent of America in 1844. [110]

However, over the last decade also the US encountered very strong intervening into the entrepreneurship arena generating policy-driven markets. But, the mental framework of policy concerning entrepreneurship and innovation is very simplistic (in the US and Germany). Take biofuels or biorefineries:

A government agency puts into place an alternative feedstock program, identifies molecules, and provides funding; national labs or research centers develop the organism; and then a private company works with development partners to move the R&D project through to commercialization.

For instance, US President Bush outlined his plan to offer tax credits, subsidies, and federal research support to fuel a drive for (bio) ethanol that would move the nation "beyond a petroleum-based economy and make our dependence on Middle Eastern oil a thing of the past."

Bush's support for ethanol and his mix of energy, economic, and electoral policies have been continued by President Obama, particularly the push for fuels made from cellulosic feedstocks. They have provided billions of dollars to support cellulosic ethanol R&D and biorefinery construction. But despite the money and talk, no commercial cellulosic ethanol biorefinery is operating in the US [Johnson 2010].

President Obama has explicitly called for government funding to be used as a tool to promote the next great companies. And the National Economic Council announced a "National Innovation Strategy" in which "the government should make sure individuals and businesses have the tools and support to take risks and innovate." [Bandyk 2009]

> Generally, for the (renewable) energy and CleanTech field there was a growing interaction and collaboration between federal and state governments and authorities and the country's entrepreneurs.
>
> The potential payoff for entrepreneurs in these fields was big enough that it is worth spending time and money. And it cannot be excluded that one or the other entrepreneur can be a beneficiary of favoritism.
>
> For policy-driven markets it is important for NTBFs looking for or being dependent on grants to have persons in their leadership team early on with experience and preferentially established contacts to the political world on the federal, state and county/communality level.
>
> The job this personnel has and the related importance emerges as "VP of Regulatory Affairs" (Gevo, Table I.99), "SVP of Government Relations" (Solazyme, Table I.90) or as a role and responsibility assigned to a particular SVP (Bluefire, Necy Sumait) or even hiring a former US state governor like Renmatix (B.2).

There is some evidence that the increasing role of the federal government has forced many entrepreneurs in the US to emphasize lobbying, politicking, and jumping through administrative hurdles (cf. SiGNa, ch. 1.2.2; B.2). Similar effects – more under the surface – can also be envisioned in Germany.

During a conference in Washington D.C. in 2009 one panelist, Jonathan Wolfson, CEO of biofuel firm Solazyme (Table I.90), was somewhat surprised he found himself in D.C. “In the 50, 60s, and 70s, entrepreneurs did not come to Washington,” he said. The difference in culture is stark between the world of politics and the world of entrepreneurship. “Silicon Valley is a meritocracy. Best business strategy wins,” said Wolfson, whose company Solazyme recently just won a contract with the Department of Defense to develop clean biofuel produced from algae for the US Navy, as well as a Department of Energy grant to build a biorefinery [Bandyk 2009].

And Wes Bolsen of the cellulosic ethanol company Coskata (Table I.99), a speaker at the conference, added. “Washington, D.C. has become the new Wall Street when banks aren’t lending.” But many attendees expressed skepticism that any one entrepreneur with a great idea could attract the federal government’s attention – and wallet – without significant political connections.

But despite all this investment in high-tech companies, venture capitalists who work in the field were skeptical that the money will find the new drivers. “Is {the stimulus money} a good use of tax dollars? Maybe,” said John Backus, managing partner at a VC firm. “But will it spur innovation? No.” One concern is that the brand-new innovative companies will get left out.” “The government doesn’t know how to work with 20-person companies,” said Backus. “Most cleantech money in stimulus won’t go to the startups. It goes to the defense contract giants.” [Bandyk 2009]

And Bandyk [2009] continued: Those who have won contracts with the government are much more comfortable with the growing collaboration between Washington and the nation’s entrepreneurs. “Government has always had a major role in the energy industry,” Wolfson said. In the case of his firm, government might be necessary. Getting production of his company’s algae-based biofuels off the ground will require significant investment – over $100 million for one plant – that the capital markets simply cannot supply right now, he said. But he did not want to be a beneficiary of favoritism – Wolfson said that after government investment gets the ball rolling, the market should take care of the rest. “Policy should be driven by ends and be technology-agnostic,” he said.

For entrepreneurs like Wolfson, the potential payoff is big enough that it is worth spending time and money away from Silicon Valley to take the risk that the Obama administration will not pick their technologies.

For instance, expenses for lobbying for Sriya Innovations Inc. (connected to RenMatix in A.1.1.6, B.2) amounted to $30,000.00 in 2010 for Provisions in pending energy and climate legislation (S.1462, S.2877, Draft APA by Kerry / Lieberman) related to Biofuels.

In the face petro-oil drying out and soaring energy demands over the past decade, more than 50 countries, including the US and Germany, have been scurrying to implement policies to integrate biofuels into the transportation infrastructure in the face of a number of pressing needs – national energy security, a sustainable agricultural sector, job creation in the rural economy, and reduction of carbon dioxide emissions to curtail climate change.

"The biofuels business globally would not exist if it weren't for the mandates." Thanks to the policies, global biofuel production has gone from about 4 billion gal in 2000 to more than 26 billion gal in 2010 [Mukhopadhyay 2011e].

A recent detailed analysis [Mukhopadhyay 2011e] culminated in the conclusion that *government mandates have shaped the market but not always for the best due to unsustainable production* (Box I.1).

It has becoming clear that biofuels will not solve all the problems proponents had hoped they would solve, but many countries are still rushing headlong as though they will, with policies that, experts say, are doing both harm and good. Governments need to pause, step back, and take a more nuanced and sophisticated view of biofuels, taking into consideration their sustainability and social costs – and, furthermore, should be aware of *systemic effects* they may induce (Box I.1) – as painfully encountered by the recent Great Recession and the financial crisis.

A particular questionable role of government for CleanTech is observed in the US looking at massive bankruptcies of solar, photovoltaic firm Solyndra, Inc. (ch. 4.3.5.2), battery NTBF A123 (ch. 3.2.1) or most recently electric car manufacturing NTBF Fisker Automotive (founded in January 2005). Fisker is the US Government's "biggest public loss since the infamous Solyndra Solar debacle." [Koetsier 2013] Fisker was one of the largest US venture capital backed companies ever.

Based on exclusive documents PrivCo [2013] outlined how a "billion dollar startup became a billion dollar disaster" and "2 of America's smartest VCs – Kleiner Perkins & New Enterprise Associates & others to lose over $1 billion dollars in 'The Largest Venture Capital Investment Debacle in U.S. History.'"

The PrivCo Fisker Papers released "never before seen original government documents regarding the Department of Energy's $529 Million loan to Fisker Automotive, definitively proving loan underwriting that no rational lender would have ever undertaken, waiver after waiver from the D.O.E. after Fisker missed covenants of the Loan, and the subsequent concerted effort by the Loan Programs Office to cover up and obfuscate the unraveling of Fisker and the inevitable erasure of U.S. taxpayer collateral that funded the Loan."

Explicit descriptions of the DOE $529 million loan issue on the "Fisker case" are also given, for instance, by Koetsier [2013], Chernova and Ramsey [2013] and Vlasic [2013]. "The untested Fisker loans totaled $529 million, more than the company had initially requested, and an amount that encouraged private backers to chip in more funds." [Chernova and Ramsey 2013] (See also Figure I.34 for encouraging private backers in policy-driven markets.)

"Fisker has become – to lawmakers and others – the Solyndra of the electric car industry." "Fisker, with its technical problems, management turmoil and mounting losses, offers a cautionary tale in the fiercely competitive arena of alternative-fuel vehicles and of government subsidies for start-up businesses." [Vlasic 2013]

A.1.1.6 The Shift from Biofuels and Co-Products to Biobased Chemicals as the Primary Target of Entrepreneurship

Back to the Agricultural Future for Chemical Innovations:
What's a rerun?
[Runge 2006:563

A Different Context

An overview of the biofuels industry situation by 2010 is given by Wikipedia. [111] As outlined in the previous sub-chapter there emerged a clear shift of many "promising" biofuel NTBFs' business models from biofuels to biobased chemicals. Here, the type of advanced biofuels (fuel not made from food-like feedstocks such as corn sugar) was cellulosic ethanol [Bomgardner 2011b].

With so much land devoted to raising livestock feed the focus of biofuels startups could also be on feed. For instance, animal feed is a lucrative business for the US, with China importing it at rates of up to 50 cents a lb. If methods can be developed to break down plant cell walls to get the sugars for biofuels while saving the proteins for animal feed, biofuel sources, such as corn or algae (Table I.90), can provide fuel and feed [Mukhopadhyay 2011e].

Also the role of the US chemical giants DuPont (Table I.83) and Dow Chemical (Figure I.179) in biofuels has been tackled above.

In the area of polymers and plastics DuPont followed a *stepwise crossover strategy* for biobased products. This means for chemical products produced by a proven process by several components the overall transition to a fully biobased product proceeds through substituting the components separately by already available biobased components.

For instance, DuPont used petroleum-derived propanediol (PDO) to produce some 10,000 metric tons of Sorona per year. DuPont's Sorona® 3GT is a copolymer designed to be made from corn-derived 1,3-propanediol and petroleum-derived terephthalic acid (Sorona: polytrimethylene terephthalate (PTT) polyester). DuPont then started construction of a large-scale propanediol fermentation facility in collaboration with carbohydrate processor Tate & Lyle. Intermediates for DuPont Sorona® polymer would then use Bio-PDO™ to get a partially biobased copolymer [Runge 2006:583] – but have to wait for availability of biobased terephthalic acid.

The enormous and looming challenge facing biofuels companies, of which none have actually gotten far enough with the research process to confront, is scaling-up to the enormous requirements of the transportation fuel market and getting the costs down to achieve pump parity. Without those two achievements and the necessary capital, all of these firms remain science and technology experiments. Additionally, for the bioengineering route to biofuels a genetic breakthrough has nothing to do with a (production) breakthrough when scaling-up to large-scale production.

Things are different if biomass from food-like feedstocks, such as corn or sugarcane, is taken into consideration. For instance, *NTBFs with a biorefinery approach* with food-related ethanol as a basic chemical for biobased polyethylene or polypropylene may find strategic partners in the Brazilian plastics industry (cf. Figure I.174).

Braskem SA, the largest (petro-)chemical company in the Americas by production capacity and among the top ten largest in the world, initiated a five-year project with Danish enzymes manufacturer Novozymes to work on a new sugarcane-based route to polypropylene. Braskem has already synthesized polypropylene from sugar-based ethanol.

In 2009 Braskem was constructing a 200,000-metric-ton-per-year plant to make polyethylene from ethanol, planned to be completed in 2010. And it signed contracts to sell the "green" polyethylene to the global packaging giant Tetra Pak [Tullo 2009]. And in 2011 Braskem started up the 200,000-metric-ton ethanol-based polyethylene plant in Brazil [Tullo 2011].

For instance, German LANXESS does not only target biobased butylrubber through its cooperation with Gevo (see above), but also ethylene propylene diene monomer (EPDM) synthetic rubber. It plans to use ethylene derived from the purely renewable resource sugarcane. Braskem shall supply the bio-based ethylene via pipeline to LANXESS' existing EPDM plant in Brazil. It will be the first form of bio-based EPDM rubber in the world [Specialchem4polymers 2011] and will be sold under the brand name Keltan Eco.

Activities of the chemical industry until 2005/2006 are described by Runge (The Chemical Industry in a Biobased Economy [Runge 2006:567-571], White (Industrial) Biotechnology in a Biobased Chemical Industry [Runge 2006:571-578] and Research, Development and Innovation with Renewable Resources in the Chemical Industry – Green Chemistry [Runge 2006:578-590]).

The concept of a “biobased chemistry” including biobased plastics is not new. In the 1920s in the US there was a strong movement in that direction under the name “*chemurgy*,” but was ultimately stopped by political interference of the petrochemical industry [Runge 2006:565-566].

Some early entrepreneurial activities in biobased chemical solvents were reported in the 1990s [Runge 2006:860-861]. The bio-oriented small- and medium-sized enterprises (SMEs) often targeted areas that were niches. The “green solvents” niche at that time was a generic thrust in line with general trends to replace organic solvents due to regulatory pressures. The emphasis was on oxygenated solvents, such as lactic acid esters (like ethyl lactate) and soy-based solvents (methyl soyate).

For instance, Vertec Biosolvents Inc. (assumed to be founded in 1997 and currently having five issued US patents) produced environmentally friendly solvents made from ethyl lactate derived from farmer grown corn and soybeans. Vertec BioSolvents offered also environmentally-friendly ink cleaners. Vertec Biosolvents, in particular, was planning to replace NMP (N-methyl pyrrolidone, Figure I.187), a powerful organic solvent with broad solubility for resins and high chemical and thermal stability.

Currently Vertec Biosolvents manufactures and sells biosolvents and formulations (blends) based primarily on four major ingredients – ethyl lactate, fatty acid methyl esters (soy methyl esters), d-limonene and ethanol for a variety of specialty applications in industrial and agricultural markets targeting replacement of petrochemical solvents in use, even NMP and hydrocarbon solvents.

Another firm which, by the mid of the 1990s, focused early on green oxygenated solvents was Diversified Natural Products, Inc. (DNP), an industrial biotechnology company organized into two divisions, Biobased Fuels and Chemicals, and Gourmet and Functional Foods. Its main product was succinic acid.

DNP addressed another niche which comprises “short chain diacids” called “building block chemicals” (see below) that can serve as key feedstock for future biorefineries (for instance, adipic acid = hexanedioic acid HOOC-(CH2)4-COOH or succinic acid = butanedioic acid HOOC-CH2-CH2-COOH) or even long-chain diacids. DNP received investment from several Japanese venture firms including Toyota Tsusho Corp., a sister company of Japanese automaker Toyota Motor Corp. [Runge 2006:860-861].

The origins of the DNP’s succinic acid business go back to 1995, when it was established by a company called Applied CarboChemicals. The company operated under this name until 2003, when it was restructured, refinanced and renamed Diversified Natural Products. The company subsequently expanded its activities into other fields; succinic acid became just one of its businesses [ORNL 2010].

In 2006, Diversified Natural Products established a collaborative R&D effort with Agro Industrie Recherches et Développements (ARD), the R&D subsidiary of a French agricultural consortium led by Champagne Cereales. The focus was on disuccinate esters as “green solvents.” Over the next two years the partners scaled-up succinic acid production to 80,000 liters and developed an economical, aqueous based isolation and purification process [ORNL 2010].

DNP Green Technology was established in 2008, when all succinic acid assets, including all intellectual property, contracts, joint venture interest and employees, were spun off from DNP. Following the spin off, the company's shares were distributed to Diversified Natural Product's shareholders, making DNP Green Technology a standalone legal entity with no ties to Diversified Natural Products.

A joint venture called Bioamber SAS between ARD and its US partner DNP Green Technology was established in 2008, resulting from the R&D partnership between its two shareholders. BioAmber targeted succinic acid production. The existing organism for production, originally funded by the DOE in the late 1990s, was further developed and scaled-up, and optimized at the large-scale manufacturing facility in France [De Guzman 2011b]

ARD industrialized its laboratory procedure and invested €21 million in an industrial demonstration facility with a capacity of 2,000 tons per year. BioDémo enjoyed also financial support from the General Council of the Marne Département (€1.25 million), the Champagne-Ardenne Region (€1.25 million) and the ERDF (€2.5 million). Diversified Natural Products contributed its intellectual property portfolio.

DNP Green Technology fully executed an exclusive license agreement for three patents invented solely by Argonne National Laboratory and jointly by Oak Ridge National Laboratory (ORNL) or Argonne National Laboratory (ANL), one patent specifically was on “*Mutant E.coli Strain* with increased succinic acid production” [ORNL 2010]. Their process uses a strain developed particularly to produce succinic acid, with wheat-derived glucose currently being used as the substrate.

In 2009, DNP Green Technology completed a $12 million financing with a group of institutional investors led by Sofinnova Partners, a European venture capital firm, and including, for instance, also the Japanese Mitsui & Co. Venture Partners. In 2010 DNP Green Technology acquired 100 percent of the shares of its BioAmber joint venture from ARD. Concurrent with the acquisition of the joint venture, DNP Green Technology changed its name to BioAmber Inc. Siclae, a leading European agricultural group and the principal shareholder of ARD, became a shareholder in BioAmber through the transition [ORNL 2010].

BioAmber owns or have exclusive rights to specific microorganisms, chemical catalysis technology and a unique, scalable and flexible purification process. BioAmber manufactures its bio-succinic acid in a facility using a commercial scale 350,000 liter fermenter in Pomacle, France, which was used to refine its process and issue a claim to make cost-competitive bio-succinic acid. The purpose of the Pomacle plant is to showcase the production technology, which is available for license by other parties.

As expected, BioAmber has set up a "veterans approach." Its management team consists of experienced professionals, possessing on average over 25 years of relevant experience in scaling-up, manufacturing and commercializing chemicals, gained at large companies or entrepreneurial startups.

Since its creation in 2008, BioAmber executed its strategy by several business partnerships and had successfully commissioned an industrial scale production facility. It has moved down the value chain through its acquisition of Sinoven Biopolymers, which produces modified PBS (Figure I.187), an innovative biodegradable polymer. BioAmber has licensed DuPont's hydrogenation catalyst technology to make bio-based 1,4 butanediol (BDO). The major uses of BDO (Figure I.187) are in the production of tetrahydrofuran (THF) and polybutylene terephthalate (PBT).

Recently, BioAmber has signed an agreement with Mitsui to jointly build a facility in Sarnia, Ontario, that is expected to produce bio-succinic acid and bio-BDO with a total capacity of 34,000 metric tons of bio-succinic acid and 23,000 metric tons of bio-BDO at full capacity.[Bomgardner 2011g].

The Sarnia plant will be operated by BioAmber's new subsidiary Bluewater Biochemicals, and will have initial capacity of 17,000 tons per year by 2013. This capacity will increase to 35,000 tons/year by 2014 and will then use next-generation yeast developed by Cargill and successfully used and commercialized by Cargill in lactic acid production.

The Bluewater Biochemicals subsidiary was specifically created as a Canadian legal entity that will own and operate the Sarnia plant. The Sarnia plant investment was supported by government grant/loans [De Guzman 2011b]. The Sarnia plant will initially use corn kernels as a sugar source for *E. coli* fermentation. But the switch to the engineered yeast licensed exclusively from Cargill will mean producing succinic acid from hydrolyzed agricultural wastes, such as corn stover [Ritter 2011].

In 2011 BioAmber formed a number of partnership, for instance, with Mitsubishi Chemical Corp. (MCC) of Japan to produce bio-succinic acid for MCC's joint venture company PTT MCC Biochem. The joint venture will manufacture and market bio-polybutylene succinate (PBS). BioAmber plans to have its succinic acid facility located next to PTT MCC Biochem's 20,000 tons/year PBS plant in Thailand.

Since its inception, BioAmber raised an aggregate $76.1M from private placements of equity securities and convertible notes. It expected to spend around $200 million per plant on construction and start-up operating costs for facilities in Canada and Thailand [De Guzman 2011b]

In 2011 *BioAmber filed for an IPO* hoping to raise up to $150 million with the US Securities & Exchange Commission. BioAmber said it has made 221 metric tons of biobased succinic acid at its facility in Pomacle, France, *but has yet to book any sales.* Instead, the firm touts its strategic partnerships with potential succinic acid buyers.

BioAmber's filing acknowledges that it faces *tough competition in the nascent biobased succinic acid market* both from startups and from established companies. [Bomgardner 2011g; De Guzman 2011b].

In 2013 BioAmber announced the pricing of its initial public offering of 8 million units consisting of one share of common stock and one warrant to purchase half of one share of common stock at $10 per unit, before underwriting discounts and commissions which means it would raise $80M at $10. But its per-share stock price fell from $10 to $8 in its first five days of trading. In the first quarter of 2013 BioAmber posted small sales of about $330,000 and posted a loss of $9.6 million for the quarter [PlasticsNews 2013].

In 2012 BioAmber set a strategic collaboration with the German firm LANXESS in the field of plasticizers to show that bio-succinate esters are viable alternatives to phthalates, which have come under scrutiny for their potential toxicity. It has also completed its Series C round of financing with net proceeds of $30 million involving existing investors and LANXESS.

Many advanced biofuel startups have been diversifying into the biobased chemicals sector given the higher potential profits for chemicals versus biofuels. For most of all, it appears to be quicker (but not easier) to get into the chemicals sector especially if you are looking into drop-ins as long as you have partners who know *the chemical industry's well-oiled system*. For instance, above the cooperation of US Gevo and German chemical firm LANXESS was described concerning their isobutanol/isobutene efforts to produce renewable butylrubber.

But many startups perceive opportunities to focus essentially on chemicals or more generally biorefineries and to execute their strategies, *seeking alliances with chemical firms* which are making considerable investments into what they call "sustainable chemistry," "green chemistry" or "CleanTech Chemistry."

For instance, OPX Biotechnologies (see below), founded in 2007 and emphasizing "good chemistry," and chemical giant Dow Chemical announced a collaboration to develop an industrial scale process for the production of biobased acrylic acid from renewable feedstocks. The global petroleum-based acrylic acid market is estimated to be $8 billion and growing 3 to 4 percent per year. Acrylic acid is a key chemical building block used in a wide range of consumer goods including paints and coatings, adhesives, diapers and detergents [Bomgardner 2011h].

Simultaneously with the increasing interest of the chemical industry and the disappointment of VCs with biofuels venture capital is re-directed towards startups targeting biobased chemicals, intermediates, resins and plastics for the petrochemical branch.

There was a golden age, from the late 1930s through the mid 1960s, when the chemical industry invented and commercialized most of the polymers we use today

[Runge 2006:411-424]. Those early plastics were so successful that it has become difficult to launch newer polymers in the marketplace.

Furthermore, an established infrastructure of resin producers, plastics converters, and processing machinery makers is dedicated to multi-million-ton-per-year applications. Many companies have since unveiled ambitious plans to establish new resins, but the history of the plastics industry is littered with their failures. "But some companies are still inventing polymers much like their counterparts did in the old days: by coming up with novel chemistry and then sorting out where it will be useful." [Tullo 2011]

The most promising way for plastic success lies in a global view, from design up to finished devices. Design integrating multiple functions is by far the most important aspect followed by compounding integration [112] preferentially supported by modeling and simulation.

> Any to-be entrepreneur addressing biobased chemicals or plastics is advised to read the story of Patrick Gruber, innovator of polylactic acid (PLA) plastics at US giant Cargill and now CEO of Gevo (Table I.99) [Benda 2003]. A further very lucid article by C. Benda (http://www.cargill.com/news/00_08_cd.htm: *Mission Possible*! Cargill News International) is unfortunately no longer accessible on the Web.

New polymers have rather long scale-up and gestation periods. And this applies also to biobased plastics, for instance, those derived from Ingeo polylactic acid [Tullo 2011; Runge 2006:129-130,581,756].

NatureWorks, a subsidiary of US giant Cargill, opened its Ingeo polylactic acid (PLA) plant in 2002 with 70,000 metric tons of capacity. At that time the company had high hopes merely because bioplastics were still new. Converters started experimenting with it for nearly every conceivable application. But, it had to cultivate markets. Simultaneously PLA took advantage from improvements in blending and multilayer technology and, finally, over the past two years, NatureWorks' sales have grown by more than 25 percent annually.

Entrepreneurship and Intrapreneurship in Drop-In Building-Block Chemicals

Riding a wave

> Over the last two to five years two strong trends have emerged – not just for existing biofuel NTBFs to jump on, but also for the foundations of new firms. They relate to shifts of advanced bioalcohol NTBFs, whether bioethanol or biobutanol, to producing *non-food related C6 and C5 sugars* and/or *drop-in biobased chemicals* ("*CleanTech Chemistry*"). In particular, there is a rush for low-cost *non-food industrial sugars* as reflected, for instance, by Bluefire's wholly owned subsidiary SucreSource (Table I.86) and Virdia (Table I.99)!

Since 2006 a new wave of findings of biotechnology companies, primarily in the US, has changed the landscape that people had about the field. Driven by big players and, to a large portion, venture capital new biotech companies propose to replace petrochemicals originated *raw materials* by chemically identical products originated from biomass feedstocks. Based on generic technologies this family of biobased products is currently called *"drop-in" technologies* because of addressing markets of existing and identical or very similar products.

Biobased chemicals appeared in the focus, not just as an expression of a "green" attitude and sustainability, but the chemical industry is also looking for ways to attenuate generally the impact of cost fluctuations in fossil-fuel feedstocks (Box I.26).

According to Lux Research biobased chemicals and materials was expected to grow to $19.billion in 2016 as its global capacity jumps 140 percent. Lux Research said that it has listed down 151 global facilities and their intended operational dates, products and capacities. These capacities are expected to climb to 9.2 million tons in the next five years [De Guzman 2011c],

In particular, the reports says [De Guzman 2011c]

> *Bioplastics* will slow down in terms of expansion though capacity is still expected to grow 57 percent from 2011 to 2016. From 2006 to 2011, bioplastics have experienced explosive growth of 1,500 percent to a current aggregate capacity of 470,000 tons, and a 10.9 percent share of all bio-based materials.
>
> Cellulose polymers and starch-based plastics remain dominant but their share of total capacity will slide from 45 percent in 2011 to 21 percent in 2016. Cellulose polymers and starch-derived materials still rule because they are durable, strong and easily biodegradable: They have been widely used in high-performance plastic coatings, buttons and yarns, and even early LEGO bricks.
>
> By 2016, there will be *consolidation* – both within sectors of biobased materials manufacturing, and regionally, as *leaders buy up technologies and access to feedstock*. Momentum derived from existing capacity – ethanol from sugarcane being converted to ethylene and propylene, for instance – will influence regional specialization.

Related *VC-based startups* bring a unique set of technical core strengths, processes and long-term business approaches again by a *"veterans" management team* to deliver a consistent, readily convertible sugar feedstock that can compete on price and quality with crude oil feedstock for petroleum fuels and chemicals and food-based industrial sugars from corn and sugarcane.

Within a *biorefinery concept* which is heavily supported in the US and Germany by policy [Runge 2006:849-873] and which embraces fuels (energy), chemicals and materials referring to non-food biomass as feedstock there are *a sugar platform and a syngas platform* [Runge 2006:865] and the two main process options to use are the (sugar-oriented) *biochemical platform* and the *thermochemical platform*. The technology platforms will be featured by industrial biotechnology, materials technology, and reaction and process design.

Notably, there is a pioneer plant demonstration to produce cellulosic ethanol by syngas fermentation, Coskata (Table I.99), which was basically co-founded by Argonne National Laboratory scientists. This means, there may be combining both platforms.

In the US a list of chemicals was created by industry and academia that considered the compounds' compatibility with existing petrochemical processing, technical complexity of the syntheses from biomass, known market potential, and other factors. A shortlist of 30 compounds was selected and from among those compounds a final 12 *top-tier compounds* that can be produced from plant sugars were chosen [Runge 2006:871].

Of the hexosen (C6 sugars) glucose derivatives offer most potential for *top-tier building-block chemicals* as key feedstocks [Runge 2006:867-868]:

- Lactic acid (cf. the graph "chemicals from lactic acid" [Runge 2006:868]);
- 3-hydroxypropionic acid (3-HP; cf. [Runge 2006:249-251,584,871-872], in particular, the graph "3-Hydroxypropionic acid (3-HP) – A New Chemical Intermediate Platform" on page 249)
- Succinic acid.

Further notable building-block chemicals for the chemical industry include acrylic acid and 1,4-butanol (1,4-BDO). Acrylic acid ($H_2C{=}CH{-}COOH$) may be derived from 3-HP. 1,4-butandiol is one of the many compounds that can be obtained from succinic acid by established chemical transformation (Figure I.186).

> Different from a start in biofuels to enter the chemical industry a deep knowledge of the broad application spectra of the targeted offerings is necessary, and also the current and anticipated regulatory environment.

Novel chemicals' market barriers are well known: product registration, performance and costs versus established products, switching costs, long market introduction cycles to name a few.

But, *drop-in chemicals will experience quite different challenges compared with novel chemicals, converging to a large part around the cost of goods* – that is levels of innovativeness, conservatism, and risk aversion of potential customers (ch. 4.2.1.1). Hence, for biochemicals-oriented startups a "*veterans approach*" has turned out to be mandatory with managers having not only broad experiences in the chemical industry, but also being appropriately connected in related networks.

Execution of a startup's strategy (commercialization) requires intense *partnerships* with firms from the sugar and/or biotechnology-oriented nutrition industry and, very importantly, the *rather conservative chemical industry* in terms of large and giant chemical companies, particularly from the US, Germany, The Netherlands and Japan.

For instance, 1,4-butanol from succinic acid provides a typical example for the highly competitive situation for entrepreneurship/intrapreneurship in biobased chemicals, particularly with regard to economies of petroleum-based chemicals (Figure I.187).

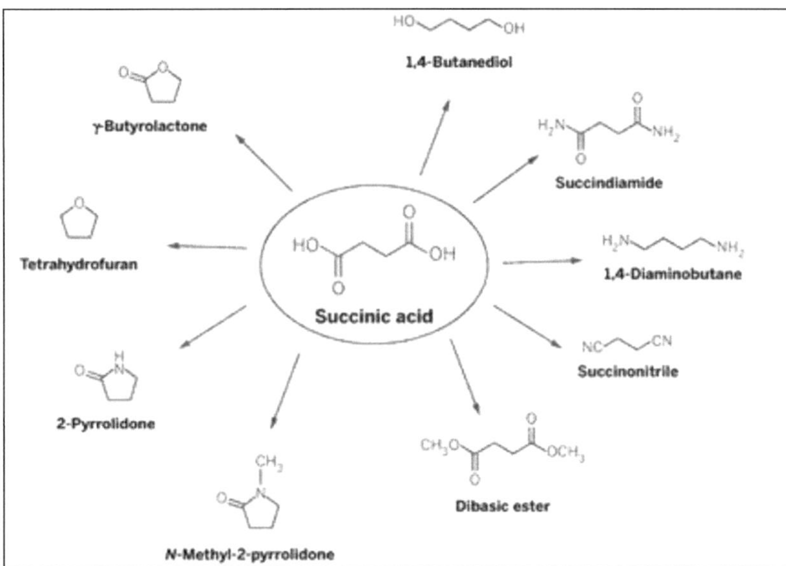

Figure I.186: Chemicals and intermediates derived from succinic acid.

US firm Genomatica, Inc., which in 2011 filed for an initial public offering (IPO) of stock worth up to $100 million [Bomgardner 2011f; Admin 2011f], is known for making as a first product 1,4-butanediol (1,4-BDO) by feeding sugar to an engineered strain of *Escherichia coli*. But, in 2012 Genomatica withdrew its $100m IPO – “in light of current market conditions.” [De Guzman 2012d].

On the basis of demonstration-scale tests with sugar processor Tate & Lyle, Genomatica said it can produce the intermediate at lower cost than petroleum-based processes. Its second product made from renewable feedstock shall be butadiene [Bomgardner 2011a]. In line with numerous examples from the biofuels scene, in 2012 Genomatica pushed back the timeline for commencing commercial-scale production of renewable butanediol from 2012 to 2013 [Lane 2012q].

Concerning renewable butadiene Genomatica clashes with Amyris (Table I.99 and below text), but more importantly its focus must be on butanediol made from biobased succinic acid which is offered by many other firms. Among the largest emerging applications of bio-succinic acid is the production of “green” 1,4-butanediol (BDO). Furthermore, consideration of just butanediol and producing it at lower cost than petroleum-based processes will not suffice: The overall cost of the customer in a system of other products related to succinic acid and BDO is relevant (Figure I.25, Figure I.187).

Moreover, by mid of 2012 LanzaTech (Table I.99) signed a joint development agreement with one of the world-leading nylon producers INVISTA focused on bio-based butadiene. According to the agreement, INVISTA and LanzaTech will collaborate on projects to develop one-step and two-step technologies to convert industrial waste gas carbon monoxide (CO) into butadiene. Initial commercialization is expected in 2016. The collaboration will initially focus on the production of butadiene in a 2-step process from LanzaTech's CO-derived 2,3-butanediol (2,3-BDO). A direct single step process will also be developed to produce butadiene directly through a process of gas fermentation.

Butadiene is a key intermediate chemical used by INVISTA in its proprietary butadiene-based adiponitrile (ADN) production technologies. ADN is a critical intermediate chemical used in the manufacture of Nylon-6,6.

Currently LanzaTech runs a 15,000 gal/year pilot facility at a steel mill in New Zealand that produces ethanol and 2,3-BDO from waste carbon monoxide gas. In Shanghai, China, LanzaTech's 100,000-gallon-per-year demonstration plant uses waste gases from a Baosteel steel mill to produce ethanol.

The issues of introducing and ultimately replacing biobased succinic acd and/or 1,4-butanediol to a large chemical firm can be lucidly illustrated looking at the specific situation of BASF.

The world's largest chemical company BASF with its "Verbund"-approach [Runge 2006:369-370] can interconnect biobased succinic acid and 1,4-BDO with its production streams for petroleum-based derivatives and products with production sites all over the world as given in Figure I.187. This provides a huge potential for *crossover strategy* for products containing various amounts of biobased components and thus a great potential for cost management and price settings. While mostly succinic acid is currently made mainly from fossil-derived maleic anhydride that provides the basic four-carbon backbone BASF can utilize its proprietary acetylene platform.

Furthermore, contrary to its peers and other large chemical firms, such as US firms DuPont or Dow Chemical, BASF has an oil business including exploration through its 100 percent subsidiary Wintershall Holding GmbH [Runge 2006:63-64,586]. It is the largest crude oil and natural gas producer in Germany (turnover of €12.1 billion, $15.8 billion) in 2011, net profit €1.1 billion). Wintershall is a big revenue contributor to BASF. Hence, BASF does not depend fundamentally so much on oil price swings as other chemical companies do.

Combining bio-BDO and bio-succinic acid opens up the possibility of greener biopolymers, such as polybutylsuccinate (PBS), which is used in biodegradable packaging films and disposable cutlery. Another potential market is in polyester polyols and polyurethanes, currently dominated by the use of adipic acid as a precursor. Companies are looking at replacing the six-carbon adipic acid with four-carbon biosuccinic acid, providing the costs become comparable, because adipic acid production is a messy process that produces a lot of carbon dioxide [Taylor 2010].

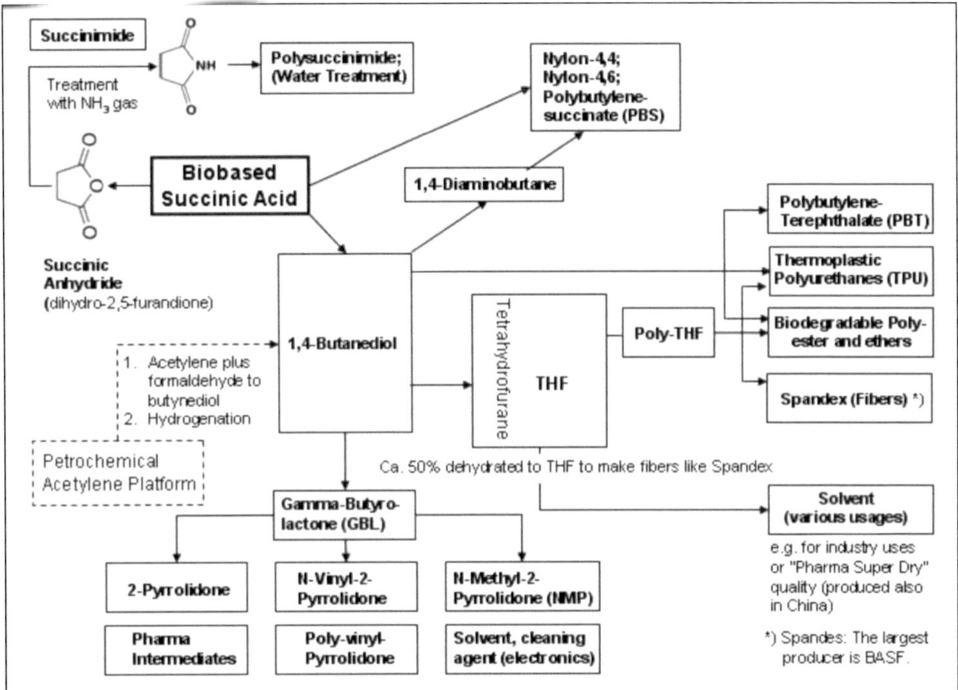

Figure I.187: The role of succinic acid for BASF in its production Verbund environment.

Currently the majority of polyamides (PAs) of the Nylon-type is on C4 and C6 components (succinic or adipic acid). But interest in polyamides based on α,ω-diacids (long-chain diacids) and diamines (C10– C18) emerged. There was already interest in "green" long-chain α,ω-diacids (with 9 or more carbon chain atoms) rather than the C4 or C6 diacids before 2007. The German firm Cognis, acquired in 2010 by BASF (now BASF Personal Care and Nutrition GmbH), was particularly active in this field targeting polyamides, polyesters, and polyurethanes. A series of Nylons were synthesized and tested using octadecanedioic acid (C18 diacid) made from Cognis's biofermentation process.

Cognis had developed a proprietary strain of the yeast "*Candida Tropicalis*" which efficiently *oxidizes natural based fatty acids produced from vegetable, animal or tall oil sources to produce the novel diacid momomers. Candidas Tropicalis* was developed to oxidize terminal ("end") methyl groups on the molecules efficiently into terminal carboxylic acid groups. Thus an alkane or renewable monobasic fatty acid feedstock can be oxidized to a dicarboxylic acid [Runge 2006:861].

Interestingly, Wallace Carothers, who was the first to develop polyamides (Nylon-6,6) at DuPont in the 1930s [Runge 2006:414-418], insisted that Nylon-5,10 is better than Nylon-6,6 or Nylon-6 of BASF (then I.G. Farben), the latter two of which are the most common for textiles and plastics. Carothers had wanted to bring Nylon-5,10 to commercial status, but was unable to, because he could not identify an economically efficient production process [Ravenstijn 2011].

BASF has developed a 100 percent bio-based PA-5,10 with performance suitable for automotive applications. However, Bio-PA-5,10 is rather expensive and thus limits its applications. Dutch firm DSM works on PA-4,6 grade expected to replace metal parts under the hood, such as turbo diesel systems components.[113] DSM's high-heat polyamide Stanyl 4,6 was one of the few successful polymer introductions since the 1980s.

Succinic acid is currently only a niche product, with the 30,000 tons produced a year creating a market worth $225 million. Market research firm Frost & Sullivan believes the market will expand six-fold to 180,000 tons by 2015, thanks largely to the introduction of bio-succinic acid. There is a rather strong competition in the bio-succinic acid arena.

The emerging bio-succinic acid market in particular shows a very competitive environment with small and large players – from Europe, the US, Japan, China and Thailand.

Apart from BioAmber discussed above at least four other groups are gearing up to develop commercial capacity for bio-succinic acid. The companies investing in bio-succinic acid clearly believed these projections are reasonable, given that collectively three of them intended to bring over 140,000 tons of capacity online by 2012 [Taylor 2010]. According to the BASF/Purac JV (see below) the main drivers are expected to be bioplastics, chemical intermediates, solvents, polyurethanes and plasticizers.

US-based Myriant Technologies is among the companies which see the main potential for bio-succinic acid as lying in the BDO market. As BioAmber also Myriant (motto: "Chemistry Refined ...Naturally") filed for an IPO in 2011 worth up to $125 million. Myriant's filing disclosed that it was planning a 220 million-lb-per-year bio-succinic acid facility in China in partnership with China National Bluestar, in addition to a smaller plant to be built in Louisiana. This plant in Louisiana should have a capacity of 15,000-ton succinic acid and should be built by 2012 with the help of $50 million in Department of Energy funding.

Founded in 2004 as BioEnergy International, Myriant focused on the production of renewable biobased chemicals using a proprietary biocatalyst platform. Myriant is the exclusive licensee of technology from the University of Florida and has since expanded its intellectual property portfolio with internally generated patents, patent applications and a scientific knowledge base. In 2009 bioethanol specialist BioEnergy International span out Myriant as an independent company and incorporated all of its biobased chemicals business and intellectual property.

The other players include very large chemical firms. Dutch DSM and the French firm Roquette Frères, through their Reverdia joint venture, expected their 10,000-metric-tons plant in Spinola, Italy, to be online in the second half of 2012. Finally, Mitsubishi Chemical Corp. (MCC) has been making PBS from fossil-based succinic acid [Taylor 2010]. It has developed its own process for making bio-succinic acid from biomass, although no details of its process are known.

The company was formerly collaborating with Ajinomoto on the project. Mitsubishi wanted to follow the typical "crossover strategy" and wanted to target the market being developed by the (then) Dow-Cargill JV NatureWorks for PLA. [Runge 2006:861]. Mitsubishi currently has also a joint venture with the Thai company PTT (PTT MCC Biochem) to develop bio-succinic acid-based PBS and also BioAmber envisions a partnership with that firm.

Owing to BASF's traditional raw material base interconnected with research, applications, and product driven business approaches it is not surprising that a leading white biotechnology project is the production of succinic acid (Figure I.187; cf. also Quantifying the BASF "Knowledge Verbund" – Figure I.188). In 2009, BASF and Purac announced they will form a joint venture to produce up to 25,000 metric tons of the intermediate in Barcelona (Spain) by 2013 planning already a world-scale plant with a capacity of 50,000 tons [Taylor 2010; Purac 2011].

Purac is a subsidiary of the Dutch firm CSM, a global player in bakery supplies and food ingredients and preservations. Purac (revenue ca. €400 million, 1,100 employees) is active in a variety of markets, with a focus on natural food ingredients, lactic acid, biogases, chemicals and biobased monomers for PLA [Taylor 2010; Purac 2011].

The JV will make bio-succinic acid using a BASF-developed bacterial strain (*Basfi succiniproducens*) which can process a wide variety of C3 (glycerin), C5 and C6 (glucose) renewable feedstocks, including biomass sources. Using a fully equipped fermentation and down stream purification plant the partners will demonstrate the economical production of succinic acid on industrial scale; carbon dioxide (CO2) will be used as a raw material and fixed during the highly efficient fermentation process (cf. BioAmber's process).

The BASF/Purac JV followed closely a stringent typical scale-up process (Figure I.8). Critical steps of the jointly developed production process have been validated in several successful production campaigns. The resulting volumes were used to evaluate the market. In particular, the giant BASF can simultaneously provide a very large in-house test field.

"After successfully testing the BASF in-house applications we are now able to make large volumes available for external customers." In view of the risky situations of existing or to emerge bio-succinic acid startups as competitors BASF emphasized that "The goal is to globally provide a high product quality and offer security of supply to the customers." [Purac 2011] This means BASF is very serious about its inroads into the bio-succinic acid intermediate – and that does not bode well for startups.

Differently to the startups BioAmber and Myriant which rely on cost models for just the production of bio-succinic acid the BASF/Purac JV works together to achieve manufacturing cost levels by empirical in-house and outside tests for making biobased succinic acid competitive for a systemic context of a wide variety of novel applications.

Correspondingly, BASF and Purac are establishing a joint venture for the production and sale of biobased succinic acid named Succinity GmbH with headquarters in Düsseldorf, Germany which should be operational in 2013. "We know from many discussions with customers and samples we sent them that the demand for biobased succinic acid for example for biodegradable plastics is set to grow faster and more strongly than expected earlier," said the President of Purac [BASF 2012c].

Generally, the demand for succinic acid is anticipated by Succinity to grow strongly in the years ahead, driven mainly by bioplastics, chemical intermediates, solvents, polyurethanes and plasticizers and Succinity to take advantage from all of these fields. This plant, having commenced operations in March 2014 with an annual capacity of 10,000 metric tons of succinic acid, will put the new joint venture company in a leading position in the global marketplace. This is complemented by plans for a second large-scale facility with an annual capacity of 50,000 metric tons of succinic acid to enable the company to respond to the expected increase in demand. The final investment decision for this facility will be made following a successful market introduction.

Additionally in 2013 BASF planed to begin also production of 1,4-butanediol based on renewable feedstock (renewable 1,4-BDO) using the patented process of Genomatica utilizing a license agreement allowing BASF to build a world-scale production facility. The one-step fermentation process is based on sugars as a renewable feedstock. And concerning non-food sugars as a renewable feedstock BASF has linked itself to Renmatix (see below). Furthermore, "initial lifecycle analyses show that Genomatica's Bio-BDO will require about 60 percent less energy than acetylene-based BDO" (cf. Figure I.187 and [Bomgardner 2011i]).

Genomatica will continue to advance its patented renewable BDO production process while BASF will produce renewable BDO, which shall be available in the second half of 2013 for sampling and trials. "We are pleased to cooperate with BASF, the leading global BDO manufacturer with a worldwide manufacturing and sales network and many years of market experience," said Christophe Schilling, Chief Executive Officer of Genomatica, and continued: "This agreement highlights Genomatica's commitment to delivering innovative process technologies to the global chemical industry." [BASF 2013]

The starting materials for the production of conventional petrochemical BDO are natural gas, butane, butadiene and propylene. BASF currently produces BDO and BDO-equivalents (Figure I.187) at its sites in Ludwigshafen, Germany; Geismar, Louisiana; Chiba, Japan; Kuantan, Malaysia; and Caojing, China, and has an annual capacity of 535,000 metric tons. BASF has recently announced the intention of building a BDO complex in China with a capacity of 100,000 annual metric tons [BASF 2013].

Biobased chemicals and plastics as a new wave is about to emerge. But, according to BioAmber's CEO Huc for entrepreneurship the wave will have a serious threshold for startups: "A lot of large chemical companies are still looking at bio and saying 'prove it.'" "In the sphere of succinic acid, if any of the early players fail it will undermine the credibility of the whole industry." [Taylor 2010]

> Obviously there is a "Catch 22" situation: Green startups need large capital and big companies waiting for proven startups.

While succinic acid addresses the "4-component" in Nylon derivatives adipic acid addresses the "6-component" which also brings up the "BASF-factor" for startups. BASF, the innovator of Nylon-6 in the 1930s from ε-caprolactam [Runge 2006:416-417] and the world's largest manufacturer of caprolactam, is generally a leading manufacturer of carboxylic acids. Its product portfolio reaches from monocarboxylic acids to dicarboxylic acids like adipic acid and fumaric acid to name the most important ones. Correspondingly BASF is one of the leading manufacturers of polyamide intermediates and polyamides with production sites all over the world. The BASF process for making adipic acd is by direct oxidation of cyclohexane using air only [ChemSystems 2010].

US startup Verdezyne, Inc. ("Green Chemistry by Design") is developing a yeast platform to optimize metabolic pathways, microorganisms and fermentation processes for the conversion of sugars to biofuels (bioethanol from C6 and C5 sugars) and biobased chemicals and plastics by proprietary metabolic pathway engineering tools. In particular, by the end of 2011 it opened its first pilot plant to produce adipic acid, the key component of Nylon-6,6. The company said "the plant will be used to demonstrate scalability of their process, validate their cost projections and generate sufficient quantities for commercial market development."

Verdezyne was founded in 2005 as CODA Genomics, a University of California at Irvine spin-out that used computational technology to design genes for the research world. In 2008 it transitioned from being a lead synthetic gene provider to pharmaceutical and industrial enzyme businesses to focus on fermentation pathway engineering for renewable fuels and chemicals. As expected it currently follows a "veterans approach" for its executive management with personnel having 25+ years of experience. Among investors in Verdezyne BP Alternative Energy Ventures and DSM Venturing BV are notable in this context.

CODA Genomics (Computationally Optimized DNA Assembly) was founded by members of UC Irvine's Intitute of Genomics and Bioinformatics as an LLC. In 2007/2008 CODA overhauled its core business strategy, recruited a new CEO, William Radany, along with a new management team, changed its name, and moved its headquarters from Orange County to Carlsbad, CA, near San Diego [Bigelow 2009c].

The business model emphasized 1) the company to look for having core expertise around developing a process and 2) validating out that process in a 10 liter laboratory scale fermentor. "After validation, it requires scale-up and that's where partnership with chemical companies has to come in." [De Guzman 2009]. Verdezyne seems to have kept its strong research orientation. Asked about its driver we read: "An exceptional R&D team." [Admin 2011g].

Foundation was actually by "stage-oriented entrepreneurs" as is Pamela Contag of Cobalt Biofuels (Table I.96). Both co-founders Rick Lathrop (Professor of computer sciences, then 51 years old) and Wes Hatfield (Professor of microbiology and molecular genetics, UCI School of Medicine. Professor of chemical engineering, UCI Samueli School of Engineering. Director, UCI Computational Biology Research Laboratory, then 65) had already founded other biotechnology firms. Hatfield's favorite quote that characterizes their stage-oriented entrepreneurship is: "Do good science and leave management to professionals." [Stewart 2006]

Originally, Verdezyne focused on fatty acid distillates or soapstocks from the oil seed processing industry (a by-product of soybean processing) [De Guzman 2009; Admin 2011g]. Verdezyne needed help with breaking down cellulosic (and hemi-cellulosic) materials into the sugars to tackle converting grass, straw, sugarcane stalks and other such tough plant material into chemicals. Interestingly in this context, one of BP's other big moves into biofuels came with its purchase of the lignocellulosic biofuels business of Verenium (Table I.84) [St. John 2011]. As investors in Verdezyne BP represents the bioethanol side, DSM is interested in the chemical side.

From the beginning of chemistry and chemical endeavors plant oils played a key role as a raw materials for the chemical and then also the cosmetics and nutrition industries. The surfactants (soaps!), detergents and oleochemicals industries existing since ages provide a strong bridge into biofuels (biodiesel) and a biobased chemical industry [Runge 2006:252-256, 563].

Relevant plant-based oils contain large varieties of fatty acids and their esters. Olefin metathesis technology has emerged as key for converting biobased oils to industrial chemicals, feedstock and consumer products [Runge:865-867]. Correspondingly, we encounter considerable entrepreneurial activities in this segment.

Through the acquisition of German firm Cognis the chemical giant BASF [Runge 2006:188] is now back-integrated to a large extent also in renewable oleochemicals. Similarly, US "food and feed giant" Cargill plays a key role here [Runge 2006:244].

Industrial Non-Food Sugars versus Petroleum as a Feedstock and Raw Material

Fundamental components of a business model of biobased chemicals oriented start-ups are producing and selling cellulosic industrial sugars and a synergistic back-end to proprietary chemical companies to produce high value products.

From an economic point of view cellulosic industrial sugars to be used for key intermediates to create the biochemical products and biofuels are a raw material *commodity* and, consequently, the business really is a cost game. And the competitive situation is envisioned as follows:
"In the end, the lowest cost providers will be the winners, and maybe a couple or three will be there. There won't be twenty." (Renmatix, B.2).

BASF is back-integrated in oil and gas as key raw materials for its intermediates. And BASF's $30 million investment in 2012 in US cellulosic sugar developer Renmatix (B.2) signals the German chemical firm intends to expand its feedstock sources and raw materials especially for its renewable chemicals and materials portfolio [De Guzman 2012b; BASF 2012a; Fehrenbacher 2012].

Renmatix (derived from **Ren**ewable **Mat**erials) claims itself to be the current lowest-cost producer of industrial sugars, the building blocks of renewable ("green") chemistry, utilizing non-edible biomass as feedstock. Mike Hamilton, the chief executive of Renmatix, said in an interview that the startup plans to build a facility by 2014 that will ship *sugar that can compete in cost with Brazil's sugarcane crop, the global benchmark for the commodity* [WOC 2012].

The end products in the two-step Plantrose™ process of Renmatix are C5 (xylose) and C6 (glucose) sugars, and optionally lignin. Basically, the process follows a patent protected "*supercritical fluid hydrolysis*" technology as well as patent-protected (supported liquid membranes, SLM) separation technology.

Supercritical fluids for use in processing biomass are used as mixtures. These include water, carbon dioxide and ethanol at selected temperature and pressure intervals, for instance, above the critical points for ethanol and carbon dioxide but at a temperature and/or pressure below that of the critical point for water, etc. Furthermore, the Nano Carbonic Solvothermal Technology (NCST) provides methods for generating micro- or nano-structured raw materials and performing biomass and particularly cellulose hydrolysis (Renmatix, B.2).

Apart from Renmatix there have emerged a number of pure-play sugar technology developers and manufacturers, such as Bluefire's wholly owned subsidiary SucreSource (Table I.86) and Virdia (formerly HCL CleanTech) (Table I.99) which changed its emphasis away from biofuels. Also Sweetwater Energy launched in 2006 in the US under the name SweetWater Ethanol, LLC, belongs to that category. It intends to advance a decentralized business model it developed to allow farmers to produce ethanol from crops right on their farms.

On the other hand, London, Ontario-based Comet Biorefining Inc., founded in 2009, produces its cellulosic sugar as syrup which has high glucose concentration. The Comet cellulosic sugar process uses a two stage process to activate cellulosic biomass, followed by conversion to glucose at very low enzyme loading. Co-products are used for energy production [Sims 2012]. It has demonstrated its cellulosic sugar technology at pilot scale and is currently scaling up to commercial applications. However, it was not saying much about the company's technology. What it said is that the firm's process uses fewer enzymes to break down biomass than in competing processes (Renmatix, B.2).

Having sugars next to come is the conversion of these into various products. In this line BASF announced by mid of 2012 a collaboration with BioTork, LLC of Gainesville, FL. According to its Web site after six months of a pilot study, both firms are going into a combination of their complementary approaches to strain development to improve the efficiency and resulting economics of biochemical production processes.

Created in 2008, BioTork LLC is a biotechnology company developing certain microbial strains for the industrial production of biobased polymers and green chemicals. BASF has been conducting intensive research on the use of microorganisms for the production of proteins, enzymes, vitamins and other high value and low cost chemicals.

> In their natural environment, microorganisms generally synthesize these chemicals only to meet their own requirements for survival. The challenge faced by chemical companies is to push these microorganisms to produce these chemicals faster, in much larger quantities, and under industrial conditions that are different from the microorganisms' natural environment. This is the only way to use microorganisms for commercially viable production of chemical products.

In addition in 2010 BASF and Solix Biofuels (now Solix BioSystems) (A.1.1.4) started a collaboration demonstrating the BASF commitment to generate growth from industrial biotechnology and algae representing an addition to BASF's technology portfolio as they offer the potential to produce a number of *specialty chemicals* and products.

Apart from "sugar entrepreneurs" and existing firms focusing on bioplastics and chemical products firms are also using 3-hydroxypropionic acid (3-HP) as a chemical intermediate platform to produce acrylic acid (and esters), acrylamide and acetonitrile, 1,3-propanediol, and malonic acid esters [Runge 2006:248-249]. Acrylic acid is a key chemical building block used in a wide range of consumer goods including paints, coatings, adhesives, diapers and detergents.

At first, in the US Codexis and food and feed giant Cargill cooperated and announced a breakthrough in developing a novel microbial process that will convert corn sugar to 3-HP [Runge 2006:248-249]. Currently Cargill and its Danish partner Novozymes are assumed to be planning to release their technology available for licensing within the next couple of years [De Guzman 2012c].

By August 2012 BASF joined the Cargill and Novozymes cooperation to develop the process for conversion of 3-HP into acrylic acid. All three firms have signed an agreement to develop technologies to produce acrylic acid from renewable raw materials. Presently, acrylic acid is produced by the oxidation of propylene derived from the refining of crude oil. BASF is the world's largest producer of acrylic acid and has substantial capabilities in its production and downstream processing. BASF plans initially to use the bio-based acrylic acid to manufacture superabsorbent polymers [BASF 2012b].

The three companies bring complementary knowledge to the project. Novozymes is the world-leader in industrial enzymes. BASF and Cargill are global leaders in their industries. Together this trio is uniquely positioned.

The large French specialty chemicals firm Arkema will look at the direct conversion of glycerin to acrylic acid, as well as the conversion of glycerin to acrolein (propanal) and use of conventional technology to oxidize acrolein to bioacrylic acid. The process is not new; the Japanese firm Nippon Shokubai has developed catalysts for conversion of glycerin to acrylic acid.

Arkema is the third-largest player in the huge global acrylic acid market after number one BASF and Dow Chemical as number two. As early as 2004, Arkema had been working on a method to make acrylic acid from renewable resources. Hence, startups in the field have to encounter heavy competition in the bioacrylics area with petroleum-based acrylics [Reisch 2010].

Dow has a rather small biotechnology portfolio compared with BASF and DuPont, but they were among the first to invest in bioplastics via the firm NatureWorks, the JV with Cargill. But Dow sold its stake at a time that the 100,000 tons plant was almost idle in 2005 [Runge 2006:129-130,245].

In 2011 Dow and startup OPX Biotechnologies, Inc. (OPXBIO) signed a joint development agreement to prove the technical and economic viability of an industrial-scale process to produce acrylic acid using a fermentable sugar feedstock with equal performance qualities as petroleum-based acrylic acid, creating a direct replacement option for the market. If collaborative research is successful, the companies will discuss commercialization opportunities that could bring biobased acrylic acid to market in three to five years [OPXBio 2011]. In 2011 OPXBIO raised $36.5 million in an equity financing round C. This added to a total of more than $53 million OPXBIO raised so far with venture investtors.

Dow is focusing on the use of sugar feedstock and the conversion process of sugar to bioacrylic acid while OPXBIO is focusing on its microbe using its “Efficiency Directed Genome Engineering” (EDGE™) platform, as well as developing the 3-HP bioprocess. OPXBio uses microorganism to biosynthesize 3-HP by fermentation of sugar and subsequent dehydration of the 3-HP to acrylic acid. Both Dow and OPXBIO will jointly fund the development, demonstration and commercialization of bioacrylic acid. OPXBIO's rebuttal against the Cargill technology is that OPXBIO claims to have a lower-cost biobased route (and also the competitive to petroleum-based route) [De Guzman 2012c].

OPXBIO was founded in 2006/2007 and follows currently a “veterans” approach for its management team. For instance, Charles R. (Chas) Eggert, the current President and CEO, has more than 30 years of experience in the global specialty chemical industry. He began his career with Monsanto Company, progressing through roles in technology, manufacturing, business development, marketing and general management.

Michael D. (Mike) Lynch, MD, PhD – the Chief Scientific Officer and Co-Founder – is still on board. He received an AB in anthropology as well as a BS and MA in biomedical engineering from Washington University in St. Louis, followed by a PhD in chemical and biological engineering from the University of Colorado in 2005 and an MD from the University of Colorado Health Sciences Center in 2007. He has nearly a decade of research experience in the life sciences, including the fields of molecular biology, protein thermodynamics, microbiology, and metabolic engineering, and is the primary inventor behind OPXBIO's platform technologies.

In 2006, Lynch started OPX Biotechnologies, and successfully raised three rounds of financing after demonstrating the concept's viability. In 2009 the company had more than 30 employees. After launching with $1 million in seed capital in 2006, OPX Biotechnologies closed funding rounds in October 2007 and April 2008 of $1.3 million and $2.6 million, respectively, before closing on an impressive $17.5 million round in 2009 [INITIAL LIGHT BULB 2009].

OPXBIO's goal is to compete with petroleum-based chemicals and fuels on both quality and price. OPXBIO has developed and piloted the microbe and bioprocess that will produce its first renewable chemical product, BioAcrylic. In 2011, based on pilot-scale development, it announced that it has achieved the commercial bioprocess performance and cost goals for BioAcrylic. The company plans to diversify its product portfolio into the fatty acid and acrylamide sectors. But both products are still in the early phase of development [De Guzman 2012c].

OPXBIO has scaled up its bioacrylic acid production to 3,000 liter fermentation (equivalent to 60,000 lbs/year) at the demonstration plant in Lansing, Michigan, owned by MBI, a non-profit organization owned by Michigan State University (MSU) Foundation. OPXBIO said the company plans to have a second demonstration plant with a capacity of 600,000 lbs/year in 2013. A commercial plant with a capacity of 100 million lbs/year is expected by 2015 [De Guzman 2012c].

Like in previous market hypes (US) *investors seemed to be ready to invest in almost any biotech company provide they follow the mainstream which this time is the "drop-in" technology.*

But a recent Wall Street Journal article (of September 2013) referring to OPX entitled "Biotech Firm Tests Investors' Patience" emphasized it may be a test case of private investors' appetite for risk with investing in related startups. Specifically, OPX "Chief Executive Charles R. Eggert has said the company doesn't expect to generate any commercial revenue until at least 2017. That is hard math for some investors."

A.1.2 William Henry Perkin and Industry Genesis in the Last Third of the Nineteenth Century

In Runge [2006:397-402] the Berlin (Prussian) Blue innovation from 1704 and the related birth of inorganic chemistry has been mapped against current notions, concepts and thinking of innovation and entrepreneurship. Furthermore, in the same way key aspects of William Henry Perkin's synthetic dye innovation in 1856 in the UK by serendipity has been roughly mapped to current concepts of radical innovation [Runge 2006:295]. Therefore, it is of interest to inquire in more details into the entrepreneurial aspects of Perkin's dye innovation (Figure I.87).

There is a myriad of literature on the synthesis of the iconic dye mauveine by (Sir) William Henry Perkin as a major landmark in the history of science and technology, as it led to the establishment of the synthetic dye industry and further development of organic chemistry. But, apart from the emphasis on history of industry, Perkin and the dye industry provide a wealth of generic features of entrepreneurship in the environment of entrepreneurial capitalism. Furthermore, it can be viewed as an example for one of the first research-based startups (RBSUs) in a broad sense. Actually Perkin was a "university drop-out."

In Table I.100 the combined entrepreneurship and innovation concepts are displayed focusing on industry genesis [Runge 2006:266- 269, 274-276; 293-296; Ball 2006] [114].

Table I.100: Current entrepreneurship and innovation concepts and processes reflected already during the middle of the nineteenth century by William Henry Perkin.

Scientific and Socio-Economic Context	There was a scientific vision of the famous German Justus von Liebig in line with societal attitudes and convictions. We "believe that tomorrow or the day after tomorrow someone will discover a process ... to make the wonderful dye of *madder* or helpful *quinine* or morphine from coal tar." [Runge 2006:293]

Table I.100, continued.

Idea and Serendipity	Following Liebig's vision August Wilhelm Hofmann at the Royal College of Chemistry based in London targeted the demand of the market to have an antimalarial drug (quinine). Ball [2006] describes the chemical rationales at that time for selecting a specific route to synthesizing quinine. William Henry Perkin, a student of A. W. Hofmann of 18 years, had been directed to make the anti-malarial drug quinine based on material from coal tar by an oxidation route. Initial trials with an envisioned starting material failed, but when using the coal tar product aniline things changed dramatically. "The resulting black sludge dissolved in methylated spirits, and the resulting solution was a beautiful purple."
Revealing the Opportunity	Perkin "stumbled" over an unexpected result, serendipity. Silk dipped in this solution took on the same royal hue. Perkin grasped that his purple solution could be used to color fabric. And Perkin took the recklessly bold move of quitting his studies to exploit the opportunity.
Opportunity Evaluation Early Assessment by a Potential Customer	Perkin realized that this coloring matter had the properties of a dye and resisted the action of light very well thus making it the world's first synthetic dye. He quickly grasped that his purple solution could be used broadly to color fabric. Perkin *changed the project direction* – he wanted to exploit the *first* synthetic organic dyestuff based on *abundantly available feedstock.* He sent some specimens of dyed silk to a dyeing firm in Perth, Scotland, which expressed great interest provided that the cost of the cloth would not be raised unduly. "It was to his credit, and luck, that he sought out the advice of Robert Pullar, the owner of a well-regarded dye works in Scotland. Pullar encouraged the eighteen year old Perkin to manufacture more dye, and told him that if the dyed fabric would remain colorfast and not fade in the sun, Perkin would be a very wealthy man." [Nelson 2002]
Securing Intellectual Property Rights	Referring to this situation, Perkin filed for a patent in August 1856, while he was still only 18.
Venture Financing and Formation	*Against* Hofmann's recommendation Perkin believed in his business idea and convinced his father to invest in his idea and borrowed his father's life savings. With the help of his father and brother (*3F financing*), Perkin set up

	a dye factory in 1857 (Perkin & Sons) at Greenford Green, near Harrow, for mass production of the first synthetic organic dye – mauveine – on a six-acre site near the Grand Union Canal, not far from London. The *location* was selected for cheap transportation of coal-tar from London, the by-product for the emerging gas lighting infrastructure.
Commercialization Issues	Inventing the dye was one thing, raising enough capital for manufacturing the dye in quantity cheaply, adapting it to cotton, getting acceptance from commercial dyers, and creating demand for it in the public was something else. Perkin was active in all of these areas. In a whirlwind of activity, he got his father to put up the capital, his brothers to partner in the creation of a factory, (as a response to Robert Pullar's remark) he *invented a mordant* (a pre-dyeing treatment) for cotton, became a one man technical service operation, and publicized it in the marketplace.
Production; New-to-the-World Products	Utilizing the cheap and plentiful coal tar that was an almost unlimited by-product of London's gas street lighting the dye works began producing the world's first synthetically dyed material in 1857. Initially there were *difficulties*. Since aniline was not readily available, it had to be produced at the factory from benzene. Manufacturing of synthetic dyestuff also revealed large *needs for inorganic basic chemicals*, such as sulfuric acid and alkalis, caustic soda, lime and soda ash.
Unexpected Market Success Stepwise Market Entry via Customer Segments	Already historic in its very founding, the company received an unexpected commercial boost from the Empress Eugenie of France when she decided the new color flattered her. In short order, mauve was the necessary shade for all the fashionable ladies of France. The product *met immediately a market with high purchasing power* based on fashion and "life style," and only *later expanded into the large end-user markets* via the textile industry.

Table I.100, continued.

Competition Hyped Industry Genesis	Ten years after Perkin's discovery of (synthetic) mauve organic chemistry was perceived as being exciting, profitable, and of great practical use. Many other dyes and new firms followed, and Perkin & Sons was soon facing stiff competition from manufacturers in England and France. By the mid-1860s, the German (later giant) companies Bayer, Hoechst and BASF were already in business making dyes, as were Ciba and Geigy in Switzerland [Runge 2006:266-269] (and Figure II.20 in Runge [2006:275]). *Britain and France dominated the dyestuff industry till ca. 1870* to then encounter a dramatic decline through the new players from Germany. The important lesson learned is that the companies and industry of the country the innovation originated in do not necessarily win. A similar situation occurred currently: First the US leadership in the photovoltaic (solar cells) industry was overtaken by Germany, and currently both the US and Germany are behind China – and China having overcome both previous leaders. Perkin discovered and marketed also other synthetic dyes.
Profits and Harvesting	Over the next few years, Perkin found his research and development efforts increasingly eclipsed by the German chemical industry, and in 1874, he sold his factory and retired from business, already a very wealthy man at the age of 35. He devoted the rest of his life to research in pure science. For instance, he became particularly interested in Faraday rotation [115] and produced over 40 papers on this topic. After Perkin's retirement from the industry he remained active in his field in other ways, such as being secretary of the Royal Chemical Society in 1869 and he became president in 1883. He also sat on the boards of several scientific journals.
Multidimensional Innovation Success:	*Industrial*: created and/or stimulated new industries (the organic color industry of coal tar dyes and pharmaceuticals); *Scientific*: stimulated organic chemistry and the search for a better understanding of the structure of molecules

It should be noted that malaria is still a very serious problem. The search for an anti-malaria drug is currently repeated by Amyris Biotechnologies (Table I.99).

By 1914 Germany dominated the world of synthetic dyes by ca. 85 percent. As described by Runge [2006:266-269] Murmann [2003] has attributed the exorbitant rise and superiority of German dye companies largely to a *co-evolutionary*, self-reinforcing development of *higher education, university research – the S&T system – and the industry system* – by strong *political involvement* outlined in detail by Streb [1999].

However, Streb [1999] made some important additions to the co-evolutionary explanation pointing out some important other external drivers and an innovative marketing and sales strategy of the German firms. On the one hand, it was both decreasing imports of natural dyes during the German-French War of 1870/71 and government's demand for dyed tunics which accelerated the innovation of coal tar dyes in Germany (cf. also the. Prussian/Berlin Blue innovation and its role for the Prussian army).

On the other hand, it is little noticed that the German producers of coal tar dyes also owed their success to the two new *marketing* strategies "*customer consulting service*" and "*customer training*." Furthermore, the German chemical firms of the late 19th century had both technological and economic innovation capital. And innovation of coal tar dyes occurred in cooperation with the textile industry in the second half of the 19th century [Streb 1999].

According to Streb [1999] German chemical firms, such as BASF, established so-called "Coloristische Abteilungen" ("Dyes Departments") to *generate new markets for product innovations* in the field of coal tar dyes. These were responsible for implementing new marketing strategies in the textile industry. Therefore, the Coloristische Abteilungen were affiliated with the industrial research laboratories and *filled with both commercial and technical staff.*

Chemists provided *customer consulting service* who did not only know the special characteristics and performance of new coal tar dyes but were also trained to demonstrate how to dye and print textiles. These chemists did understand the problems and the "language" of textile producers. *Before sales* they explained textile producers how to apply the new coal tar dyes in production plants and also provided technical help in cases of actual processing problems ("*after sales service*").

Around 1900 the Coloristische Abteilungen developed *customer training*. The Coloristische Abteilungen used customer consulting service and customer training to gain textile firms as long-term buyers by technological knowledge transfer. They taught employees of textile firms how to handle the latest techniques of dying and printing. Of course, they informed the trainees only about their own products inducing preferences.

Obviously their one-year training was advantageous for textile firms. The chemical firms on their part won the *loyalty of future customers*. In our times firms these marketing strategies are recalled when playing the game of technological cooperation for various areas by "Technical Service Centers."

"The enormous expansion of universities, technical universities (in German Technische Hochschulen) as well as the many research institutes formed after the 1880s was orchestrated from the desk of Friedrich Althoff, who all handled professorial appointments at Prussian universities and Technische Hochschulen, between 1882 and 1907, serving under five successive ministers.
Because he shared the vision that broad scientific and technological research and education would be of immense benefit to society, he was a key ally in the efforts of the dye industry to expand educational facilities. Furthermore, given his unique control over the direction of the Prussian university system and Prussia's trend-setting role for other German states, the dye industry would have to form an alliance with him if they wanted to be successful at all during his long tenure. The German dye industry employed three strategies to upgrade its supply of scientific and engineering talent that could staff its firms: 1) use collective organizations to mobilize support, 2) lobby parliament directly, 3) create private-public academic partnerships." [Streb 1999]

There are also mentioned some more factors contributing to the exorbitant rise of the German synthetic dye industry. The development of a very broad variety of dyes could be based on many *platform technologies* (chemically different types of dyes) [Runge 2006:268].

According to the patent law of 1877 the chemical industry could only protect processes in Germany, contrary to the situation in France or the UK where also products (substances) could be protected by a patent. Therefore during the acquis the new law had less protection impact than expected by the firms. And it is argued that, as no one could afford to rest on its monopoly, the German patent law "outright drove the companies into innovation competition." [Hoffritz 2013]

A.1.3 Structures and Issues of Current University-Industry Relationships

Emphasizing technology transfer (ch. 1.2.6.3) there is a broad diversity of approaches and models for university- and public research institute-industry relationships. Diversity does not only result from the involved units of two partners, but also from the number of different partners and the more or less active role policy may play. There is a pentuple of partners: industry firms (from medium-sized to giant), universities of different types, national laboratories or research centers, other public research organizations and policy. Usually, the relationships can be characterized as *project-like* goal- and time-related endeavors, which are rather firm-specific concerning preferences.

> Persons working in particular university-industry or public research institute-industry organizations may take the experience they gained here (Figure I.64) as a springboard for technology entrepreneurship.

Examples in the US for biofuels of such relationships with "Big Oil" companies are described above (A.1.1; Table I.83). For instance, chemical giant DuPont got $9.0

million for a partnership with the biofuel startup Bio Architecture Lab from the US Department of Energy's Advanced Research Projects Agency-Energy (ARPA-E). Chevron had a biofuels development agreement with the National Renewable Energy Laboratory (NREL) and also an agreement with Texas A&M University (four-year period). It formed research arrangements with Georgia Tech, the University of California, Davis and the Colorado Center for Biorefining & Biofuels, which is a consortium of the Energy Department's NREL, three major Colorado universities and other private companies (Table I.83).

Royal Dutch Shell established a joint venture Cellana LLC in 2007 with HR BioPetroleum to build a pilot facility for growing marine algae and producing algal oils that can, in turn, be used to make biofuels. HR BioPetroleum Inc., headquartered in the State of Hawaii, is a developer of large-scale microalgae production technology. It is a University of Hawaii, School of Ocean and Earth Science and Technology based company. Additionally, an academic research program would support the project. The program would include scientists from the Universities of Hawaii, Southern Mississippi and Dalhousie, in Nova Scotia, Canada; with professional management (Table I.89).

Industry calls for proposals for collaborative research networks (like US firm HP); the German government may call for development projects across the value system (in Germany "Verbundprojekte" – "joint projects").

For technology entrepreneurship university-industry relationships based on exchange of R&D personnel or sharing R&D personnel in a dedicated organizational unit, such as a laboratory or a firm (Figure I.41), can play an important basis *for potential entrepreneurs to gain experience and first insights into the business world* (ch. 2.1.2.4), but also *may reveal opportunities* for to-be entrepreneurs. Generally, the personality- or personnel-oriented R&D-industry relationship types are as follows.

1. Sponsorship of a professorship or a network of professors for given research or technology directions by industry
2. Exchanging research personnel in both directions
3. Sharing personnel in joint firms (or joint projects)
4. Sharing research personnel at a firm's site (laboratories) or on the campus.

More in the US than in Germany sponsorship of individual university professorships mostly refer to sponsorship by (wealthy) individuals or foundations. But sponsoring professorship or research units involving several professors mostly from different disciplines has become also of interest to industry as a mode for creating knowledge and technology transfer.

For instance, BP provided $500 million over ten years to establish a dedicated biosciences energy research laboratory attached to a major academic center. BP funding for the Energy Biosciences Institute (EBI) is led by Berkeley and includes additionally Lawrence Berkeley National Laboratory and the University of Illinois at Urbana-Champaign (Table I.83).

German firms have historically had very close ties to academia (Box I.3; A.1.2). For the German industry it is quite common that people from industry hold teaching positions or professorships at universities. Current university-industry-relationships do not only follow common tracks, but, often with support from policy, try new approaches. For instance, the German Karlsruhe Institute of Technology (KIT, Table I.20) established the concept of "*Shared Professorship.*" The concept is based either on a 1:1 research institute-industry relationship or a 1:many relationship.

In one case capable young scientists are given the opportunity to gain experience in research and industry (basically for a period of four years) in order to facilitate their later decision in favor of a university or an industry career. The model for a shared professor means working half of the time period at industry and the other half of the time at the KIT. This close-to-industry professorship is to enhance permeability between the KIT and industry by a *talent transfer in both directions* – with benefits for both partners, the KIT and industry [KIT 2008].

In another case [KIT 2009] the shared professorship involves several enterprises as industry partners. For instance, a cooperation designed for a period of five years will comprise KIT, Bayer Technology Services, BASF SE, and Roche Diagnostics, each industrial partner contributing a quarter of the funds – apart from a professorship – the setup of an institute-overlapping thin-film-technology platform [KIT 2009].

Thin Film Technology (TFT) deals with the setup and properties of thin layers and the devices and process technology required for their production. The thickness of the layers varies between a few micrometers and a few nanometers. A particularly promising new market is Organic Electronics with organic photovoltaics. In this field, TFT mainly focuses on polymer solar cells and hybrid solar cells, which means, on polymer solar cells with inorganic nanoparticles. Other projects cover medical diagnosis test strips, coatings and varnishes as well as functional thin layers and structures for thin-film batteries and optical foils.

A related approach to intensify industry and science and universities relationships was followed by German Henkel AG & Co. KGaA. In 1998 Henkel and "to-be-professors" and researchers from a university worked in dedicated projects for a restricted period in the industrial research environment [Runge 2006:689-690]. A specific description involving the chemical industry in Germany is described by Runge [2006:687-692].

A very special situation of technology transfer in the context of entrepreneurship is the private-public-partnership (PPP) firm founded by a university and a large company. These firms do not only do research, but develop marketable products. Here, a legal entity is set up between the two partners.

Henkel AG & Co. KGaA set up a biotechnology and cell physiology PPP-firm Phenion GmbH & Co. KG with the Johann Wolfgang Goethe-Universität in Frankfurt am Main and merged it with Henkel AG & Co. KGaA effective January, 1 2009. Furthermore, it set up a new materials firm Sustech GmbH & Co. KG on the campus of the Technical University of Darmstadt [Schweinberg 2007].

The research company SusTech Darmstadt with an appropriate legal form was established by Henkel, the Technical University of Darmstadt, and five professors from different disciplines and different universities – with Henkel having a majority stake in it and concentrating on management and commercialization (Table I.101). The start phase of the PPP-firm was supported by the German Federal Ministry of Education and Research (BMBF).

Table I.101: Organizational setup of the PPP-firm Sustech GmbH & Co. KG.

Henkel Henkel AG & Co. KGaA	**Technical University of Darmstadt**	**Five Professors from Various Organizations**
60%	10%	30%
Finances Utilization, Exploitation Management Legal and Patents Chairman of Advisory Board: CTO of Henkel	Offices, Laboratories, Infrastructure Foundation: 2000 Acquired by Henkel: Sep. 2008 Employees: 16 (Sep. 2008) 20 (2006)	Colloids, Emulsions (U Saarbrücken). Polymers, Surfaces (U Aachen, RWTH), Biomineralization (MPI Desden), Particular Systems (HGF FZK Karlsruhe, now KIT), Modeling (TU Darmstadt)

When it started Sustech had an international team of 30 scientists that should develop new materials, systems and products. It was set up to enable the fast conversion of innovative ideas into economically usable products and processes. The emphasis was on utilizing the practical potential of nanoparticles to tackle the widespread problem of sensitive teeth. Founded in 2000 Systech was acquired by Henkel in Spetember 2008.

A similar PPP-approach is also found in the UK, for instance, for ionic liquids technology Scionix Ltd. Scionix is set up as a joint venture between the University of Leicester and Genacys Ltd., a wholly owned subsidiary of the Whyte Group Ltd., Britain's largest privately owned chemical company (A.1.5 and Scionix in Bioniqs Ltd. – B.2).

Chemical giant BASF follows basically two approaches for sharing research personnel, at a firm's site (laboratories) or on the campus.

Combining the creative freedom and rapid exchange of ideas unique to the academic environment with the resources of a giant company, BASF wanted to create a new paradigm for productive academic-industrial research, with all the benefits of both worlds. The concept relies on bringing academic and industrial researchers physically as close as possible together in one laboratory.

In 2003 it opened a laboratory at the Institut de Science et d' Ingenierie Supramoleculaires (ISIS), Louis Pasteur University, Strasbourg (France). The ISIS group was headed by a researcher from BASF. Along with BASF's own expertise with nanostructured polymeric materials, a multidisciplinary international team of postdoctoral researchers provided a wide range of scientific backgrounds, from supramolecular complexes to polymer/layered silicate nanocomposites, to sol-gel condensation of highly porous silica networks, to the design and use of automated reactor systems for high-output polymer chemistry [Runge 2006:691-692].

The BASF lab at Louis Pasteur University in Strasbourg (ISIS) is specialized in supramolecular chemistry, developing synthesis pathways for synthetic foams with nanometer-scale pore sizes. These nanopores prevent cell gas molecule collisions, and in this way reduce heat conduction in the foam to less than half of that observed with conventional materials. The nanofoam is designed as an insulating material for refrigerators, buildings, cars and even planes. It will reduce energy consumption and save materials, thus benefiting the environment.

Furthermore, since 2006 BASF and Heidelberg University run a Catalysis Research Lab (CaRLa) [116] led by BASF in the Technologiepark Heidelberg (Technology Park Heidelberg) devoted to homogeneous catalysis. The laboratory was funded by both partners and by the State Government of Baden-Württemberg. Furthermore, the Chemistry Department of the University of Heidelberg has a catalysis research area specially funded by the German Research Foundation DFG ("Sonderforschungsbereich," SFB 623 (*Collaborative Research Centre)*: Molecular Catalysts: Structure and Functional Design).

CaRLa is led by a Steering Committee with representatives of the University of Heidelberg and BASF. It has six postdocs from the University and funded by the University. Also six postdocs and a senior researcher from BASF as managing head of the laboratory are financed by BASF. The laboratory's proximity to Heidelberg University as well as to BASF's global Research Verbund with its Ludwigshafen based Research Headquarters offers ideal conditions for outstanding catalysis research and for a swift transfer of technology. The joint laboratory has become a prime research location attracting catalysis researchers from around the world.

Another approach of BASF concerning innovation and technology transfer is the company InnovationLab GmbH (iL) [117] inaugurated in September 2006, a "Joint Innovation Lab" (JIL). Here, BASF experts are collaborating with partners from industry and academia on materials and device structures from the field of organic electronics. The researchers at the JIL are currently focusing on the areas "Organic Light Emitting Diodes" (OLEDs) and "Organic Photovoltaics" (OPV) and also fuel cells.

Referring to the photovoltaics value system (Figure I.11, Figure I.12) iL reflects an innovation strategy of BASF turning to a higher level in the value system. It is funded equally by science and industry. It is conceptually embedded in the so-called "Top

Cluster Organic Electronics" including printed electronics supported and awarded as an excellent cooperation network by the Federal Ministry of Education and Research (BMBF) comprising twenty-five firms, universities and research institutes working in the field on organic electronics.

The Universities of Heidelberg and Mannheim (near Heidelberg) have 40 percent and 10 percent stakes, respectively, and industry hold the other with 8,33 percent each for the chemical firms BASF SE, Merck KGaA and Freudenberg & Co., Heidelberger Druckmaschinen AG (the printing machine giant), Roche Diagnostics GmbH und software giant SAP AG. The initiative is part of the German Federal Government's high-tech strategy and the OLED initiative is one of it.

The business model of iL aims to establish research platforms for various key technologies. Organic Electronics represents a first step. Each platform has three central elements: applied research, supporting young talents and services, such as acquisition of public funds. The activities of each platform are controlled and managed by independent management.

If researching partners have no interest in commercializing inventions of the research platforms of iL the inventions will be assessed by the unit "Transfer and Incubation" with regard to opportunities and, in case, they will be further developed to reach a market within a "virtual enterprise." iL will not go for commercialization within its own structure. Commercialization is intended by either bringing the project back to one of the research partners or by a spin-off as a new firm and thus iL may act as an incubator.

When looking for "New Instruments for Promoting Innovation," the origin of the "Top Cluster Organic Electronics" was the German Ministry of Education & Research (BMBF) which established a competition for high-level technology clusters. But it did not specify categories or established preconceived boundaries as to the kinds of technologies that were eligible. In 2007, it published a call for entries in the first of a total of three rounds of the German Top Technology Cluster competition. The prize was roughly $260 million for a maximum of five clusters. Out of thirty-eight applicants twelve finalists were eventually invited to make a 10-minute presentation to the prize jury.

Two of the five winners were from the Heidelberg – Rhine-Neckar Rivers (RN) region (Figure I.50). A tight focus helped winning the prize. [Short 2009a]. The organic electronics cluster – Forum Organic Electronics – is managed by InnovationLab GmbH (iL) with a primary focus on products such as luminous wallpaper and solar absorption coatings for energy-efficient heating and cooling. [Short 2009a]. The other cluster BioRN emphasizes biotechnology [Short 2009b].

The heavy focus of BASF on worldwide research cooperations is illustrated in Figure I.188. As similar structures are found also for other German large to giant industrial firms it stresses the concept of a "networked economy" for the German innovation system (Figure I.51).

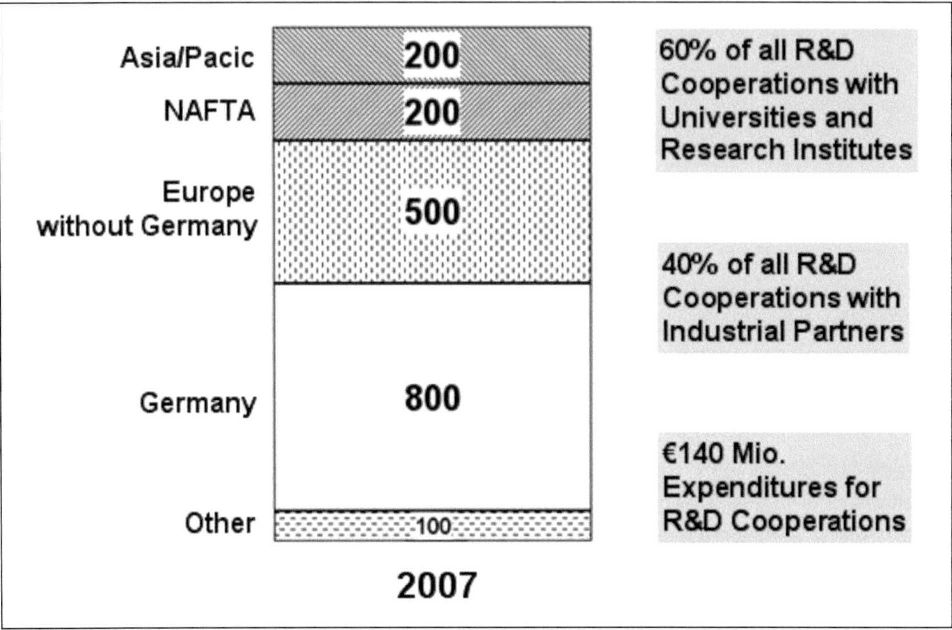

Figure I.188: Quantifying the BASF "Knowledge (now Science) Verbund": 1,800 R&D Cooperations [Jahn 2008].

For German university-industry relationships and technology transfer based on sharing personnel more concepts have been reported, such as "project houses" [Runge 2006:556-557, 575-576] or "science-to-business-centers" [Runge 2006:575; Dröscher 2008].

Moreover, for the chemical industry, in particular, a joint program "Academia-Industry-Exchange (in German Akademia-Industrie-Austausch – AIA) has been set up by the German Chemical Society and the (German) Chemical Industry Association to bring (for two to eight weeks) academics into industrial R&D laboratories or industrial researchers into laboratories of universities or public research institutes. Experiences of an academic in a so-called Innovation Concept Lab of German Merck KGaA in Cambridge (Mass.) have been reported recently [Schneider 2010].

PPP structures similar to those in Germany are also found in the US. The broad areas of industry-academia alliances concern essentially *"entry" into new or emerging technologies*, for the chemical industry, for instance, ionic liquids, nanotechnology, "white and green biotechnology" and biobased or green chemistry or co-evolutionary areas, such as electronic chemicals.

For instance, DuPont had intense cooperative agreements with MIT which started in 2000 with about 30 specific research programs in biotechnology. Each research project at MIT was assigned a DuPont Liaison and projects were regularly reviewed at DuPont. The DuPont MIT Alliance (DMA) was renewed in 2005.

Originally as a five-year $35 million investment, the alliance should receive another $25 million from DuPont to continue funding through 2010. This 10-year, $60 million commitment made the DMA the largest corporate R&D investment at MIT. In the second stage, the alliance planned to expand beyond biobased science to work with nanocomposites, nanoelectronic materials, alternative energy technologies and next-generation safety and protection materials [Runge 2006:690, 691].

Furthermore, DMA also provided an opportunity for DuPont to collaborate with MIT's Sloan School of Management to define *new business models for these emerging technologies*. Another aspect of DMA concerned idea generation. Since its inception, the DuPont MIT Alliance has also asked for *proposals* from the MIT community that draw upon the science, engineering and business expertise at MIT to extend DuPont's reach in the areas of biology, genetics, bioinformatics and catalysis .

Another way for academia-industry relationships targeting compelling topics is used by the US firm HP. Its Innovation Research Program (IRP) [118] is administered by the HP Labs Open Innovation Office, which is responsible for enabling strategic collaborations with academia, the government and the commercial sector to produce mutually beneficial, high-impact research.

HP Labs' IRP is designed to create opportunities at colleges, universities and research institutes around the world for collaborative research with HP. It offers awards in the range of $50,000 to $75,000. Each year IRP sends out open calls for proposals. It is designed to *create opportunities for breakthrough collaborative research with HP*. Proposals will be judged on their potential scientific and societal impact, as well as the caliber of the principal researchers, the availability of matching funds for the project and the quality of the proposed research plan.

Discussing issues of university-industry relationships we can refer again to BASF. Fundamentally, industry at the moment might benefit from getting something into the market quickly, but it also might destroy the free-ranging activity of the professor(s). It is the question over the degree of influence the industry partner firm could exercise over projects chosen. While not opposed to university-industry partnerships to address technical challenges such partnerships might compromise researchers' independence and commitment to the public interest, such as a "green" environment, *unless the program had a clear organizational structure* [Reisch 2007a].

In the US in recent times the partnership between industry and universities has been weakened over difficulties associated with negotiating IP rights in research contracts. Largely as a result of the lack of federal funding for research, American universities have become extremely aggressive in their attempts to raise funding from large corporations. But industry feels that it takes too much time, effort, and money to negotiate an agreement.

Typically at present, negotiating a contract to perform collaborative research with an American university takes one to two years of exchanging emails by attorneys, punctuated by long telephone conference calls involving the scientists who wish to work together. All too often, the company spends more on attorneys' fees than the value of the contract being negotiated. This situation has driven many large companies away from working with American universities altogether, and they are looking for alternate research partners [Johnson 2005; Reisch 2007b].

In 2007 BASF and Harvard University's Office of Technology Investment agreed to form the BASF Advanced Research Initiative. With $20 million from BASF, the five-year program would initially support 10 postdoctoral students and other Harvard researchers, primarily in the School of Engineering & Applied Sciences. However, the initiative would also draw on a network of faculty and students in labs throughout Harvard.

Under the agreement, BASF will have the opportunity to further develop discoveries and innovations. But Harvard faculty investigators reserved the right to distribute and publish any discoveries from the initiative. Although it involves two high-profile names, the announcement of this program set off no obvious alarms, perhaps because it had many of the hallmarks of traditional university-industry research initiatives and involved no debate over public policy. Its focus was on research leading to new products. BASF would decide on the projects it will fund and had pledged to work with Harvard on applying fundamental research to new product development [Reisch 2007a; Reisch 2007b].

A.1.4 Foundation and Development of SAP AG in Germany

German SAP AG is the largest software enterprise in Europe and the fourth largest software enterprise by revenues in the world as of 2009 (behind Microsoft, IBM and Oracle). Its revenues over the last years amounted to €11.6 billion in 2008 (€10.6 billion in 2009 during the Great Recession and €12.5 billion in 2010).

The company now is best known for its SAP Enterprise Resource Planning (SAP ERP) software covering accounting, controlling, distribution, purchasing, production, storage and inventory, and human resources. Describing the foundation and development of SAP AG rely on selected literature [SAP-1; SAP-2; Nonnast 2006, Anonymus 2004; Breuer 2009] with information relevant for entrepreneurship.

In April 1972 five computer experts employed at the IBM office in Mannheim (Germany) founded their own firm with the simple name "Systemanalyse und Programmentwicklung" SAP ("*Systems Analysis & Program Development* in Data Processing"), using the simplest legal form (GbR) for a firm in Germany. They rented rooms in a building in Mannheim close to the IBM office. And as they told, these rooms were usually left empty, occupied by just a secretary, to receive calls from customers as the team was off developing software with customers [Nannost 2006].

At that time (mainframe) computers were like big cabinets with many switches and buttons. These mainframes were run by "operators" through stack processing generating a sequence of programs with a meaningful order. Input of data and commands via keyboard monitors did not exist. And IBM did not only provided hardware, but delivered also customized software programs worked off by the computer, for instance, for accounting or payroll.

By the early 1970s, many in the computer industry realized that an affordable video data entry terminal could supplant the ubiquitous punched cards and permit new uses for computers that would be more interactive. At universities (at least in Germany) there was no computer science or informatics, just courses in programming with ALGOL 60 or on analogue calculations.

Dietmar Hopp, an engineer (diploma in telecommunication technology) who after graduation started to work at IBM as a software developer, and then system consultant and account manager and his assistant Hasso Plattner, also an engineer (diploma in telecommunication technology) and having started at IBM as software developer, were engaged in an IBM program for order processing – based on stack processing - for a customer Imperial Chemical Industries (ICI) in Germany, the once British chemical giant, then split and now no longer existing.

Hopp and Plattner, both from the University of Karlsruhe, suggested to ICI that the task could be done much smarter with monitors. ICI agreed and IBM developed an order processing system using monitors which then was much requested also by other customers.

Both developed the *idea of standard software* for accounting whose work processes proceed largely identical in firms. And both were ready to develop a corresponding system for IBM. However, something which is not so rare in big firms happened. Infights with regard to the project between the Mannheim office and the IBM Headquarter in Germany emerged – and Hopp and Plattner were frustrated.

Simultaneously, ICI wanted to have a monitor-based solution for procurement, inventory and invoice-checking. Hopp and Plattner recognized the opportunity: provided ICI with what it wanted and got agreement to distribute the created programs with their own firm. Both convinced other IBM colleagues, mathematician Hans-Werner Hector, graduated physicist Klaus Tschira and economist Claus Wellenreuther, business administration and accounting specialist who also thought about standardization of accounting software, to join the foundation team.

SAP was founded with capital of the founders and ICI was their very first client in 1972. By the end of 1972 there were nine employees and revenues in the first year amounted to DM620,000 (€310,000) leaving a small profit. Their vision and business idea was to *develop standard application software for real-time business processing*, in particular, standardization and unification of all kinds of software that is applied in firms (which later was extended to provide enterprise software applications and support to businesses of all sizes globally). More of their entrepreneurial and innovative approach is described under the heading “value innovation” (ch. 4.3.6, Table I.78).

The five ex-IBMers actually acquired the technology from which they based their software platform from IBM itself which got it from Xerox as a swap deal for a contract Xerox had with IBM. It was called the SDS/SAPE software then and IBM gave the software rights to the five engineers in another swap deal, now, for the five’s stock ownership of about 8 percent.

After a short while they came out with the very first financial accounting software. This was the seed from which other components were developed to create the system known as the SAP R/1 where “R” means real time. Thus came to fruition the founding engineer’s vision of developing the standard software systems for real time business data processing [SAP-2].

During the first years the founders were simultaneously consultants, developers and salesmen who could complement each other by various personal characteristics. Success factors included that SAP standard ERP software filled a market niche, they had customers and marketing was unnecessary; development was done with the customers and internationalization soon took place. Simultaneously, the founders created a corporate culture of trust and customer-orientation as the basis. For employees they were always addressable. Key was a corporate environment free from fear and accepting error and failure so that innovation could thrive.

The founders focused more on development than sales, but average yearly growth rate in terms of revenues in the first five years remained very strong with ca. 60 percent. Further development was with ups and downs. A serious obstacle occurred when IBM announced its own accounting software. SAP could not sell any accounting software for one year.

SAP lived on software introduction service [Breuer 2009]. And we learn: “In the first half of 1975 we did not grasp any new order.” Five years after foundation SAP was still a small firm with 25 employees and DM3.8 million (€1.9 million) revenues [Nannost 2006]. *It took SAP approximately ten years to achieve €10 million in revenues (Figure I.189) which simultaneously represented the inflection point when linear growth changed to exponential growth.*

In 1977 the legal form of the company was converted to a limited liability company (“GmbH,” LLC) and the name was changed to “Systeme, Anwendungen, Produkte in der Datenverarbeitung” (SAP GmbH; Systems, Applications and Products in Data Processing) and it moved its headquarters to Walldorf near Heidelberg.

The breakthrough was induced again by the former employer of the team. Early in 1979 IBM launched a new mainframe computer. The new type 4300 was faster by a factor of four compared to the most powerful computers existing so far on the market. Furthermore, the price of the 4300 series was just a quarter of the other high power computers.

And SAP had the right software R/2 which could map many business processes. In the 1980s, SAP released SAP R/2 which boasted of a better stability compared to its predecessors. It also started becoming multilingual and had become multi-currency to accommodate the needs of their international customers.

Only one year later half of Germany's top 100 industrial firms were SAP's customers. SAP's significant growth became more noticed. In 1986 SAP's revenues exceeded DM100 million (€50 million) [Nannost 2006]. By 1995 the global chemical industry had the largest percentage of SAP installations and accounted for 40-50 percent of SAP's annual revenues and Chemical Market Reporter called SAP the "ERP standard among large chemical companies." [Runge 2006:241]

The original foundation team worked together roughly for ten years. In 1982 Claus Wellenreuther left the firm. In 1988, SAP GmbH transferred into SAP AG (a stock-based corporation by German law), and public trading started on the stock exchange by the end of the year. Hans-Werner Hector became responsible for SAP's business in the US. Until its IPO SAP financed its growth by own profits.

After fifteen years in 1987 SAP achieved revenues of €77.7 million with 468 employees, and €1.38 billion with 6,857 employees in 1995, after 23 years (Figure I.189). For comparison, it took Microsoft nine years (until 1983) to reach $50 million in annual revenues and fifteen years to cross $1 billion (1989) (cf. [Bhidé 2000:16] and Figure I.144, Figure I.157).

What was the basis of SAP's success? Contrary to Microsoft or Oracle which, during their early phases, relied essentially on products, tools (programming languages and computer operating system or relational databases), SAP was focused on *business processes* emphasizing industrial customers. And, according to Dietmar Hopp, there was an additional situation: "Simply, at the right time we had the right idea" ("Wir hatten einfach zur richtigen Zeit die richtige Idee").

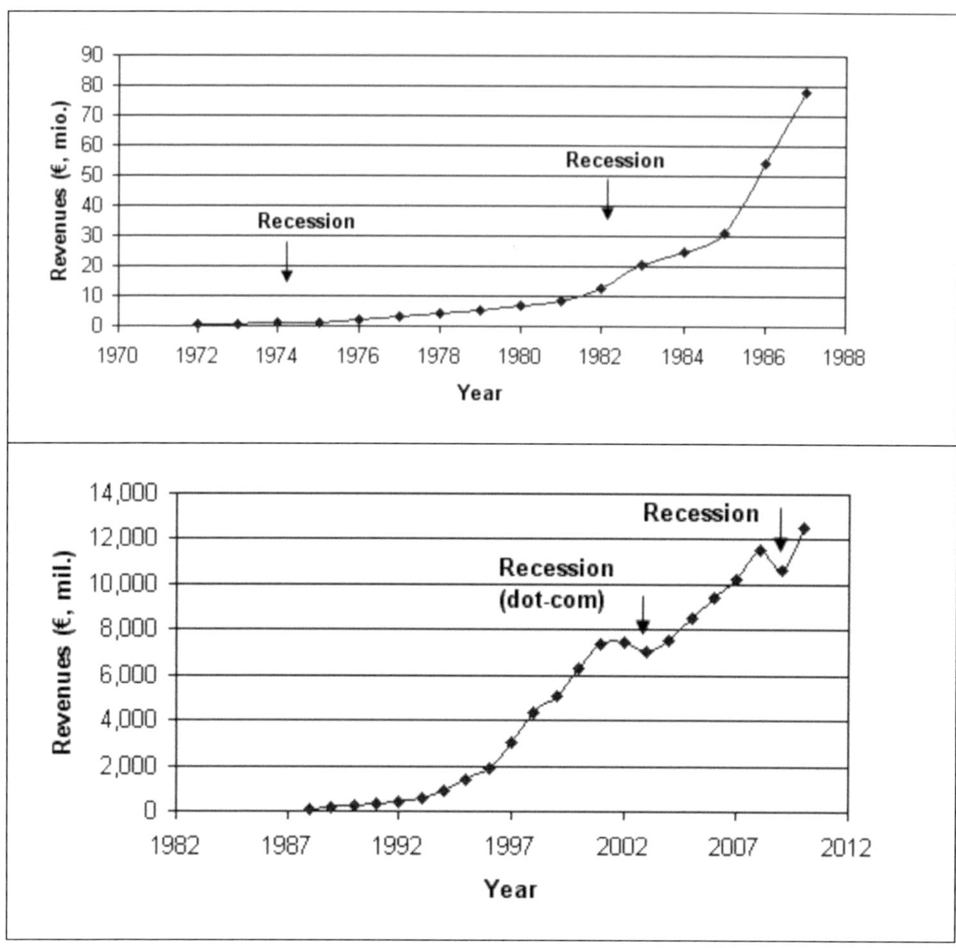

Figure I.189: Developments of revenues of German SAP AG from its foundation.

A.1.5 Entrepreneurship Cases Referring to Ionic Liquids

Referring to the field of ion liquids (ILs) provides exemplary insights into aspects of the birth of (largely) "economic markets" (Table I.15) out of science and the approaches of entrepreneurs to grasp corresponding opportunities based on a *new technology* – which means essentially a *technology push* approach (ch. 1.2.5.1, Figure I.26).

But contrary to biofuels and biobased chemicals (A.1.1) venture capital did not show up here as a financial source. Related cases tackle entrepreneurship perceived as a *disruptive innovation* based on a *platform technology* with broad applicability for many markets. Furthermore, the evolved area after ca. five years had a *limited number of players*.

We shall focus on the fates of two university spin-outs (RBSUs) and one NTBF, two from Germany and one from the UK, over the first 8-10 years of their existence. Only one survived the entry into the new technology area (IoLiTec GmbH, B.2), one was purchased by a large firm (Solvent Innovation GmbH, B.2) and one went bankrupt (Bionics Ltd., B.2). The Bioniqs case contains also a larger discussion of another university spin-out in the UK, Scionix Ltd. founded in 1999, which still exists.

	Solvent Innovation GmbH	**IoLiTec GmbH**	**Bioniqs Ltd.**
Founded	1999	2002/2003	2004
Current State	Purchased by Merck KGaA in 2007, integrated 2008	Still operational; has US subsidiary	Dissolved 2011 in Nov.; liquidation > 12 months

Concerning entrepreneurship it is notable that IoLiTec has been founded by a former employee of Solvent Innovation.

Ionic liquids for broad chemical, pharmaceutical and biotechnological, industrial and research applications emerged only by the end of the 1990s though they were known for decades. They are salts (with negatively charged anions and positive cations), but they are not solid as commonly salts are but liquid at "low" temperatures (usually considered to melt near or below 100 degC) and are relatively low viscous. However, only a melting point below 80 degC allows a broad substitution of conventional organic solvents by ionic liquids – in principle.

Replacing an organic solvent by ionic liquids can lead to remarkable improvements regarding reactivity and selectivity. In many cases the proper choice of the cation/anion-combination allows an optimization of the ionic liquid solvent, especially for a reaction under investigation. Ionic liquids have been used to dissolve not only simple organic compounds, but also enzymes, polymers, even coal and nuclear waste.

The role of ionic liquids as a solvent is not restricted to chemical reactions; it can also be used for extraction reactions. The extraction of metals from different sources will play an important role in times of depleting resources. Selected ionic liquids show a high extraction capacity for some metal ions which will make these ionic liquids interesting solvents.

Researchers from the University of Leuven, Belgium, have used ionic liquids to separate the rare earth metals neodymium and samarium from transition metals like iron, manganese and cobalt – all elements that are used in the construction of permanent rare earth magnets, which are found in electronic devices ranging from hard drives to air conditioners and wind turbines. Hence, recycling old magnets with ILs provide opportunities, so that rare earth metals can be re-used in electronics [Farrell 2013].

China has almost a monopoly concerning rare earths. This degree of dependence gives many Western governments an uneasy feeling, especially when the materials are so crucial to high-tech defense projects. Recycling is at least a partial solution to the supply-risk problem.

China has 37 percent of the world's accessible reserves, according to the British Geological Survey, followed by the former Soviet Republics that make up the Commonwealth of Independent States, then the US and Australia. But China supplies about 96 percent of the world's rare earth elements (REEs). Many green technologies are heavily dependent on the REEs, especially wind turbines and hybrid cars; each Toyota Prius hybrid car is reported to contain as much as 1kg of neodymium in its motor and 10-15 kg of lanthanum in its battery.

“Although less than 1% of rare-earth elements are recycled currently, 20% of global demand could be met in this way. By combining mining and recycling the western world could become largely independent of China in the future.” [Farrell 2013]

Ionic liquids are also specified as “*designer solvents*.” The choice of the cation or anion can affect other salt properties, including density, viscosity, and water stability and miscibility. Tailor-made ionic liquids were becoming increasingly important.

Ionic liquids form two-phase reaction systems with many organic product mixtures. In this way, *simple product separation* by phase separation and *easy catalyst recycling* is possible. Moreover, the lack of vapor pressure allows destillative separation of the product from the ionic catalyst solution without formation of azeotrops. In some cases the catalyst is even stabilized by the ionic liquid during distillation. Currently a growing variety of ionic liquids is becoming commercially available, a development that has fed the surge of research using these unorthodox liquids [Runge 2006:538-540].

One of the most widely heralded features of ionic liquids is the virtual absence of vapor pressure. Ionic liquids are “the ultimate non-volatile organic solvents.” A great deal of attention has consequently been given to the use of ionic liquids as “green” replacements for volatile organic compounds (VOCs), which are being subjected to increasingly stringent regulations [ICB Americas 2004].

By theory about 10^{18} different ionic liquids are conceivable. The numbers of ca. 500-1,000 products for R&D and 10-20 different commercially available industrial scale products [Schubert 2008a] seem to be just a start for the new technical field. This ratio also demonstrates that *design of ionic liquids is a game with gigantic numbers and combinatorics*.

Though being rather fragmented into many relatively small markets this totally new technical area attracted not only some notable startups and NTBFs (Solvent Innovation GmbH and IoLitec GmbH in Germany, Scionix Ltd. and Bioniqs Ltd. in the UK), but also large chemical companies, such as the German firms Merck KGaA (which acquired recently Solvent Innovation GmbH), Cytec and Covalent Associates in the US as well as the chemical giant BASF in Germany.

Ionic liquids can be viewed as a *generic* and *platform technology* (Table I.12, Table I.51) with a very large number of commercially relevant applications as is shown by IoLiTec GmbH founder T. Schubert [2008a]. He lists 26 specific applications clustered into 6 different general areas. A corresponding slide in a presentation (available with the author) was published by IoLiTec already in 2004 [Schubert 2004], one year after its foundation. A corresponding list of applications was later also published by Short [2006].

Having such a *broad choice set of opportunities* requires careful selection and setting priorities. And the setting of priorities means usually making hard choices among conflicting (sub)goals.

But publicizing and illustrating the broad spectrum of applications for ionic liquids was not only meant to show what choices IoLiTec could make from existing options and which ones it finally made. The spectrum of applications was additionally made public to prevent any entrant to patent a particular application and thus constrain IoLiTec's further expansions into other applications of interest to them.

In search for commercial applications of ionic liquids and the necessity to reveal further opportunities ionic liquids startups had to closely watch results of the intense research efforts in the field. In this line startups focused on building broad and intense networks with academia and, in particular, to get leading scientists as members of their Advisory Boards.

Markets

The gross market of ionic liquids comprises research and development in academia and industry as well as components or auxiliaries for technical devices and machines as well as solvents for special technical processes (functional fluids).

Most industry observers reckoned that the chemical industry's interest in ionic liquids, mainly as a solvent, was kick-started only in 1999 by Solvent Innovation GmbH, founded in Cologne, Germany focused on the development, production and marketing of ionic liquids. Soon Degussa AG (now Evonik Industries) through its subsidiary Creavis Technologies & Innovation gained an interest in Solvent Innovation (B.2). Also in 1999 in the UK Scionix Ltd. (in Bioniqs Ltd. – B.2) was established.

"The purpose of the strategic partnership that we have entered with Solvent Innovation is to obtain ionic liquids as a new product category for large-scale production applications." According to a Degussa spokesperson around 2004 the company aimed to produce and market ionic liquids as specialty chemicals, particularly for technical applications, such as pigment-sensitized solar cells, high-capacity batteries, fuel cell membranes, plastics additives and special functional coatings. Degussa was rumored already to have commercialized a hydrosilylation reaction using ionic liquids [ICB Americas 2004].

Chemical giant BASF SE entered the field in 2002 and in the same year also the formation of IoLiTec GmbH occurred (in Cologne/Aachen, Germany).

Most of the applications of ion liquids that do come to fruition in 2006 were estimated to be relatively small. “Everybody is looking for the blockbuster – but it won’t happen. There will be $1 million here and $5 million there, a lot of different uses in a lot of different markets.” It was expected that in this way the worldwide ionic liquids business could add as much as $50 million annually in a very short time [Short 2006].

Sales projections by market entrants were understandably still vague in a business that was less than a decade old. For his startup IoLiTec (B.2) Schubert [2004] estimated the total market for R&D environments in 2004 to be €2.5 million per year and distribution to be via chemicals’ catalogs of dedicated firms, such as German Merck KGaA or Sigma-Aldrich. These firms sell (and produce) its chemical and biochemical products and kits that are used in scientific and industrial research, biotechnology, pharmaceutical development, the diagnosis of disease, and as key components in high technology manufacturing. On the other hand, Schubert envisioned applications for sensors (for the detection of moisture, the use of gases and dangerous materials) to reflect a total market of €25 million per year.

Most ion liquids may have *very high prices*, €400 - €2,000 per kilogram. Economies of scale for ionic liquids would not exist soon, and their expense would slow their adoption. Costs, however, may be surmountable.

It was pointed out that “the price might look bad in the beginning, but it is always price-to-performance that is important.” If the performance of an ionic liquid is 20 times that of the material it aims to replace, for example, a customer would need much less of the ionic liquid. Furthermore, one hears “*You always need to help a customer differentiate a product* from a competitor’s or make a technology leap.” “If the improvement is only incremental, you’d better forget about it.” [Short 2006]

Experts fast dampened expectations arising from early research into ionic liquids – particularly when it came to forecasting widespread applicability. “Most of the ionic liquids that academia is playing with are new chemicals. They are not listed in regulatory framework. They can be used in research but can’t be used in large quantities without *being registered with the authorities around the world*. And that, will involve considerable amounts of time and money.” [Short 2006]

“In Degussa,” the German large specialty chemicals company renamed to Evonik Industries, “researchers have some degree of freedom to work on potential projects and to be innovative. But once something becomes a bigger, controlled project, questions come up: Toxicity? Raw materials? Availability? Listing? These questions do kill some projects.”

The willingness of large companies, such as Evonik, to encourage the use of ionic liquids will be a key to their widespread adoption.

More specific ionic liquids market segments and their magnitudes are described in the context of the cases of Solvent Innovation GmbH (B.2) and IoLiTec GmbH (B.2).

After 2005 large or giant global German firms like Evonik (Degussa), Merck KgaA or BASF looked for large-scale applications of ionic liquids and had increased their in-house production capacities to the multi-ton level for selected applications.

Some ionic liquids are not miscible in organic solvents, a property that BASF made the foundation of its BASIL technology. For example, the company produces alkoxy-phenylphosphines at multi-ton scale by reacting phenyl-chlorophosphines with alcohols [ICB Americas 2004].

Merck KGaA began working with ionic liquids for battery applications already in the mid 1980s, though the project was ultimately dropped, said its Urs Welz-Biermann. In 1999, the company was one of the co-founders of the QUILL network (Solvent Innovation GmbH, B.2), in which it took a fairly passive role, following developments and considering how it might use ionic liquids. In 2002, however, Merck restarted its own ionic liquids program. "We decided we would try to sell compounds and see if there was interest," said Mr. Welz-Biermann.

Merck began a Web site offering a list of compounds, at that time over 250, many of them based on building blocks patented by the company. "We've put a lot of effort into analytics," he adds. "We're doing the business a different way from other companies-not just putting together a catalog of compounds, but also giving specific data like melting point, solvation, etc., to help customers begin working with new compounds." [ICB Americas 2004]

Merck made most of its ionic liquids in-house at its headquarter in Darmstadt, Germany. Large volumes were not a problem, according to Mr. Welz-Biermann – a new multi-purpose facility built for the company's liquid crystal business was also available for manufacturing ionic liquids in the hundreds of kilograms.

While large firms, such as BASF SE and Merck KGaA, have multi-purpose plants for large production levels economies of scale as, for instance, found for biofuels (A.1.1) or solar cell manufacturing (Figure I.154), this did not exist for NTBFs and their ionic liquids. Without economies of scale the related little reductions of expense and price of ionic liquids also slowed adoption of their offerings.

"The success of ionic liquids will not necessarily be equated with large-scale chemistry," noted Prof. Robin D. Rogers, at the University of Alabama, Tuscaloosa. As most applications will be small, usually only a couple of tons of ionic liquid per application per year, a micro-reactor that can quickly be configured to produce different types of ionic liquids on a kilogram-per-day scale could be an advantage. For IL NTBFs scaling-up via micro-reactor technology (MRT) has turned out to be the *technology of choice for small and mid-sized companies to face the challenges of scale-up* (IoLiTec GmbH, B.2).

In their early phases ionic liquids startups usually relied on joint production alliances with large firms (mostly Merck KGaA) to produce larger volumes of ILs, say more than 1 metric ton. But concerning the issue of scaling-up startups also considered limiting themselves concerning production capacities.

For instance, Claus Hilgers, the co-founder of Solvent Innovation GmbH said [Short 2006]: "We could extend up to 50 or 100 metric tons, but that's it. We won't go beyond that. We are positioned between the global players and the small guys." If his company needs significantly larger quantities, he added, it would work with Degussa (Evonik), BASF, or another large company to actually produce the compounds.

Entrepreneurial Startups in Ionic Liquids

The *industry's concept of what ionic liquids can do* has evolved significantly over the last years. Advanced materials and functional compounds, such as high-performance lubricants, thermal fluids, and dispersion of nanoparticles, became major directions.

In 2010 the ionic liquids markets worldwide was forecasted to reach $3.4 billion by 2020 from 300 million that year. Catalysis and synthesis are seen as the biggest applications by value. German and US companies lead the market and developments with a share today of 70 percent [Helmut Kaiser Consultancy 2010].

Generally, IL startups could take advantage from the *general interest in the field* in academia *and* industries – at least in Germany. A serious entry of startups into commercialization began around 2005.

Around 2006-2007, for instance, to push growth in *promising technical directions* and development of the markets by sales professionals Solvent Innovation looked to catch €2 million of investment capital (B.2) and IoLiTec looked for €3.5 million investment capital. In parallel, *scientific research* on ionic liquids progressed with fast pace. A host of known academic groups was busily expanding the limits of what is known about ionic liquids.

As a consequence, IL startups must *keep knowledge about the developments* of new classes of materials, including their applications and potential markets.

Computer-supported prediction of the performance and properties and simulation of ILs turned out to be mandatory to manage the myriad of continuously created new data and also to respond to *customized solutions for clients*. IoLiTec and Bioniqs run (ran) related computer systems. The issue here is often a trade-off between properties, for instance, hydrophobicity (water-repellent; tending to repel and not absorb water), thermal stability and price versus biodegradability and corrosiveness [Sahin and Schubert 2012].

> "If a new material is to be accepted as a technically useful material, the chemists must present reliable data on the chemical and physical properties needed by engineers to design processes and devices." – Lowell A. King, Pionier of Ionic-Liquids-Research

The startups had to continuously adapt to their environments by changing their *business models* and *organizational structures*, having sufficient *financial resources* and establishing *networks* with appropriate universities and public research institutes and *cooperation* with industrial partners which are potential customers and participate in corresponding *project consortia* – utilizing public R&D and public financial support.

They had to continuously work on *reducing the price* levels of the offerings, but simultaneously *keeping the quality*. For ILs quality is associated with purity, and purity usually translates into performance of the ILs. *Collaborative projects* are a preferred method of *introducing materials to the market place as an optimized process solution*.

Furthermore, in the sense of *"technology push*" (ch. 1.2.5.1) and commercialization of the technology startups had to *develop their markets* and fight for market share. If ILs occur in the market as a *new technology* or if ILs appear as an *enhancing* or *generic technology*, fight is against other, often well established technologies in the market.

The above requirements represent important factors and conditions against which the fates of the startups mentioned in the introduction can be discussed, considering additionally the role of the Great Recession.

Solvent Innovation GmbH

The *RBSU* Solvent Innovation GmbH (B.2) survived the Dot-Com Recession around 2001, as did Scionix Ltd. in the UK.

Solvent Innovation GmbH (SI) was founded in 1999 by Claus Hilger and later Prof. Peter Wasserscheid, a worldwide renowned pioneer and expert in ionic liquids, as a spin-out of the Technical University (RWTH) Aachen in Germany.

By January 2008 the German firm Merck KGaA :took over SI with ten employees at that time and integrated it into Merck's Performance & Life Science Chemicals unit, but it continued to operate as "Merck Solvent Innovation GmbH." In this regard Solvent Innovation shared the fate of another spin-out of the Technical University (RWTH) Aachen, Puron AG, which also was acquired by a large firm after a couple of years ([Runge 2006:95-96]; Table I.41, Figure I.73).

Before foundation of Solvent Innovation in 1999 availability of ionic liquids in commercially relevant quantities did not exist. A small number of systems for laboratory experiments could be purchased from catalog firm Sigma-Aldrich [Wagner and Hilgers 2008]. Furthermore, Cytec, Acros and other "catalog firms" had supplied the laboratory market with ionic liquids for years.

Having the adequate entrepreneurial mind-set Claus Hilgers went in 1998 to the Technical University (RWTH) of Aachen. Under the leadership of Dr. Peter Wasserscheid a working group emerged that dealt with the synthesis and applications of ionic liquids and Hilgers performed his doctoral thesis in this group. At that time industry became more interested in ILs. And as they were not commercially available, Wasserscheid's workgroup received an increasing number of requests for samples.

Talks with industry professionals indicated that *people wanted the ionic liquids and were willing to pay for ILs*. When development of demand reached a certain level, Wasserscheid und Hilgers recognized the opportunity for a business and founded Solvent Innovation GmbH.

Such an entrepreneurial situation starting already with industrial or academic customers is often observed with RBSUs, such as WITec GmbH (B.2), Attocube AG (B.2), Nanion Technologies GmbH (B.2) in Germany or Cambridge Nanotech (B.2) in the US.

The company's founders were both pioneers in the development and application of new ionic liquids with enhanced efficiency. The most significant research results of Wasserscheid and Hilgers were combined to form a unique technology platform, the AIMFEE™ technology (Advanced Ionic Materials for Enhanced Efficiency), which was seen as the basis and a powerful tool for numerous potential applications in life science and chemical synthesis as well as catalysis and material science.

SI's *technology was protected by a number of patents* (or patent applications, respectively).

In the early days Solvent Innovation viewed itself as a partner for systems solutions in the field of ionic liquids rather than only a producer. I*t also offered custom synthesis of specialties and contract research.* Early customers included the big names in the chemical and petrochemical industry.

For foundation and further developments SI followed the typical entrepreneurial path of a German RBSU.

Solvent Innovation did not need external financing during its startup thrust phase (ch. 4.3.2; Figure I.125), its first three to four years of existence [Hilgers 2006]. On the one hand, the founders could utilize the laboratories and the infrastructure of the university. Furthermore, after Hilgers' scholarship for his thesis ended the program PFAU ("Programm zur Finanziellen Absicherung von Unternehmensgründern aus Hochschulen") of the State Government of Northrhine-Westphalia secured his cost of living.

The program financed founders of the state universities for a maximum of two years by a quasi-salary. Hence, Hilgers did not need to make revenue, but could concentrate on developing the business. Hilgers was supported by PFAU for the period July 1, 2000 until June 30, 2002.

The time of the PFAU scholarship was used essentially for generating a technical Proof-of Concept (PoC), developing concepts for financing and distribution and market tests. During that period also two new processes were developed with the claim to reduce production cost by 35 percent and increasing quality and SI submitted these as patent applications.

Solvent innovation could not finance growth which required also a new location by sufficient own revenues (profits) and, hence, looked for an investor. As there where already contacts with the German large specialty chemicals firm Evonik Industries (named Degussa at that time) Hilgers succeeded in getting Evonik on board as a third (minority) partner for the GmbH (LLC) in 2003. It complemented its growth financing deal together with capital from a public investment organization.

In February 2004 Solvent Innovation moved to its new site at the Biocampus Cologne, a sort of technology park. Since 2004 Solvent Innovation manufactured its products in Cologne and had an option on the neighboring building so that its site could be extended without problems.

Business orientation was driven essentially by the fact that published research results indicated that the unique character of ionic liquids could open up new "solutions" for catalysis and organic synthesis emphasizing the "green" character of *ionic liquids for chemical processes* in terms of

- replacing volatile organic solvents,
- minimizing the consumption of catalyst,
- enhancing the overall activity and selectivity of chemical processes.

In order to meet *rising market demands for ionic liquids in larger quantities*, in 2005 Solvent Innovation increased its capacities distinctly with the acquisition of a new 100 l plant. Together with the already existing 25 l plant and two 20 l reactors at that time Solvent Innovation had an annual production capacity of more than 5 metric tons.

Between 2005 and 2007 within Solvent Innovation a *business re-orientation* emerged. The industry's concept of what ionic liquids can do had evolved significantly. Therefore, Hilgers changed the business model. Over time it had turned out *that positioning and commercialization of ionic liquids as a replacement of organic solvents for syntheses were not sufficient for distinct growth.* Solvent Innovation should no longer be viewed as only a producer of solvents – mainly used by academic research

SI turned to *materials and functional compounds* – for industrial customers.

And there was a new business model. Organizationally, SI could expect its impressive list of persons on the Advisory Board to be helpful for the re-orientation. Products would be sold directly via a *catalog business* or via *distributors*. SI then focused on the fields of

- Separation
- Analytics
- Organic Synthesis
- Enzymatic Biocatalysis
- Electrochemistry
- New Materials.

This shift of emphasis away from the solvent aspect was associated with addressing a *different type of customers*. And the firm had to learn that for the market of functional materials the times from first contact to applications with the existing customers were significantly shorter. Until an industrial customer replaces a solvent in a running process an extremely long time will pass. Other applications of ionic liquids can be implemented within one or two years [Hilgers 2006].

Solvent Innovation strove for becoming *a systems and solutions provider*; it offered the complete portfolio of *customer services, joint development, consultancy*, and so on.

Solvent Innovation offered two kinds of products [Wagner 2006a, 2007]:

- Platform products – pure ionic liquids
- Integrated products – finished formulated products and masterbatches.

A masterbatch is a product in which components (often pigments and/or other additives) are already optimally dispersed in a carrier material that is compatible with the main target/material in which it will be used. Integrated product classes of SI with market potentials of €300 million to €500 million were, for instance,

INNOLUBE™ High-performance lubricants and electrically conductive lubricants	INNOLUBE™ acts as an electrically conductive lubricant for a bearing in frequency-controlled motors
INNOVAC™ Liquid for vacuum pumps and compressor fluid for screw compressors	INNODISPERS™ Dispersing agents for nano-particles
INNOSTAT™ Anti-static agents for plastics and coatings	AMMOENG™ acts a dispersing agent, for instance, for homogenization of color pigments

However, it appeared that most of the products had still the status of *prototypes*. For instance, the prototype INNOSTAT™ anti-static agents, such as INNOSTAT™PU or INNOSTAT™PVC or INNOSTAT™PC targeted the polymers and plastics commodity markets of polyurethanes, polyvinylchlorides and polycarbonates which, however, are produced since decades on a million tons level relying on an established set of highly competitive suppliers of processing aids and additives.

To replace existing anti-static agents for well established and optimized manufacturing processes would mean that "technical specification of customers met," as noted by Wagner [2007] for INNOSTAT™PU, does not suffice. SI had to fight against switching costs and convenience and customers taking the risks these additives to function not only in a laboratory or 500 kg level pilot plant, but in a multi-million tons plant.

In 2006 SI targeted an institutional investor to finance finishing its "products" and the development of the market by sales professionals.

According to SI's Head of Marketing & Sales M. Wagner the biggest risk was associated with the *challenge of efficient market penetration* for the newly developed products INNOLUBE™ and INNOSTAT™.

By January 2008 the German firm Merck KGaA acquired SI. The takeover meant *acquisition of technical know-how and experience*. Furthermore, Merck obtained additional production capacities and products to access new markets with the high-performance lubricants and antistatic agents for plastics.

SI was slow in transforming science into businesses providing sufficient revenues compared with its direct German competitor IoLiTec whose most important co-founder was a former employee of SI. It lacked sufficient financial resources to drive pilot products into commercial offerings and lack of human resources to sufficiently support the successful entry into lucrative markets by technically versatile professionals.

Bioniqs Ltd.

As Solvent Innovation GmbH in Germany also in the UK the *RBSU* Bioniqs Ltd. (B.2) was founded (in December 2004) by an *entrepreneurial pair* consisting of a scientific co-worker or graduate, respectively and a professor.

It provided designs and developed proprietary ionic liquids (ILs) which aimed to facilitate and improve *biochemical and biocatalytic processes in industry*, particularly in the chemical, pharmaceutical, paper and textile sectors. It addressed a heterogeneous set of industrial processes, from bioconversions and chemical synthesis to analytics (chromatography), extraction of natural products and decontamination/cleaning.

Bioniqs was set to be profitable by the end of 2007 [RSC 2007]. However, Bioniqs went bankrupt and was dissolved in November 2011.

Founding Bioniqs had a biological origin and perspective. It was essentially *science-driven* based on attitudes and activities of Adam Walker who wanted to work at the interface between biology and chemistry. For a study for a PhD to find a way to integrate biological catalysts into the preparation of an opioid analgesic Walker joined Neil Bruce, then at the Institute of Biotechnology at the University of Cambridge, UK.

Walker realized that the intermediates in the path to that analgesic are poorly water soluble, but enzymes only function in a water-based environment. In the attempt to solve the problem Walker came across using ionic liquids as solvents. Eventually he succeeded in chemically modifying an ionic liquid to make it resemble water more closely, so that his enzymes and drug intermediates remained stable in one solvent.

Towards the end of his PhD, Walker had the business idea that *replacing water with modified ionic liquids as solvents for industrial applications* would be commercially viable – and a related startup would not seriously interfere with other IL startups operating already in the UK and Germany

And he decided to set up a spin-out company, Bioniqs, with Neil Bruce. The University of Cambridge filed the patents for their "designer solvents," but before Walker and Bruce could set up a spin-out company in Cambridge, Bruce was offered a position as chair of biotechnology at the Centre for Novel Agricultural Products (CNAP) at the University of York, UK.

Bruce and Walker moved to CNAP and they "designed an ionic liquid that would mimic water, but would not hydrolyze enzymes" – "Second Generation Ionic Liquids." They thought they can *develop tailor-made ionic liquids at a competitive price*.

They positioned their technology as an *enabling technology* (Table I.12) which means, *you can do things that you cannot do using existing processes.* Furthermore, the new company would be based upon a *strong patent portfolio* arising from work performed at the Universities of York and Cambridge.

Bioniqs aimed to *generate revenues through design and process development (royalty stream on sale of licensed products), not manufacturing* – hence, exploiting opportunity by alliance rather than competition with major IL manufacturers. Bioniqs also offered *contract research and consulting services.* Its major target *markets* were in the *pharmaceuticals and fine chemicals sectors.*

By type there were two customer segments, industrial customers (*solvents for industrial enzymes*) and customers from academia and public research institutes. But, *de facto* the consultancy element of what Bioniqs did was really helping the firm to understand and develop its own products rather than generating a stream of revenue.

To fill roles of CEO and director of operations for growing a company quite literally from scratch and *pushing new technology into almost non-existing markets*, a "technology push" situation, Walker had to master a steep learning curve.

CNAP in York took a very proactive approach to spin-outs. In the *York Science Park* as its location Bioniqs had an analytical room where ionic liquids were designed and a synthesis suite, where the resulting liquids were produced in small scale. In York they also found a partner in Amaethon Ltd., a technology commercialization company specifically created to commercialize CNAP research. *Financing* was through *own and public sources* as well as *private investors.*

IL research of the then founders of Bioniqs was also funded by the ProBio Faraday Partnership and BBSRC (Biotechnology & Biological Sciences Research Council).

In 2006 Bioniqs could take advantage of Connect, Yorkshire's Fast Invest Scheme – a program that offered technology businesses loans of up to £50,000 combined with business mentoring (at January 2007: 1 £ = 1.52 € = 1.96 $). To help Bioniqs' transformation from a purely development focus to one of sales and growth, Fast Invest allowed Bioniqs to recruit a business development manager, an experienced commercial director on a consultancy basis.

Also in 2006 Bioniqs established a partnership agreement with the large German chemicals firm Merck KGaA (Darmstadt). Through this partnership, *Merck KGaA manufactured and distributed through its catalog a selected range of Bioniqs' proprietary ammonium based ionic liquids.*("catalog business").

In 2007 the Yorkshire Forward Bioscience award for the "Young Company of the Year" was given to Bioniqs.

By 2006/2007 with ionic liquids as solvents and its application in the chemical and pharmaceutical industries Bioniqs put its focus largely on production processes via production and distribution alliances with *IL as solvents or auxiliaries* for *decontamination and cleaning* of process reactors and recycling processes as well as *extraction of natural products from biomass* and *biocatalysis.*

It hoped to *take advantage from the "green chemistry" and CleanTech trends* which emerged clearly by 2005/2006. As a differentiator, Bioniqs *positioned* its offerings on identification and design of environmentally friendly solvents that offer performance and efficiency improvements over many hazardous materials and as a *timely service* – as many conventional solvents were becoming more difficult and expensive to use due to increasingly stringent environmental and safety legislation, such as the so-called REACH registration for Europe. This would tend to require replacing substances and solvents because of their negative environmental impact.

As there was (and is) much discussion about the notion and the understanding of "green solvents" as a *marketing tool* Bioniqs introduced and promoted a green solvent certification (named "*econiqs*") in response to confusion over the reality of claims made about "green chemicals" and many novel solvents.

As a *further marketing instrument*, ahead of its 2009 product catalog, Bioniqs launched three solvent kits – "Product Catalogue Starter kit," "Low Viscosity kit" and "Hydrophobic kit." These offered a representative selection of ammonium salts and would *address researchers who are new to protic ionic liquids* (PILs).

The *cleaning business* promised to be multi-scale tons envisioning tailor-made ionic liquids for dissolving poorly soluble active pharmaceutical ingredients off the walls of reactors.

In 2008/2009 Bioniqs was successful in winning funding from the UK government to design solvents that will enable some *plastics (high performance polymers) to be re-*

cycled more efficiently. The related HiPerPol project aimed to enable polymers to be separated from plastic waste-streams.

As Bioniqs strived for assisting the customers in developing increasingly sustainable, safe and ecologically efficient working practices it used its "solventS" service to work with their customers to develop solvents which they claim are *optimized for their technical, economic and environmental performance*.

This service was made by the *high-throughput screening and design capabilities* of Bioniqs *solvent modeling software* and proprietary *database* of over 12 million solvent permutations (including both ionic and molecular solvents). The ROSETTA solvent simulation database [Newton 2009] combined advanced structure-property alignment tools with a series of databases to evaluate the performance and properties of solvents along with other requirements, such as cost and toxicity/environmental impact.

Concerning the potential of "*extraction of natural products*" for ionic liquids Bioniqs had, for instance, successfully extracted *artemisinin (*also called *artemesinin*), the anti-malaria drug precursor (Table I.99, Amyris), from both fresh and dried plant material following an *in silico* solvent design process (performed on computer or via computer simulation) from a database of some 350 proprietary ionic liquids. In 2008 Bioniqs had secured £50,000 of investment to enable continuing to fund its work with artemisinin.

But, Bioniqs' approach to artemisinin extraction turned out to be a scientific investigation rather than a recipe for implementing a real process – it was not a demonstration of the artemisinin extraction process at scale and fulfilling the commercial potential of Bioniqs' ionic liquid. The study revealed that further fine tuning can lead to the end product of an ionic liquid optimized for the needs of the real process. A set of process parameters were revealed and it was recommended that these parameters are used as the basis for a final product specification and that multi-parameter screening is used. In addition, the involvement of chemical engineering specialists was recommended (B.2).

On foundation in 2004 Bioniqs consisted of three people, but was growing to employ eight people in 2007. But the financial decline of Bioniqs towards a financial collapse was already reflected by some financial indicators, such as "cash at bank" and "net worth," comparing 2006 and 2007 data. After five years of existence, Bioniqs was no longer viable, the liquidation process started by the mid of 2009 (B.2).

There were still tremendous *issues of market entry* in terms of *cost of ILs* and *replacing existing processes* including a fight against switching cost and attitude and risk aversion of customers to implement a totally new technology – industrial customers did not want to act as the "guinea pigs."

Bioniqs Ltd remained largely a *curiosity-driven research endeavor* with (probably) meager revenue in an entrepreneurial environment relying on perceived potential or unexpected commercial opportunities without a clear identification and focus on its major markets and executing related market entry – or, at least, convincing demonstrations.

Over its time of existence Bioniqs can be assumed to have had little direct, sufficient contacts with the market and its customers. Competing with other IL startups concerning contract research and consulting services for revenue generation may have suffered from the same services offered by the more established other IL startups. The end was a state of the firm without enough cash and probably no chance for further financing due to the Great Recession.

IoLiTec GmbH (and Scionix Ltd.)

Ionic Liquids Technologies (IoLiTec) GmbH (B.2), founded in 2002/2003, has been cited already in various contexts in this book as an example to illustrate particular entrepreneurial situations. The current discussion shall focus on IoLiTec's development and position in emerging markets based on its technology push approach. IoLiTec emerged as a rather successful NTBF and the most successful of the ionic liquids startups.

IoLiTec is engaged in top value technology (TVT, Table I.1). Its founder reported several times that *IoLiTec made always a profit since its foundation*. It shows continuous growth by various indicators. For instance, after six years of existence it had 12 employees, whereas on average the number of TVT startups' employees after 6-8 years is 8 (Table I.70). The Compound Annual Growth Rate (CAGR, Equation I.10) of employees between 2005 and 2012 is 17.8 percent.

Contrary to all other IL startups which are university spin-outs (RBSUs) IoLiTec GmbH is an "academic startup" (Table I.2) and is special by four facts.

- It started with a customer, which eased getting further financings via banks.
- It comprised a team of experienced founders in the ionic liquid fields and in industry.
- The key founder Thomas Schubert worked as a "post doc" with Professor Peter Wasserscheid at the Technical University (RWTH) in Aachen (Germany), a worldwide renowned pioneer and expert in ionic liquids, and worked already for 18 months for Solvent Innovation as a Leader of Distribution, with responsibilities for technical synthesis and marketing of ionic liquids. In Aachen he also met one of the then co-founders of IoLiTec.
- The founder team fell apart rather soon, but without serious troubles. Thomas Schubert replaced one of the co-founders by an angel investor in 2004 who took over responsibilities for taxes, finances and law. In 2006 the second co-founder left IoLiTec to further pursue his scientific career at a university (with Professor Peter Wasserscheid).

The firm was born in a climate of excitement about ionic liquid technologies in Germany and also the UK (Scionix Ltd. in the Bioniqs Ltd. case, B.2) that was shared between academia and industry and embedded in a "green" attitude of society and policy. There was (and is) much support by related joint projects (ch. 1.2.6, Figure I.40, Table I.92) financed by federal and state governments and non-governmental organizations (NGOs), such as the Deutsche Bundesstiftung Umwelt (DBU; The German Federal Environmental Foundation).

DBU is one of the largest foundations in Europe. It promotes and funds innovative and exemplary projects for environmental protection. And IoLiTec's development projects represented what in Germany is called "*sustainability innovation*" (ch. 1.2.5.1; [BMBF 2007]) – or, at least, paths to sustainability innovations.

IoLiTec had to successfully respond to a number of issues associated with ionic liquids and to achieve competitive advantage.

- A *technology push* situation with very many fragmented rather small markets
- A *science-driven environment* in which technical developments by firms are continuously associated with external scientific developments. It was not to use just one key scientific effect or result which had to be commercialized as a technical solution and subjected to further developments
- Success of ionic liquids will not necessarily be equated with large-scale production
- Economies of scale for ILs do not exist yet for ionic liquids as, for instance, for biofuels (A.1.1) or solar cell manufacturing (Figure I.154). Hence, their slow reduction of expense and *high prices* will also slow their adoption
- A *special situation for scale-up* of production.

IoLiTec founders selected carefully the *location* of the firm in the city of Freiburg and its BioTechPark, having had in mind networking with academia in the science-driven field.

> Massive networking with academia, public research institutes, and industrial firms as *external resources* (Figure I.125, Figure I.170) became a typical feature of IoLiTec's further development.

During its first three to four years of existence, its *startup thrust phase* (ch. 4.3.2; Figure I.125), IoLiTec was in search for lucrative applications and markets and developing "experimentally" its business model. *Major customers were from academic and industrial research groups* [Schubert 2006a].

IoLiTec elaborated a rather large *opportunity options set* (Box I.13, ch. 5.1) of a *platform technology* from which to select the most promising opportunities for the firm. The options were related to the various broad application fields matching relevant properties of the ionic liquids technology. Furthermore, *priorities* had to be established.

The derived technology strategy and associated marketable products and strategy evolvement over time, consequently adapting the firms organization to the strategy and implementing and executing the strategy, represented the framework for IoLiTec's development over its first ten years of existence.

Evolvement of IoLiTec's organizational and network structures are presented by graphics in the case description (B.2). These reflect the increase of complexity of the firm, changing functions and roles for the leadership team and developing applications (businesses). In the case description (B.2) it is also exemplified how IoLiTec achieved a competitive advantage.

In its early days around 2004 IoLiTec's offerings had two components.

- (Bio-)analytical applications (IoLiTec offered new materials that could make the life of biochemists and scientists from other disciplines much easier.)
- Consulting and custom R&D.

Additionally, anti-static fluids – functional fluids for the use on glass-surfaces – were offered. Activities in nano-particles and sensors were planned.

Concerning *intellectual properties* (IPs) by 2005 IoLiTec had submitted eight patent applications and owned some trademarks (for instance, IoLiTive®, IoLiTherm® and IoLiSens®).

Around 2005 IoLiTec decided to focus on the following five areas for commercialization:

1. Contract R&D services
2. Special Chemistry (Ionic Liquids)
3. Sensor technology
4. Energy
5. Nanotechnology

IoLiTec ran own R&D, but contract R&D to generate revenues.

> Since then, fundamental and necessary R&D for all these areas was partially pursued by *participating in publicly funded projects*, essentially the typically German joint projects ("Verbundprojekte"; ch. 1.2.6, Figure I.40).

Joint projects cover universities, public research institutes, and small and mid-sized firms. They are often initiated out of "*competence networks*" (Figure I.39) in which IoLiTec also participated. This did not only broaden the scope of IoLiTec's network, but provided also many contacts to and cooperation with potential customers and was used to enlarge its group of Scientific Advisors. Project money always represented *an important revenue stream* for IoLiTec.

Running almost continuously R&D projects, publicly financed by the German Federal Ministry of Education and Research (BMBF), Federal Ministry for Economics (BMWi) and the Deutsche Bundestiftung Umwelt (DBU), was (and is) central for financing IoLiTec and running its own internal applied R&D.

IoLiTec *increased systematically its knowledge base* and set up *computer-supported technology intelligence* focusing on literature and patent search activities. These are the basis of knowing the state of the art und current awareness about new developments. The *systematic activity* of "current awareness" and "state-of-the-art" knowledge based on the scientific literature and patents became the basis for related databases – and to provide an IoLiTec newsletter as well as consulting activities and design of new ILs or customized ILs.

As a *marketing instrument for technology push* IoLiTec launched its free newsletter "Ionic Liquids Today" already in March 2005. It does not only provide news about IoLiTec's products and their applications and cooperation set up by IoLiTec, but reports also on scientific and technical progress in ILs.

In 2005 IoLiTec moved out of BioTechPark Freiburg to a new location in Denzlingen very close to Freiburg due to more favorable cost of needed facilities. And since 2006 IoLiTec acted also as *a distributor for other IL firms*. It made phosphonium ionic liquids of the US firm Cytec Industries available.

Within a huge joint project (called NEMESIS) IoLiTec engaged intensively itself with the scaling-up technology via micro-reactor-systems. Micro-reactor technology (MRT) is the *technology of choice for small and mid-sized companies to face the challenges of scale-up*. Another aspect during the project was the development of concepts for efficient *recycling of used ionic liquids*. One micro-reactor was set up in 2008, in 2009 IoLiTec built its second micro-reaction system.

Furthermore, micro-reactor technology was assumed to be not only the means for scale-up, but also for *quality management* and to *reduce the prices of ILs*.

By 2007/2008 IoLiTec had a growth strategy in place with a corresponding organizational structure and a related requirement of investment (Figure 4 and Figure 5 in the case description, B.2). As T. Schubert did not succeed getting finances from banks he turned to an investment firm, the "Zukunftsfonds Heilbronn" (ZFHN). The venture capital fund acquired a 30 percent stake in the technology firm and IoLiTec had to move its location from Denzlingen to that of the investor, to Heilbronn (Germany).

Preparing for the move to Heilbronn had consequences for the original plans of IoLiTec. Activities in R&D projects were reduced and its intention to establish a subsidiary in the US was delayed. Though in 2009 IoLiTec, Inc. was incorporated as a one-man-firm in the Business Technology Incubator at the University of Alabama in Tuscaloosa. But it started the operative business only in April 2010. The selection of the location in Tuscaloosa followed very rational arguments.

After Europe the US is the most important market for ILs – and simultaneously there was no IL startup. Out of a small office the IoLiTec representative is contacting North

American companies, universities and researchers interested in ionic liquids. ILs are shipped to Tuscaloosa from Germany, and IoLiTec Inc. then fills the orders for North American customers.

Having established its distribution organization for ILs, marketing and sales of ionic liquids *have remained tough as they remained expensive* due to the difficulty of manufacturing them. Much of the difficulty is associated with purification. Hence, *recycling* of ILs played an important role to support sales. But, in 2010 IoLiTec opened another revenue stream, i*onic liquids rental service*. IoLiTec claimed to be an industry first with renting ionic liquids to customers.

By the end of 2011 IoLiTec achieved a status concerning the level of progression from R&D via piloting to commercialization of offerings which is depicted in Figure 6 of the case description (B.2) and which also indicates the state of *executing the strategy*.

IoLiTec has positioned its ionic liquids in the "*low volume, high value*" segment of specialties. The diversified orientation of offerings and industry segments makes *IoLiTec rather independent from economic ups and downs* of the addressed industries. Growth of the number of employees across the period of the Great Recession (Table 2 of the case description, B.2) corroborates this.

Several metrics for IoLiTec show that the firm grew considerably and healthy and since 2006/2007 has accelerated its development. Characteristics of IoLiTec's growth are:

- *Innovation persistence* (ch. 4.2.3, Figure I.115, Figure I.127)
- Strong *customer orientation*
- *Execution* of a longer term strategy and
- *Diversification* of applications and continuous expansions of its basis of ILs as its *platform technology*.

Concerning the competitive landscape, as described in the IoLiTec case (B.2), there was no serious competition with the very few large firms active in ILs except probably with Merck KGaA which acquired the NTBF competitor Solvent Innovation GmbH in 2008. IoLiTec operates in lucrative niches and for large firms like BASF the related markets and volumes of production are too small to be of interest. IoLiTec operates complementary to the large firms.

Concerning the IL startups during its life-time Bioniqs Ltd. (B.2) has never been a competitor of IoLiTec. Bioniqs focused largely on ionic liquids as solvents and on application in just the chemical and pharmaceutical industries.

The remaining competitor from the UK, Scionix Ltd. (in case Bioniqs Ltd, B.2), shows few fields that could become competitive areas, if Scionix would leave its two key markets

- Electrochemistry (metal deposition and electropolishing, metal recovery) and
- Process technology (metal extraction including catalyst recycling).

Scionix' ionic liquids are essentially based on choline chloride promising mass product related ionic liquids which could be assumed to be applicable to large scale processes. One basic orientation of Scionix is ILs to offer *a clean way to carry out chemical processes avoiding strong acids* (removing harsh acid-based processes).

Scionix is a joint venture between the University of Leicester and Genacys Ltd. (a wholly owned subsidiary of the Whyte Group Ltd). The company was set-up in 1999 to commercialize the industrial use of a novel class of solvent systems, focusing on ILs.

Scionix pursued *an entrepreneurial process which is quite different from IoLiTec's approach*. It follows the narrow path of a university/industry cooperative organization. This Private-Public-Partnership (PPP; A.1.3) allows fundamental and applied research to be carried out at the university while providing the production, marketing and licensing capability by the private organization. The essential structure in case of Scionix attributed a dual role to Prof. A. Abbott of the University of Leicester (UK) to be a research leader at the university and as a co-founder to act as the Research Director of Scionix.

The PPP construct involves Whyte Group Ltd. which is Britain's largest privately owned chemical company and has a number of diverse activities, including manufacturing, distribution and R&D. The flagship of the group is Whyte Chemicals Ltd., one of the largest private distributors of chemicals and polymers in the UK and it also manufactures pharmaceuticals – and ultimately also ILs.

The interrelations between Leicester University and Whyte Group are mediated by Genacys Ltd. Genacys is a special external organization of the Whyte Group acting as a *corporate venturing company (CVC)* and on the basis of bringing together research ideas and entrepreneurial spirit, targets to turn early-stage technologies into separate successful corporate entities through collaborations, strategic partnerships and joint ventures.

Contrary to IoLiTec which pursues many links with industrial firms and universities and public research institutes Scionix concentrates essentially on only the University of Leicester and Whyte Group. This means, Scionix' access to external resources for research, development and commercialization and competitive strength are rather small compared with those of IoLiTec.

With regard to the IL NTBFs IoLiTec's *competitive advantage* is discussed in the case description in terms of the VRIO attributes of a resource-based view (RBV) (ch. 4.3.3,Table I.75):

A final remark concerns entrepreneurship and innovation with high priced offerings. In contrast to, for instance, German Smart Fuel Cell (SFC) AG (now SCF Energy AG) which could sell its fuel cells systems in niche markets (such as leisure, recreation vehicles, military) with customers with high purchasing power [Runge 2006:331-335] for ILs there are no such opportunities. An entry into a broad industry segment through a market with high purchasing power occurred also when William Henry Perkin (A.1.2) introduced his synthetic mauve dyestuff [Runge 2006:293-294].

A.1.6 Formalization of Structures of Founder Teams and Architectures of New Firms

Dealing with "architectures," "configurations" (ch. 4.3.2; Figure I.124) and ultimately taxonomies of new technology ventures (Figure I.128) relates to several complementary aspects of subjects or objects of a set: "identification," "distinction" and "similarity."

Such concepts play an important role to describe the development of new firms in terms of biological analogies. The discipline that provides *metaphoric explanations* of firm or industry segment growth is Developmental Biology [Runge 2006:7; Bhidé 2000:249].

Developmental biology relates to three key concepts, similarity, homomorphism and heritage versus analogy (actually function-analogy). Two structures are *homologous*, if they *look similar* and the *similarity* is due to descent from a *common ancestor* possessing the ancestral version of the part in question (Figure I.190). If two structures are *analogous* there is *no common ancestry* and the parts look similar as the pressure of natural selection has forced a convergence of structure to meet *the need for similar function*. In biology *identity* of form or shape or structure is termed isomorphism.

Fundamentally we dealt with sets of n entities, objects or states, and their representations in terms of the "Diagram Lattice" (with Young Diagrams) which are related to *partitions of an integer n* (Figure I.73, upper right). This means decompositions of n into a sum of integers and is related to permutation algebra to deal with structures of founder teams and discuss *issues of coordination* (Figure I.72 – Figure I.74).

In essence, there was a proposition that structures and attributes can be associated with meaning. For instance, in Figure I.73 we have introduced a binary differentiation of weak and strong *coupling of (structural) subjects* based on the subjects' attributes. Sharing the same attribute has been *interpreted* as "strong coupling" requiring much less efforts of coordination – for instance, by reference to effects of "boundary spanning" (ch. 1.2.3).

Using Young Diagrams, in a similar way as coordination, one could also represent and discuss *social ties* of the members of a founder team (social coupling, Figure I.71).

Furthermore we used *partial orders* of Young Diagrams (Figure I.74) which are related to moving boxes within a Young diagram and illustrated that in Figure I.73 (lower right). The WITec GmbH case (Figure I.73, lower right) provides a special example for dealing with *homology* without changing the number n of objects due to a common ancestor (here the column with three boxes).

In discussing development of new ventures Young Diagrams [119] provide also a mode of (structural) relations of relevance for mappings and representations of the current subject of interest. That is, the relations generated by adding or removing a box providing *relations for build-up processes*.

> From a systems point of view in the realm of entrepreneurship we are dealing with small numbers of components (2-30), often numbers between 1 and 8-12. For systems of such smallness (the lowest level of "organized complexity") one can expect that adding or removing just one component may have a significant effect for a team and/or the firm.

To discuss structures and similarities of teams we shall understand a partition diagram of a Ruch lattice as a histogram (bar chart, Figure I.191) or, in other words, a finite discrete distribution with a partition diagram being comparable to an intra-system state.

Without needing to go into the details of (mathematical) group theory, particularly of the symmetric groups S_n and their irreducible (non-decomposable) representations (mathematically homomorphisms), Zhao [2009] provides sufficient information to be utilized for our purpose.

One needs to know that irreducible representations of the symmetric group S_n are in a one-to-one correspondence with Young Diagrams having n boxes. And a representation of the symmetric group on n elements, S_n, is also a representation of the symmetric group on (n − 1) elements, S_{n-1}.

Furthermore, one can describe a basis of each irreducible representation using standard Young Tableaux, which are numberings of the boxes of a Young Diagram with 1; 2; …; n such that the rows and columns are all increasing. For instance, the bases of the standard representation of S_3 correspond to the following two standard Young Tableaux relating to the partition (2,1) (Figure I.73) which are the only standard tableaux for (2,1).

$$\begin{array}{|c|c|}\hline 1 & 2 \\ \hline 3 \\ \cline{1-1}\end{array} \qquad \begin{array}{|c|c|}\hline 1 & 3 \\ \hline 2 \\ \cline{1-1}\end{array}$$

A general example of a standard tableau would be

$$\begin{array}{|c|c|c|}\hline 1 & 2 & 4 \\ \hline 3 & 5 & 6 \\ \hline 7 & 8 \\ \cline{1-2} 9 \\ \cline{1-1}\end{array} .$$

There are interesting connections for further discussions following essentially a *visual approach*. Suppose we have an irreducible representation in S_n and we want to find its induced representation in S_{n+1}. It turns out that the *induced representation* is simply the *direct sum* of all the representations corresponding to the Young Diagrams obtained by adding a new box to the original Young diagram! For instance, the induced representation of the standard representation from S_3 to S_4 is simply

$$\mathrm{Ind}_{S_3}^{S_4}\,\ydiagram{2,1} = \ydiagram{3,1} \oplus \ydiagram{2,2} \oplus \ydiagram{2,1,1}.$$

Similarly, the *restricted representation* can be found by removing a box from the Young diagram

$$\mathrm{Res}_{S_2}\,\ydiagram{2,1} = \ydiagram{2} \oplus \ydiagram{1,1}.$$

These relations and corresponding extensions become lucidly visualized by another approach interconnecting Young Tableaux and representations of S_n. In addition to partial order of Young Diagrams by the Diagram Lattice (Figure I.74) relations in terms of the “Young Lattice” *generalize the addition or removal of boxes*.

Following Zhao [2009] let the symbol $\lambda \nearrow \mu$ denote that μ can be obtained by adding a box to λ. To create a Young Lattice at the nth level all the Young Diagrams with n boxes are drawn. In addition, λ to connected to μ if $\lambda \nearrow \mu$., Figure I.190 displays the bottom portion (n = 4) of the Young Lattice which will extend infinitely upwards ($\varnothing$ is the NULL).

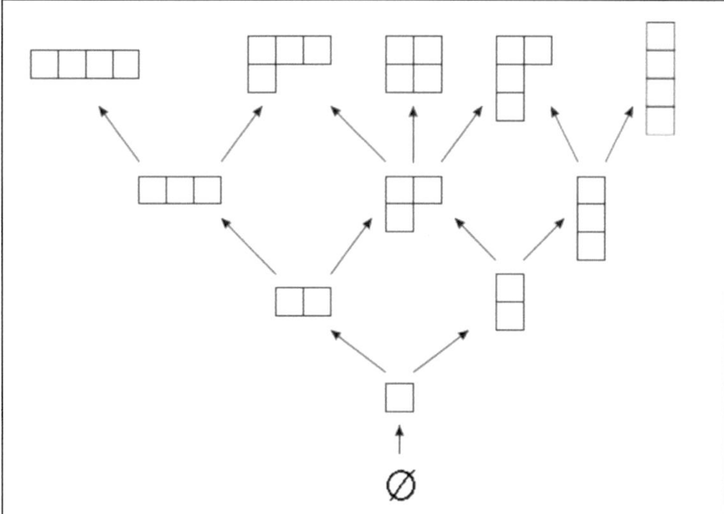

Figure I.190: The bottom of the Young Lattice [Zhao 2009].

We can think that a Young Diagram is a pile of bricks and the Young Lattice is the order in which bricks are placed.

The issue of firm development can be related to the question: What was the shape of this pile of bricks in the past and what would be a favorable shape for the pile in the future? Or could there be an adequate interpretation of the shapes when building a pile? In essence rows of the Young Lattice can represent homologies in biology.

To be useful with regard to the last question one must ask, can there be any meaning attached to the relation $\lambda \nearrow \mu$ with regard to "directions" or strategies chosen for further development of new firms?

To answer the last question one can use the example of entrepreneurship with the number of founders being $n \leq 4$ which covers almost all situations of technology entrepreneurship or we can use sources of financing (essentially $n \leq 8$).

For our purposes the discussion refers primarily to *structural characteristics*, such as looking at a founder team with three members as equivalent subjects. On secondary consideration we shall introduce *attributes of property or function* ("roles") of the structural entities, such as personal operational competencies of the individuals of founder team (Figure I.72). Adding attributes in this discussion is similar to using "colored symmetries" as displayed in Figure I.2.

If we attribute numbers to the boxes in Figure I.190 (which has nothing to do with Young Tableaux) all boxes in a row reflect only one entity (for instance, one member of a team) having a set of relevant attributes (for instance, personal operational competencies) which equals the number of the boxes in the row. In this view the outer left part of the Young Lattice (Figure I.190) means "*growth by learning*," adding an attribute to the same entity, such as learning of new subjects like a competency or chunk of knowledge.

As the extreme a single entrepreneur (Figure I.72) has all the needed personal competencies to found and develop a new firm. A lucid example coming into one's mind would be William Henry Perkin (A.1.2). In a related way concerning financial sources of a new firm one can define the attribute of a source of capital, for instance, as ten percent per box (Figure I.193).

Hence, one can view the left part of a Young Lattice with attributes, symbolized by $\lambda\ {}_{L}\!\uparrow \mu$, to put the emphasis on rows as "*centralization*" ("monopolization") of attributes and the right part focusing on columns, $\lambda \uparrow_{R} \mu$, as "decentralization" ("diversity"). In a Young Diagram the length of the leftmost column is the maximum number of non-equivalent objects/subjects. Hence, $\lambda \uparrow_{R} \mu$ may represent, for instance, adding a further person to a team with a particular useful competency or knowledge.

Furthermore, Ruch's Diagram Lattices allows a general description of comparing frequency distributions as an order relation. This can be extended to discuss a firm's organization or sub-states on the basis of Ruch's [1975] concept of the "*mixing character*," distinction of objects by relating classification to mixing character.

Accordingly, the mixing in the set of n objects is certainly at a maximum, if the number of distinguishable objects is maximal. This suggests that increasing mixing character should be defined by a *mixing process*. The *mixing character* allows comparing sets of differing mixing character. That is, a judgment can be obtained of "more" or "less mixed," at least for comparable (compatible) cases.

The **mixing character** denotes the composition of a set of partly equivalent objects or the distribution of objects among different states. Two sets are "*mixing-equivalent*," if the partition of the classes (structure) is the same for both. In Figure I.73 the founder teams of Puron AG and JPK Instruments AG would be structurally mixing-equivalent, but attributively different.

Generally, according to Ruch [1975], a statement that a set M is more mixed than another one, M', implies that the comparison must be restricted to pairs of sets such that M is obtainable by mixing together sets of mixing character of M'. We must define M to be more mixed than M' if M can be obtained by mixing sets with the same mixing character as M'. If the set finally obtained has a partition of objects which is an integral multiple of a partition of n, we may consider a corresponding set of n objects as equally mixed.

We can characterize all "mixing-equivalent" sets by a sequence of integers ν_l (including zero) in a column matrix or graphically by means of diagram-like figures in which rows with ν_l boxes are arranged along a vertical scale with indices $l = 1, 2, \ldots n$ denoting the different kinds of objects or the distribution of objects among different states.

Figure I.191 illustrates how to address *similarity* (including equality) for multidimensionality of underlying criteria. It provides an example of structurally mixing-equivalent sets of three subjects and of the representative diagram and the binary relationships which these exhibit with regard to structurally and attributive characteristics. In so far, these considerations may provide an option to discuss aspects of *heterogeneity* which is important concerning competencies or traits of members of a founder team or resources of a startup or NTBF, respectively.

Attributes refer, for instance, to personal operational competencies. Here we can differentiate *inequivalency* and *incompatibility*. However, we regard the structurally mixing-equivalent, but attributively mixing-inequivalent constellations having the same set of attributes as *similar*. For our purposes in the realm of small numbers, usually $n \leq 8$, graphics will suffice to work with issues of mixing characteristics.

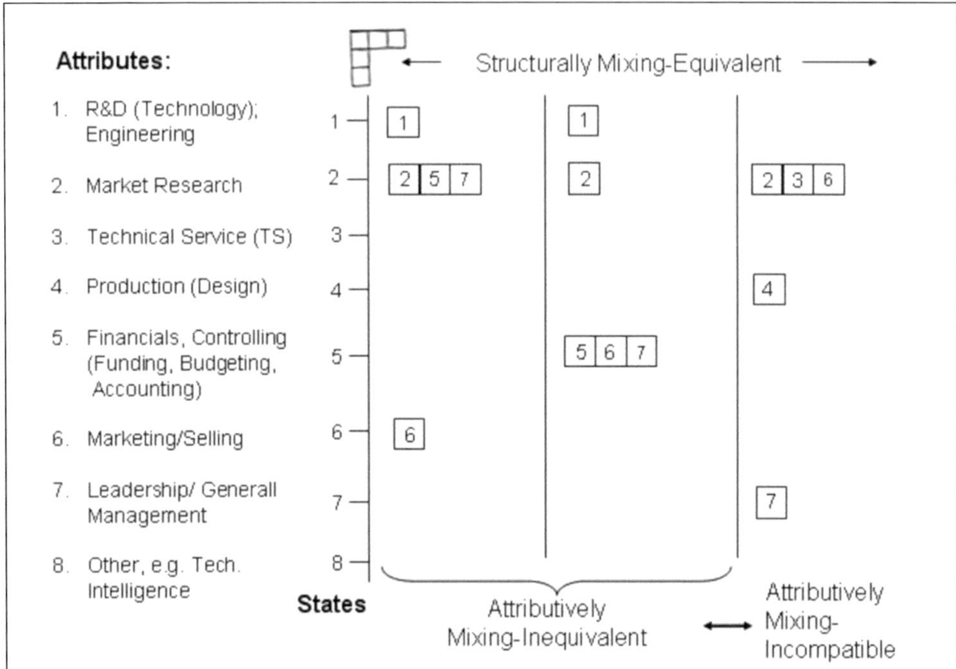

Figure I.191: Illustrating structural and attributive features with regard to their mixing characteristics referring to an entrepreneurial triple with regard to the personal operational competencies for firm foundation.

For future applications it should be noted that Young Diagrams allow also a mapping to (state) vectors which is shown in Figure I.192. The integers of the example partition can be mapped to a 2-dimensional vector. To obtain vectors with the same dimension within a Diagram Lattice one expands the 2-dimensional vector with zero elements.

Correspondingly, one could also separate structure and attributes denoting structure by a Young Diagram and attribute by a vector, respectively, reflecting the different attributes' value by a 1 and otherwise a 0 for the related vector's coordinates. For instance, for state 2 in Figure I.191 with attribute values (2,5,7) out of eight attributes one can write a binary value row vector:

(2,5,7) → (01001010)

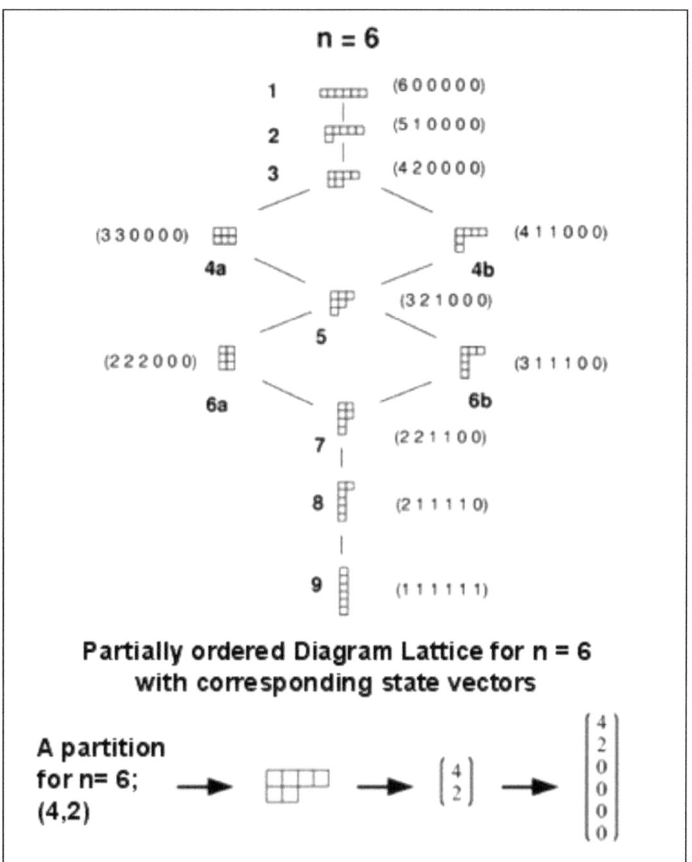

Partially ordered Diagram Lattice for n = 6 with corresponding state vectors

Figure I.192: Representation of Young Diagrams as vectors based on their partitions.

It is immediately evident that in Figure I.193 the founder teams concerning experience of the team members are equal in structure and attributes. To quantify "similarity" of entities with regard to attributes generally vector calculus can be utilized.

The related non-systemic and component-oriented *cosine similarity* may provide a rough measure. Cosine similarity is a measure of similarity between two vectors by measuring the cosine of the angle between them. The cosine of 0 is 1, and less than 1 for any other angle; the lowest value of the cosine is -1. For our case, however, the cosine similarity will be restricted to $0 \leq 1$.

The cosine of the angle between two vectors thus determines whether two vectors are pointing in roughly the same direction. In Euclidean space, the dot product of two unit vectors is simply the cosine of the angle between them. This follows from the formula for the dot product, since the lengths are both 1. Hence, we could measure cosine similarity by creating n-dimensional vectors, normalizing them to unit vectors, and then calculating the cosine similarity.

For instance, in Figure I.193 we compare operational competencies of two startups founded by an entrepreneurial triple. And we deal with eight-dimensional vectors. In the first step we construct the attribute vectors of the related Young Diagrams starting with (3,1,1) versus (2,2,2) and then add the three related vectors in the usual way to get in the first case (1,1,0,0,1,1,1,0) and 0.4472·(1,1,0,0,1,1,1,0) as the unit vector. Creating the second unit vector and multiplying both leads to cosine similarity = 0.73 for attributes – a rather high (quantitative) similarity. This, however, does not tell anything about the actual performance of the two startup teams. At best it is an indicator that both have a similar potential of performance.

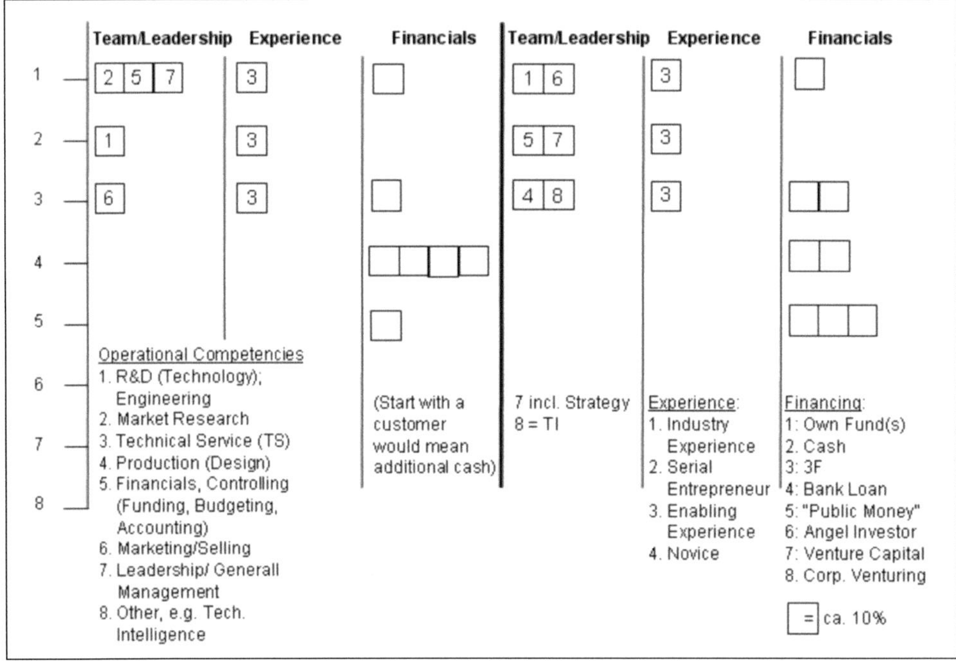

Figure I.193: Comparing architectures of two startups founded by an entrepreneurial triple focusing on selected resources.

One can extend this approach to deal with finances as another key resource of a startup. For the financial attributes we assume for simplicity equally weighted financial sources quantified by financial contributions expressed in multiples of 10 percent. Figure I.193 allows comparisons between the financial states of two startups. However, an immediate question arises concerning the underlying attributes, the origins of the money and the consequences for the founder(s) in terms of control over the startup.

For instance, debt and equity financing (Table I.28) will mean incompatible attributes if equity means also external control of the firm by VCs. Hence, it would not allow comparing startups with related financial sources. The financial states of the two new firms in Figure I.193 show compatible financial attributes which puts them into the same class of startups (Table I.74). They are structurally mixing-inequal.

For quantitatively assessing rough cosine similarity we give characterization of the financial sub-states by vectors with non-binary-valued coordinates (Figure I.192). This would mean (row) vectors (1,0,1,4,1,0,0,0) and (1,0,2,2,3,0,0,0) for the two firms and attributive cosine similarity = 0.70.

The outlined approach may be used to visualize architectures or even configurations of startups (ch. 4.3.2) which means *the firm's genetics* for a particular state in time. For instance, in Figure I.193 the initial configuration of two NTBFs or the startup thrust phase was displayed referring particularly to *resources* (knowledge/competencies, business experiences or foundation experience, respectively, and financial endowment).

A more elaborate visualization of architectures may connect structures and attributes to block forms of their dimensions as given in Figure I.194. It is envisioned that the firm is in the startup thrust phase (ch. 4.3.2) and having a customer for its foundation (initial configuration) means generating cash as part of the financial structure.

The NTBF's architecture given here shall not be discussed in detail. It should be noted how further resources may be integrated into such a diagram, such as classes of technology underlying the firm's offerings. And network partners and type of networking activity indicate how to tackle the firm's environment to represent also configurations of new ventures in a related manner.

Finally, also relationships between the firm's offerings, business activities and its revenue model (Table I.3) can be visualized by such block forms. And if we focus on the early stage of startups with say less than 15 employees one can envision also to deal with organizational states in terms of specialization as discussed in Table I.69 or given by their value chains (Figure I.7) including an advisory board.

Variations on a Theme

The preceding outlines focused on identifying *generic* relationships, where the attribute "generic" will be viewed as an "operational definition": the limited or even unlimited number of structural variations based on one particular selection of structural features/units or one particular function (application).

For entrepreneurship from a GST and permutation perspective corresponding variations (Figure I.194) can be related metaphorically to music, to V*ariations on a Theme*.

In music, variation is a formal technique where "content" is repeated in an altered expression (theme-and-variation form) and is recognizable by the addressee. Variation forms show up as "free-standing" pieces for solo instruments or ensembles, or can constitute a movement of a larger piece. Most jazz music is structured on a basic pattern of theme and variations.

Most famous and very illustrative for the current context are Johann Sebastian Bach's Goldberg Variations (BWV 988) and the first movement of the Piano Sonata in A Major (K. 331) or the finale of the Quintet for Clarinet and String Quartet in A Major of Wolfgang Amadeus Mozart.

These examples have been chosen as there are lucid demonstrations available referring to perception of variation on the *acoustical* and *visual* levels (inspecting the sheets of music) and to get a feeling for the "systemic nature" of the underlying patterns [Denk 2012].

Furthermore, the topic allows also to see that composers use pieces of other composers to generate variations, for instance, Johannes Brahms wrote "Variations on a Theme by Haydn." All this establishes interconnections to Faltin's [2007] article entitled "Founding successfully – The Entrepreneur as an Artist and Composer."

The Goldberg variations do not follow the melody of the aria, but rather use its bass line (and chord progression) and even without any music education and even if you cannot read music or have little feelings for music variations of the baseline are easily grasped by inspecting visually the notes or/and hearing them and recognizing the four eight notes sequences as "similar patterns."
(http://en.wikipedia.org/wiki/Goldberg_Variations)

> In music variation can concern a number of elements. In its most literal sense, for music a *melody* is a sequence of pitches (in German Tonhöhe) and durations of tones and a *theme* is the material, usually a *recognizable melody*. A *chord* is a sound created by a set of three or more different notes from a specific key that sound. A *rhythm* is the variation of the length and accentuation of a series of sounds or other events – also a matter of execution.

Hence, *variation* is a formal technique where material is altered during repetition: *reiteration with changes.* And changes may involve, for instance, the theme/melody, chords or rhythm or a contrast of major/minor mode.

Similarly, the realization of the *systemic character of a founder team* depends on the "measuring device" to bring it up. *Viewing* three music notes arranged vertically upon each other on a music note sheet may realize three individual "music note components." However, pressing on a piano the three related piano keys, the "chord." one *hears* a tone which do not allow most of us to discern the individual constituting tones of the notes – it is a "whole" (Figure I.193).

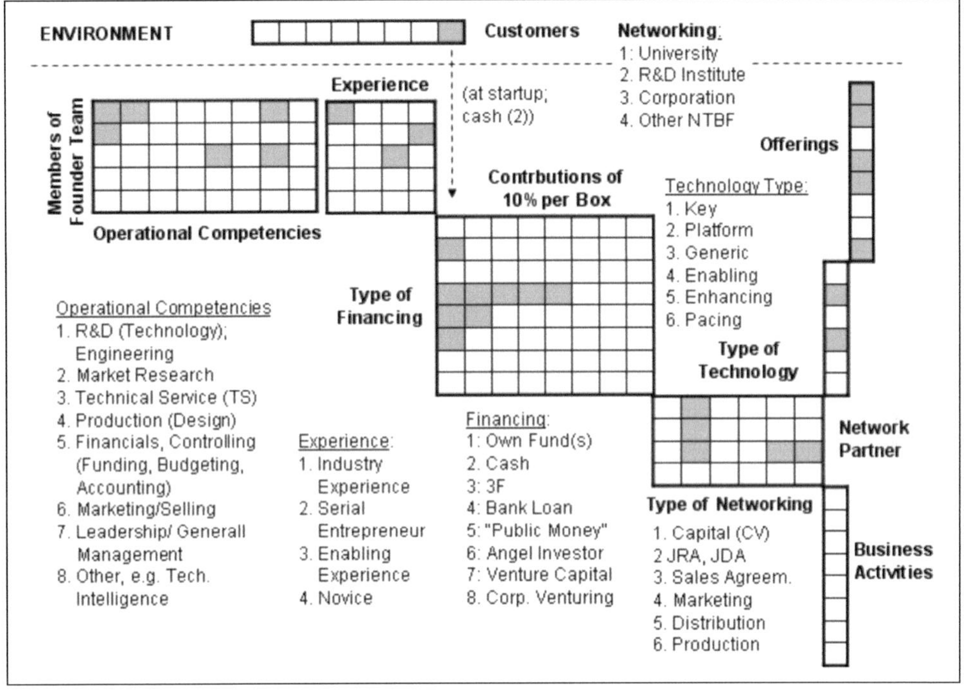

Offerings:
1. Product(s)
2. Tech. Processes
3. Services
4. Consulting
5. Contractual R&D
6. Analytics, Tests
7. IPR, Licenses
8. Knowledge

Early Phase Business Activities:
1. Business Development
2. Product/Process Development
3. Application Development
4. Project Management
5. Purchasing, Procurement
6. Selling Products/Processes
7. Managing Patents and Licenses
8. Stakeholders Relationships

Figure I.194: Architectural outline of an NTBFs startup thrust phase as block diagrams using related dimensions of involved entities (business activities not specified).

Chords can be related to collections of attributes of a single entity, such as a person with personal traits, experiences, attitude towards risk etc. or a financing structure of a firm (as seen in Figure I.193 and Figure I.194).

Correspondingly, for the core-shell model the "core-entrepreneur" or the new venture (Figure I.16) chords can be viewed to be transformed to particular extended chords (via interactions with super-systems) and variations may occur by changing the original chord (initial architecture) by additional systemic "notes" and accentuation (strengths of interactions of the systems) establishing configurations.

Entrepreneurship then is shaped by *environmental rhythm* (ch. 1.2.1) of the shells which provide the relevant exogenous variables (parameters), the drivers from the set of given shells in Figure I.13, for firm development over time.

A.1.7 Special Networking Effects for Entrepreneurship: The "PayPal Mafia"

According to Wikipedia [120] the "PayPal Mafia" is an informal term for a network of American business people, actually (serial) entrepreneur-investors, centered in Silicon Valley, who were founders or early employees of *PayPal* before founding a series of other technology companies. The PayPal Mafia are often credited with inspiring Web 2.0 and for the re-emergence of *consumer-focused Internet companies* after the dot-com bust of 2001. Some commentators consider these credits to be exaggerated or partly mythologized.

This group of serial entrepreneurs and investors represents a new, but special generation of wealth and power. In some ways they are classic characters of Silicon Valley, where *success and easy access to capital breed ambition and further success* [O'Brien 2007].

PayPal grew to serve the broader *market of electronic currency*, used in particular for online auctions. It went public in 2002 and was bought by eBay later that year for $1.65 billion, after eBay gave up on its own competing service, BillPoint. A number of serial entrepreneurs from the group worked with each other in the following years to form new companies, venture funds, and to make private equity investment in each other's companies, particularly in the field of social networking. Figure I.195 shows which companies dealt with in this treatise can trace their ancestries to PayPal.

It is not uncommon for a company's employees to leave and start successful new companies of their own after their old company is acquired. PayPal's former employees launched more successful companies in a shorter time than almost any other company in history. Apart from the networking effect the diversity of skill-sets among the former employees ensured that the social group has a full range of financiers, engineers, designers, operations experts, marketers and others available to help each other start new companies.

The group's name for itself, "PayPal Mafia" was already a minor cultural meme, but gained wide exposure due to a 2007 article in Fortune Magazine [O'Brian 2007] that used the term in its headline and featured most of the members posed at San Francisco's Tosca Cafe in gangster outfits. Members included and startups are listed in note 120.

Peter Thiel and PayPal's Development

At the center of the network there is Peter Thiel with a story of a money manager extraordinaire and special personality and the foundation of PayPal. While at high school, Peter developed into both a math genius and a chess prodigy. He achieved and has maintained his US Chess Master rating up to the present.

He picked nearby Stanford, deciding to major in philosophy, and became a free-market libertarian, believing that people should be permitted to do as they wished, assuming they did not impinge on the freedoms of others [Thomas 2010].

Following Thomas [2010], after graduating in 1989, Thiel decided on law school at his alma mater and earned his JD from Stanford. He clerked as a corporate lawyer for several organizations, but after a couple of months he resigned because of total boredom. Thiel needed the excitement and thrill of "the deal" to keep his considerable professional juices flowing. Finance has always been more his thing.

He decided to polish his totally theoretical investment skills by joining the firm of CS Financial Products, now part of the Credit Suisse Group. He quickly decided it was time to follow his own map, not someone else's. In California, he somehow raised $1 million from friends and family, beginning his first macro fund, Thiel Capital Management. With no experience, Peter faced daily struggles to raise funds and investors, but by 1998, he did have more than $4 million in his management portfolio.

This was about the time Thiel had the fateful meeting with Max Levchin, a Ukrainian-born American computer scientist. While he was conducting a finance lecture at Stanford, the young software engineer, Max Levchin, dreaming about an Internet startup, walked into his class by chance.

During a later meeting for breakfast of the two, Levchin asked if Thiel would invest in his idea to offer a secure method of allowing handheld computers to communicate. Peter liked the idea. Originally believing he would have a short-term relationship Thiel eventually froze his fledgling hedge fund career, dedicating the next four years to the new company – which eventually grew into PayPal. Thiel even joined the new venture as a co-founder and its CEO in December 1998, but together Thiel and Levchin set out to "*create the new world currency*." [O'Brien 2007]

A staunch libertarian, Thiel figured a Web-based currency would undermine government tax structures. Getting there, however, would mean taking on established Industries – commercial banking, for instance – which would require financial acumen and engineering expertise [O'Brien 2007].

Thiel had invested $240,000 in the new company after his meeting with Levchin. Only eight months after PayPal's IPO it was sold to the giant auction site eBay for $1.65 billion and Thiel realized a very attractive payout. Only 34, he rode off into the sunset with $60 million [O'Brien 2007].

Only weeks after the PayPal sale to eBay in 2002, Thiel decided to found a hedge fund firm, Clarium Capital Management, LLC, in his apartment. And, indeed, he became a successful hedge fund manager in the US. In 2005 Thiel started a San Francisco-based venture capital investment firm, The Founders Fund, with fellow members. Peter Thiel and the firm's six partners have been founders of or early investors in numerous companies, such as Facebook, PayPal, Napster, and Palantir Technologies. Founders Fund launched four suites of funds by 2010 with more than $1 billion in aggregate capital under management [Thomas 2010] [120].

The amazing success of PayPal was not without problems and serious issues. Shortly after PayPal began fulfilling the dream of its becoming the "new currency for the world economy," Thiel had to face a number of challenges. Russian hackers managed to pirate millions of dollars from the new venture by cribbing credit card numbers. Credit card processing companies claimed that PayPal was in violation of their regulations. Customer-service complaints flooded the phone lines and in-boxes and were often dealt with by simply not answering the phone or doing a mass deletion. Louisiana temporarily banned PayPal from doing business in the state; MasterCard threatened to pull the plug because of the high number of chargebacks [Thomas 2010; O'Brien 2007].

At one point, PayPal had enough funding to survive only another two months while still losing around $10 million a month. Shortly after, because of what was then the dot-com zenith, Thiel tried to raise money for PayPal, then valued at around $500 million by VC's even though losing money at a rapid rate. NASDAQ had just broken its own record, hitting 5,048, and the majority of investors thought the dot-com phenomenon would last forever [Thomas 2010].

Thiel already realized what the market came to know shortly thereafter: There was very little substance to the majority of dot-com "superstars." Nonetheless, he capitalized on the opportunity presented to him, worked feverishly to locate interested VCs and jumped through numerous obstacles to close the deal so quickly and rose $100 million to fund these hard times for PayPal. The closing for the deal was March 31, 2000. This was critical as the very next day the NASDAQ began its famous freefall [Thomas 2010].

Thiel established PayPal as the leading company to handle purchase payments over the "Net" and was successful at branding his company as the expert in this area. As a result, he began positioning the company for an IPO, which he registered only weeks after the World Trade Center ("9/11") tragedy. Again his strategy was successful.

Thiel was not always successful. He missed the opportunity to invest in the highly successful YouTube, bought by Google for $1.65 billion, (Thiel said, "It just kind of fell through the cracks." [Thomas 2010]). But he was very high on Facebook, founded by Mark Zuckerberg in 2004. Thiel backed the startup and advised Zuckerberg to relocate to Silicon Valley. Calling Thiel his mentor, Zuckerberg did just that. He said goodbye to Harvard University, where he was a student, and headed west (Box I.15).

In the context of the expected IPO of Facebook (in 2012) Facebook's first outside investor Peter Thiel led a $500,000 investment in Facebook in late 2004. He has 44.7 million shares that could be worth more than $2 billion. Accel Partners, whose principal partner, Jim Breyer, invested in the startup seven years ago (Box I.15), holds

201.4 million shares. Accel could have a thousandfold return on some of its investment [Bilton and Rusli 2012].

PayPal's Corporate Culture

For PayPal's development Thiel and Levchin had to bring in several hundred employees to what would become PayPal. They signed up more than 20 million users and burnt $180 million in funding before breaking even and selling out to eBay. The eBay deal indicated a remarkable factor of PayPal's success – obviously unique corporate culture and networking conditions. Most of PayPal's key employees left eBay, but they stayed in touch. [O'Brien 2007].

It is hypothesized by O'Brien [2007] that PayPal's success comes back to the early hires. Thiel and Levchin began recruiting everyone they knew at their alma maters. "It basically started by hiring all these people in concentric circles," Thiel rememberd. "I hired friends from Stanford, and Max brought in people from the University of Illinois."

They were looking for a *specific type of candidate*. They wanted competitive, well-read, multilingual individuals who, above all else, had a proficiency in math. Thiel and Levchin also wanted workaholics who were not MBAs, consultants, frat boys, or, God forbid, and jocks (an American term for a stereotypical male athlete). In other words, they were *looking for people like themselves*. For instance, Levchin lives to work. The company was male-dominated; nearly all employees were young men. [120]

"The difference between Google and PayPal was that Google wanted to hire PhDs, and PayPal wanted to hire the people who got into PhD. programs and dropped out," said Roelof Botha, PayPal's onetime CFO. "Most of them were very introverted anyway," Levchin recalled. "They'd come in, eat crappy food all day, and sleep under their desks." [O'Brien 2007]

Thiel's leadership style was as unconventional as his worldview. All employees, not just managers, were made aware in detail of company finances, performance etc. His management at PayPal (at least, pre-IPO) was the all-hands open-book session. Customer logs, revenue flow, fraud losses, burn rate: He showed it all for every employee to see. This access to information, coupled with the lack of offices, created a flat structure where any idea could win the day.

Company decisions were made according to reasoned arguments ("Good decision-making flows out of details") rather than executive experience. It was allowed and even encouraged for low level employees to criticize executive decisions and lobby for their own positions.

Reid Hoffman, a former executive VP, loved PayPal's meritocracy. "The group was very analytical," says Hoffman. "It was all about, 'Here are my arguments; here's my perspective.' You could never say, 'In my experience,' because experience wasn't there as a variable."

And O'Brien [2007] continued that the former COO David Sacks further opened the culture by establishing a no-unnecessary-meetings policy. He became a meeting cop. Anytime he saw a closed-door discussion happening, he had sit in for three minutes. If he considered the meeting to be valueless, he would declare it adjourned. Sacks recalled how the lack of meetings helped create a culture of many workers and few managers. Prestige was measured "by how few people there were above you who could prevent you from doing what you wanted to do."

Not everyone liked the PayPal vibe. Chief among the dissenters was Elon Musk. Musk who came to PayPal not through Levchin/Thiel's regime but during the company's merger with his Internet bank, X.com. Despite having perhaps the greatest entrepreneurial streak of all the PayPal mafia (see last paragraphs), Musk was purged from PayPal like some kind of toxin. Soon after the merger, Thiel resigned. Musk became CEO of the combined company and decided it was time for a technological overhaul. Specifically, he wanted to toss out Unix and put everything on a Microsoft platform.

That may sound innocent enough to laypeople but not to Unix zealots like Levchin and his team. A holy war ensued. Musk lost. The board fired him and brought back Thiel while Musk was on a flight to Australia for his first vacation in years. "That's the problem with vacations," Musk deadpanned. It was not just Musk; anyone who did not mesh with the Levchin/Thiel culture ran into trouble. X.com had a number of people from the banking industry who did not last long. And that awkwardness turned into total dysfunction and warfare. Most X employees ended up leaving or getting fired.

The infighting eventually stopped. It had to, because there were too many other issues to deal with. PayPal problems and losses were multiplying as described above.

Networking – The Soil of Entrepreneurship

By then, the PayPal Mafia was well established. PayPal's founders encouraged *tight social bonds among friends* who continued to trust and support one another despite their relatively short time together at PayPal. They call upon one another when they need money or advice – and when they need both, they go to Thiel, who seems to be at the center of it all (Figure I.195).

Many of PayPal employees were mining new territory. Chad Hurley, Steve Chen and Jawed Karim founded the ever more successful video sharing Web site YouTube, Inc, finally selling it to Google for $1.65 billion.

The now-famous Max Levchin founded Slide, a popular photo-sharing website. Reid Hoffman, Executive VP, started the successful LinkedIn Corp. for business networkers (ch. 3.4.2.1, B.2), while Vice President Jeremy Stoppelman began Yelp, helping people find restaurants, entertainment, businesses and shops in their local area [Thomas 2010]. Slide was sold to Google in August 2010 for $182 Million and, in August, Levchin joined as Vice President of Engineering. In August 2011, Google announced it was shutting down Slide, and that Levchin was leaving the company.

And both YouTube and Yelp learned a valuable lesson from PayPal: The *first idea is not always the best*. YouTube started as a video dating play. CEO Hurley rememberd his PayPal days as an education in business. When he arrived in California with a degree in art from Indiana University of Pennsylvania, building a successful company seemed like something other people did. “You never think it could happen to you,” said Hurley. “But seeing Peter and Max and the guys come up with ideas and seeing how to make things work gave me a lot of insight. You may not have a business degree, but you see how to put the process into effect. The experience helped me realize the payoff of being involved in a startup.” [O’Brien 2007].

Yelp, Inc. is a company that operates yelp.com, a social networking, user review, and local search Web site. Yelp.com had more than 54 million monthly unique visitors as of late 2010. Yelp was one of three projects, including Adzaar and Slide, to come out of the San Francisco incubator, MRL Ventures. The project arose out of research into the local services market by David Galbraith, who worked with Jeremy Stoppelman on the early stages of the project. Stoppelman and Russel Simmons, both of whom were early software engineering employees at PayPal, spun the service off as a separate company. After an aborted start as an email recommendation service, Yelp launched its namesake Web site into the San Francisco market in October 2004. [121]

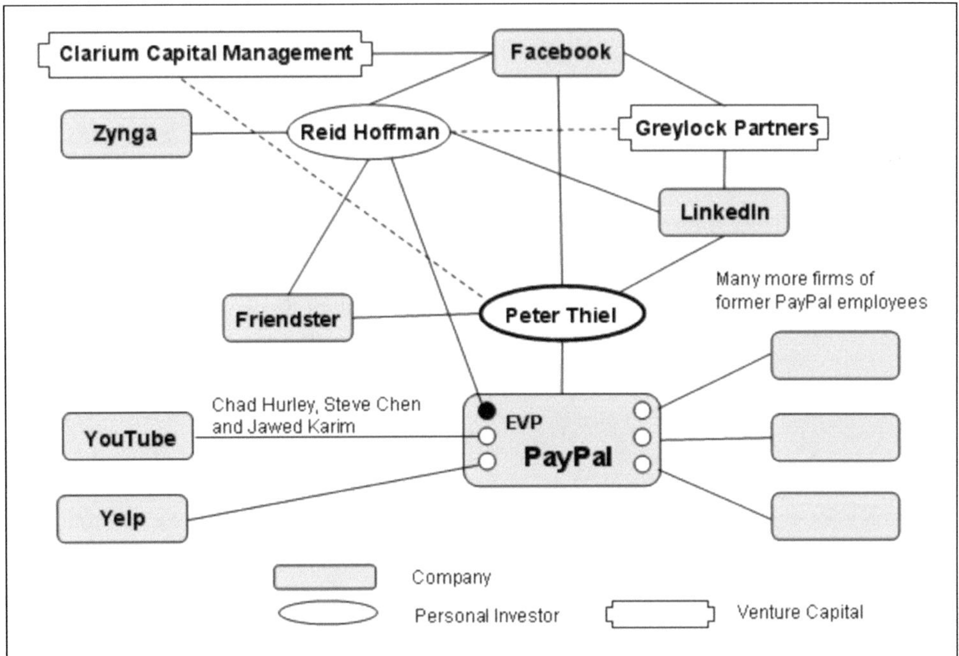

Figure I.195: Significant interrelationships of selected persons and new firms involng the “PayPal Mafia.”

In the context of the entrepreneurship of the PayPal context the aspect of "negative thinking" is notable.

Many successful businesspeople reject the idea of setting firm goals, but assume that sometimes the best way to address an uncertain future is to focus not on the best-case scenario but on the worst ("negative thinking") [Burkeman 2012].

"Positive thinking," by contrast, is the effort to convince yourself that things will turn out fine, which can reinforce the belief that it would be absolutely terrible if they did not. Prof. Saras Sarasvathy interviewed 45 *successful entrepreneurs*, all of whom *had taken at least one business public* ("serial entrepreneurs"). Almost none embraced the idea of writing comprehensive business plans or conducting extensive market research.

They practiced instead what Sarasvathy calls "effectuation." Rather than choosing a goal and then making a plan to achieve it, they took stock of the means and materials at their disposal, then *imagined the possible ends*. Effectuation also includes what she calls the "affordable loss principle." [Burkeman 2012].

With an emphasis on technology entrepreneurs and in the context of PayPal in this book a short addendum on *Elon Musk* is required. [127] Elon Musk (born 1971), who was born in South Africa and came to the US to study at the University of Pennsylvania received a BS in physics and BA in economics from the University of Pennsylvania, was not only a co-founder of PayPal. Before PayPal's sale to eBay he was the company's largest shareholder, owning 11.7 percent of PayPal's shares.

Drawing inspiration from innovators such as Thomas Edison and Nikola Tesla, Musk considered three areas he wanted to get into that were "important problems that would most affect the future of humanity," as he said later, "One was the Internet, one was clean energy, and one was space."

In June 2002, Musk founded his third company, Space Exploration Technologies (SpaceX) of which he is currently the CEO and CTO. SpaceX develops and manufactures space launch vehicles with a focus on advancing the state of rocket technology.

NASA selected SpaceX to be part of the first program that entrusts private companies to deliver cargo to the International Space Station (ISS). In December 2008, SpaceX was awarded a $1.6 billion NASA contract for 12 flights of their Falcon 9 rocket and Dragon spacecraft to the ISS, replacing the Space Shuttle after it retired in 2011. In seven years, SpaceX had designed the family of Falcon launch vehicles and the Dragon multi-purpose spacecraft from the ground-up.

Musk was also co-founder and head of product design at Tesla Motors, where he led development of the Tesla Roadster, the first production electric sports car. Musk's interest in electric vehicles extends long before the creation of Tesla. He originally went to Silicon Valley to do a PhD in Applied Physics and Materials Science at Stanford, where his goal was to create ultracapacitors with enough energy to power electric cars.

Musk provided almost all of the capital for Tesla's first two funding rounds and continued to invest in every subsequent financing round. As a result of the financial crisis in 2008 and a forced layoff at Tesla, Musk agreed to assume the additional responsibility of CEO.

Tesla Motors after having built an electric sports car and having shipped over 2,200 vehicles to 31 countries expected to be in production with its four-door Model S sedan by July 2012. In addition to its own cars, Tesla sold electric powertrain systems to Daimler for the Smart EV and Mercedes A Class, and to Toyota for the upcoming electric RAV4. Musk was also able to bring in both companies as long term investors in Tesla.

Finally, Musk provided the initial concept for the firm SolarCity, founded in July 2006 by brothers Peter and Lyndon Rive, where he remains the largest shareholder and chairman of the board. SolarCity is the largest provider of solar power systems in the US. His cousin Lyndon Rive is the CEO and co-founder. Musk's underlying motivation for funding both SolarCity and Tesla is to help combat global warming.

The aerospace industry reached a milestone in May 2012, when Space Exploration Technologies Corp. (SpaceX) became the first private company to dock a spacecraft with the ISS. The accomplishment also helped buttress plans by the NASA to pay private companies to transport cargo and crew to the orbiting station. But SpaceX was assumed to face huge challenges to turn its achievement into a thriving, long-term business – one that could help birth an industry of privately funded space ventures [Pasztor 2012].

The major question is whether Musk and his management team can transform SpaceX from a boutique development outfit into a low-cost, relatively high-volume production house, said aerospace industry officials and space experts. "Will they be able to reliably repeat this, and do it at the price they promised?" asked a former senior NASA official [Pasztor 2012].

APPENDIX B

B.1 Background Information on the NTBF Selections

Concerning NTBF selection we differentiate providing information on startups or other firms only in a specific context of the book's content or on those firms which were selected intentionally and elaborated as cases. The majority of selected cases cover critical early histories of mostly successful entrepreneurs whose firms had survived and grown through 8 – 12 years. This alone would reflect a "survival bias" and "hindsight bias" for the selections. Therefore, we looked also at some cases that showed NTBFs' failures to survive. Admittedly, many of them were wiped out during or after the Great Recession.

Concerning inter-cultural and socio-economic effects the emphasis is on German and US startups and existing firms with some firms from the UK allowing differentiating situations of culture in the US and UK which both follow Anglo-American capitalism for their economic systems.

Rationales for selection of cases according to technical subjects are given in Table I.1; for software/Internet startups the focus was on social networks (ch. 3.4).

Further selection criteria for cases referred to

- "Competitive groups" to inquire into competitive strategies, but also to provide more insights into related industry segments the new firms are operating in and
- "Strategic groups, particularly the so-called German "Hidden Champions" (ch. 4.1.1) and outstanding representatives (such as Enercon GmbH, Prominent Group and SAP AG).

Finally, on secondary thoughts, NTBFs were also selected to cover a broad range of modes of financing by the various sources available to technology ventures (ch. 1.2.7).

The focus of cases of new ventures with three or more representatives was

- Biofuels and biobased chemicals/materials
- Lighting/Optoelectronics (LED/OLED)
- Nano coatings/films
- Nano-tools/Scientific Instruments
- Ionic Liquids
- Specialty/fine chemicals and polymers and plastics
- I&CT = Informatics (HW/SW)/Consumer/Web Services/ incl. Bio- and Chemin-formatics.

Out of the global ca. 350 "plant-based biomass" firms and 200 algae-related startups we have, for instance, selected ca. 40 NTBFs for detailed discussions in the text, not as cases, but rather detailed in a specific context, the *biofuels industry* (A.1.1). Most of these NTBFs are "promising" biofuel firms. For their identification reference was made to Biofuels Digest which published on the Web the "50 Hottest Companies in Bioenergy" Here, "hottest" does not mean "best," "biggest" or "most significant" – it means the companies that are, in the readers' judgment, the *most worthy of attention*.

Selection of biofuels firms focused, furthermore, on representatives for biomass type, used conversion technologies and their interrelations, entrepreneurial history, interconnections with agricultural, oil, chemical and automotive industries, financing models and diversity of business models.

The selection of "successful" technology startups relied on assessments of NTBFs regarded as "technology pioneers," "most promising," "fast or strong growing," "high expectation" etc. expressed in terms of being awarded or nominated in contests for prestigious national or international awards and prizes provided by international NGOs, national governmental representatives, technical or business magazines, entrepreneurship-oriented for-profit firms like savings banks in Germany or big industry firms or consulting firms.

Approximately 80 percent of founders invited giving guest lectures within the author's Technology Entrepreneurship curriculum [1] were either rewarded with or nominated for a highly prestigious award or prize, respectively.

The most important of these assessments often referred to in this book (following [Bhidé 2000]) is the ranking of *Inc.* Magazine which is for the people who run growing privately held companies. The magazine publishes an annual list with rankings of the 500 (5000) fastest-growing private companies in the US, the "*Inc.* 500." or 500/5000 list categorized by industries. In addition to rankings by growth rates, since 2009 in *Inc.* Magazine a new ranking will also determine which small businesses have generated the most jobs. Most industry classes of *Inc.*, however, do not belong to the industries covered by technology entrepreneurship.

Another US magazine in this category with a focus on practice is "Fast Company." Through identification of the very creative individuals sparking change in the marketplace the magazine and Web site intends to uncover best and "next" practices, and thus may help new leaders work smarter and more effectively.

In any Systems Design (ch. 5.1) one is faced with the problem of determining the extent to which certain variables can be quantified and measured (ch. 4.1; Box I.17). Therefore, when making selections of NTBFs, we did not only utilized the various organizations which publish lists and/or rankings to select "successful" firms, but looked also into the basis, the criteria of their assessments, to gain insights into what features and data are regarded as relevant. The major sources for NTBF selection and their criteria are given and discussed in Box I.17 and sub-chapter 4.3.6.

B.2 List of NTBFs and Other Companies Surveyed by the Author

Startups/NTBFs investigated in detail for or during the preparation of the treatise as cases in separate documents are listed in Table I.102. The corresponding case documents that were publicly available when this treatise was published are listed in subchapter B.3.

Sharing entrepreneurial stories widely does not only create impacts for learning-by-example of founders or to-be founders but may also provide impacts on national policy issues.

The cases of the NTBFs will refer usually to no more than the first twelve years of their existence.

The typical gross structure of the cases is as follows:

- The Technology and the Market
- The Entrepreneur(s)
- Awards and Publicity
- Business Idea, Opportunity and Foundation Process
- Innovation Persistence, Expansion and Diversification
- Vision/Mission, Business Model and Risks
- Intellectual Properties
- Key Metrics
- Competition
- References and Notes.

Table I.102: Case documents prepared specifically for this book.

Firm Name (Country) [1, 5)]	Type of Technology	Industry / Customers	Prod. [2)] (yes/no)	VC [3)] (yes/no)
Bioniqs Ltd (UK) 5, 6)	Ionic liquids	Various industries, industrial and academic research	No	No
IoLiTec GmbH (DE) 5)	Ionic liquids	Various industries, industrial and academic research	Yes	No
Solvent Innovation (DE) 4)	Ionic liquids	Various industries, industrial and academic research	Yes	No

Table I.102, continued

Xing AG (DE)	Social networks (for professionals)	Industry, universities, research institutes	No	No
LinkedIn Corp. (US)	Social networks (for professionals)	Industry, universities, research institutes	No	Yes
Gameforge AG (DE) (Flaregames) 5)	Games on com-puters and mobile devices	Social network, consumers	Yes	Yes
Zynga, Inc. (US)	Games on com-puters and mobile devices	Social network, consumers	Yes	Yes
hte AG (DE) 4, 5)	Hard-/software; high throughput screening	Industry	Yes	Yes
CeGaT GmbH (DE) 5)	Genomics, bioinformatics, high throughput screening	Academic research, consumers, industry	No	Yes
Nanion Technologies GmbH (DE) 5)	Nano-tools, high throughput	Academic and industrial research: medicine, pharmaceutical industry	Yes	No
Nanofilm LLC (US)	Nano-coatings, nano-films	Industry, consumers	Yes	No
Nano-X GmbH (DE) 5)	Nano-coatings, nano-films	Industry, consumers	Yes	No
Nanopool GmbH (DE) 5)	Nano-coatings, nano-films	Industry, consumers	Yes	No
Industrial Nanotech, Inc. (US)	Nano-coatings, nano-films	Industry, consumers	Yes	No
InovisCoat GmbH (DE)	Multi-layer coatings, photographic films	Industry, consumers	Yes	Yes
WITec GmbH (DE) 5)	Nano tools, scientific instruments	Academic, in-dustrial R&D	Yes	No

JPK Instruments AG (DE)	Nano tools, scientific instruments	Academic, industrial R&D	Yes	Yes
Cambridge Nanotech, Inc. (US) 6)	Nano tools, scientific instruments; ADL	Academic, industrial R&D	Yes	No
Attocube AG (DE) 4)	Nano tools; scientific devices and instruments	Academic, industrial R&D	Yes	No
Vitracom AG (DE) 5)	Image processing; shop monitoring	Industry, communalities	Yes	Yes
ATMgroup GmbH (DE) 5)	Process, automation, measuring and inspection technologies, industrial image processing	Industry	Yes	Yes
SiGNa Chemistry, Inc. (US)	Fine chemicals, reagents, hydrogen	Academic, industrial R&D	Yes	?
ChemCon GmbH (DE) 5)	Fine chemicals, APIs, CRO	Academic, industrial R&D	Yes	No
ASCA GmbH (DE)	Fine chemicals, APIs, CRO	Academic, industrial R&D	Yes	No
Polymaterials AG (DE) 5)	Polymer, plastics, compounding	Industry	Yes	?
Novaled AG (DE) 4))	OLED, Displays, liighting	Industry	No	Yes
Zweibrüder Optoelectronics GmbH (DE) 4, 5)	LED, Lighting	Professional customers, consumers	Yes	No
MineWolf AG (DE/CH) 4)	Mine cleaning	Military, Governments, NGOs	Yes	?
Torqeedo GmbH DE)	Electromobility	Consumers	Yes	Yes
SunCoal GmbH (DE)	Bioenergy, biomass-to-coal	Industry, communalities	Yes	Yes
Renmatix, Inc. (US)	Biobased chemicals and biofuels	Industry	Yes	Yes
Enercon GmbH (DE)	Wind power, wind turbines	Industry	Yes	No

Quiet Revolution, Ltd. (UK)	Wind power, wind turbines	Professional customers, consumers	Yes	Yes
TimberTower Gmbh (DE) 5)	Wind power, wind turbines	Industry, Professional customers	Yes	Yes
SkySails GmbH & Co. KG (DE)	Wind power, energy efficiency, shipping	Industry	Yes	Yes
Marrone Bio Innovations, Inc. (US)	Biopesticides	Farmers, consumers	Yes	Yes
Heppe Medical Chitosan GmbH (DE)	Raw materials, natural products	Industry, chito-san for phar-maceuticals, cosmetics	Yes	Yes
Aluplast GmbH	PVC windows	Industry, professional customers and consumers	Yes	No
KWO Kunststoffteile GmbH (DE)	Plastics	Industry	Yes	No
Nanosolutions GmbH (DE) 6)	Nano-pigments, in ink jet printers or fountain pens	Consumers, industry	No	No
Zoxy Energy Sys-tems AG (DE) 6)	Batteries	Industry	Yes	Yes
MnemoScience GmbH (DE) 6)	Special plastics (with shape memory)	Industry, medicine	Yes	Yes

1) DE = Germany; 2) firm with production or anticipated production; 3) early financed by private venture capital including corporate venturing 4) acquired by a large firm; 5) presentations available on the Technology Entrepreneurship Web; 6) insolvent or bankrupt by 2012.

The startup of some firms is described in individual sub-chapters including Perkin & Sons (UK, founded in 1856, A.1.2), SAP (DE, founded in 1972, A.1.4) and PayPal (US, founded in 2000, A.1.7, ch. 4.3.2). The situation of four startups in ionic liquids with an emphasis on a technology push situation and the competitive constellation is presented in sub-chapter A.1.5.

Larger discussions in context are provided for the birth of Microsoft Corp., Cisco Systems, Inc. and Q-Cells AG.

Startups/NTBFs presented in tabular form or tackled in large text blocks in particular contexts with or without associated figures or in text boxes ("short stories") are given in Table I.103.

Table I.103: Startups/NTBFs tackled by larger descriptions in the text.

Firm Name (Country) [1)]	Technology	Firm Name (Country) [1)]	Technology
Verenium Corp (US)	Biofuels	Cobalt Biofuels, Inc. (US)	Biofuels
Virent Energy Systems, Inc.	Biofuels	Cilion, Inc. (US)	Biofuels
BlueFire Ethanol Fuels, Inc. (now BlueFire Renewables)	Biofuels	Range Fuels, Inc. (US)	Biofuels
CHOREN Industries GmbH (DE)	Biofuels	Mascoma Corp. (US)	Biofuels
Süd-Chemie AG (DE)	Biofuels	Coskata Energy, Inc. (US)	Biofuels
Bio Methanol Chemie Nederland BV (BioMCN) (NL)	Biofuels	Gevo, Inc.	Biofuels
ZeaChem, Inc. (US)	Biofuels	LS9, Inc.	Biofuels
Cellana, LLC (US)	Biofuels	Amyris Biotechnologies, Inc. (US)	Biofuels
Solazyme, Inc. (US)	Biofuels	HCL CleanTech Ltd. (now Verdia) (US)	Biofuels
Algenol Biofuels, Inc. (US)	Biofuels	KiOR Inc. (US)	Biofuels
Cyano Biotech GmbH (DE)	Biofuels	LanzaTech (US/NewZealand)	Biofuels
Sapphire Energy, Inc. (US)	Biofuels	Nanophase Technologies Corp. (US)	Nanotechnology

IGV GmbH (Institut für Getreideverarbei-tung) GmbH (DE)	Biofuels	Scionix. Ltd. (UK) 2)	Ionic liquids
Bioprodukte Prof. Steinberg GmbH" (BPS) DE	Biofuels	Google, Inc. (US)	Search engine and electronic media
Solix Biofuels (now Solix BioSystems) , Inc. (US)	Biofuels	First Solar	Photovoltaic, solar cells
ButylFuels, LLC (US)	Biofuels	US LED Ltd. (US)	LED, Lighting
Green Biologics Ltd. (GBL) (UK)	Biofuels	Albeo Technologies, Inc. (US)	LED, Lighting
OHB AG (DE)	Aerospace	MetroSpec Technology, LLC (US)	LED, Lighting

1) DE = Germany, 2) more detailed in Bioniqs case document.

NTBF cases including historical firms' foundations' dealt with in the author's previous book [Runge 2006] and updated as appropriate are listed in Table I.104.

Table I.104: Other case stories of firm foundations by Runge [2006].

Firm Name (Country, Foundation Year) [1)]	**Original Type of Technology**
BlueBioTech GmbH (DE, 2000)	Algae based additives to nutrition, cosmetics. Fine chemicals
Nanogate AG AG (DE,1999)	Nano-coatings, nano-films
SFC Energy (Smart Fuel Cell) AG (DE, 2000)	Fuels Cells (Direct Methanol Fuel Cells, DMFCs)
Puron AG (DE, 2001) 1)	Water treatment, membranes
Prominent GmbH (DE, 1960)	Water treatment, pumps, filters, chemicals
Osmonics, Inc. (US, 1969) 1)	Membrane separations, water treatment
vH&S GmbH (DE, 1971)	Aerospace, instruments, devices

Closure Medical Corp. (US, 1971; 1997) 1)	Adhesives, medical adhesives
Perkin & Sons (UK 1856)	Organic dyes
Bayer,AG (DE, 1863)	Organic dyes, pharmaceuticals
The Dow Chemical Company (US, 1897)	Chemicals
Avery Dennison (US, 1935)	Adhesives, pressure-sentive adhesives (PSA)
Henkel AG & Co. KGaA (1876)	Washing, adhesives
The Eastman-Kodak Company (US, 1899)	Photography, apparatus and films
Röhm & Haas (DE, 1907); Rohm & Haas (US)	Chemistry
Perstorp Holding AB (SE, 1881)	Chemistry
The Prussian (Berlin) Blue Endeavor (DE, 1704)	Inorganic dyes/pigments

1) Acquired by a large firm

B.3 Publicly Available Case Documents of Companies

Table I.105 provides the list of documents of NTBF case stories of those firms of Table I.102 that are published together with the treatise (on the KIT EnTechnon Web site – Downloads). Other documents will follow regularly over 2014/2015.

For firms marked with an asterisk corresponding presentations of founders are also available on the Technology Entrepreneurship Web. [1]

Table I.105: Published case stories of firms.

Firm Name (Country) 1)	Firm Name (Country) 1)	Remarks Concerning the Cases
IoLiTec GmbH *)	Gameforge AG	The pairs Gameforge vs. Zynga and Xing vs. LinkedIn allow comparing Germany vs. US IoLiTec, Bioniqs (including Scionix) and Solvent Innovation is a competitive group of startups
Solvent Innovation GmbH	Zynga, Inc. (US)	
Bioniqs Ltd. (UK)	Xing AG	
Nanion Technologies GmbH *)	LinkedIn Corp. (US)	
Novaled AG *)	Nanopool GmbH 2)	

1) German firms if not stated otherwise; 2) as a presentation.

Glossary

This glossary is an alphabetical list of terms in a particular broad area of knowledge with the definitions for those terms. The list includes terms that are either newly introduced or specialized and aims to cover several domains of knowledge accounting for the many scientific and technical disciplines which are relevant for technology entrepreneurship.

It is also a reference to notions in the text for which definitions are used that are sometimes differently understood in various scientific disciplines or sometimes have slight differences in a particular discipline to have the basis for common understanding among disciplines.

Cross-references to other terms in the glossary are indicated in the list by italics face.

Apart from the many possibilities to look up definitions, notions and terms on the Web the work of Dorf and Byers [2007] provides a glossary directed toward business administration vocabulary.

Advisory board (or board of advisers):
A group constituted to provide advice and contacts ("networking") to a venture by members with distinct skills and knowledge – and by level of reputation of its members it may also add credibility to a venture.

Agent:
Participants who play a role in achieving the objectives or changes of a system; attendees of the various programs.

Ascribed value (or imputed value):
A systemic category which is generated by consensus or common interests and behavior of a social group in attributing value. It refers usually to valuating or perceiving value of a current situation or object or specifically a firm and the expectation that it will provide future socio-economic benefit – often related to (technology) speculation.

Asset:
Something with economic value that an individual, corporation or country owns or controls with the expectation that it will provide future benefit.
Assets are acquired to increase the value of a firm or benefit the firm's operations.

Backward-integration:
Means the situation in which on a corporate basis a plant or business (of a firm) is interrelated to an *upstream* plant or processing facility for producing its offerings (cf. *forward-integration*).

Book value:
The net worth (net asset value) of a firm according to accounting rules, calculated by total assets minus intangible assets (patents, goodwill) and liabilities.

Breakeven:
Is the point at which cumulated income (revenues, sales) equals loss (expenses, cost).

Brand:
A combination of name, sign or symbol that identifies the goods or services sold by a firm.

Business model:
A business model is an organization's core logic for creating value; a hypothesis how to create value for all its stakeholders.

Business plan:
A business plan is a structured document that includes a current and projected description of a new venture or a business of an existing company, its offerings like products and/or services and related market(s) and how the business will achieve its goals in a particular environment.

Cannibalization:
Processes of introducing offerings, mostly products that will compete with, usually even replace, existing products of a firm.

Cash flow:
Means the transfer of cash into or out of a business, project or financial product (note that the word cash is used here in the broader sense, where it includes bank deposits). It is usually measured during a specified, finite period of time.

Complement:
A product that improves or perfects another product.

Certainty:
Under certainty there is complete knowledge of the value of the outcomes and of the occurrences of the states (of the system).

Closed-loop system:
A self-regulated system is called a closed-loop system and it has its output coupled to its input.

Cluster:
Here, a cluster is a network of interconnected companies and/or organizations with spatial proximity of the nodes (organizational components) and similar or related activities of the nodes.

Communication:
Covers interactions of entities (people or things) and behavior; with regard to people it may relate to actions to work together for a given goal, purpose, function or effect.

Competitive advantage:
An advantage that a firm has over its competitors; how a firm could gain and sustain an advantaged position at potential customers. There can be many types of competitive advantages including the firm's cost structure, product offerings, distribution network and customer support.

Cognitive framework:
Cognitive psychology explains human behavior broadly by examining the "cognitive frameworks" that are used to interpret (how to see) the world and change attitude, behavior and activities accordingly.

A (holistic) system of reciprocation which is a relation of mutual dependence or action or influence over time. A cognitive framework of an individual appears as an internal mental counterpart of a state which is induced by perceiving a situation as the external counterpart.

Complexity:
Complexity in the context of systems expresses a condition of the number of components in a system and the numerous forms of relationships among the components. The numbers may be very large or even infinite, but remain enumerable in the mathematical sense.

Configuration:
In the context of a system, particularly a firm, its situation or state, respectively, characterized by endogenous (system-internal) variables combined with exogenous (external) variables (called "parameters") of supersystems. All the factors are interdependent and interacting, such that their effects may be enhanced or diminished.

Control:
The purposive influence or enforcing power toward a predetermined goal involving continuous comparison of current states to future goals ("is" versus "shall" assessment).

Coordination:
For living systems comprising components and/or subsystems coordination together with sub-ordination in the system exists or is developed to produce an output or outcome or achieve a goal. It is seen as a source of competitive advantage.

Core competency:
It is the one thing that a company can do better than its competitors; an area of specialized expertise that is the result of orchestrating complex streams of technology and work activities and processes, including building and keeping unique relationships with customers, suppliers, research, development or marketing partners, and operational agility or unique business practices.

Customization:
Provision of a product or service designed to meet a customer's or user's preferences or specifications.

Decision-maker:
A decision-maker is someone who is internal to a system and who can change the performance of the parts. Responsibilities for the guidance of the system toward achievement of its objectives are with decision-makers, managers and agents.

Directed Evolution:
Describes a set of *techniques* for the iterative production, evaluation and selection of variants of a biological sequence, usually a protein or nucleic acid (http://dbkgroup.org/direvol.htm);
Directed evolution allows exploring enzyme functions never required in the natural environment and for which the molecular basis is poorly understood. Its purpose is, for instance, to produce useful biocatalysts. With directed evolution one now has the abil-

ity to tailor individual proteins as well as whole biosynthetic and biodegradation pathways for biotechnology applications.
Francis H. Arnold Research Group. Evolution, Synthetic Biology, Protein Engineering, Biocatalysis, Biofuels.
(http://www.che.caltech.edu/groups/fha/, http://www.che.caltech.edu/groups/fha/).

Discovery:
Describes a novel observation or finding of something already existing, often of a natural phenomenon or effect of a (natural) product.

Driver:
Is a relevant endogenous variable or exogenous parameter of a model which provides sufficient power ("strength") or influence to explain (and probably "predict") a system's state and development. Drivers are those combinations of factors which suffice to determine an observable response of the system.

Dominant Design:
Means that, after a technical innovation and a subsequent era of digestion and progressive developments in an industry, a basic architecture of a product or process becomes the accepted market standard.

Due diligence:
A process of gathering and verifying facts, data and information in plans or purposive documents, such as a business or project plan or description of a firm to be acquired, before making a commitment to the terms of an investment, or firm merger or acquisition.

Entry barrier (barrier of entry):
A factor that keeps a firm from entering an industry or a market.

Dynamic capability:
The ability to build and develop firm-specific capabilities and, simultaneously, to renew and re-configure the firm's competencies in response to key factors and conditions of the environment.

Economics:
Economics is the social science that analyzes the production, distribution, and consumption of goods and services (http://en.wikipedia.org/wiki/Economics) and sharing of information. Information may have economic value because it allows individuals to make choices that yield higher expected payoffs or expected utility than they would obtain from choices made in the absence of information.
Economics is "the science which studies human behavior as a relationship between ends and scarce means which have alternative uses." The subject thus defined involves the study of choices as they are affected by incentives and resources (http://en.wikipedia.org/wiki/Scarcity).

Efficacy:
A measure of the extent to which a system contributes to the purposes of a high-level system of which it may be a subsystem.

Entity:
An entity is something that has a distinct, separate existence, though it need not be a material existence. In general, there is also no presumption that an entity is animate. Entities are used in system developmental models that display communication and internal processing of, say, documents compared to order processing (http://en.wikipedia.org/wiki/Entity.

Equifinality:
Means a system can reach the same final state from different initial conditions and by a variety of paths (the ability to reach a goal from myriad ways and beginning at various locations). For open systems this option of finding equally valid ways is the expression of equifinality.

Equivocality:
The state or quality of being ambiguous in meaning or capable of double interpretation. It is viewed as the existence of multiple and conflicting interpretations about an organizational situation or situations where multiple meanings information or information patterns exist among people (striving for the same objectives).

Environment:
In a systems approach any system is viewed in relation to all other systems larger than and interfacing with itself. Such a "Whole System" comprises all the systems deemed to affect or to be affected by the problem at hand. Within a Whole System the environment is defined as comprising all the systems (subsystems and supersystems) over which a decision-maker of a given system has no control.

Environmental rhythm:
The environment within which the entrepreneur operates may have certain regularities or patterns. An environmental rhythm exists when patterns in the environment vary over time with some regularity. Its recognition, however, is perhaps not simple or easy.

Epistemology:
Is the theory of knowledge and the branch of philosophy concerned with the nature and scope (limitations) of knowledge focusing on 1. What is knowledge? 2. How is knowledge acquired? 3. What do people know? 4. How do we know what we know? (http://en.wikipedia.org/wiki/Epistemology).

Escalation:
An expression of a binary relation originating with a special phenomenon or event and showing up either as a significant increase or rise, lifting something's extent, volume, number, intensity or scope stepwise to a higher level or a corresponding decrease.

Established business ownership:
Percentage of population aged 18-64 years who are currently owner-manager of an established business, for instance, owning and managing a running business that has paid salaries, wages, or any other payments to the owner for more than 42 months (according to Global Entrepreneurship Monitor).

Exit:
The way investors or entrepreneurs get their money out of a venture making a significant profit.

Expert:
A person with considerable experience (and knowledge) in a certain field – someone who knows the most serious mistakes one can make in a field and how to avoid them.

Factor market:
A market where the factors of production (conversion) are bought and sold, such as the labor markets, the capital market, the market for raw materials and the market for management or entrepreneurial resources.

Feedback:
The regulatory mechanism of *closed-loop systems*. It is the modification or control of a process or system by its results or effects; output of an action is "returned" (fed-back) to modify the next action.
"*Negative* feedback" is a process in which an initial change will bring about an additional change in the opposite direction. A "*positive* feedback" is a process in which an initial change will bring about an additional change in the same direction. In positive feedbacks, a small initial perturbation can yield a large change which is *self-reinforcement*. A produces more of B which in turn produces more of A.

Feedforward:
It is a control mechanism that can be measured but not controlled. The disturbance is measured and fed forward to an earlier part of the control loop so that corrective action can be initiated in advance of the disturbance having an adverse effect on the system response.
In industrial processes when some output of an earlier step is fed into a step occurring down the line; self-fulfilling prophesy: if people believe the stockmarket is going to rise, their purchases drive up the stock prices thus creating the very situation they believed will happen

Finality:
A term used to describe the goal-seeking nature of systems, that is, achieving a predefined future state. Open systems have equally valid alternatives easy of attaining the same objectives from different initial conditions (cf. *equifinality*).

Forward-integration:

The situation in which on a corporate basis a plant or business (or a firm) is interrelated to a *downstream* plant or processing facility for producing its offerings (cf. *backward-integration*).

Function:
Means that something is used for; serve: serve a purpose or *role*; a form/structure or activities of a system to achieve a particular purpose or goal or a specific subsystem to contribute to purpose/goal achievement of the supersystem, usually associated with an order of processes (cf. also the division/department/function of a firm).

Gatekeeper (Boundary Spanner):
Someone acting as an interface, if "differences" between intervening parts are too large to allow direct contact and communication between the parts. For instance, a "technical gatekeeper" interconnects various scientific and technical disciplines or corporate-internal and external research.

Genetic Engineering:
Genetic engineering, recombinant DNA technology, genetic modification/manipulation (GM) and gene splicing are terms that apply to the direct manipulation of an organism's genes.
Genetic engineering uses the *techniques* of molecular cloning and transformation to alter the structure and characteristics of genes directly.
(http://en.wikipedia.org/wiki/Genetic_engineering)

Genome:
A genome is the total of all an individual organism's genes. Thus, "genomics is the study of all the genes of a cell, or tissue, at the DNA (genotype), mRNA (transcriptome), or protein (proteome) levels." (US Environmental Protection Agency).

Genomics:
A branch of genetics that studies organisms in terms of their *genomes* (their full DNA sequences) – (http:// wordnetweb.princeton.edu/perl/webwn);
a branch of biotechnology concerned with applying the *techniques* of genetics and molecular biology to the genetic mapping and DNA sequencing of sets of genes or the complete genomes of selected organisms, with organizing the results in databases, and with applications of the data (as in medicine or biology).
(http://www.merriam-webster.com/dictionary/genomics)

Goal: – *Objective*
A goal is where you want to be; it focuses on a qualitative statement. It is a broad, general, tangible, and descriptive statement. It does not say how to do something, but rather what the results will look like.
Some common business goals are: being always profitable, achieving sustainable competitive advantage, customer loyalty, etc.

Goodwill:
In law and accounting, an intangible asset constituting a value over and above the valuation of the tangible assets of the business, and representing all benefits derived from the distinctive location, trademarks, credit rating, reputation, and patronage of the business.
On the sale of a business, a charge usually is made for the goodwill as one of the assets. Sometimes goodwill may be sold by itself without the transfer of any other assets; for example, a business that is moving to another locality may sell the right to use its name and to occupy its former premises.
(http://encarta.msn.com/encyclopedia_761567903/Goodwill.html)

Hard system: – Soft System
A characterization of a system with regards to determining the extent to which certain variables can be quantified and measured and lines of reasoning. "Hard" systems usually will admit formalized reasoning processes where logico-mathematical and analytical-mechanistic derivations, causality and quantitative approaches using the "Scientific Method" are prevalent. Typical domains cover physical sciences, engineering and chemistry and systems are usually treated as "closed" ones.

Human-activity system:
A social system and open system dealing primarily with *goal-seeking subjects*. It is a *purposeful* system directed toward the achievement of a final state, the *goal* – usually a man-machine system with "subjects" and "objects" and characterized by exhibiting *organized complexity*.

Implementation:
Is the use or adoption of change; the actions of accomplishing some goal or executing some order; to put into practical effect; executing given procedures.

Impact value (in German Wirkwert)**:**
A value that is not directly associated with a tangible or intangible entity *per se*, but emerges in a particular situation or constellation and may be related to a particular time period. (Shakespeare, Richard III: "A horse, a horse, my kingdom for a horse!)

Indicator:
In social sciences the decision how to relate measurement to a particular "observable" ("metrics") is often associated with its operational definition, an indicator assumed to reflect a variable.

Influence:
A capacity to change the behavior of other individuals, to get them to do something that they would not otherwise do.

Information asymmetry:
A state of social group interactions where one or several individuals of the group has more or better information than the other ones; or, at least, one party has information relevant for a subject under consideration whereas the other(s) do not. This could lead to imbalance for decision-making or power which may become counter-productive for the group.

Intangible:
Nonmaterial: lacking material qualities, and so not able to be touched or seen (http://encarta.msn.com/encnet/refpages/search.aspx?q=intangible);
Antonym – Tangible: able to be realized, capable of being given physical existence tangible – financial benefits.

Intelligence:
"Intelligence is knowledge and foreknowledge of the world around us – the prelude to {Presidential} decision and action" (US CIA Factbook on Intelligence).

Interdisciplinarity:
Interdisciplinarity "is based on the integration of ideas from across fields and directed towards a common goal. In this regard it is essential that those involved have a fundamental understanding of the core concepts of the area, its research traditions or themes and the basic questions under consideration."
A multi-disciplinary approach – that is often confused with interdisciplinarity – generates little or no cooperation between areas.
(A Moral Tourism Industry? Release: Jul. 29, 2010.
http://www.hotelmule.com/html/72/n-3072-3.html

Intervening variable:
In social science intervening variables ((“latent variables”) are *hypothetical* internal states (constructs) that are used to explain relationships between observed variables, such as independent and dependent variables. They are not real things; they cannot be seen, heard, or felt. They are *interpretations of observed facts*, not facts themselves. But they create the illusion of being facts. Typical examples include personality, traits, memory or learning.

Leadership:
Is a process of social *influence* in which one person (or a coherent group) can enlist the aid and support of others in the accomplishment of a common goal. Leadership is reflected by *a purposive collective or group process* and is ultimately concerned with fostering change directed toward some future end or condition which is desired or valued. Leadership is about people (and “doing the right things”).

Learning Curve:
A learning curve shows the rate of improvement in performing a task as a function of time, or the rate of change in average cost (in hours or €/$) as a function of cumulative output.
It is a (graphical and/or mathematical) representation of the common sense principle that the more one does something the better one gets at it (the more times a task has been performed, the less time will be required on each subsequent iteration).

Management:
Management is a process and the art, or science or practice, of setting and achieving *objectives* utilizing and coordinating appropriate *resources* including people in order to attain them with least cost and minimum waste which means attaining the best return on such resources by getting things done efficiently. Management is about business results and processes (and “doing the things right”).

Measurement:
Is the assignment of numbers (or numerals) to represent attributes (properties). Numerals possess order only because of arbitrary assignment or mere convention. One of the first requirements of measurement is the determination of the appropriate scale in which the attribute in question could be mapped. Prevalence of measurement scales differs for hard and soft sciences.

Mental model:
Mental models are representations (for instance, connected information), about a particular topic or subject. Mental models include not only cognitive information, but also feelings and motives in regard to the particular topic. The subject of mental models often involves aspects of the self (for instance, self-concept in regard to spelling), and aspects of the world (for instance, beliefs about a competitive firm).

A mental model is an explanation in someone’s thought process for how something works in the real world; it reflects conscious or subconscious perceptions of reality.

Metabolic engineering:
Metabolic engineering is the practice of optimizing genetic and regulatory processes within cells to increase the cells' production of a certain substance. Producing beer, wine, cheese, pharmaceuticals, and other biotechnology products often involves metabolic engineering (http://en.wikipedia.org/wiki/Metabolic_engineering).
Means targeted and purposeful alteration of metabolic pathways found in an organism in order to better understand and use cellular pathways for chemical transformation, energy transduction, and supramolecular assembly.
(http://www.metabolicengineering.gov/)

Network: – *Cluster*
An interconnected system of entities denoted as "nodes" (such as people, firms or things) irrespective of distance. Interconnections may be "hard" like electric lines in a computer network or "soft" via human relationships, interactions and communication as for a *social network*.

New Technology-Based Firm (NTBF):
An entrepreneurial organization with the goal to actively create, develop, and/or commercialize offerings based on technology and/or research, particularly innovative products, processes, applications and services, which is no more than 12 years in operation and which is usually still led by the original founder or founder team or, at least, one member of the founder team.

Objective:
An objective or *goal* is a projected state of affairs that a person or a system plans or intends to achieve – a personal or organizational desired end-point in some sort of assumed development. Many people endeavor to reach goals within a finite time by setting deadlines (http://en.wikipedia.org/wiki/Objective_(goal));
but, an objective is a specific, measurable, actionable, realistic, and time-bound situation that must be attained in order to accomplish a particular goal. Objectives define the actions that must be taken within a time period to reach the goals. For example, if an organization has a goal to "grow revenues," an objective to achieve the goal may be "introduce 2 new products by the third quarter (Q3) of 20xy." Other examples of common objectives are, increase revenue by x% in 20xy etc.
(http://www.fastcompany.com/blog/dan-feliciano/lean-six-sigma-rock-star/do-you-know-difference-between-goal-and-objective)

Open system:
An open system has an *environment* with which it has inflows and outflows, for instance, of material, energy, information – or people. It possesses other systems with which it relates, exchanges and communicates (for instance, shares information). All systems with living components are open systems. In particular, man-machine systems with "subjects" and "objects" are open systems.

Opportunity:
A timely and favorable juncture of circumstances providing a good chance for a successful venture [Dorf and Byers 2007:28] (in German often translated as "unternehmerisches Handlungsfeld").

Organizational learning:
Comprises the acquisition, application, and mastery of new information and intelligence, tools and methods that allow more rapid decisions and improvement of those processes which are critical to the success of an organization.

Organized complexity: – *Complexity*
Organized complexity has as the main feature that there are only finite, relatively small numbers of components and relationships in the system. Organized complexity is what we usually encounter dealing with new firms (with "small" numbers of employees, say number is < 30).

Outcome: – *Result*
An outcome is any result or consequence, good or bad, intended and unintended, desired or undesired.

Parameter:
An observable or measurable factor exogenous to a system forming one of a set that influences or defines the conditions of the system's operation in the sense of a variable.

Partnership:
A *legal* partnership is created when two or more people work together with a view to make a profit. Legal partnership means that partners are jointly and separately responsible for all the partnership's debts and liabilities.

Performance:
Is the quotient of Actuality (A) and Potentiality (P) when A corresponds to the current achievement of a system using existing resources and constraints and P is what could be achieved by developing resources and removing constraints (Performance ~ A/P). Performance is a relation between "what is" and "what could be," or verbalized "more with the same or even more with less." Change of productivity is an indicator of performance. In essence, in the current context performance can be related to the first derivative of *productivity*.

Plan: – *Strategy*
When you know *what you want to do* and exactly *how to do it*. A plan is characterized by knowing what the next step will be.
Each step is designed by taking into account the next step.

Positive feedback*:* – *Self-reinforcement*
Positive *feedback* occurs in a feedback loop when the mathematical sign of the net gain around the feedback loop is positive. That is, positive feedback is in phase with the input, in the sense that it adds to make the input larger. Positive feedback is a process in which the effects of a small disturbance on a system can include an increase in the magnitude of the perturbation. Positive feedback tends to cause system instability.
(http://en.wikipedia.org/wiki/Positive_feedback)

Procedure:
A specified course of action intended to achieve a result; a *prescribed* process; a *structured* process by an explicitly given, often documented order of activities according to steps, rules and execution conditions;
how to execute a process, is repeatable.

Process:
Ordered activities to achieve a goal, purpose or function; steps may be sequential (finite, prescribed steps (for 1 to n) or not prescribed, parallel, may be branched (if ...then; case x = 1, ...case x = n), looping (do ...while) or facultative (do X out of Ys).

Productivity:
An economic measure of output per unit of input. Inputs include labor and capital, while output is typically measured in revenues, for instance, revenue per number of employees. Capital and labor are both scarce resources, so maximizing their impact is always a core concern of modern business (cf. *performance*).

Program:
A program is coded or prearranged information (actually "instruction") that controls a process (or behavior) leading it toward a given end; it is a planned sequence and combination of activities designed to achieve specified goals, also a planned series of future events.

Program structure:
Is a classification scheme relating the activities of an organization according to function they perform and the objectives they have been designed to meet. The program structure may cut across formal organizational (and other) boundaries.

R&D intensity (research intensity):
The proportion of R&D expenses in relation to the overall revenues (sales) of the firm in percent.

Recursion:
In mathematics a recursive definition (or inductive definition) is used to define an object in terms of itself. Most recursive definitions have three foundations: a base case (basis), an inductive clause, and an extremal clause.
The base case satisfies the definition without being defined in terms of the definition itself. The factorial function n! ($0! = 1$, $n! = n\cdot(n - 1)\cdot \ldots 2\cdot 1$) is a typical example.
(http://en.wikipedia.org/wiki/Recursive_definition)

Red Tape:
"Red tape is excessive regulation or rigid conformity to formal rules that is considered redundant or bureaucratic and hinders or prevents action or decision-making. It is usually applied to governments, corporations, and other large organizations."
(http://en.wikipedia.org/wiki/Red_tape)

Reflexivity:
In social sciences a circular relationship between cause and effect. A reflexive relationship is bidirectional; with both the cause and the effect affecting one another in a situation that renders both functions cause and effect.

Research-Based Startup (RBSU):
Also called an academic spin-out; is mostly viewed as a new for-profit and independent company based on the findings of a member or by members of a research group at a university or public research institution.

Resource:
A source of aid or support that may be drawn upon when needed.

Result:
A result is the final consequence of intended actions or of events
(There are explicit references to actions/activities or events).

Regulation:
In the context of control rather than law means that the interrelated subjects and objects constituting a system must be regulated in some way so that the *goal* can be achieved. Regulation implies that deviations must be detected and corrected (cf. *feedback*).

Revenue:
Sales of offerings of a firm after deducting all returns, rebates, and discounts.

Revenue model:
Specifies by which kinds of offerings the firm will earn *revenue* to make more money than it spends. In business, a revenue model is generally used for mid and long-term projections of a company's profit potential and operation.

Risk:
In situations of risk the decision-maker knows the value of the outcomes and the relative probabilities of the states (of the system). For a given situation it relates to hazard.

Role:
A function: the actions and activities assigned to or required or expected of a person or group or thing ("acting as"); "the function of a manager";
a role is a set of behaviors, rights and obligations conceptualized in a social situation.

Routine:
A course of action to be followed regularly, not necessarily by an explicitly defined or prescribed given order;
a series of steps followed in a regular definitive order to accomplish something; also a set of instructions designed to perform a specific task; a standard *procedure.*

Scale-up:
Is the process of transfer of materials from preparation in the lab ("lab scale") to large scale or mass production or from small models and prototypes of machines, devices or vehicles to actual size machines/devices/vehicles and their production in large numbers as for automobiles. Here often science is connected with development and engineering to ultimately "Production/Manufacturing."

Self-reinforcement:
It represents a mechanism, a *positive feedback* mechanism, by which a system's output or state is enhanced or brought into a more favorable situation.

Positive feedback is in phase with the input, in the sense that it adds to make the input larger. Positive feedback is a process in which the effects of a small disturbance on a system can include an increase in the magnitude of the perturbation. (http://en.wikipedia.org/wiki/Positive_feedback)

Self-employment:
In the technological context self-employment refers to a restricted aspect of entrepreneurship related essentially to an autonomy orientation ("be one's own boss," perceived freedom).

Securities:
Securities are financial instruments that can be traded freely on the open market and representing ownership (stocks), a debt agreement (bonds), or the rights to ownership (derivatives); they are broadly categorized into debt securities (such as banknotes, bonds and debentures) and equity securities, for instance, common stocks; and derivative contracts, such as forwards, futures, options and swaps. (http://useconomy.about.com/od/glossary/g/securities.htm)

Serendipity:
Is finding something unexpected and useful while searching for something else entirely.

Social network:
In its simplest form, a social network is a system of specified ties between individuals as network nodes being observed or studied concerning relationships and social interactions ("edges"), such as friendship, kinship, common social values or interest, relationships of beliefs, knowledge or prestige.
Functionally, it is currently understood as an Internet-based service that allows interacting with others. They play a critical role in socialization into norms and determining the way problems are solved, organizations are run, and the degree to which individuals succeed in achieving their goals.

Soft system: – *Hard System*
A characterization of a system with regards to determining the extent to which certain variables can be quantified and measured and lines of reasoning. "Soft" systems, typically covering biology and life sciences, psychology, cognitive, behavioral and social sciences, focus more on qualitative approaches, perceived causal relationships, intuition, discontinuities with low level of replication, etc. They are usually treated as "open" systems.

Spillover effect:
In economics neglecting the effects of one system upon another one is often referred to as "spillover effects." In systems theory it is related to *sub-optimization*.

Spin-off:
Is a new organization or entity directly formed by a split from a larger one, such as a new company formed from a large firm being still governed by the parent company.

Spin-out:
Is a firm formed when an employee or group of employees leaves an existing entity to form an *independent* startup firm. This can refer to a university or a research institute, directly or mediated by a business incubator. Spin-outs typically operate at arms length from their parent organizations (formally and legally independent, but usually

with certain ties) and have independent sources of financing, products, services and customers.

Strategy: – *Plan*
When you know what goal you want to achieve, but you are not sure exactly how to do it. A strategy is characterized by not knowing what to do at the next step until you have results from the previous step.
Each step of a strategy is realistically influenced by what was learned from the previous step.

Stakeholder:
A stakeholder (in German Einflussnehmer, Anteilsnehmer) is a person (or a group) who has a stake or interest in the outcome of a system's activities, operations and conversion processes, but also one who is or may be affected by a firm's projects;
all the parties that have an interest, financial or otherwise, in a company, including shareholders, creditors, bondholders, employees, customers, management, the community, government and even media.

Sub-optimization:
Sub-optimization refers to issues of improving or even optimizing the performance of open systems. Optimization is only possible for closed systems! Open systems can, at best, only be partially optimized – we have sub-optimization. Moreover, optimizing the subsystems does not guarantee that the total system optimum is reached, whereas the optimization of the total system (if it could ever be reached) does not guarantee that all the subsystems can be optimized at the same time. It is related to *spillover effects*.

Supply chain:
The supply chain (or *value system*) is a "supplier-to-end-users *value chain.*" Specifically, *a supply chain is a system* of organizations, people, technology, activities, information and resources involved in moving an offering, product or service, from supplier to the "end-user customer".

Synthetic Biology:
The design and construction of new biological parts, devices, and systems, also the re-design of existing, natural biological systems for useful purposes.
(http://syntheticbiology.org/).
It is a new area of biological research that combines science and engineering in order to design and build ("synthesize") novel biological functions and systems.
Engineers view biology as a technology. Synthetic Biology includes the broad redefinition and expansion of biotechnology, with the ultimate goals of being able to design and build engineered biological systems that process information, manipulate chemicals, fabricate materials and structures, produce energy, provide food, and maintain and enhance human health and our environment.
One aspect of Synthetic Biology which distinguishes it from conventional genetic engineering is a heavy emphasis on developing foundational technologies that make the engineering of biology easier and more reliable.
(http://en.wikipedia.org/wiki/Synthetic_biology).

Task:
A task is an activity or set of activities (that might be defined as part of a process) that needs to be accomplished within a defined period of time.
Skill is the ability to perform a task.

Tacit knowledge:
Tacit knowledge (as opposed to formal or explicit knowledge) is a kind of knowledge mainly ingrained in people rather than documented or encoded and represents correspondingly an issue for "knowledge or technology transfer" or also "licensing" of technology.
With tacit knowledge, people are often not aware of the knowledge they possess or how it can be valuable to others. It is usually gained through personal experience in particular fields and environments of activities (cf. *learning curve*).

Tacit technology:
Is not codified or not documented practical knowledge of and experience with technical fields of people and, hence, an important competitive advantage of a firm and part of its core competencies – as long as the firm can keep the people. Tacit technology is often brought to bear as and when it is required (cf. *resource*).

Technique:
Represents an applicable element of a *technology*. Techniques constitute what is also called instructional (practical) knowledge. Like any recipe they comprise essentially instructions that allow people to "produce" or "re-produce," respectively. A technique is a *procedure* used to accomplish a specific activity or *task*.

Technology:
Technology has more than one definition, but generically it refers essentially to the body of know-how about the means and methods of producing tools, goods or services. Technology comprises a system of application-oriented statements about means and ends. Technology comprises often a set of *techniques*. Correspondingly, this notion does not require an interconnection to science; it may related to "art, skill and craft" and "useful arts" to create some value. Current definitions often refer explicitly to science, in particular when focusing on "high technology."

Technology implementation:
Technology implementation means selecting the *techniques* to target a given goal related to tools, goods or services.

Teleology:
Teleology anticipates future existence of systems. As an analytical method it is associated with purpose. Related to *finality*, it represents an antithesis to causality and linear thinking in terms of causes and effects, which is prevalent in natural sciences.

Trade-off (or tradeoff):
A trade-off (or tradeoff) is a situation that involves losing one quality or aspect of something in return for gaining another quality or aspect. It implies a decision to be made with full comprehension of both the upside and downside of a particular choice. Trade-offs are regarded as "compromises" or exchanges which decision makers must effect when all their *objectives* cannot be carried out at the same time. As the extreme one has to be sacrificed totally at the expense of the other, for instance, either quality or low price, but not both.

Trait:
Traits are multiple, thematically-related personality features that collectively reflect the operation or characteristic functioning of a particular area of person, a characteristic way in which an individual perceives, feels, believes, or acts.
For example, the trait of intelligence describes the level of functioning of broad areas of the cognitive system. Traits typically emerge from many diverse contributors.

Transcriptomics: – *Genome*
The study of the complete set of RNA transcripts produced by the genome (transcriptome) at a given time.

Uncertainty:
Under uncertainty, the values of the outcomes may be known but no information on the probability of (occurring) effects or events is available.

Utility:
In the context of systems theory utility can be used as valid guides for *decision-making*. Utility is assigned expected utility values to choices and represents, to a certain degree, the behavioral (psychological) characteristics of decision makers, who are faced with choice situations under risk [Gigch 1974:107].
In economics, utility is a measure of the relative satisfaction from, or desirability of, consumption of various goods and services. Given this measure, one may speak meaningfully of increasing or decreasing utility, and thereby explain economic behavior in terms of attempts to increase one's utility. (http://en.wikipedia.org/wiki/Utility)

Value chain:
A value chain is a sequence of activities that a firm operating in a specific industry performs in order to deliver something valuable, such as a product or service.

Value proposition:
Is a statement how customer value will be created summarizing why a customer should buy an offering (product or service). This statement should convince a potential customer that one particular product or service will add more value or better solve a problem than other similar offerings (if they exist).

Value system:
It is the network of organizations and the value producing activities involved in the production and delivery of an offering. The value system consists of value adding components which correspond to supplier/channel-customer bunches. It is an interconnection of processes and activities within and among firms that creates benefits for intermediaries and end-users (consumers).

Verbund (German):
The Verbund principle enables a firm to add value as one company through efficient use of its internal and external resources.

Verbundprojekt (according to the German BMFB ministry):
It is a pre-competitive, division of labor and cooperation of several independent partners from industry and academia with independent contributions to the solution of a research and development task. (Ein Verbundprojekt ist eine vorwettbewerbliche,

arbeitsteilige Kooperation von mehreren unabhängigen Partnern aus Wirtschaft und Wissenschaft mit eigenständigen Beiträgen zur Lösung einer Forschungs- und Entwicklungsaufgabe.) – usually translated as "cooperative project" or joint project."

Verbund system:
A Verbund system creates efficient value chains that extend from basic input right through to high-value-added offering like products. The Verbund principle extends beyond production and technology to embrace the firm's employees and also their interconnections of knowledge, experience and expertise to other internal or external people (Employee Verbund, Technology Verbund, ..., Customer Verbund).

Window of Opportunity:
Is a short time period during which an otherwise unattainable opportunity exists.

Zeitgeist:
The Zeitgeist (spirit of the age or spirit of the time) is the intellectual fashion or dominant school of thought which typifies and influences the culture of a particular period in time. According to the German philosopher Georg Wilhelm Friedrich Hegel (1770-1831) "no man can surpass his own time, for the spirit of his time is also his own spirit." (http://en.wikipedia.org/wiki/Zeitgeist)
In an extension of this view, systems theory would allow to use the notion also in a restricted sense, such as how to do business at a particular time period and location (region, country) or "prescriptions" how to innovate or found a firm.

Zeitgeist – Goethe:

Faust – 575-577	Translation *)
„Was ihr den Geist der Zeiten heißt, das ist im Grund der Herren eigner Geist, in dem die Zeiten sich bespiegeln."	The spirit of the ages, that you find, In the end, is the spirit of Humankind: A mirror where all the ages are revealed. What you call "spirit of the ages" Is after all the spirit of those sages In which the mirrored age itself reveals.

*) http://www.poetryintranslation.com/PITBR/German/FaustIScenesItoIII.htm; http://goethe.holtof.com/faust/Faust_I_02.htm.

Acronyms

ALD	Atomic layer deposition
AUTM	Association of University Technology Managers
B2B	Business-to-business
B2C	Business-to-consumers
B2G	Business-to-government
B2P	Business-to-public
BMBF	Bundesministerium für Bildung und Forschung (Federal Ministry of Education and Research)
BMVh	Bundesministerium der Verteidigung (Federal Ministry of Defense)
BMWi	Bundesministerium für Wirtschaft und Technologie (Federal Ministry of Economics and Technology)
BOS	Balance-of-System (Cost)
CAPEX	Capital expenditures
CEO	Chief Executive Officer
CFO	Chief financial officer
CdTe	Cadmium telluride
CFD	Computational fluid dynamics
CHP	Combined heat and power plant (in German Blockheizkraftwerk – BHKW)
CIGS	Copper indium gallium diselenide
CMP	Chemical mechanical planarization
COO	Chief operating officer
CoP	Communities of practice"
COTS	Commercial-off-the-shelf (products, components)
CRADA	Cooperative Research and Development Agreement
CSF	Critical success factor
CSO	Chief-science-officer

CTO	Chief-technology-officer
CV	Corporate venturing
CVC	Corporate venture company
DFG	Deutsche Forschunsgemeinschaft (German Research Foundation)
DIY	Do-it-yourself
DOD or DoD	(US) Department of Defense
DOE or DoE	(US) Department of Energy
DPMA	Deutsches Patent- und Markenamt (German Patent and Trade Mark Office)
DSO	Days Sales Outstanding
EEG	(Germany) Erneuerbare Energie Gesetz – Renewable Energy Act
EERE	(US) Office of Energy Efficient and Renewable Energy
EGC	Entrepreneurial growth company
EH&S	Environmental health and safety
EISA	(US) Energy Independence and Security Act of 2007
EMEA	European Medicines Agency
EPA	(US) Environmental Protection Agency
EPC	Engineering, procurement and construction
EPO	European Patent Office
ERP	Enterprise resource planning
ESA	European Space Agency
ESTCP	(US) Environmental Security Technology Certification Program
EU	European Union
EU27	European Union comprised of 27 members
EV	Electric vehicle
F2P	Free-to-play
FCV	Fuel cell vehicle
FFE	Fuzzy Front-End

FFV	Flex Fuel Vehicle
FhG	Fraunhofer Gesellschaft (Society)
GAAP	Generally Accepted Accounting Principles
GDP	Gross Domestic Product
GDR	German Democratic Republic
GEM	Global Entrepreneurship Monitor
GMA	General morphological analysis
GmbH	Gesellschaft mit beschränkter Haftung (cf. LLC)
GMO	Genetically modified organism or object
GP	General Partner(ship)
GST	General Systems Theory
HAP	Hazardous air pollutant
HR	Human Resource(s)
HAWT	Horizontal axis wind turbine
HTS	High throughput screening; high-temperature superconductor
HVT	High value technology
HW	Hardware
I&CT	Information and communication technology
IC	Intellectual Capital
ID	Identification (mark, sign, signal, code etc.)
IHK	Industry und Handelskammer (Chamber of Industry and Commerce)
IP	Intellectual property
IPO	Initial Public Offering
IPR	Intellectual property right
IT	Information technology
JD	Juris Doctor (degree), Doctor of Jurisprudence
JDA	Joint development alliance
JRA	Joint research alliance

JV	Joint venture
KAI	Kirton Adaptive Innovative (instrument)
KDT	Knowledge discovery in text databases
KIBS	Knowledge-intensive business service (firm)
KIT	Karlsruhe Institute of Technology (Germany)
LCF	Lignocellulose feedstock
LED	Light emitting diode
LLC	Limited Liability Company (cf. German GmbH)
LP	Limited Partner(ship)
M&A	Mergers & Acquisitions
MBI	Management buy-in
MBO	Management buy-out
MBTI	Myers-Briggs Type Indicator
MD	Medicinae Doctor (Doctor of Medicine)
MIT	Massachusetts Institute of Technology
MMOG	Massively Multiplayer Online Game
MPG	Max Planck Gesellschaft (Society)
MPI	Max Planck Institut
NAFTA	North American Free Trade Agreement
NAICS	North American Industry Classification System
NASA	(US) National Aeronautics and Space Administration
NASDAQ	National Association of Securities Dealers Automated Quotation
NBD	New Business Development
NGO	Non-governmental organization
NIH	National Institutes of Health; not invented here
NPD	New Product Development
NREL	(US) National Renewable Energy Laboratory
NSF	National Science Foundation (of the US)

NTBF	New technology-based firm
NVCA	(US) National Venture Capital Association
NYSE	New York Stock Exchange
OLED	Organic light emitting diode
OPEX	Operational expenditures
OPV	Organic photovoltaic
OTC	Over-the-counter
P2P	Pay-to-play
PARC	Palo Alto Research Center
PDA	Personal digital assistant
PoC	Proof-of-concept
PPP	Private-public-partnership
PR	Public Relations
PV	Photovoltaic
RBV	Resource-Based View
RES	Renewable energy source
RFS	Renewable Fuel Standard (of US EPA)
RI	R&D intensity (or research intensity)
ROI	Return on investment
S&T	Science & Technology (System)
SBA	Small Business Administration (in the US)
SBIC	Small Business Investment Company
SBIR	Small Business Innovation Research
SME	Small and medium-sized enterprise
SOL	Society for Organizational Learning
SSBIC	Specialized Small Business Investment Company
STVP	Stanford Technology Ventures Program
STTR	Small Business Technology Transfer

SW	Software
SWOT	Strengths, Weaknesses, Opportunities and Threats
Syngas	Synthesis gas
TBS	Technology-based service
TIM	Technology and Innovation Management
TOI	Timing of industrialization
TS	Technical service
TFT	Thin film transistor
TVT	Top value technology
UIRC	University-industry-research centers
USDA	US Department of Agriculture
USP	Unique selling proposition; Unique selling point
USPTO	US Patent and Trademark Office
VAWT	Vertical axis wind turbine
VC	Venture capital (or venture capitalist)
VOC	Volatile organic compound
WWI	World War I
WWII	World War II

INDEX

Indexing is rather detailed including hierarchies following Figure I.1, Figure I.5 and the Table of Contents. This is intended to facilitate working with this book, such as "answering questions", preparing a presentation, a course and reading list or looking for background for a research issue.

A page number for an index entry may be the start covering a range of following pages associated with that index term.

For the index term "Definition/Explanation" the page number of the key definition will be given in **bold** face.

For an index term in a figure or a company description in a sub-chapter, table or box the related page number may be given in *italics*. A company name and directly associated page in italics means a detailed description of the company in a box, sub-chapter or table.

Company Index

Subject Index